Traceability in Chemical Measurement

Paul De Bièvre · Helmut Günzler (Eds.)

Traceability in Chemical Measurement

Springer

Prof. Dr. Paul De Bièvre
Duineneind 9
2460 Kasterlee
Belgium

Prof. Dr. Helmut Günzler
Bismarckstrasse 11
69469 Weinheim
Germany

Library of Congress Control Number: 2004114857

ISBN 3-540-43989-7 **Springer Berlin Heidelberg New York**

Springer is a part of Springer Science+Business Media
springeronline.com

Printed in Germany

Cover design: *design & production* GmbH, Heidelberg
Production: LE-TEX Jelonek, Schmidt & Vöckler GbR, Leipzig

Printed on acid-free paper 52/3111/YL - 5 4 3 2 1 0

Preface

Not many terms covering concepts in measurement have circulated over the last ten years in the chemical measurement community around the world so intensely as the term 'traceability'. It appears in the title of CITAC (Cooperation on International Traceability in Analytical Chemistry) since 1993. It is addressed almost yearly in Workshops of EURACHEM (A Focus for Analytical Chemistry in Europe). Documents of ILAC (International Laboratory Accreditation Cooperation) require it to be used in the process of accreditation. Standards and Guides of ISO (the International Organisation for Standardization) mention them frequently and insistingly.

In short, everybody talks and writes about 'traceability' (because everybody talks and writes about 'traceability'?).

The 2nd edition of the International Vocabulary of General and Basic Terms in Metrology, VIM2, (1993) defines it as the 'property of the result of a measurement or the value of a standard whereby it can be related to stated references, usually national or international standards, through an unbroken chain of comparisons all having stated uncertainties'.

Over the years the problem had arisen that the term 'traceability' became more and more ambiguous because it was used for many different traceability concepts such as traceability of a sample (sample traceability), traceability of a document (document traceability), traceability of an instrument (instrument traceability) or -most important- traceability of a measurement result (measurement traceability). The VIM2 definition clearly meant it to be related to a measurement result.

The revised edition of the VIM (VIM3), will probably fine-tune the term for traceability of a measurement result to be named 'metrological traceability'. It is also likely that this definition is improved to read something like 'property of a measurement result relating the result to a stated metrological reference through an unbroken chain of calibrations or comparisons each contributing to the stated measurement uncertainty'.

Metrological traceability of chemical measurement results means the establishment of a relation to a stated metrological reference (a 'trace'). This can be the definition of a measurement unit which, of necessity, must go through a practical realization or (better: an embodiment) of that definition. But in case of operationally defined measurands (no units), metrological traceability can be to the result of an (internationally) agreed measurement procedure, or to the quantity value[1] carried by a measurement standard such as a certified reference material. All of these metrological traceabilities must be realized through an 'unbroken chain of calibrations or comparisons'. The chain ensures that the metrological traceability of a measurement result has been established to a metrological reference which must be stated. Only when measurement results are 'traceable' to a *common* metrological reference, is their direct metrological comparability possible, i.e. is their ability assured to be comparable.

This anthology contains 56 outstanding papers on the topic 'Traceability', published in the Journal "Accreditation and Quality Assurance" since its inception, but mostly in the period 2000–2003. They reflect the latest understanding of the concept 'measurement traceability' -or lack thereof- and possibly some rationale(s) for the answer to the question why it is important to integrate the concept of measurement traceability into the standard measurement procedures of every analytical laboratory.

For one thing, the wide variety of opinions reflected in the papers demonstrates that we have not yet achieved a common understanding of the concept 'traceability' and therefore not yet international understanding based on a concept which is unambiguously understood in the same way by everybody. Thus the international discussions will (have to) go on for some time because agreement must be reached. Measurement traceability (metrological traceability) is a cornerstone property of any measurement result. Only measurement results which are traceable to a stated common metrological reference (such as a measurement unit), are directly 'comparable'. 'Comparability' of results is essential in any border-crossing context, whether that is the estimate of the monetary value of goods, based on measurement results, or the rejection of goods based on measurement results for toxic substances contained in the goods, or when comparing results of clinical

[1] quantity (German: 'Messgrösse', French: 'grandeur de mesure', Dutch: 'meetgrootheid') is not used here in the meaning 'amount', but as the generic term for the quantities we measure: concentration, volume, mass, temperature, time, etc., as defined in the VIM.

measurements in case of international business and leisure travel. At least as important is the fact that proper evaluation of measurement uncertainty is only possible after metrological traceability has been established, i.e. after the 'trace' or 'track' has been decided by the analyst along which (s)he will organize the plan of the measurement in order to make sure that metrological traceability to a common metrological reference would be in place. That is needed because the measurement uncertainty in a measurement result can only be evaluated by combining the uncertainty contributions generated by every step along the metrological traceability chain.

This anthology hopefully is of benefit to both the producers and the users of results of chemical measurements: the basic concepts and the basic thinking in measurement are the same for both. Only their measurement uncertainty will differ.

Prof. Dr. P. De Bièvre
Editor-in-Chief
Accreditation and Quality Assurance
Kasterlee 2004-04-02

Contents

Contributors

Hassan Y. Aboul-Enein
Biological and Medical Research Department, MBC-03, P.O. Box 3354, Riyadh 11211, Saudi Arabia

Freddy C. Adams
Department of Chemistry, University of Antwerpen (UIA), Universiteitsplein 1, B-2610 Wilrijk, Belgium

Des Alphonso
Centre for Optical and Environmental Metrology, National Physical Laboratory, Teddington, Middlesex, TW11 0LW, UK

Elena Amico di Meane
Istituto di Metrologia "Gustavo Colonnetti" del C.N.R., Strada delle Cacce 75, I-10137 Torino, Italy

Boris Anisimov
The National Physical Laboratory of Israel (INPL), Givat Ram, Jerusalem, 91904, Israel

V. P. Antipin
All Russia Institute for Light Alloys, Bldg. 2, Gorbunova Street, 121596 Moscow, Russia

Paul Armishaw
National Analytical Reference Laboratory, AGAL, Clive Road 3, Cottesloe, 6011 Western Australia, Australia

Charles M. Beck II
National Institute of Standards and Technology, Gaithersburg, MD 20899, USA

Brian Belanger
Visiting Committee on Advanced Technology (Retired), National Institute of Standards and Technology, Gaithersburg, MD, 20899–1004, USA

William Bell
Centre for Optical and Environmental Metrology, National Physical Laboratory, Teddington, Middlesex, TW11 0LW, UK

Chris Brookes
Centre for Optical and Environmental Metrology, National Physical Laboratory, Teddington, Middlesex, TW11 0LW, UK

Lutz Brüggemann
Department of Analytical Chemistry, UFZ Centre for Environmental Research, Leipzig-Halle, Permoserstrasse 15, 04318 Leipzig, Germany

Mirella Buzoianu
National Institute of Metrology, Sos. Vitan-Bârzesti 11, 75669 Bucharest, Romania

Philippe Charlet
Bureau National de Métrologie – Laboratoire National d'Essais, 1 rue Gaston Boissier, 75724 Paris Cedex 15, France

Ludmila Dabrowská
National Institute of Public Health, Šrobárova 48, 100 42 Praha, Czech Republic

Paul De Bièvre
Institute for Reference Materials and Measurements, European Commission – JRC, B-2440 Geel, Belgium

Hansa D'Souza
Centre for Optical and Environmental Metrology, National Physical Laboratory, Teddington, Middlesex, TW11 0LW, UK

Steluta Duta
National Institute of Metrology Sos., Vitan-Barzesti 11, 75669 Bucharest, Romania

René Dybkaer
Department of Standardization in Laboratory Medicine, Frederiksberg Hospital, H:S Copenhagen Hospital Corporation Øster Farimagsgade 5, DK-1399 Copenhagen, Denmark

Steve L. R. Ellison
LGC Limited, Teddington, Middlesex, TW11 0LY, UK

John Fleming
LGC Limited, Queens Road, Teddington, Middlesex, TW11 0LY, UK

D. Glavič-Cindro
"Jožef Stefan" Institute, Jamova 39, 1000 Ljubljana, Slovenia

Irena Goldfeld
The National Physical Laboratory of Israel (INPL), Givat Ram, Jerusalem, 91904, Israel

Brian Goody
Centre for Optical and Environmental Metrology, National Physical Laboratory, Teddington, Middlesex, TW11 0LW, UK

A. A. Grigorieva
All Russia Institute for Light Alloys, Bldg. 2, Gorbunova Street, 121596 Moscow, Russia

Bernd Güttler
Physikalisch-Technische Bundesanstalt (PTB), Bundesallee 100, 38116 Braunschweig, Germany

Gill Holcombe
LGC Limited, Queens Road, Teddington, Middlesex, TW11 0LY, UK

Paul E. Holland
Centre for Optical and Environmental Metrology, National Physical Laboratory, Teddington, Middlesex, TW11 0LW, UK

Lars Jorhem
National Food Administration, Box 622, 751 26 Uppsala, Sweden

Robert Kaarls
Netherlands Measurements Institute, AR, 2600 Delft, The Netherlands

Mary M. Kimberly
Division of Laboratory Sciences, National Center for Environmental Health, Centers for Disease Control and Prevention, Atlanta, GA 30341–3724, USA

Barnard King
Analytical Measurement Consultant, 25 Bolton Gardens, Teddington, Middlesex, TW11 9AX, UK

B. King
National Analytical Reference Laboratory, 1 Suakin Street, Pymble, NSW 2073, Australia

Matjaz Korun
"Jožef Stefan" Institute, Jamova 39, 1000 Ljubljana, Slovenia

Jan Kučera
Nuclear Physics Institute, 250 68 Řež u Prahy, Czech Republic

Ilya Kuselman
The National Physical Laboratory of Israel (INPL), Givat Ram, Jerusalem 91904, Israel

Richard Lawn
LGC Limited, Queens Road, Teddington, Middlesex, TW11 0LY, UK

Wang Linzhen
National Research Center for Certified Reference Materials, 100013 Beijing, People's republic of China

Markus Lipp
Monsanto Company, 800 North Lindbergh Boulevard, St. Louis, Missouri, 63167, USA

Randie R. Little
Departments of Child Health and Pathology, University of Missouri School of Medicine, Columbia, MO, 65212, USA

M. Máriássy
Slovak Institute of Metrology, Karloveská 63, 842 55 Bratislava, Slovakia

A. Marschal
Bureau National de Métrologie – Laboratoire National d'Essais, 1 rue Gaston Boissier, 75724 Paris Cedex 15, France

A. Mathiasová
Slovak Institute of Metrology, Karloveská 63, 844 55 Bratislava, Slovakia

Roderick G. Millar
National Analytical Reference Laboratory, AGAL, Clive Road 3, Cottesloe, 6011, Western Australia, Australia

Martin Milton
Centre for Optical and Environmental Metrology, National Physical Laboratory, Teddington, Middlesex, TW11 0LW, UK

Yoshito Mitani
Centro Nacional de Metrologia, Mexico, Col. Centro, Apdo. Postal 1-100, QRO. 76000 Queretaro, Mexico

Mathias M. Müller
Institute of Laboratory Diagnostics, Kaiser Franz Josef Hospital, Kundratstrasse 3, 1100 Vienna, Austria

Gary L. Myers
Division of Laboratory Sciences, National Center for Environmental Health, Centers for Disease Control and Prevention, Atlanta, GA 30341–3724, USA

Bernd Neidhart
Philipps-Universität Marburg, Hans-Meerwein-Strasse, 35032 Marburg, Germany

Ioannis Papadakis
Institute for Reference Materials and Measurements, Joint Research Centre – European Commission, Retieseweg, B-2440 Geel, Belgium

H. Steffen Peiser
National Institute of Standards and Technology 638, Blossom Drive, Rockville, MD, 20850, USA

Alejandro Perez-Castorena
Centro Nacional de Metrologia, Mexico, Col. Centro, Apdo. Postal 1-100, QRO. 76000 Queretaro, Mexico

Margherita Plassa
Istituto di Metrologia "Gustavo Colonnetti" del C.N.R., Strada delle Cacce 73, I-10136 Torino, Italy

Gary Price
P.O. Box 57, Menai, NSW 2234, Australia

Han Qiao
National Research Center for Certified Reference Materials, 100013 Beijing, People's republic of China

Paul Quincey
Centre for Optical and Environmental Metrology, National Physical Laboratory, Teddington, Middlesex, TW11 0LW, UK

Stanley D. Rasberry
National Institute of Standards and Technology, 31 Longmeadow Drive, Gaithersburg, MD, 20899, USA

William P. Reed
National Institute of Standards and Technology, Gaithersburg, MD, 20899, USA

Wolfgang Richter
Physikalisch-Technische Bundesanstalt (PTB), Bundesallee 100, 38116 Braunschweig, Germany

Rod Robinson
Centre for Optical and Environmental Metrology, National Physical Laboratory, Teddington, Middlesex, TW11 0LW, UK

Mike Sargent
LGC Limited, Queens Road, Teddington, Middlesex, TW11 0LY, UK

Michela Sega
Istituto di Metrologia "Gustavo Colonnetti" del C.N.R., Strada delle Cacce 73, I-10135 Torino, Italy

Lothar Siekmann
Institut für Klinische Biochemie, Universitätsklinikum der Universität Bonn, Sigmund-Freud-Str. 25, 53127 Bonn, Germany

Ilona Šperlingová
National Institute of Public Health, Šrobárova 48, 100 42 Praha, Czech Republic

Peter Spitzer
Physikalisch-Technische Bundesanstalt, Bundesallee 100, 38116 Braunschweig, Germany

Christoph Tausch
Philipps-Universität Marburg, Hans-Meerwein-Strasse, 35032 Marburg, Germany

Philip Taylor
Institute for Reference Materials and Measurements, European Commission – JRC, B-2440 Geel, Belgium

*Milo*ò *Tichý*
National Institute of Public Health, Šrobárova 48, 100 42 Praha, Czech Republic

Ian Uprichard
Centre for Optical and Environmental Metrology, National Physical Laboratory, Teddington, Middlesex, TW11 0LW, UK

B. Vodenik
"Jožef Stefan" Institute, Jamova 39, 1000 Ljubljana, Slovenia

Emil Völkert
Chairman of EDMA WP Standardisation, Roche Diagnostics GmbH, Sandhofer Str. 116, 68305 Mannheim, Germany

L. Vyskočil
Slovak Institute of Metrology, Karloveská 63, 842 55 Bratislava, Slovakia

Andrew Wallard
Bureau International des Poids et Mesures, Pavillon de Breteuil, 92312 Sèvres, France

Wolfhard Wegscheider
Institute of General and Analytical Chemistry, University of Leoben, Franz-Josef Strasse 18, 8700 Leoben, Austria

Alexander Weisman
Chemagis Ltd., P.O.Box 9091, Tel-Aviv 61090, Israel

R. Wennrich
Department of Analytical Chemistry, UFZ Centre for Environmental Research, Leipzig-Halle, Permoserstrasse 15, 04318 Leipzig, Germany

Alex Williams
Chairman EURACHEM Working Group on Measurement Uncertainty, 19 Hamesmoor Way, Camberley, Surrey, UK

Peter T. Woods
Centre for Optical and Environmental Metrology, National Physical Laboratory, Teddington, Middlesex, TW11 0LW, UK

Adolf Zschunke
Bundesanstalt für Materialforschung und -prüfung (BAM), Richard-Willstätter-Strasse 11, 12489 Berlin, Germany

Accred Qual Assur (1996) 1:3–13

P. De Bièvre
R. Kaarls
H. S. Peiser
S. D. Rasberry
W. P. Reed

Measurement principles for traceability in chemical analysis

Received: 30 June 1995
Accepted: 30 June 1995

Paul De Bièvre (✉)
Institute for Reference Materials and Measurements, European Commission – JRC, B-2440 Geel, Belgium

Robert Kaarls
Nederlands Meetinstituut, NL-2600 AR Delft, The Netherlands

H. Steffen Peiser
638, Blossom Drive, Rockville, MD 20850, USA

Stanley D. Rasberry and William P. Reed
National Institute of Standards and Technology, Gaithersburg, MD 20899, USA

Abstract By the definition of the mole as a base unit for amount-of-substance measures within the International System of Units (SI), chemists can make chemical measurements in full compliance with established metrological principles. Since the mole requires exact knowledge of the chemical entity, which is often neither available nor of practical relevance to the purpose of the measurement, the SI units of mass or length (for volume) are unavoidable in the expression of results of many chemical measurements. Science, technology, and trade depend upon a huge and ever increasing number and variety of chemical determinations to quantify material composition and quality. Thus, international harmonization in the assessments of processes, procedures, and results is highly desirable and clearly cost effective. The authors, with relevant experience and responsibilities in Europe and America, have found some consensus in the interpretation of the metrological principles for chemical measurements, but believe open discussion should precede wide implementation by chemical communities. In fostering this dialogue, this paper shows, for instance, that more precise interpretation of the definitions for "traceability," "calibration," and "validation" is needed for present-day chemical measurements. Problems that face scientists in making measurements do not all vanish just by adherence to the SI. However, such compliance can improve communication among chemists and metrologists.

Key words Traceability · Mole · Definition · Measurements · Chemical metrology · Calibration · Validation · Reference materials

Introduction

Science, technology, and commerce require rapidly rising numbers and types of measurements that for good reasons can be trusted [1–4]. Worldwide acceptance of measurement results requires reliable, traceable, and comparable measurements for reduction of costs, efficient production processes, subsequent use of measurement data, realization of fair-trade conditions, and for internationally recognized and accepted laboratory accreditations. Physical measurements made in accordance with the International System of Units (SI), which was introduced under the Convention of the Meter (with status of an International Treaty), have satisfied many of these needs [5, 6]. Such measurements typically rely on a comparison of the measured quantity in the item concerned with the same quantity in a "standard." Chemical measurements are usually not made by comparison with an equivalent chemical "standard." Chemical measurements are not yet widely made in terms of the SI unit of amount of substance, the mole [7]. This paper will explore the possibilities for bringing a stronger metrological foundation to chemical measurements and will specifically describe a role for reference materials in the traceability of chemical measurements to the SI [8].

Amount-of-substance measurements

Most chemists will agree that the majority of chemical measurements are, or could be, expressed as amount-of-substance measurements. When appropriate, they will in this paper be so described. However, whereas mass or length (volume) measurements at the smallest attainable uncertainty do not generally require a detailed understanding of the material whose property is quantified, amount-of-substance measurements require reference to the exact composition of the measured entity, to interfering impurities, and to the material – by composition, mass, or volume – within which that entity is measured.

In many chemical measurements one neither knows nor, at the time of measurement, wishes to know the exact composition of the matrix. To give an example, a metallurgical firm will receive ore shipments measured by mass in kilograms. Representative samples in the seller's and receiver's laboratories are measured for quality by the amount of substance of a specified metal element or compound per given mass of ore. It is unnecessary and far too complex to attempt amount-of-substance measurements on all components of the bulk. In exactly the same way, a food laboratory might measure the amount of substance (say lead) in orange juice in milligrams per liter (per cubic decimeter). The charm of the SI system lies in a coherence, which makes it possible to express all measured quantities in a combination of base and derived units [9].

Thus, whereas chemists have historically expressed analyses mostly by mass per mass, or as convenient percentages, or by mass per volume, they could express their measurements in amount of a specific substance per mass (mole per kilogram) or per volume. In cases such as pure materials and gases, mole per mole can be used. A percentage statement, or one in parts per thousand, million, or billion, is possible, though not recommended. In the SI system, as originally visualized, such dimensionless numbers as results of measurements are not favored. The quantitative result of any measurement should be expressed by a number "multiplied" by the appropriate unit associated with the measured quantity. As is further discussed below, this original preference proposed for the International System does not fit well with much of current practice in chemical measurements.

Towards harmony in amount measurements

There is no doubt that chemical measurements are and must be widely used in science and research, technology, engineering, and agriculture, as well as in regulatory issues, including boundary crossings, health control, environmental assessment, and commerce. A vast number of chemical measurements is made every year. Ever more will be needed for reasons of increasing complexities in human interactions with the environment [4, 10, 11]. For many measurements worldwide – such as ozone levels in cities and the upper atmosphere – it is necessary to maintain anchor points with long-term stability. More generally, all equivalent measurements should be made in harmony with each other [2], even when the practically needed and achievable reproducibility [9] has to be superior to the best attainable uncertainty in measurement relative to "true value." The relation to true value, however, remains the ultimate test for quality of a measurement [12]. At present it is a rather widely accepted opinion that, even when the relation to the true value is elusive, chemists in different laboratories equipped to make repeatable measurements can still make them comparable to one another by the use of a reference material (RM) [13–15]. The correctness of this concept will be discussed later.

The use of the mole

We seek to understand the reasons why chemists tend not to express their measurements by the mole, the SI unit of amount of substance, which is said to have been introduced at their request and which is appropriate for many chemical measurements. Some of the background has been discussed previously [7, 16, 17]. Here we hope to discuss:

1. Why and to what extent we advocate a coherent implementation of a wider use of "amount of substance" by chemists
2. Why the use of the mole itself does not solve pressing common problems in chemical measurements
3. Why certified reference materials can meet many, but not all, needs of chemists
4. How we hope a consensus either exists or can be achieved regarding the traceability of measurements to SI
5. Why RMs are necessary to promote harmony among chemical measurements worldwide.

The nature of chemical measurements

Measures of a mass, a length, or a time are not dependent on the composition and constitution of the material. By the definition of the mole, need exists for amount-of-substance measurements to specify the entity among possibly many types of entities in the material under consideration. Amount-of-substance measurements are highly dependent on the composition and constitution of the material. Chemical measurements fall into four groups:

1. Measurements that can be expressed as a mole/mole ratio, the most basic measurements in chemistry, are typified by processes which react, interact, blend, or replace a described amount of substance A with a described amount of substance B. Included are solution concentration measurements when all solutes are known in a known solvent. Note especially that these measurements are independent of the magnitude of the unit mole. Note also that if these measurements are made by mass or volume determinations, the uncertainties in the corresponding atomic or molecular mass values must be taken into account.
2. Measurements that can be expressed as a mole/kilogram or mole/liter ratio are the most commonly made and are typified by a described amount of substance of compound A in an unspecifiable amount of substance B. Note that for these measurements the uncertainty of and relation to the unit mole, just as those applicable to the kilogram or meter, are involved.
3. Measurements that can only be expressed as kilogram/kilogram or kilogram/liter are unusual because they involve amounts of substances of unknown composition. Instances of this type are not really rare. Examples are particulates in air and condensed-ring compounds in tar. Chemists can be reassured that no mention of the mole is made or needed for expressing the results of such measurements.
4. Measurements that are described directly in terms of multiples and submultiples of the kilogram, the liter, or the mole are the measurements that provide the underpinning of chemical measurements in science, technology, and trade. They are typified by calibrations or validations of values of weight sets, reference materials, or instruments, as well as by determinations of the magnitude of the unit mole of a specifie compound (from the quotient of that compound's mass divided by that of a single ^{12}C atom), or of the Avogadro Constant.

Measurements for which reproducibility is more easily obtained than accuracy

Practical chemical measurements are commonly more precise than accurate. By that statement, we mean that the uncertainty of a measurement relative to the true value expressed, in either the mole or the kilogram, is greater than the range for repeated measurements in the same or even different laboratories at different times or by different operators under different environments.

By contrast, satisfactory practical mechanical, electrical, optical, and thermal measurements are often made adequately for the purpose at hand, even if less accurate than corresponds to the optimum achievable uncertainty relative to "true value" expressed under SI. Routine measurements in these fields can thus be expressed conveniently in terms of the relevant SI unit to an uncertainty determined principally by the uncertainty of the practical measurement in the "field."[1] Harmony among most physical and engineering measurements can be achieved to the uncertainty of the measurement in the field by traceability of all measures to the SI unit without invoking an intermediate "standard"[2] or RM.

In physical science there are occasionally instances where measurements need to be more reproducible than the lowest achievable uncertainty relative to the true value in SI units. Chemists, not just occasionally but as a rule, must achieve traceability of measurements by use of some standard, a reference material, a reference instrument, or a reference method [18]. The spread of these measurements made in different laboratories is often required to be smaller than the uncertainty with respect to true value. Nevertheless, one should state any such measurement in moles along with an assessment of the quality of its reproducibility. Such a statement will be different depending on its applicability within a laboratory, between laboratories, for a given method and environment, or in relation to an RM. When an RM is used, one must also include its often larger uncertainty of traceability to the SI unit. This uncertainty of the value of the RM must be included in the total uncertainty of the unknown.

Some important issues in all metrologyy

When discussing traceability of both physical and chemical measurements, one must be clear from the outset on the following conditions applicable to any measurement or measurement capability.

The type of measurement

First, we must specify the type of quantity: a base quantity such as temperature (a property that is coupled to a base unit within the SI), a derived quantity such as pressure (a property coupled to a derived unit, being

[1] The use of "in the field" is intended without detriment to measurements made in laboratories other than those whose main concern is the traceability link to the true value and the SI.

[2] Although for physical measurements one often speaks of various kinds of "standards," there is a functional difference, but no sharp distinction, in current usage between, say, a transfer standard and an RM.

the quotient of two or more base units within the SI), or even a quantity such as a hydrogen-ion concentration (a property that by convention is not commonly coupled to the SI, although perhaps it should be).

The relevant range of measurement

Measurements are needed over a total of many more orders of magnitude of a quantity than any one measurement methodology or instrument can achieve. For electric current, the measurement in a nerve fiber near 1 nA will differ from one applicable to a gigantic TA current in a magnet laboratory. At the two ends of the measurement range there is a non-trivial need to relate any "standard" in the smallest or highest range to the applicable SI unit itself. Needed amount-of-substance measurements, too, may range over more than 12 orders of magnitude.

The uncertainty statement

The uncertainty applicable to a measurement contains components for repeatability and reproducibility [9, 19, 20], caused in part by variability of measurement-relevant parameters. The uncertainty also depends on the individual making the measurement, the laboratory facilities used, and the environment during the measurement. Without some quality control over measurements, statements on relevant traceability can have little meaning. Such controls provide a laboratory with confidence in its operators and credibility to the outside.

Often of general interest is the reproducibility of measurements when operator, equipment or environment is not the same. One must commonly distinguish clearly between uncertainties applicable to measurements at different times (called repeatability [9]) and those made in different places (called reproducibility). A statistical analysis of homogeneity may be needed whenever a measurement is made on a representative sample from the object to be evaluated.

The similarity principle of metrology

In metrology in general, the closer the similarity between two specimens, the smaller the relative uncertainty of the measured difference between them and the easier it is to make a reliable measurement. Thus, by the use of suitable "standards," measurements in the "field" can become highly reliable and far less demanding and costly.

By this similarity principle it is possible to measure precisely and relatively easily small differences from an amount, or ratio of amounts, given by a "standard." RMs thus become very attractive vehicles for measurement traceability and quality. However, there is an associated problem: good reproducibility of comparisons between pairs of similar specimens is liable to mislead and, in practice, often causes underestimations of total uncertainties through failure to consider additional, large error sources.

Fitness for purpose

The achievement of smaller uncertainties than needed is usually uneconomical. In a practical way, realistic uncertainty assessments in relation to true requirements lead to economically sound planning for measurements to be fit for their intended purpose.

The classical measurement pyramid

A simple view of measurement services pictures the International Bureau of Weights and Measures (BIPM) with its international prototype kilogram and the seemingly perfect constants of physics at the peaks of huge pyramidal systems for all types of measurements, each with many levels. The first level below the apex lists the realizations of units at a number of national metrology institutes, passing on slightly more uncertain measurements to a much larger number of laboratories, which in turn service lower-tier measurement laboratories, until at the very bottom of the broad-based pyramids the workbenches receive calibrations that have become a little more uncertain at every intervening level. That system is simple to understand and works well for most industrial and legal services and for the control of small-scale markets, for which the step-by-step losses from impressive accuracies near the appropriate apex level are tolerable. An inverted pyramid may also become useful for illustrating traceability [7].

For chemical measurements, a possibly preferable system is illustrated in Fig. 1. Various possible forms of realization of traceability are given. They range from virtual lack of traceability to a fully "SI-bonded" measurement. The authors tentatively use the term "SI-bonded" to indicate a direct realization of the SI unit, as opposed to being traceably linked by way of measured values. Any user laboratory must seek a reference laboratory that is capable of providing measurement links of the adequate uncertainty and that provides the direct bond to the SI, if that is needed. The reference laboratory can in turn choose the traceability quality that it wishes to maintain, with the responsibility of fulfilling the corresponding competence requirements.

Fig. 1 Traceability Schemes for Field Measurement values (FM)

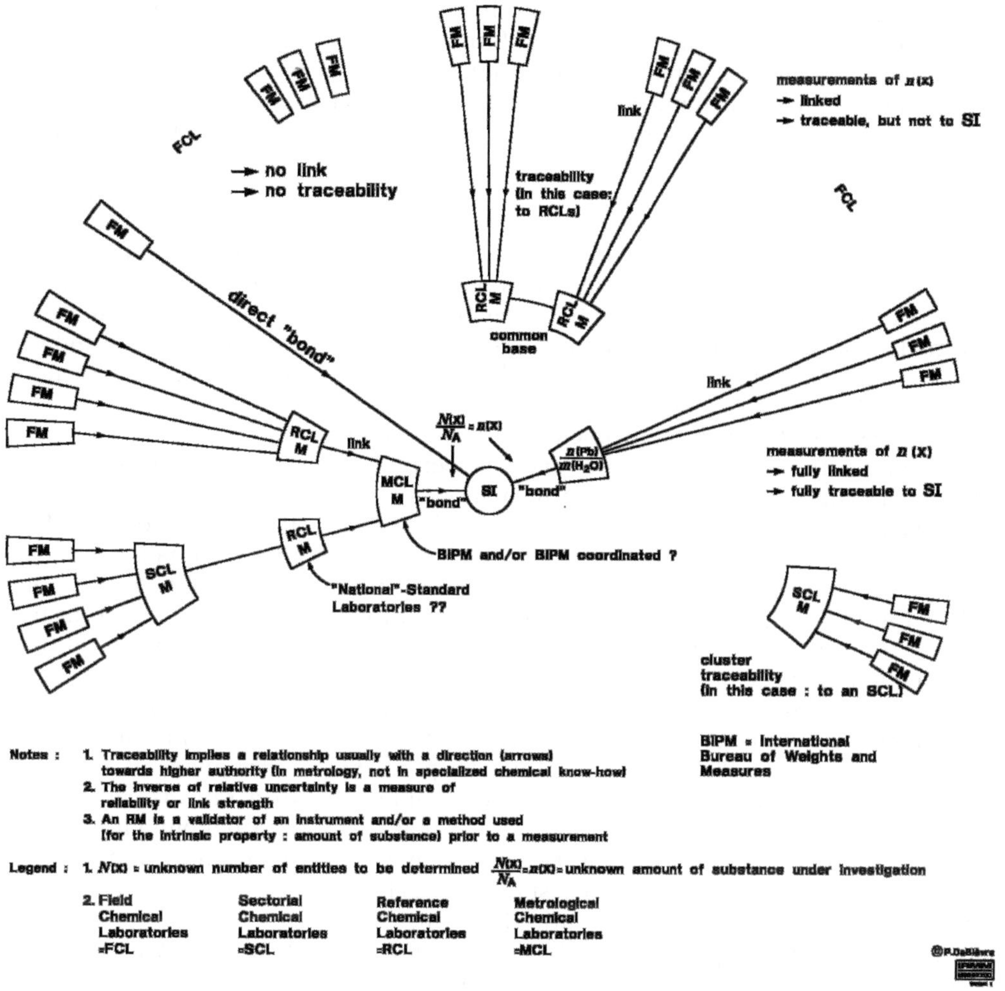

M = Measurement values

In modern high-technology situations, however, very high reproducibilities are frequently required. A good metrological system must provide means whereby any measurement station can have access to the highest needed level of the system.

Some important issues for the wider introduction of metrological concepts into chemistry

The above features are common to all measurements. However, some chemical considerations do not have a clear equivalent in physical measurements.

The diversity of chemical measurements

Whereas the types of measurements in physics and engineering do not exceed the numbers of base and derived units of the SI, chemical measurements are virtually infinite, equal to the number of chemical elements and compounds. Whereas the magnitude of, say, mass is defined independently of the entity for which it is measured, the amount-of-substance determination is made specifically relative to one entity. This situation should not lead to confusion, but some chemists fear that it might. For instance the "mole of nitrogen" is not defined until it is said whether reference is made to N_2 or N.

The word "mole"

Other potential difficulties for chemists arise from differences among molecular, molar, and the historic meaning of "mole" in chemistry [16, 17]. Some find the mole unsuited as a base unit in SI because, they argue, it is just a number of entities. Others find the use of "amount of substance" awkward, especially when the entity – for instance an ion – is not generally regarded as a substance. However, true and relevant some of these objections to current nomenclature and definitions are, a consensus is most unlikely to be reached on any related change in the foreseeable future. Discussion on such a change here is therefore not relevant to more immediate opportunities for a useful consensus in amount-of-substance measurements.

The matrix effect

Whereas the measurement of, say, mass depends little on the character (e.g., density) of the object for which it is made, the measurement of amount of substance is strongly dependent on the matrix in which the entity resides. Chemists have always been concerned with interferences, but the general problem has become more important with the introduction of many powerful analytical-chemical instruments, the performance of which depends not only on specific physical properties of the entity to be measured, but also on the matrix within which the entity is contained. Chemists may wish for RMs for all entities to be measured in all kinds of matrices of interest to technology or trade. However, the production of every RM is a time-consuming expensive process. Chemists are thus faced with the unending problem of available resources imposing severe practical limits to the number of RMs that can be produced in conflict with the wide range of matrices of interest. Consequently, a most important contribution that basic chemical science must make is in the development of matrix-independent methods of measurement [21]. The challenge is to separate the one entity to be measured from the influence of all other entities in a mixture. By widespread abilities to do so, metrology in chemistry will reach its most desirable aim to make accurate amount measurements related to the mole unit. In the future, chemical metrology should be directed at the basic science on RMs whose link to SI is strong and at field methods whereby specimens can be compared reliably with the RMs, independent of matrix [7, 22].

SI recognizes derived units (products or quotients of base units)

Whereas the measurement of, say, mass can be stated as a fraction of a total mass (e.g., mass of a sample in a bottle), the amount of substance of a given entity can usually not be stated as a fraction of all amounts of substance in a material. One typically does not even know or care about all the other entities, and one certainly does not generally wish or need to analyze the material in terms of all its constituents. The SI system permits and widely encourages coherently associated units. The substance of interest should, where possible, be expressed in the SI unit, the mole. The other substances, the amounts of which are of no immediate interest to the determination, are quantified in terms of SI units such as the kilogram that do not distinguish entities.

Do physicists use the mole?

Geophysicists generally describe the composition of the universe or of the earth by mass percentages. They could use the mole, the amount of terrestrial substance of, say, aluminum. In the very processes leading to the birth of the elements, amount ratios are of prime interest. The end amount of Al would be expressed in mole per average terrestrial kilogram.

Measurement by ratio

Proponents of the SI for chemistry must consider that proportionality is deeply embedded in chemical thinking.[3] Many of the potentially most reliable analytical techniques – for instance isotope-dilution mass spectrometry – yield ratios in the first place. In complex series of ratio measurements the uncertainty propagation is more straightforward than when sums and differences from standards – such as for mass determinations – are involved. Consistent with the use of SI, the value of a ratio is called a "measurement" when numerator and denominator are multiplied by a unit and the related uncertainties have been evaluated.

Uncertainty about the nature of the entity to be measured

Chemists may not exactly know what is the entity they wish to measure in a material. A common example is moisture, say in grain. There are known to be continuous levels of strengths of chemical bonding of the water molecule in products. Mass loss on heating is routinely used to determine moisture in grain, but may cause error by including in the measured loss volatile compounds other than water and will also depend on the method used, principally on the temperature and time of drying. In giving the result in mole of H_2O per kilogram, one cannot assure that it was free water in the grain, where some of it was present as a different chemical entity. The same may apply to a metal element, say aluminum in an alloy. The user may well be interested in whether a mole of Al per kilogram refers to total aluminum or just the metallically bound – as opposed to oxide – aluminum. The result obtained in a measurement will then depend on the measurement

[3] By contrast, many measurements are initially additive, as is true for mass, time interval, and length.

method that is used. The use of amount-of-substance measurements can neither help nor hinder the chemist's need to carefully distinguish significant entity differences such as those due to chemical bonding and molecular association in a material.

Some vague usages of terms in measurement processes [23]

It is quite common for the chemical community nowadays to use the terms "calibrate" and "calibration" for any process that converts an observed value into a more reliable result, which is then called "corrected," "true," or "calibrated." We must also concede that RMs are sometimes used that do not have a matrix closely similar to that of the sample. To make matters worse, uncertainties associated with that situation are generally ignored. Insofar as the chemical community is aware of these problems, the call goes out for more and more RMs in appropriate matrices beyond available capabilities to produce reliable RMs. In order to arrive at rational conclusions on these issues, it is necessary to examine closely and to understand the proper role of "calibration" and "validation" procedures. In the following paragraphs we describe our views and hope that others will endorse them.

What constitutes a measurement?

A measurement of a specified property in an "unknown" material is a quantitative comparison by ratio or difference made of that property between a reference standard or reference material and the unknown or between relevant settings in an instrument, preferably in the appropriate unit for the quantity under investigation, provided:

a. Measurements of the relevant type and range, at the site where the measurement is made, are subjected to reliable uncertainty assessment.

b. The result (difference or ratio) is proven to be a known function of the true difference or ratio, or appropriately corrected for non-linearity, usually by means of a set of RMs.

c. The comparison applies only to a constituent part of either or both the RM and the "unknown," and the comparison is:

i. proven to be independent of the matrices,

ii. based on knowledge that the matrices are precisely similar, or

iii. quantitatively evaluated for variability with matrix

d. The result is given with its uncertainty including those caused by possible lack of linearity and by the above criteria applied to RMs involved.

Under these conditions, the comparison constitutes a measurement, and the value given of the property in the "unknown" has been determined.

Chemists will have an important reservation concerning this understanding of what constitutes the uncertainty of a measurement. Physicists and engineers may not, but chemists often are subjected to major sampling, stability, blank, and contamination errors. Chemists should include them in their total uncertainty estimates. The distinction between the measurement uncertainty and the degree to which the measured sample fails to represent the relevant larger bulk needs to be debated and discussed for consensus and understanding.

What is a calibration?

Let us begin with the ISO definition [9]: A calibration is a "set of operations that establish, under specified conditions, the relationship between values of quantities indicated by a measuring instrument or measuring system, or values represented by a material measure or reference material, and the corresponding values realized by standards." Applied to amount measurements, the "standards" would then be the values assigned to the RMs (of defined composition) at the stated uncertainty relative to the true value of the property, expressed in SI units, or relative to an internationally recognized, certified standard RM for the relevant property, range, and matrix composition.

An instrument or system is said to be calibrated for amount measurements only if, within a specified range, a value versus signal (response) curve has been evaluated against RMs including two near the ends of the range. At the present time, it is unfortunately quite common to use the term "calibration" to describe any process which converts a single observed measurement into a more reliable result.

What is a "validation"?

An RM can validate a measurement procedure (including the measurement instrument) [13] if, prior to its use for an unknown sample, it has been shown to give:

1. A quantitative response for the quantity (in the relevant range) to be measured
2. A response with a defined and acceptable repeatability
3. A response with a defined and acceptable reproducibility over changing times and measurement conditions
4. A defined and acceptable estimate of their overall intrinsic uncertainty

Traceability for chemical measurements

ISO in its vocabulary for metrological terms [9] defines traceability as follows: "property of the result of a measurement or value of a standard whereby it can be related to stated references, usually national or international standards, through an unbroken chain of comparisons all having stated uncertainties". Thus, the term does not apply directly to laboratories, but should be applied to the results of chemical amount-of-substance measurements. Every link in the traceability chain should consist of comparisons that are measurements in accordance with the above proposed meanings, which include the validation of measurement procedure by RMs. A measurement therefore often has strong links to internationally accepted RMs, but may be only weakly connected to the SI unit. For comparability among measurement laboratories, the strength of the link must be adequate to assure equity in trade. Weakness of the relation to SI may thus be acceptable, but the metrologically minded chemist will be disposed to aim for strongly linked reference measurements, methods, and instruments. They are based on simpler concepts with greater permanence and would be more easily understood by the wider public. Other definitions of traceability have been described [24–26].

Not all chemical measurements are, or should be, traceable to the mole. We have seen instances where the unit of mass was the proper SI unit for a quantitative measurement of a material of unspecified entities. There are chemical measurements that are not, but probably should be, referred, and preferably be traceable, to the SI unit. Color is used either simply as a qualitative attribute not subject to a measurement, or it is measured quantitatively by some spectrometry, where it may inevitably be subject to high uncertainties from both the measurement itself as well as from theory, such as the Lambert-Beer Law, but well understood in relation to SI.

The description of the relation of a measurement to an SI unit encounters a basic difficulty when the desired meaningful measurement result is a ratio, as in many chemical determinations. The magnitude of the unit then becomes irrelevant. Chemists err when they claim that the inaccuracy of their weight set relative to the international prototype is a component in their uncertainty budget. The self-consistency of their weight set is of course of paramount importance. Since that would include tareweights, internal balance weights, and sensitivity weights, the advice to use weights calibrated against the international kilogram is still generally good.

The quality of ratio measurements seems not to be concerned directly with the SI unit. The only essential condition is that the unit for the numerator be the same as that for the denominator. Traceability requirements for many amount-of-substance measurements, therefore, appear to concern not the unit mole, but a standard measured ratio, preferably between pure defined substances in one RM. Nevertheless, the authors propose that by consensus it shall be a rule for all measurements, where a choice could be made, that it shall fall on the SI unit.

Unusual are measurements for which a direct link to the mole is useful. We should probably not talk about traceability in that connection, because that term is defined as a relation between measured values. An acceptable chain of measurements for compound X of established purity, containing element E that has isotope ^{i}E and that would establish a link to the mole, then would take one of the following general routes: the amount of substance $n(\mathrm{X}) \rightarrow n(\mathrm{E}) \rightarrow n(^{i}\mathrm{E}) \rightarrow n(^{12}\mathrm{C})$; or $n(\mathrm{X}) \rightarrow n(\mathrm{E}) \rightarrow n(\mathrm{C}) \rightarrow n(^{12}\mathrm{C})$. The ratio of atomic masses $m(^{i}\mathrm{E})/m(^{12}\mathrm{C})$ is also involved in the definition, but that ratio is known with a negligible uncertainty compared with the other links in the chain. Clearly, only in a few instances will laboratories attempt to execute such a chain of measurements for a link to the SI unit. Is it fear that such a difficult process is involved in every chemical analysis that has kept so many chemists from using the mole as the way to express chemical measurement values? Or is it just habit and the convenience of a balance that subconsciously links amount of substance to amount of mass?

Laboratory accreditation

For laboratory accreditation, based on ISO guide 25 [27] and the EN 45001 standard, as well as for certification, based on the ISO 9000 series of standards [1], it is required that measurement and test results be traceable to international, defined, and accepted physical and physicochemical standards [28]. This requirement includes the use of conventionally expressed quantities and units in conformity with the SI [29]. It also includes the proper use of the concept of measurement uncertainty. All these are necessary conditions for reliance on the measurement results of another laboratory. Accreditation is granted when a laboratory has demonstrated that it is competent and capable of working in the above-mentioned sense. Technical trade barriers then fall away, and the needs and requests from industrialists, traders, and the general public can be met in the interest of open and fair trade, health, safety, and the environment.

For amount-of-substance measurements we include kilogram mass units, which are linked to the amount-of-substance unit in SI by the atomic-weight values. The latter differ greatly in uncertainty for different chemical entities, but are always available, with the best estimates by current knowledge of their uncertainties, through the International Union of Pure and Applied Chemistry [30, 31].

Reference materials

In the above sections we have already illustrated some of the characteristics and uses of RMs. A more formal definition of RM by ISO is [9]: "material or substance one or more of whose property values are sufficiently homogeneous and well established to be used for the calibration of an apparatus, the assessment of a measurement method, or of assigning values to materials." Extraordinary care in the production of RMs [15] is essential for effective, harmonized chemical measurements. Special features of certified RMs are carefully explained by that ISO document [9, 15] and their designation as measurement standard specifically authorized.

One may be inclined to suppose that for each type of chemical measurement there is a need to build a measurement system based on the pyramid concept [7, 32]. For the practicing chemist, however, this would be seen only as an unhelpful imposition. Previously discussed limitations of such a pyramid system would apply equally to the use of RMs. In addition, there is a major difficulty due to the previously discussed differences between RM matrix and sample matrices. Whereas for extrinsic measurements the composition of an RM or other traveling standard is of little or no concern, intrinsic amount-of-substance measurements are generally affected by the internal composition, structure, and texture of the RM.

The limited number of reliable RMs that can be prepared and made available leads to the use of possibly inappropriate RMs. When the matrix in a sample differs from that of the RM, reliable comparison may be very difficult. A provision for the support of critically important and accurate bench level measurements is needed. In such situations there is a better alternative: from the bench level a specimen with typical matrix properties is sent to a laboratory having competence appropriate for providing a "reference measurement." That value is communicated back to the "bench" where it provides a certified value – a kind of in-house RM – for comparison with routine sample measurements. Thus, the concept of reference measurement emerges as being equally as important as that of the RM. Chemical science has no other choice, since the combined output of RM-producing institutions could not possibly accommodate all the rapidly diversifying demands for all measurands in all matrices of interest.

In order to establish traceabilities of measurements, we advocate the structure shown in Fig. 1 where many types of linkage can be found, including but not limited to those terminating in SI.

A system for describing types of candidate chemical materials for RMs

We would also advocate the optional use of a descriptive materials system for candidate RMs. Firstly, we should have categories depending on the chemical nature of the materials (Table 1). Secondly, we should agree on RM classes dependent upon their degree of traceability (Table 2).

The isotopic composition of an element in a specimen can be established and expressed in abundances – that is amount-of-substance fractions or moles of iso-

Table 1 Categories of reference materials, determined by their chemical nature

Category	Kind of material	Description and criteria in terms of mateiral composition
A	High purity	Pure specified entity (isotope, element, or compound) stoichiometrically and isotopically certified as amount of substance, with total impurities <10 μmol/mol
B	Primary chemicals	As above, but with limits of <100 μmol/mol
C	Pure	One constituent >950 mmol/mol
D	Matrix	Matrix with one or more major constituents >100 mmol/mol
E	With minor constituents	Minor constituents in matrix <100 mmol/kg
F	With trace constituents	Trace constituents <100 μmol/kg
G	With ultra trace constituents	Ultra trace constituents <100 nmol/kg
H	Undefined	Entities unspecified or undefinable

Table 2 Classes of reference materials determined by their traceability

Class	Description and criteria in terms of traceability to SI
0	Pure specified entity certified to SI at the smallest achievable uncertainty
I	Certified by measurement against class 0 RM or SI with defined uncertainty by methods without measurable matrix dependence
II	Verified by measurement against class I or 0 RM with defined uncertainty
III	Described linkage to class II, I, or 0 RM
IV	Described linkage other than to SI
V	No described linkage

tope per mole of element – by comparison to synthetic mixtures of enriched isotope class 0 RMs of that element.

Certification of an elemental class 0 RM can be performed by metrology laboratories having the best scientific procedures for the establishment of traceability routes to the SI system. For every such RM the cost in facilities and experts' time is very high and in practice cannot easily be balanced against sales. Only a long history of the laboratories' reliability and their free and open discussions of problems, coupled with energetic self-criticism, will reassure the scientific and technological communities. Metrological quality, not cost and economy, should be the prime concern of operators within such laboratories.

All other classes of RMs are needed in much greater number and diversity. They are therefore of much greater potential interest commercially. Intercomparison between similar RMs is always helpful. Only one class IV or class V RM should be made available by consensus for a certain purpose, so that all laboratories are encouraged to make their measurements comparable to others through just one RM.

Validation of a measurement procedure including an instrument can be performed with an RM of class 0, I, or II, but only if differences in matrix or impurities are specified, small, and of proven limited influence on the uncertainty. The uncertainty of the RM relative to true value or the mole may be larger than the link between the measurements on the material and the RM. Traceability between measurements can be achieved with the help of all classes of RMs, but requires a clear statement on uncertainty. Traceability to the mole, if not by direct realization of the mole, can be established only by class 0 RMs. Their relation to the unit mole must be established by way of atomic-weight determinations or by direct atomic mass comparisons with carbon 12 atoms.

An example of a class I RM is an RM for which the amount of substance of an element has been measured by isotope dilution against a class 0 RM, provided the measurement has been shown to be in accordance with basic laws [7, 22, 33] of chemistry and physics.

Conclusions

Reliable chemical measurements in future will depend on more RMs with direct links to the SI as well as on RMs of greater diversity than are available now. Chemical science will be assisted by clear consensus definitions of traceability, certification, and validation, as well as by a widely accepted system for describing RMs by material composition, degree of traceability, uncertainty, quality, and purpose. Ultimately, chemists, physicists, and engineers benefit from adherence to the well-grounded and well-established discipline of metrology under a coherent system of units.

Acknowledgement The authors thank two reviewers of the manuscript, Robert Watters and Walter Leight, who provided numerous valuable suggestions.

References

1. International standards for quality management (1992) ISO Series 9000, Geneva
2. King B (1993) Analyst 118:587–591
3. Rasberry SD (1993) Fresenius J Anal Chem 345:87–89
4. De Bièvre P (1993) Int J Environm Anal Chem 52:1–15
5. Quinn T (1995) Metrologia 31:515–527
6. Page CH, Vigoureux P (1975) The International Bureau of Weights and Measures 1875–1975. National Bureau of Standards, GPO Washington
7. De Bièvre P (1994) Fresenius J Anal Chem 350:277–283
8. Terms and definitions used in connection with reference materials (1992) ISO Guide 30, 2nd edn, Geneva
9. International vocabulary of basic and general terms in metrology (1993), 2nd edn. ISO, Geneva
10. De Bièvre P (1993) Quimica Nova 16:491–498
11. Lamberty A, Moody J, De Bièvre P (1996) Accred Qual Assur 1:(submitted)
12. Cali P, Reed WP (1976) National Bureau of Standards, Special Publications, vol 422, pp 41–63
13. Marchandise H (1993) Fresenius J Anal Chem 345:82–86
14. Reed WP (1995) Fresenius J Anal Chem 352:250–254
15. Quality system guidelines for the production of reference materials (1995) ISO Guide 34, Geneva (in press)
16. De Bièvre P, Peiser HS (1992) Pure Appl Chem 64:1535–1543
17. McGlashan ML (1995) Metrologia 31:447–455
18. 5th International Symposium on Harmonization of Internal Quality Assurance Schemes for Analytical Laboratories (1993) Association of Official Analytical Chemists, Washington D.C.
19. Guide to the expression of uncertainty in measurement (1993) BIPM, IEC, IFCC, IUPAC, IUPAP, OIML ISO, Geneva
20. Kaarls R (1981) Metrologia 17:73–74
21. De Bièvre P (1995) Updated from original German, Ch. 7. In: H. Günzler (ed) Akkreditierung und Qualitätssicherung in der Analytischen Chemie, pp 131–156, Springer, Heidelberg (in press)
22. De Bièvre P (1993) Anal Proc 30:328–333
23. Management and quality assurance vocabulary (1986) ISO 8402, Geneva
24. Belanger BC (1980) Standard News 8:22–28
25. Belanger BC (1983) In: 37th Annual ASQC Quality Congress Transactions. American Society for Quality Control, Boston, pp 337–342
26. Rasberry SD (1983) In: 37th Annual ASQC Quality Congress Transactions, American Society for Quality Control, Boston, pp 343–347

27. General requirements for the competence of calibration and testing laboratories (1990) ISO Guide 25, 3rd edn, Geneva
28. Cigler JL, White RV (eds) (1994) National Institute of Standards and Technology Handbook 150
29. Quantities and units (1993) ISO Standards Handbook, Geneva
30. Atomic weight of the elements 1993 (1994) Pure and Appl Chem 66:2433–2444
31. De Laeter JR, De Bièvre P, Peiser HS (1992) Mass Spectrom Rev 11:193–245
32. Taylor JK (1993) National Institute Standards and Technology Special Publication 260-100, pp 1–102
33. De Bièvre P (1990) Fresenius J Anal Chem 337:766–777

Accred Qual Assur (1997) 2:168–179

Paul de Bièvre
Robert Kaarls
H. Steffen Peiser
Stanley D. Rasberry
William P. Reed

Protocols for traceability in chemical analysis

Part I: Definitions and terminology

Received: 12 January 1997
Accepted: 31 January 1997

P. De Bièvre (✉)
Institute for Reference Materials and Measurements, European Commission – JRC
B-2440 Geel, Belgium
Tel.: 32-14-571-605; Fax: 32-14-591-978

R. Kaarls
Netherlands Measurements Institute, 2600 AR Delft, The Netherlands

H. S. Peiser, retired from
National Institute of Standards and Technology, 638, Blossom Drive, Rockville, MD 20850, USA
Tel.: 1-301-762-6860; Fax: 1-301-762-1173

S. D. Rasberry · W. P. Reed, retired from National Institute of Standards and Technology, Gaithersburg, MD 20899, USA

Abstract The authors propose definitions and terminology for protocols on traceability links, generally to the international system of units, for specific chemical-analytical measurements in accordance with recognized principles of science. These definitions and terms could be useful in science, technology, commerce or law. A chain of such links leads from a measurand in a sample up to a unit in the International System of Units or, if unavailable, to a value on an internationally recognized measurement scale. The quality of such a chain is quantified by combining all recognized uncertainties estimated for all its links. These uncertainties of the measured values arise from many potential error sources. The protocols should give details of specific uses of reference materials, measuring instruments and standard measurement methods.

Introduction

This publication is the second of three contributions on traceability in chemical analysis. The first was published in this Journal [1] and deals with the general principles, whereas the third is planned chiefly to present examples, but also to suggest implementation procedures, to assess comments from chemical groups and to introduce possible modifications of concepts and definitions [2, 3].

The second contribution we present in two parts. In this first part we discuss definitions and terminology, mostly from recognized sources [2–9]. Some ideas in this article go beyond established international understandings; they are introduced for debate and possible refinement. The terms used here are responsive to the fundamental concepts under which chemical analysts can formally substantiate and record a traceability link. A chain of such links should lead from the value of a quantity in a sample up to a unit in the International System of Units (SI) [5] or, where that is not possible, up to a unit on an agreed and conventional measurement scale.

We address chiefly individuals or groups of analysts who aim to originate a protocol, that is a document recording the procedures for a specific link. That protocol establishes scientifically reliable measurements for the benefit of equity in trade and industry, as well as for legal interpretations of scientific realities.

A protocol must deal with the quality of the link based upon carefully estimated uncertainties [6, 7] from all foreseen error sources that remain after due precautions have been taken and after significant corrections have been applied where possible. The combined uncertainties of all links in the chain of links will then define the quality of a link to SI[1] or to some other rele-

[1] As has become customary [4], we use "the traceability (or the link) to SI" meaning "the traceability (link) to an appropriate unit or units in the SI"

vant scale unit. This quantitative assessment of quality of traceability to SI from combined uncertainties is not inconsistent with the metrological term of "accuracy" [2, 6]. It differs from popular meanings of "accurate" such as "free from error" and "highly exact". A traceable measurement may be adequate for its intended purpose, yet be inferior compared with the optimum achievable.

Fundamental understandings

By "protocol for *traceability*" [2][2] we here mean a documented *record* of a relationship, consisting of a "*link*", or *chain* of links, emanating from the *value* of a *quantity*[3]. The value is obtained by a *measurement* applicable to the *measurand,* the property of an entity in a *sample* which may consist of a pure material or incorporate the entity in a *matrix.* Each such traceability link is established for a stated chemical purpose and asserted by virtue of that measurement, which relates the forementioned value to another value in a *reference material* (see Fig. 1 for an outline of the use of RMs in typical chemical analyses) or to the response of a calibrated instrument (see Fig. 2). This measurement is carried out in a *responsible laboratory* using planned and described procedures, in a *validity interval* (time period) for a specific type of quantity (such as a concentration or other material property), within a limited *range* of magnitude of the quantity measured. The measurement is characterized, in part, by an observed *repeatability* and invariably by a substantiated estimated *uncertainty* [6] (including especially any arising from matrix effects), which is the sole indication of *quality* of the traceability relationship for each link or, when duly combined, for a chain of links. Thus, the uncertainty becomes the quantified indication of quality for the measured value itself. Wherever possible, the value of that measurement is ultimately made traceable to an SI *unit* (or units) [2, 5, 8], through *realizations* of those units. If not possible, the final link is made to a unit in an internationally recognized scale.

[2] Traceability is defined as follows [2]: "property of the result of a measurement or the value of a standard whereby it can be related to stated references, usually national or international standards, through an unbroken chain of comparisons all having stated uncertainties"

[3] When using "quantity" [3] we will refer to a property, such as mass, length, amount of substance, or speed of light. We will not use "quantity" for describing a portion or bulk of a material, a chemical, or a sample, but rather consistently use "amount" of a sample or of rubber etc. Furthermore, we will try to distinguish such a general use of "amount" from "amount of substance" which is an SI base quantity requiring specificity of entity in terms of its chemical formula

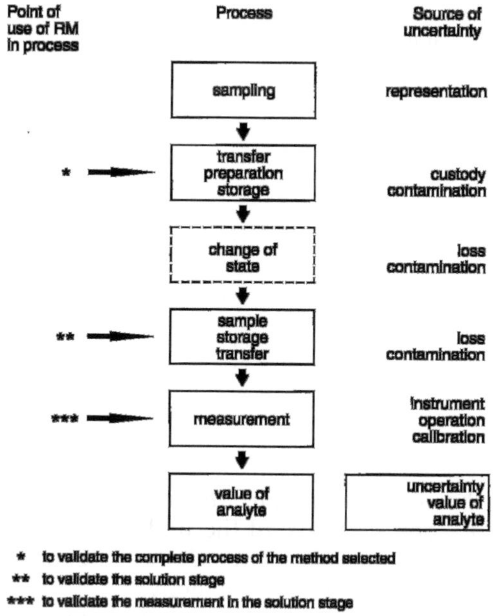

Fig. 1 Schematic of typical chemical analysis

Underlying concepts and definitions

The record

The record of a traceability protocol is a written document that may serve: in science and technology as reference; in product control as procedural "written standard"; in environmental comparisons as precept [10]; in trade and industry as basis for agreement, especially where border crossings are involved; and in courts of law as a means to judge whether specified limits are met[4].

The sample

For a chemical measurement the history and homogeneity of the sample must be established. Subsamples from the original sample are drawn for measurement. Although the selection of a sample in the "field" is important and often of concern to analysts, this paper deals only with the sample as delivered to the laboratory[5]. Two kinds of questions there can profoundly affect the uncertainty associated with the measurement [12]:

[4] An elaboration on how a duly recorded traceability protocol might be used outside the professional chemical arena is not the subject of this paper

[5] Questions of chain of custody of samples or "trackability", a term recently proposed [11], and due representation of the population are not addressed in this article

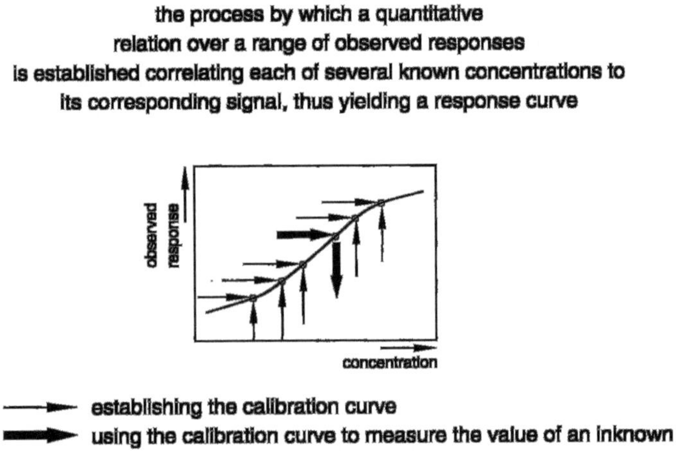

Fig. 2 Instrument calibration for analytical measurements

1. Is the sub-sample on which the measurement is made representative of the undivided sample?
2. Does the measurement of the intended quantity meet the measurement objective?

An important issue, for example, may be whether the measurement value is to apply to a specific sample or to the bulk of its source material. When surface contamination is the purpose of the measurement, total trace-element measurements could be misleading. Thus, surface sampling may have to be part of the protocol. Conversely, measurement methods that give results preferentially for surface layers may not be representative of the bulk of the subsample.

The measurand

The measurand is the selected property to be quantified by measurement of a constituent in a sample. For most analytical measurements, one prefers to select a quantity that is invariant on division of the material. Mass, amount of substance, volume etc. do not remain invariant on division and so are unsuitable for the characterization of a material.

Temperature remains invariant on division, but is unsuitable for characterizing a material because of the dependence of that base SI quantity on the external environment. To a lesser extent, external temperature and pressure conditions affect volume, too. In chemistry, the commonly used ratio of amount of substance (of a stable[6] entity) to mass of the material, in which it is uniformly contained, is not only invariant to contamination-free division, but is also independent of external environmental conditions, as long as they ensure the stability of the entity. In this respect amount-of-substance concentration per mass of material is suitable for characterizing a material, but not unique. Mass per heat capacity has similar invariance on division of the material and independence upon the environment. Concentration by mass per amount of substance has the additional unique property for a pure substance that these two base quantities are related by the molar mass of the entity. Thus, the analyst has the option of measuring either of the base quantities and from it deriving the other base-quantity value and hence the concentration with an appropriate increase in uncertainty.

Issues involved in the choice of measurand may be related to the purpose of measurement, which should be precisely and unequivocally identified and stated in the protocol.

The measurement

The measurement – a quantified comparison by difference or ratio between two values of the same quantity – shall conform to well-accepted principles of the chemical profession and good measurement practice [12–14].

The value

The value is the numerically expressed magnitude of a quantity either in a material or indicated by a calibrated instrument (see Fig. 2). An uncertainty must be associated with every value. Its important estimation is discussed in detail in the section "The uncertainty" below (see also [6, 8]).

Wherever possible, every value and every uncertainty that is cited in a protocol or quoted in the implementation of a protocol shall be expressed in a unit of the SI (with or without prefix) associated with a number commonly called the numerical value [5]. Thus, the value and its uncertainty are multiples or fractions of that SI unit. If an SI unit for the relevant quantity is not in

[6] The meaning of "stable" as here intended does not always indicate constancy of value. For radioactive materials, for example, a quantified change with time is here understood to be "stable"

common use, the value could be associated with an internationally accepted measurement scale or on the scale of a written procedure, perhaps involving an in-laboratory prepared reference material that is properly identified. In some cases, even a commercially available stock solution may serve this purpose. The long-term constancy of such values is an important issue, to be considered in the use of any reference (see section "The validity interval" below). Possible non-negligible instabilities of a reference value used in a protocol must be mentioned and appropriately taken into account.

The unit

In the physical and engineering sciences, the metric International System of Units for measurements [5] has become widely accepted throughout the world. An intended characteristic of all SI units is their unsurpassed stability and their independence of location and time. That characteristic of the SI contributes decisively its unique appeal to measurement science and technology. In technology, the use of SI is gradually gaining acceptance over many customary, especially non-metric, units. Chemists have no problem in using SI for mass comparisons, because of the convenience and sensitivity of analytical balances and the universal acceptance of the kilogram as the unit for mass[7]. Chemists also use SI units for measurements of volume, temperature, and some other quantities, but tend to avoid the use of the mole for amount-of-substance measurements, even when the chemical entity is well defined and although that quantity is the most meaningful for the consideration of chemical formulae, reactions, kinetics, and energy. In such situations the use of the mole in protocols should be expected. Traceability to SI, however, can be claimed relative to any SI unit that is appropriate for the measured quantity. That SI unit can be a base or a derived unit, or even a unit temporarily accepted for use within SI. When traceability is planned to derived units or composed of products or quotients of other SI units, it is often operationally necessary to achieve traceability to these SI units separately. This is the situation for the quantity of greatest interest for characterization of any material by chemical measurement: concentration measured in mole per kilogram or mole per cubic meter. Most laboratories may routinely maintain traceability to the SI units of mass and length at lower uncertainties than is needed for many protocols for chemical analytical measurements.

Whereas mass can be quantified irrespective of the intrinsic nature of a sample, amount-of-substance quantification requires the explicit chemical description of the entity pertaining to the measurand [15]. A description in mole units may be inappropriate for specific purposes when the composition of the chemical entity lacks specificity and would impose an objectionable molar mass or volume uncertainty. This limitation applies frequently in quantifications for purposes of trade, such as when describing amounts of polymers (with large uncertainty of molar mass and additional difficulty in defining its mean), a lithium compound (of variable isotopic composition), sodium carbonate (of undefined hydration), iron oxide (of undefined Fe valency), chlordane (composed of several molecules), nutritional fibers (vague definition), and numerous other substances.

The mole is associated with a specific chemical entity as defined by its chemical formula [15]. Its structural formula, isotopic composition, isomeric form, crystal structure, or chirality may have to be given in order to completely specify the entity of interest. The achievable uncertainty of amount-of-substance measurement is limited by that of its apparent molar mass. This consideration affects not only measurements on entities with variable molar mass, but those on pure substances. It is related to the traditional and important concern about purity.

Ratio measurements in analytical chemistry will often relate values in different units for the numerator and denominator. The most commonly used ratios between SI units are summarized in matrix form in Table 1 (adapted from [5]). Confusion may result when values are stated in different SI quantities. The pharmaceutical industry, for example, is careful to distinguish values in mg/g from values in mmol/g. It is important to retain the two units used in expressing a ratio such as mol/g. Differential measurements are often made, obtaining a ratio of ratios for which the numerator and denominator are generally expressed in multiples of identical pairs of units, e.g. (g/L)/(g/L). For such a ratio of ratios no great harm is done by stating, for instance: "the con-

Table 1 Frequently used ratios of SI units

Quantity	Amount of substance	Volume	Mass
Symbol of quantity	n	V	m
Name of SI unit	mole	(derived) cubic meter	kilogram
Symbol of SI unit	mol	(derived) m^3	kg
	mol/unit of quantity	m^3/unit of quantity	kg/unit of quantity
Unit of quantity/mol	mol/mol	m^3/mol	kg/mol
Unit of quantity/m^3	mol/m^3	m^3/m^3	kg/m^3
Unit of quantity/kg	mol/kg	m^3/kg	kg/kg

[7] In fact, the kilogram is the SI unit most likely to be redefined in the near future. If so, it will be in terms of an atomic-scale mass rather than by an artifact

centration in the numerator equals a (dimensionless) number of times the concentration in the denominator." An unlike pair of units for a ratio of ratios should be avoided, otherwise it becomes imperative to identify explicitly all four units for a meaningful description of the measurement.

On some occasions, protocols may involve SI units of time, electric current, thermodynamic temperature, or luminous intensity. These units are also base units of the SI. Traceability to SI can even refer to realizations of derived SI units, such as those for energy, pressure, and amount of electricity. Solubility per unit pressure may be quoted in $(mol/m^3)/(m \cdot s^2/kg)$ or in $(mol/m^3)/Pa$, but should not be written as $mol \cdot s^2/(m^2 \cdot kg)$ [5, 20], that is: not in reduced form relating to units of quantities not actually measured.

There are chemical measurements for which the decision to use the kilogram or the mole as SI unit depends on the type of deduction that is intended to be made from the measurement. Such examples could arise in polymer studies, in alloying, in isotropic displacements, in assessing electronically active impurities, in effects from variations in isotope abundances, or in those arising from chemical binding states. When documenting a formulation for an industrial reaction process, the use of mass proportions is appropriate even when an entity is known, because weighing devices alone are likely to be available for preparing the needed mixture. In ionic crystals and the alumino-silicates of the earth's crust, especially when dealing with their solid solutions, the concept of a molecular entity has little relevance. Although quantification under those circumstances is best achieved in terms of mass, certain amount-of-substance ratios represent important features of such materials. For instance the ratio of quadrivalent to trivalent ions in feldspars gives meaningful descriptions of attributes of rocks. For abundance of the elements on earth or in space, the common use of kilogram per tonne should with advantage be replaced by the amount of substance per kilogram or by the less common 'Cosmic Abundance Units' (atoms per 10^6 Si atoms).

The dalton is not accepted within SI [9]. It is perceived as a molecular mass of a specific species. In protocols, molar mass, relative molecular mass or unified atomic mass units should be substituted. The last of these, in conjunction with the SI mass unit, is currently acceptable with an added relative standard uncertainty of 10^{-6}.

Realization of an SI unit

A value – whether based on a specific material or on the output of a detector – when traced by a single link to a multiple or submultiple of an SI unit, at a stated low uncertainty, without requiring intermediate standards, reference materials, or significant empirical correction factors, is a realization of that SI unit. The conditions that are involved in a realization of an SI unit include all those involved in the recently proposed definition of a primary method of measurement [4]. Measurements made by a primary method are in principle realizations of an SI unit. For other realizations of an amount of substance, the entity must be defined and the purity of the material containing the entity must be determined. Conceptually, every entity might be deemed to require its specific realization of its mole.

Reference materials (RMs)

In all but a very few chemical measurements, use is made of reference materials (RMs) with appropriate pedigrees [17–19] (Fig. 1). Analytical-chemical RMs are generally certified by properties (such as concentrations) of entities and by values with their uncertainties, and are sometimes provided with limit values [17–19]. These values within their uncertainty ranges remain in-

Table 2 Categories of reference materials, determined by their chemical nature[a]

Category	Kind of material	Description and criteria in terms of constituent(s) certified
		Single major constituent
A	High purity	Pure specified entity (isotope, element, or compound) stoichiometrically and isotopically certified in amount-of-substance ratios with total impurities <10 μmol/mol
B	Primary chemicals	As above, but with limits of <100 μmol/mol
C	Defined purity	As above, but with limits of <50 mmol/mol
		Maxtrix types
D	With major constituents	Major constituents (in matrix) >100 mmol/kg or >100 mol/L
E	With minor constituents	Minor constituents (in matrix) <100 mmol/kg or <100 nmol/L
F	With trace constituents	Trace constituents <100 μmol/kg or <100 μmol/L
G	With ultra trace constituents	Ultra trace constituents <100 nmol/kg or <100 nmol/L
		Undefined
H	Undefined	Entities unspecified or indefinable

[a] A similar table was originally published [1] as basis for discussion. The version here presented incorporates significant changes. Further suggestions for improvements are welcome. – The proposed limits of the concentrations are arbitrary, and, for instance, preclude RM's for most organic entities in category A. Chemists may well use and state their own RM category designations

variant on division into subsamples. RMs are widely regarded and used in chemical analysis, just as are travelling and transfer standards in engineering and physics practice.

Tentative attempts have been made to categorize chemical RMs in terms of their material composition (Table 2) and to classify them by the length and strength of their traceability chain (Table 3). This classification might in future be generalized and expanded in terms of ranges of relative uncertainties.

Whereas an individual value, the result of a measurement, can be said to be "verified" by another measurement, a chemical method or procedure should be "validated", generally by success in application of that method to a certified RM. Validation remains undefined in the ISO vocabulary of terms in metrology [2]. A definition along the following lines is under discussion within EURACHEM/EUROMET: "A validation is a set of operations that establishes, under specified conditions and for a specific range, the suitability of a given measurement instrument, measurement procedure, or measurement method for a measurement of a specified quantity at a stated level of uncertainty." Validation usually is accomplished with the help of RMs and gives an opportunity to forge a link (see next section), but does not verify a given value or its uncertainty.

Table 3 Classes of reference mateirals, determined by the length and strength of their traceability[b]

Class	Description and criteria in terms of traceability to SI
0 Primary	Pure specified entity certified to SI at the smallest achievable uncertainty
I	Certified by measurement against class 0 RM or SI with defined uncertainty (no measurable matrix dependence)
II	Verified by measurement against class I or 0 RM with defined uncertainty
III	Described linkage to class II, I, or 0 RM
IV	Described linkage other than to SI
V	No described linkage

[b] This, like the preceding Table, was originally offered for discussion [1]. Currently it is not widely adopted. The authors welcome proposals for changes

By one convention, illustrated in Fig. 3, the validation of a measurement method, as replicated in the field, succeeds if the certified value of an appropriate RM falls within the estimated measurement uncertainty when the RM is measured using the method. That definition of validation does not require a mutual conformance condition in which the two values (that deter-

Fig. 3 Issues arising when considering the meaning of a certified value and its uncertainty in a certified reference material

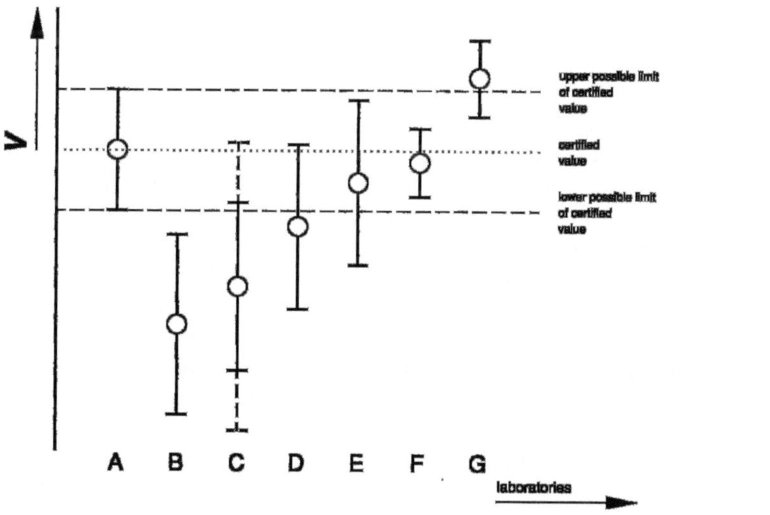

A: is the RM's certified value and its uncertainty. The latter could be used to define the range of permitted values

B: is a measurement of the RM that fails validation as normally understood. An individual protocol, however, may rule that, for a stated purpose, agreement between even wider limits (than indicated by A and B) shall be considered good enough for a purpose at hand

C: is a measurement which - despite overlap of its uncertainty with that of the certified value - would by most definitions fail validation of the method under investigation, unless the uncertainty could be shown to have been underestimated. With this uncertainty corrected (see dotted line) the RM's certified value falls within the measurement's uncertainty range, thus meeting a possible minimum criterion for validation of the method

D: is a measurement by the method to be validated which meets the validation criterion, defined under C

E: is a measurement by the method to be validated that meets the validation criterion under C and the additional possible requirement that its value shall be in the uncertainty range of the RM

F: is a measurement by a new more advanced method also traceable to SI. It does not justify discontinuation of the use of the older RM with its original uncertainty

G: is also a measurement by a new more advanced method traceable to SI. Now the continued use of the older RM might be questioned

mined at the "field" laboratory and that given by the laboratory establishing the protocol) must both lie in each others uncertainties[8].

The values in a sample can be linked to the values in RMs in several ways (see Fig. 1) [14, 16–19]. This relationship can, for example, be established:

1. Directly by a controlled comparison of values from measurements on a sample with certified values for identical entities in closely similar RMs[9], or
2. Indirectly through an instrument calibration established for values for identical entities in closely similar RMs[8]
3. In conjunction with a specified method (or procedure) of measurement [8]
4. To confirm the sensitivity of an instrument or method to detect a trace impurity or contamination.

The link

A trace ("traceability" by definition [2]) is established by a link or an unbroken, single-path (compare footnote 18) chain of links that connects by measurement two or more values of the same indicated quantity in a unidirectional order of authority. One of these values usually refers to a sample that is representative of a material embodying the measurand. The other value may be of that quantity in a reference material, or indicated by an instrument reading, or that of an SI unit (see sections "The value" and "The unit" above (to be continued in the next issue)). Links in a chain, as here discussed, can be thought of as having direction, emanating from a sample and leading progressively through higher levels of authority and perceived expertise to the SI. Each link could then be likened to a vector with the magnitude of its uncertainty. Links without directionality, between laboratories at equal level (see Fig. 1 in [1]), are of great importance to a successful measurement network but are not further discussed in this article. The chain link with the largest uncertainty is the weakest in that chain. It is not exceptional in measurement practice for the link connecting the value in a measurand to the reference chain to be stronger than the rest of the chain is to SI.

The values associated with an established traceability link are given as a difference or ratio and must have an associated uncertainty. This, combined with the uncertainty of the higher link, determines the uncertainty of thc value at the lower end of the link.

The responsible laboratory

For every link there is a qualified analyst or team of chemical specialists operating within a laboratory accepting personal and institutional responsibilities for the end result and its uncertainty, in full knowledge of the technical aspects further outlined below [12, 13].

The chemical analyst

In various sections of this paper, the authors appropriately emphasize the needed professional knowledge, experience, integrity and responsibility of the analyst. The handling of samples, the estimation of uncertainties, and the vigilance for unexpected errors also require some familiarity with statistics and possibly the help of a statistician as a consultant. However, the final assignment of uncertainties is the responsibility of the analyst who has actually performed the analyses.

The all-important manipulative skill of the analyst has yet to be underscored. No automated instrumentation or computer software can substitute for the analyst's dexterity and alert observation. Nevertheless, it must be understood that a disparity in this regard exists even between competent analytical analysts. No shame is attached to acknowledging a greater uncertainty in a given analysis than is achieved by the most experienced and the most skilled. During implementation of a protocol, an analyst, estimating a higher uncertainty for his own measurements than indicated in that document, may be demonstrating trustworthiness rather than doubt in his measurements.

Under the enormous ever-growing volume of needed analyses for production controls, environmental needs, and medical test programs, protocols have to be designed to be executed reliably by trained technicians. That protocol development, however, is and must be understood to be exclusively the proper role of the analytical chemist.

The validity interval

The validity interval is the time period during which relevant measurement operations are maintained in control with acceptable repeatabilities by each laboratory involved. This important limitation of validity of a traceability link is provided by design or could be im-

[8] An initially failed validation calls for repeat measurements or a reassessment in order to ascertain whether the uncertainties in the "field" were underestimated. Larger uncertainties may be acceptable for the purpose of the measurement. If so, the original measurement itself may be acceptable

[9] Even minor differences in matrices, however, will require determination of the significance of such differences and their effect on the uncertainties involved

posed from neglect. In order to assign a value to an RM, for example, a laboratory has to work within the validity interval for maintaining all relevant competences and procedures, such as to determine homogeneity and constancy of that RM. Thereafter, however, the measured, preferably certified, value of the RM remains valid, subject only to a validity period based on the RM's stability and requirements for its storage. These should be part of the RM's certificate with the aim of protection against contamination, temperature extremes etc. [18, 19].

The range

The range of values of the measured quantity is defined by the upper and the lower value for which the record is valid. At and between these extremes repeat measurements may not differ by more than an indicated uncertainty (see below). A zero should not be used to specify the lower end of a range. For small values of a measurand, a protocol may indicate the needed repeatability[10] of measurement or specify the smallest required detectable value.

Traceability to the SI

Wherever possible, a traceability chain of measured values terminates in an SI unit. When the base unit for mass is appropriate, this relationship is readily achieved through a mass standard, calibrated in terms of the kilogram prototype. The concept of traceability to SI has to be more carefully considered when conformity to SI depends on the SI concepts in the definition of the SI unit itself.

Basic to chemistry is the numerically simple (stoichiometric) proportion of entities in reaction and in formulae of compounds. The chemical analysts' purposes are therefore well served by comparing numbers of defined entities. The numerical value (see sect. entitled "The measurement" above) for the SI measurement of the amount-of-substance quantity fits those purposes. Historically, however, few were the analytical-chemical methods by which entities could be counted or counts of different entities could even be compared. With the nearly correct assumption that molar masses of the elements from terrestrial sources are constants of nature, chemistry made spectacular progress by measuring mass and converting to amount of substance by the factor of Avogadro's constant. Their measurements were thereby burdened by the uncertainty in that constant – which for many purposes cancels – as well as by the uncertainties in the molar masses – which do not cancel and which become significant as the total uncertainty of a measured value is reduced. For good measurement practice, protocols should therefore prefer traceability to the mole, as is stated in sects. entitled "The value" and "The unit" above.

The SI traceability statement for a chemical composition of a material cannot be completed by the traceability to the mole of one entity. The statement must include reference to another quantity, which could be a mass, a length, some other quantity, or even an amount of substance of another entity. Examples of such traceability statements for chemical composition could refer to a mole and the kilogram for the concentration, say, of a known element in an ore. The source of the element in that ore is then described in terms of the ratio of SI units mol/kg. Similarly by the SI units of mole and meter one could designate the solution of a defined organic compound, that is in mol/L. For some important chemical measurements we need to find traceability to SI for the mole of one entity as well as the mole of another entity. These moles are not identical and need separate traceability chains (see next paragraph). Measurements by mol/mol ratios are appropriate, for instance, for a trace impurity of known composition in a pure compound, or for an amount of isotope-to-element substance ratio (abundance).

The ratio measurement between the numbers of two entities establishes an amount-of-substance ratio that might satisfy the principal purpose of a chemical measurement. In measurement science, however, under the SI system, relative quantities do not fully satisfy the concepts. There remains an underlying requirement for all values to be individually traceable to the appropriate SI unit. For amounts of substance that unit is itself a number, the number of carbon-12 atoms of mass 0.012 kg. The magnitude of a given amount of substance, that is the numerical value of the SI quantity, is the number of defined entities divided by the SI unit number. It follows that equality of amount of substance is equality of the numbers of the two relevant entities.

If, in a ratio of amounts of substance, both of the two numbers of entities are traceable to numbers of carbon-12 atoms, and if the ratio between the entities is obtained by an appropriate measurement, the measurement is perfectly true to the concept of that SI unit. Calling that relationship traceable to SI is thereby reasoned and should be considered correct. Just as for mass measurements, the realization of amounts of sub-

[10] "Repeatability" [2] is defined as the: "closeness of the agreement between the results of successive measurements of the same measurand carried out under the same conditions of measurement." Repeatability should be distinguished from "reproducibility" for which the closeness of the agreement concerns the results of measurements (of the same measurand) that are not necessarily made successively and not under the same conditions, by the same method, or in the same laboratory

stance to SI could be left as a prime responsibility to national measurement laboratories.

That realization will often have larger uncertainty than relative measurements of entities by analysts. This should not surprise; similar conditions commonly apply throughout metrology. One can link the value of a good-quality gram standard to another similar standard with smaller uncertainty than that with which either can be linked to the international prototype kilogram. The designer of the protocol must carefully consider to what extent uncertainties in the realization of the SI unit cancel for the purpose of a given measurement. Not the calibration, but only the self-consistency of a mass-piece set, built-into – or external to – a balance, may enter into the uncertainty of a mass ratio. Similarly, the amount-of-substance ratio may have an uncertainty partially independent of that of the realization of the SI unit.

Summing up conclusions in this section: traceability of chemical measurements to SI involves concepts other than a direct comparison with a physical standard. Uncertainties may relate principally to values produced in the laboratory. The uncertainties of the links of two values of the same quantity to the SI unit may be larger than the uncertainty of the link between these values made in accord with the concepts of the SI.

The repeatability of a measurement

All uncertainty estimates start with that associated with the repeatability of a measured value obtained on the unknown. It is neither required for the sake of quality control, nor could it always be economically justified, to make redundant determinations of each measured value, such as would be needed for complete statistical control. Repeat measurements of a similar kind under the laboratory's typical working conditions may have given satisfactory experience regarding the range of values obtained under normal operational variations of measurement conditions such as: time intervals, stability of measurement equipment, laboratory temperature and humidity, small disparities associated with different operators, etc. Repeatability of routine measurements of the same or similar types is established by the use of RMs on which repeat measurements are made periodically and monitored by use of control charts, in order to establish the laboratory's ability to repeat measurements (see sect. entitled "The responsible" laboratory above). For this purpose, it is particularly important not to reject any outlier, unless cause for its deviation has been unequivocally established as an abnormal blunder. Rejection of other outliers leads a laboratory to assess its capabilities too optimistically. The repeatability in the "field" of a certified RM value represents the low limit of uncertainty for any similar value measured there.

When fewer than about 100 measurements of the same type are needed, the use of control charts becomes impractical. A few repeat measurements made within the routinely encountered range of relevant values is sufficient to estimate the repeatability of a single measurement. Difficulty arises only when a measurement type or procedure is inordinately time-consuming or costly to replicate. Relevant examples are: the measurement of an unusual trace constituent in a sample of minimal size, and a lengthy isotope dilution mass-spectrometric determination. The analyst is then required to depend on general experience of reliability of a method and would be wise to estimate the uncertainty with special care.

Just as the value obtained by measurement of a sample carries an uncertainty, so does the laboratory 'in-field' realization of the certified value of an RM. If the purpose of the measurement is to validate (Fig. 3) a procedure or instrument calibration, the measurement uncertainty estimated by the laboratory should include the certified value of the RM. If the measurement in the laboratory consists of determining the difference of the value in an unknown with that in an RM, the latter is taken as the reference value. Only when evaluating the uncertainty of the unknown to SI, the RM's certified uncertainty must be combined with that of the in-laboratory measurement of the unknown.

The uncertainty

Central to the protocol is the uncertainty[11], symbol u, or, if expanded, U, and u_c or U_c when combined [6]. It is expressed in the same SI units as the value V to which it refers. For propagation of uncertainties by mathematical formulae, relative uncertainties, such as U/V, are often needed.

All uncertainties are estimated and necessarily themselves uncertain. They should not be given to more than two significant figures; that is, to at most 1% of the total uncertainty. Individual smaller uncertainties thus become neglected. Uncertainties are generally given to be symmetric for positive and negative deviations from the evaluated best value[12].

---

[11] The importance of reliable uncertainties in protocols cannot be overstressed; they distinguish between insignificant differences and dangerous discrepancies

[12] On occasions, chemists designing a protocol recognize good reasons for expecting an asymmetry of likely deviations, such as for analyses of trace constituents that cannot be less than zero, of pure chemicals that cannot be more than 100%, or the molar mass of hydrogen gas (obtained by electrolysis) that cannot have less than zero content of deuterium. Under such circumstances, well-reasoned asymmetries of uncertainties may be introduced into a protocol

The protocol must present an uncertainty budget. Its components should be carefully estimated, and may be stated in standard uncertainties, but expanded uncertainties can have great utility, provided the *k* factor is carefully chosen and indicated [2, 4, 6][13]. All supposable uncertainty sources (of types A and B)[14], must be considered. Uncertainty components are concerned with contaminations, matrix effects, corrections, lack of stability or of stoichiometry, impurities in reagents, instrument non-linearities and calibrations, inherent uncertainties in standard methods, and uncertainties from subsample selection. Explicitly excluded may have to be sample selection in the "field" before submission to the laboratory and contamination prior to sample submission to the laboratory. The responsibility for adhering to the protocol's procedures, for which the planned complete uncertainty budget applies, rests with the laboratory and the analyst in charge of the measurement.

An uncertainty from a specific source may be neglected if its magnitude would have a negligible effect on the combined uncertainty value. Since that is not recorded to more than a two-digit precision, any individual uncertainty contribution of 1% or less will be disregarded[15]. A protocol for a link that includes one dominant uncertainty contribution overshadowing all others imposes an obligation on the originator or user of the protocol to consider and, if possible, devise an improvement of the protocol. Often it is convenient to split a large uncertainty into two components, one of which can be corrected by one additional measurement. This obligation will commonly arise when differences in matrix between sample and RM exist in the absence of quantitative knowledge of matrix-dependent effects on the measurement. In the absence of such knowledge, a fairly large increase of this individual uncertainty component will have to be budgeted for. In consequence, an improved protocol will often have to be devised.

Overstated uncertainties, however, are also discouraged because of loss of possibly useful information from the measurement or because of inappropriate reduction in responsibility for the measurement.

Although every form of training includes teaching the avoidance of blunders, sometimes called spurious errors [8], these errors are unfortunately not always prevented. Nevertheless, the uncertainty budgets of protocols should not include components for possible "blunders", because the magnitude of their effects on measurements is completely unpredictable. Such uncertainties are clearly not of a scientific nature.

The responsibility for avoidance or elimination of blunders lies squarely on the shoulders of the analysts applying the traceability protocol. The best chance of finding and correcting blunders is by introducing some redundancy into measurements. Repeat measurements by the same analysts, by other analysts in other laboratories, or by other methods are potentially effective in revealing previously undetected blunders. Most analysts forestall potential blunders by wisely asking colleagues to independently verify their data and calculations up to the end result.

Even the most conscientious experimentalist in retrospect occasionally detects in a previously determined value an unanticipated error[16] source. The possibility of such fictitious errors should not be compensated for in advance by use of unduly expanded uncertainties [6]. A principal purpose of expanded uncertainties remains to provide the assurance that in a large number of similar measurements only a very small proportion will lie outside the indicated uncertainty.

Possible error sources for a chemical analysis can arise from a great diversity of necessary procedures. Analysts must rely on best-available knowledge on reaction kinetics, extraction efficiencies, solid solutions, minor element interferences, effects of matrices, etc. Analysts therefore commonly feel, more than most other scientists and engineers, reluctance in estimating realistically what their measurement errors might be. The uncertainty of the allocated uncertainties is not directly expressed in a final protocol. Yet, uncertainty of uncertainties is a great challenge in the production of every trustworthy protocol in analytical chemistry. Though that uncertainty of the uncertainty may be large, it should not lead analysts in a protocol to choose a larger value of *k*.

Unreliability of uncertainty estimates in chemistry has caused difficulties in following prescriptions of measurement science. The authors of this paper hope to add to the currently gaining consensus that trustworthy uncertainties are essential in modern science, tech-

[13] Protocols may require a *k* factor of 2, a choice preferred by some laboratories and obligatory in many accreditation programs. That $k=2$ value approximates to a "confidence limit" of 95%, used in past descriptions of "possible error"

[14] The distinction between the two categories of uncertainties (types A and B) is based on the method of their evaluation. Those of type A, but not those of type B, can be evaluated by statistical methods [6]. These categories do not correspond exactly to the former grouping of uncertainties into random and systematic. No distinction is made between types A and B for the combination of uncertainties

[15] This deduction is not reasonable, when a very large number of uncertainties, all of magnitude less than 1%, are included for one links

[16] A superior measurement may show retrospectively that an earlier measured value with assigned uncertainty was in error [6]. The initial uncertainty may, for instance, have been under- or overestimated, or the earlier value – judged by the smaller uncertainty of the later measurement – may be found outside the initially estimated uncertainty range

nology, trade, and equity in markets. Drafting or just understanding a protocol will continue to require effort, experience, and judgement, but will also offer commensurable rewards.

Matrix effects on the uncertainty

For many technological applications such as in agriculture, medicine, and geology, the entity to be measured resides in a complex matrix of other entities. Composition and texture of that matrix may affect the measurement of the measurands. A call for RMs with all kinds of matrices and applicable measurands in many concentrations is not reasonable [1]. There is an inevitable shortage of appropriate RMs with a matrix that adequately resembles that of the unknown sample, and there is little hope that this shortage will fade in time [14]. On the contrary, the ever-increasing complexity of our samples and the reduced uncertainties that will be needed will exacerbate the problem. The development and preferential choice of measurement methods that deliver matrix-independent results are more likely to lead to methodologies of wider applicability. Such methods should be seen as a major aim for chemical measurement research.

Before it can prescribe the procedure for the combination of uncertainties, the protocol must treat important matrix effects and their uncertainties. These effects are liable to be disregarded, yet often overshadow other error sources, and thus lead to underestimation of total measurement uncertainty. Ignorance of matrix influence on the measurement will often cause any realistic uncertainty estimate to be so large that the use of a method, instrument or RM would become inappropriate[17]. At other times, however, substantial improvement can be achieved by introducing additional measurements to quantify and correct for the effect of matrix differences on the measurements. The correction itself will introduce a residual small uncertainty.

[17] Retroactive but necessary enlargements of uncertainties have often explained apparent discrepancies in terms of differences that should have been expected

The quality of a link

All non-negligible uncertainties are combined by root-mean-square summations. That sum establishes the quality of the traceability link. A traceable measurement may fail as an appropriate measurement if its uncertainty is too large. Its quality will then be too low for the purpose. Conversely, it should be recognized that there are highly reproducible chemical measurements which are useful, but are not traceable to any standard.

In metrology generally, and particularly in protocols for chemical measurements, a good measurement plan should tend to eliminate as many error sources as possible and to offset effects from different error sources. Self-cancelling errors should be combined with care in the total uncertainty budget of a measurement link[18]. Most uncertainties, however, are not self-cancelling, although in the root-mean-square combination of all uncertainties they are in fact partially offset. For instance, the uncertainty applicable to a measurement in one sample cannot be held to cancel the uncertainty of a separate but similar measurement in another sample. This is especially true when the matrices of the two samples are not identical.

Whereas type A and B uncertainty components [6] are almost always added by root-mean-square additions, there are circumstances under which direct additions of uncertainties from dependent error sources are preferable. When protocols use such direct additions of uncertainties, these conditions should be clearly explained in the protocol. That information is needed for a fair inter-laboratory comparison of values and uncertainties of measurements on the same material.

Acknowledgements The authors are greatly indebted to a number of colleagues who have reviewed this paper in draft and made valuable suggestions, many of which we have been permitted to adopt freely. Included in the list of colleagues are: E. L. Garner (NIST), T. M. Plantenga (NMI), P. D. P.Taylor (IRMM), and R. L. Watters (NIST).

[18] The relative uncertainties of the difference or ratio between two values, for example, are often far smaller than the relative uncertainty of either value which would not appear in the uncertainty budget

References

1. De Bièvre P, Kaarls R, Peiser HS, Rasberry SD, Reed WP (1996) Accred Qual Assur 1:3–13
2. International Vocabulary of Basic and General Terms in Metrology (1993) 2nd edn. ISO, Geneva
3. Quantities and units (1993) ISO Standards Handbook, Geneva
4. Comité Consultatif pour la Quantité de Matière, Report of First Meeting, Bureau International des Poids et Mesures, Sèvres, April, 1996
5. Guide for the Use of the International System of Units (SI) (1995) SP 811 NIST USA
6. Guide to the expression of uncertainty in measurement (1993) BIPM, IEC, IFCC, IUPAC, IUPAP, OIML, ISO, Geneva
7. ISO Guide 30 (1992) Terms and definitions used in connection with reference materials 2nd edn. ISO, Geneva

8. EURACHEM (1995) Quantifying Uncertainty in Analytical Measurement, LGC Teddington, UK
9. Quantities, Units and Symbols in Physical Chemistry (1995) IUPAC, Blackwell Scientific Publications, Oxford
10. De Bièvre P, Lamberty A (1995) Fresenius J Anal Chem 352:255–260
11. Fleming JW, Neidhart B, Wegscheider W (1996) Accred Qual Assur 1:35
12. Prichard FE (1995) Quality in the Analytical Chemistry Laboratory. Wiley, New York
13. Sargent M (1995) VAM Bull 13:4–5
14. Uriano GA, Gravatt CC (1977) CRC Critical reviews in analytical chemistry 6:361–411
15. McGlashan ML (1994/5) Metrologia 31:447–445
16. Résolutions adoptées par la Conférence Générale des Poids et Mesures (1995), BIPM, Sèvres
17. Rasberry SD (1994) Worldwide production of certified reference materials, Status Report. ISO-REMCO, Geneva
18. ISO Guide 31 (1981) Contents of certificates of reference materials. ISO, Geneva
19. ISO Guide 35 (1989) Certification of reference materials – General and statistical principles. ISO, Geneva

Accred Qual Assur (1997) 2:270–274

Paul De Bièvre
Robert Kaarls
H. Steffen Peiser
S. D. Rasberry
William P. Reed

Protocols for traceability in chemical analysis

Part II: Design and use

Received: 23 April 1997
Accepted: 27 April 1997

P. De Biève (✉)
Institute for Reference Materials and Measurements, European Commission-JRC, B-2440 Geel, Belgium

R. Kaarls
Netherlands Measurements Institute, 2600 AR Delft, The Netherlands

H.S. Peiser · S.D. Rasberry
W.P. Reed (retired)
National Insitute of Standards and Technology, Gaithersburg, MD 20899, USA

Abstract In continuing their attempt to bring general issues concerned with trustworthy chemical measurements to review and international discussion, the authors propose basic aims and requirements for protocols of chemical-measurement procedures with traceability to the SI or, where this is not possible, to units of internationally recognized measurement scales. Documents describing such protocols could be useful in science, technology, law, or trade. Concepts and definitions for protocols have been introduced in Part I of this contribution. Part II here deals with the development and application of protocols for intended in-laboratory, commercial, national, or international recognition. Protocols deal with measurement methods, instrumentation, and the estimation of uncertainties from all possible sources of measurement errors. Uncertainties define the quality of all links in a traceability chain starting from the value of a measurand in a sample, often through a certified value in a reference material, either to the SI, or – if this is not possible – to a value on a suitable, internationally agreed measurement scale. A protocol may concern itself with the complex interplay between uncertainties, tolerances, and any limit values introduced by the set aims of specific measurements.

Key words Traceability · Protocol analysis · Reference Material · Uncertainty

Introduction

Previously the authors have brought into discussion principles for traceability in chemical analysis [1]. In this Journal is also the first part of this contribution [2] on protocols for traceability of analytical-chemical measurements. This first part is intended mainly for specialists who develop such protocols. It deals with terminology and definitions used when describing protocols for traceability[1]. These terms are mostly taken from recognized literature sources [3–7]. Analysts, who want to judge the applicability of an established protocol and to use it, will be familiar with most of these terms and find others self explanatory. They may, nev-

[1] In the International Vocabulary of Basic and General Terms in Metrology [3], "traceability" is defined as: 'property of the result of a measurement or the value of a standard whereby it can be related to stated references, usually national or international standards, through an unbroken chain of comparisons all having stated uncertainties'. It should be noted that the trace runs from one value to another value in identical units of measurement. These values are generally not numerically equal, but their trace connects them to each other by a relative uncertainty in value. It should also be understood that by "traceability to the SI" we here mean a chain of measurements to the appropriate unit value in the SI

ertheless, wish to refer to specific sections of the preceding Part I [2], but will refer mostly to the present paper (Part II), in which we here submit for international consideration and debate ideas on the design and use of such protocols. Subsequent articles are encouraged from other authors. There is a special need for published examples of protocols related to analytical method or field of application. Other topics that deserve further examination in the literature include: implementation procedures, assessments by groups of chemists, and proposals for modifications of concepts and definitions.

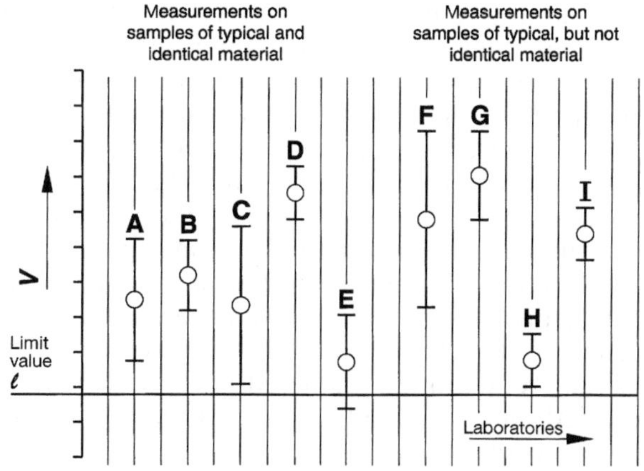

Fig. 1 Hypothetical measurement results (values with uncertainty bars ± U) obtained by different laboratories using a new protocol

Laboratory A developed and tested (on a typical sample) a protocol obtaining the value V with combined uncertainty ± U.
Laboratory B is a higher-level laboratory (a "reference" laboratory) that carried out a test run.
Laboratory C had a measurement problem: U may be unacceptably large.
Laboratory D obtained an unsatisfactory value and had a suspected problem in underestimating U.
Laboratory E had a measurement problem unless the protocol specifically permitted $(V-U)<V$ (lo).
Laboratory F had a measurement problem: U may be unacceptably large.
Laboratory G used a sample differing from that used by laboratory A, and appears to have obtained a satisfactory measurement.
Laboratory H may have inappropriately underestimated the uncertainty in order to justify accceptance of that sample.
Laboratory I obtained a low uncertainty, possible by use of a better method. Another protocol may be needed.

Use of protocols for substantiation of traceability

Here we propose the additional concepts under which analysts can formally substantiate and record their traceability link. A chain of such links should lead from the value of a quantity in a sample or reference material (RM) up to the value of a relevant unit in the International System of Units (SI)[5] or – where this is not possible – up to internationally agreed measurement scales. A protocol records specific details of scientifically reliable measurement procedures for the benefit of equity in trade and commerce, as well as for legal interpretations of scientific realities. Some ideas in this article go beyond established international understandings; these are presented for debate and possible refinement.

Types of protocols

The protocol may be intended for only one determination or, much more commonly, for a specific set of measurements, a measurement process or procedure. Protocols are then typically developed by a group of chemical specialists on behalf of professional associations or recognized organizations for specific, limited, but recurrent use. The individual analyst, under a contractual requirement, may have to use such an openly published protocol. Still, that analyst needs to be professionally convinced of the protocol's soundness and its applicability to measurands at hand (see Section "The analyst's use of a protcol" below). Confidence in the protocol is often strengthened by testing its use with a specific material, preferably a certified RM.

The quality of traceability links

A protocol must deal with the quality of the traceability link based upon carefully estimated uncertainties [4] (see also Fig. 1) from all possible error sources that remain after due precautions have been taken and after corrections have been applied, where possible. The combined uncertainties of all links in the relevant chain of links to an SI unit or to some other relevant, internationally agreed scale will then define the quality of that chain. This quantitative assessment of quality of traceability to the SI is not inconsistent with the metrological term "accuracy" [3]. It differs from popular meanings of "accurate" such as "free from error" and "highly exact". A traceable measurement result may have a considerably larger uncertainty than the optimum achievable, but may still be adequate for its purpose. Every traceable measurement result, nevertheless, has an "accuracy" quantified by its uncertainty [4].

When using a protocol for a measurement, laboratories will have varying success in achieving the intended reliability. A limit value shown in the schematic Fig. 1 may not be involved. Other features illustrated are more universal, such as the fact that different samples will yield different values within a defined range. Only their uncertainties are strongly interrelated by the protocol, which may restrict the allowable maxima in components of the uncertainties. Very rarely will an analyst using the prescribed method improve upon the uncertainties estimated by both the initiating laboratory (under A in Fig. 1) and by a laboratory that is higher in the authority chain to the SI (under B in Fig. 1). When an analyst is tempted to make such a claim, that low uncertainty estimate should be re-evaluated. With advancing science or technology, newer measurement methods with inherently lower estimated uncertainties are to be expected, but, for significantly different new methods, new protocols are required.

Some increase in uncertainties over those estimated by the initiating laboratory will commonly be permitted by the protocol. Laboratories implementing the protocol, perhaps under semi-routine conditions, cannot in practice devote the same time and care as that devoted by the initiating analyst in full awareness of the relevant responsibility of the initiator.

The analyst's use of a protocol

For the great majority of chemical measurements, a chemical analyst works in the assigned chemical laboratory by self-reliant professional procedures (see part I [2], section III.10,11). A protocol in no way compromises the relevant freedom and obligations of applying professional knowledge, experience, integrity, and judgment. For many reasons, such as convenience in contractual understandings, a formally recognized protocol supporting the analysts' work and explaining traceability of measurements may be helpful, advisable, or essential. Such formally recognized protocols may be in the form of standard methods developed and published by recognized bodies such as IUPAC, WHO, ISO, CEN, AOAC, ASTM, etc. Description of uncertainty estimates is an essential feature of every protocol.

An individual analyst in most instances is concerned only with links from values in samples to similar values in certified RMs [9–13] or from calibrated instruments. When using an existing protocol, chemists must carefully follow all procedures, minimize errors, such as those resulting from contaminations, compled with the uncertainties in estimating these contaminations. Rigorously correct use of a protocol and aggressive self-criticism in its application constitute a professional challenge.

In a situation where an analyst cannot justify the use of a protocol for a task at hand, but is required by high administrative authority to proceed, the analyst has the duty to record the reservation in all relevant reports by clearly describing all discerned additional limitations and uncertainties associated with the use of the protocol.

The development of protocols

The protocols for links to higher-level RMs and, generally, to the SI must be supplied by the producer of those RMs, or by a recognized (metrological) reference laboratory.

The additional responsibility of developing the protocol for even the first link to values in RMs or instrument readings, obtained by reference procedures or primary methods of measurement, requires wide-ranging experience at times gained only by the critical scrutiny of past results. Even if a planned new protocol is not destined for formal recognition by some authority or organization, like AOAC, its publication may be desirable for examination, appraisal, and potential use by others.

Many protocols must be responsive to complex issues of uncertainties[2] in relation to tolerances and acceptabilities of samples and results of measurements. Measurement qualities affect risks of rejecting acceptable samples or measurements or accepting those which are unsatisfactory. That balance is often influenced by upper or lower limit values, $V(\mathrm{up})$ or $V(\mathrm{lo})$. Occasionally, both types of limit value are indicated and need to be separately treated by the protocol. Specific limit values are often set from technological, commercial, or legal considerations (see Fig. 1) and not by the analyst in the design or implementation of a protocol. The interpretation of the measurement results, however, depends on the protocol clearly stating the relationship between observed values, uncertainties, and limit values. One simple – by no means the only – method for the protocol to lay down that relationship is the following:

When an upper limit value $V(\mathrm{up})$ applies, the largest acceptable measured value, $V(\max)$, is given by: $V(\max) + U \leq V(\mathrm{up})$. Similarly, for a lower limit value, $V(\mathrm{lo})$, the lowest acceptable measured value, $V(\min)$, is given by $V(\min) - U \geq V(\mathrm{lo})$. For a near-normal distribution of V, when $U = 2u_c$, the true[3] value will transgress the limit in very few of those measur-

[2] For instance, when more than one chain of links is maintained between the value in a sample and an SI unit, it might be possible to reduce the uncertainty even below that of the least uncertain single-path chain. That reduction would have to be carefully evaluated based not only on the uncertainties of all the links in the dual path, but especially on the independence of measurements in the alternative paths. Such a complex consideration is not recommended for protocols as here discussed

[3] The authors suggest that the well-established concept of "true" value [3, 6] be retained in the context of a hypothetical error-free value, rather than accepting the notion that the value of a measurand is the error-free value [7]. A clear distinction between the value as measured (as normally understood) and the hypothetical error-free value seems to have practical merit.

ements[4] for which V is on the permitted side of a boundary value.

In many commercial situations and under some regulatory requirements, trade-offs arise between measurement costs and the aim of low uncertainty in measurement. Clear relevant statements associated with the protocol are needed. For example, whereas normally U must be made equal to or smaller than a set tolerance, a protocol may for technical and economic reasons explicitly permit a less stringent requirement (see Fig. 1).

All protocols, especially those involving limit values that may be of legal significance, must include a procedure for maintaining confidence in the reliability of the measurements. Full statistical control of all measurements on all unknown samples cannot be achieved under normal cost restraints. Yet, redundancy of measurements is an important criterion of measurement science. Therefore, for every type and every range in which measurements are made, the experimental design should specify a minimum number of repeat measurements and their relative independence in time, place, equipment, and operator. Statistically [14] more than a few repeat measurements (without any change of measurement conditions) do not add significantly to their benefit[5]. A protocol may require a further evaluation when a measurement difference between identical samples is found in excess of, say, $2U$.

Protocols for in-laboratory reference materials

In some instances chemists will prepare in-laboratory RMs, generally by dissolution of a pure compound in a solvent to predetermined concentrations or even by dilution of commercially prepared stock solutions. The uncertainties connected with such RMs and relevant procedures must be understood by the analyst and entered into the uncertainty budget for the measurement of the unknown. The analyst then assumes responsibility for the uncertainties of the link of the RM value to the SI, all expressed in the relevant SI units, even when relative uncertainties are used.

[4] The basis for this statement is the following: if all measured values corresponded to the maximum (or minimum) permitted and if Gaussian distribution applied, 5% of the measured values would result in false acceptance. In any real situation, the measured values would differ from the maximum (or minimum) permitted values. Those nearer to the limit value would be rejected anyway, and of those further from the limit value substantially fewer than 5% would result in false acceptance

[5] For a Gaussian distribution the uncertainty gain achieved by a sixth repeat measurement over five is only 1%

Protocols for establishing traceability links for reference materials for use in other than the originating laboratory

RMs, preferably certified by reputable and accredited laboratories, provide vital contributions to the science of chemistry, to industry in its quest for quality assurance, and to commerce for consensus product characterizations. All such RMs should have designated values bonded to the units of the SI, where possible through national metrology laboratories [9–13].

Industry should be encouraged to produce fully traceable RMs. Professional organizations, working with experienced laboratories, should likewise feel motivated to develop and distribute RMs to the "field". The principles by which traceability of their values is established for such RMs to the values of Class 0 or Class 1 primary RMs (See Part I, Table 2) differ in no way from those described in the earlier sections for traceability from any measured sample value to that of an RM. However, the organizations and laboratories concerned should take very seriously the great responsibilities assumed when distributing any RM to other laboratories [12], possibly by commercial sale. Good accompanying descriptions of such an RM and of its proper certification are certainly a moral and often a legal responsibility [11]. Knowledge of the RM's homogeneity, the traceability of its value to that of higher RMs, the relevant uncertainties, and its exact matrix description are all essential. Sample protocols illustrating the use of the RM in the field should be developed, made available and possibly published. The information might be helpful for users of these and other RMs and the effective employment of standard analytical methods and instruments.

The quality of chemical measurements for industry and commerce will often benefit by the following procedure: a reference measurement laboratory offers itself to measure a typical "field sample" and thereby provides a reference value. Subsequently, that sample is sent back to the field laboratory and there can be used for validation of measurements. This approach is often appropriate because that RM has a matrix typical of the material analyzed in that field laboratory.

Guidance by the International Bureau of Weights and Measures for traceability links to Class 0 or Class 1 reference materials established at national levels

All considerations described in the previous section apply. The full traceability to the SI may vary in its difficulty from simple routine to a full research undertaking. At the national level, Class 0 or Class I RMs (see Part I, Table 3) may be certified for specific chemical entities, but this can clearly not be realized for all compounds.

For many units of the SI other than the mole, the BIPM has long guided participating national laboratories and taken an active role in international comparisons of values of units and scales. With the establishment in 1994 of the Consultative Committee for Amount of Substance, and following the Resolution adopted by the 1995 General Conference of Weights and Measures of the intergovernmental Convention of the Metre [15], it is hoped that BIPM will undertake a similar role for the mole, and thereby become recognized for guidance of chemical measurement values at the ends of the traceability chains. The concept of amount of substance measured in moles applies to all specified entities, including all chemical compounds. Practical measurements in moles can be entity-dependent in their uncertainties, when made through measurements of other quantities, such as that of mass. BIPM, in close coordination with OIML (Organisation Internationale de Métrologie Legale), IUPAC (International Union of Pure and Applied Chemistry), ISO (International Organisation for Standardization) and other international organizations, have the opportunity to assist progress towards wide-ranging acceptance of the concepts of traceability to the SI for chemical measurements.

Acknowledgements The authors are greatly indebted to a number of colleagues who have reviewed this paper in draft and made valuable suggestions, many of which we have been permitted to adopt freely. Included in the list of colleagues are: E.L. Garner (NIST), T.M. Plantenga (NMI), and P.D.P.Taylor (IRMM).

References

1. De Bièvre P, Kaarls R, Peiser HS, Rasberry SD, Reed WP (1996) Accred Qual Assur 1:3–13
2. De Bièvre P, Kaarls R, Peiser HS, Rasberry SD, Reed WP (1997) Accred Qual Assur 2:168–179
3. International Vocabulary of Basic and General Terms in Metrology (1993) 2nd edn. ISO, Geneva
4. Guide to the expression of uncertainty in measurement (1993) BIPM, IEC, IFCC, IUPAC, IUPAP, OIML, ISO, Geneva
5. Quantities, units and symbols in physical chemistry (1995) IUPAC, Blackwell, Oxford
6. EURACHEM (1995) Quantifying uncertainty in analytical measurement. LGC, Teddington, UK
7. Guide for the use of the International System of units (SI) (1995) SP 811 NIST USA
8. Quantities and units (1993) ISO Standards Handbook, Geneva
9. Rasberry SD (1994) Worldwide production of certified reference materials. Status Report ISO-REMCO, Geneva
10. ISO Guide 30 (1992) Terms and definitions used in connection with reference materials 2nd edn. ISO, Geneva
11. ISO Guide 31 (1981) Contents of certificates of reference materials. ISO, Geneva
12. ISO Guide 34 (1996) Quality system requirements for reference material producers. ISO, Geneva
13. ISO Guide 35 (1989) Certification of reference materials – General and statistical principles. ISO, Geneva
14. Natrella MG (1966) Experimental Statistics. US Government Printing Office, Washington D.C.
15. Résolutions adoptées par la Conférence Générale des Poids et Mesures (1995) BIPM, Sèvres

Accred Qual Assur (2003) 8:46–52
DOI 10.1007/s00769-002-0571-y

René Dybkaer

Metrological traceability in laboratory medicine

Received: 1 August 2002
Accepted: 22 November 2002

Based on a lecture at an IUPAC Seminar, EC JRC Institute for Reference Materials and Measurements, Geel, BE, 2001–12–18

R. Dybkaer (✉)
Department of Standardization in Laboratory Medicine,
Frederiksberg Hospital,
H:S Copenhagen Hospital Corporation,
Copenhagen, Denmark
Fax: +45 38 16 38 79

Abstract The establishment of a reference examination system necessary for metrological traceability of the many types of sophisticated examination result in laboratory medicine is a daunting task, which has been made mandatory by the EU Directive on in vitro diagnostic medical devices and the requirements for accreditation. Following a definition of examinand and allowed examination uncertainty, a dedicated calibration hierarchy is established from stated reference through alternating reference examination procedures and calibrators providing a traceability chain from examination result to the reference, often a definition of a measurement unit. The various types of possible calibration hierarchy are outlined in EN ISO Standards. Recent efforts by national and international stakeholders to establish a global reference examination system have led to the creation of a Joint Committee on Traceability in Laboratory Medicine with the International Committee for Weights and Measures, International Bureau of Weights and Measures, International Federation of Clinical Chemistry and Laboratory Medicine, International Laboratory Accreditation Cooperation, and World Health Organization as the principal promoters. This structure will identify reference procedures, reference materials, and reference laboratories, and seek support for further prioritised and coordinated development of the system.

Keywords Calibration hierarchy · Comparability of results · Joint Committee on Traceability in Laboratory Medicine · Metrological traceability · Reference examination system

Introduction

The size of laboratory medicine has become quite considerable, comprising tens of thousands of laboratories in Europe alone – not counting near-patient examination sites in clinical departments, hospital wards, and general practitioners' offices. The expenses for all these activities are around one-twentieth of the costs of health services in industrialized countries, and the global market of in vitro diagnostic medical devices is now about €25 thousand million a year.

Such numbers and the often crucial importance to the patient of correct examination results have naturally led to a demand for reliable and transferable service.

The implementation of IQC (Table 1) and EQAS is not sufficient to achieve comparable results between laboratories as appears from the outcome of many EQASs where various examination procedures for a given type of property often show significantly different distributions of results. This is an unacceptable situation because patient data are increasingly being transferred between different parts of a health care system and between such systems. Classically, the solution is to impose standard examination procedures, but this stifles innovation and comparison over time.

The modern answer is to create a complex reference examination system with the following exemplified interrelated elements:

Table 1 Abbreviations

AdvaMed[a]	Advanced Medical Technology Association
APLAC	Asia Pacific Laboratory Accreditation Cooperation
APMP	Asia-Pacific Metrology Programme
BCR	Community Bureau of Reference
BIPM[a]	International Bureau of Weights and Measures
CAP[a]	College of American Pathologists
CASCO	ISO Committee on Conformity Assessment
CCQM[a]	Consultative Committee for Amount of Substance: Metrology in Chemistry
CDC[a]	Centers for Disease Control and Prevention, US
CEN	European Committee for Standardization
CGPM[a]	General Conference on Weights and Measures
CIPM[a]	International Committee for Weights and Measures
COOMET	Euro-Asian Cooperation of National Metrological Institutions
CRM	Certified reference material
DTR	Draft Technical Report, ISO
DGKC[a]	Deutsche Gesellschaft für Klinische Chemie e.V.
DIS	Draft International Standard
EA	European co-operation for Accreditation
EC[a]	European Commission
EC4	European Communities Clinical Chemistry Committee
ECBS[a]	Expert Committee on Biological Standards, WHO
EDMA[a]	European Diagnostic Manufacturers Association
EN	European Standard, CEN
EQA	External quality assurance
EQALM[a]	European Committee for External Quality Assurance Programmes in Laboratory Medicine
EQAS	External quality assessment scheme
EU	European Union
EUROMET	European collaboration on measurement standards
FDA[a]	Food and Drug Administration, US
FDIS	Final Draft International Standard, ISO
ICSH	International Council for Standardization in Haematology
IEC	International Electrotechnical Commission
IFCC[a]	International Federation of Clinical Chemistry and Laboratory Medicine
ILAC[a]	International Laboratory Accreditation Cooperation
IQC	Internal quality control
IRMM[a]	Institute for Reference Materials and Measurements, EC-JRC
ISO	International Organization for Standardization
IUPAC	International Union of Pure and Applied Chemistry
IUPAP	International Union of Pure and Applied Physics
IVD MD[a]	In vitro diagnostic medical devices (industry)
JACR[a]	Japanese Association of Clinical Reagents Industries
JCCLS[a]	Japanese Committee of Clinical and Laboratory Standards
JCTLM[a]	Joint Committee on Traceability in Laboratory Medicine
JRC	Joint Research Centre, EC
LGC[a]	Laboratory of the Government Chemist, GB
MTI	Measurements and Testing, Infrastructures, EC
NACC	North American Calibration Cooperation
NCCLS[a]	National Committee for Clinical Laboratory Standards, US
NIBSC[a]	National Institute for Biological Standards and Control, GB
NIST[a]	National Institute of Standards and Technology, US
NMi[a]	Nederlands Meetinstituut, NL
NMI[a]	National metrology institutes
NORAMET	North American Cooperation in Metrology
OIML	International Organization of Legal Metrology
prEN	draft European Standard
PT	Proficiency Testing
PTB[a]	Physikalisch-Technische Bundesanstalt, DE
SI	International System of Units
SIM	Sistema Interamericano de Metrología
TC	Technical committee, CEN or ISO
WHO	World Health Organization

[a] Representation at the Symposium on Traceability in Laboratory Medicine, BIPM, 2002-06-09/11, leading to the creation of JCTLM

- Standardization ISO/TC 212, CEN/TC 140
- Certification ISO/TC 176
- Accreditation by ISO/IEC CASCO, ISO/TC 212, CEN/TC 140, ILAC, EA, APLAC, and NACC
- Metrological institutes coordinated by BIPM, EUROMET, APMP, NORAMET, COOMET, and SIM
- Reference measurement laboratories, preferably organized in networks
- Dedicated industries providing examination systems, including reagents, organized in, e.g., EDMA and AdvaMed
- Reference examination procedures
- Examination standards MTI, IRMM, NIBSC, NIST, LGC, and industry
- Professional scientific organizations, e.g. IFCC, ICSH, and IUPAC
- EQAS, organized in EQALM
- Professional curricula by IFCC and EC4

In various ways this multitude of elements all have a role to play in the effort to obtain globally and time-independently reliable and comparable examination results through metrological traceability. This desirable goal has recently been changed into a requirement by two developments. First, the essential requirement of the EU Directive on in vitro diagnostic medical devices to assure traceability of values assigned to reference materials [1] and, second, the trend to have medical laboratories accredited according to the International Standards ISO 17025 [2] or 15169 [3] which require that measurement results be metrologically traceable.

Metrological traceability

The current definition of traceability in metrology reads 'property of the result of a measurement or the value of a standard whereby it can be related to stated references, usually national or international standards, through an unbroken chain of comparisons all having stated uncertainties' [4–6.10].

This definition has been fashioned by physicists to cover only results for quantities, i.e. properties having magnitude, thus comprising differential and rational quantities, and hesitantly including ordinal quantities.

The modifier 'metrological' is used advisedly to distinguish the present concept from (historical) traceability of, e.g., materials, instruments, and documents.

The definition could be changed to cover all types of examination results, including those for nominal properties.

To obtain metrological (or examinational) traceability for the examination results of a given type of property, several aspects must be considered.

Definition of examinand

To decide on the reference end of the examinational traceability chain, it is necessary primarily to define that which is to be examined. According to the IUPAC/IFCC Recommendation 1966 [5], the designation of a property must comprise identity of system and any pertinent component(s), together bearing the property, and the kind-of-property involved. Standardized templates for such designations are currently found in an IFCC/IUPAC database comprising over 10000 entries [6]. In some cases, the examination procedure becomes an integral part of the definition of the examinand.

Required examination uncertainty

Depending on specified medical needs, the allowed combined examination uncertainty, also called an analytical performance goal, should be decided upon a priori because its magnitude influences the choice of metrological traceability chain. Guidance on how to choose combined uncertainty is given in the upcoming ISO/DTR 15196 [7].

Calibration hierarchy

Metrological traceability, according to its definition, is a property of a measurement result or quantity value, and the trace is towards a 'stated reference' through a metrological traceability chain. Its links and relations between them have to be established a priori in the opposite direction from the chosen reference towards the measurement result. This structure is termed a calibration hierarchy.

Calibration hierarchy in physics

The concepts of calibration and especially of metrological traceability were elaborated by physicists as mentioned above. The reference – or top of the calibration hierarchy – preferably is the definition of an SI unit, which is realized or embodied as a primary measurement standard. By direct comparison, the quantity value of a secondary measurement standard can be established. Subsequent comparisons may furnish quantity values of reference measurement standard, working measurement standard, and routine measurement standard with which the object carrying the measure and is compared to obtain its measurement result – which then retrospectively is metrologically traceable to the SI unit. The primary measurement standard, as the definition of metrological traceability says, is preferably an international or national measurement standard.

Calibration hierarchy in chemistry

Although many elements of measurement in chemistry are by nature physical, such as those involving mass, volume, time, temperature and spectral absorbance, the calibration hierarchy in chemistry is seldom described as a series of 'comparisons' between measurement standards. Rather, a measurement procedure points to a measuring system performing a measurement which assigns a quantity value and measurement uncertainty to a calibrator – itself a type of measurement standard – which serves to calibrate the next measuring system, operated according to a second measurement procedure, and so on.

This principle can be implemented in various ways depending on the available elements.

Full calibration hierarchy

The currently optimum reference is the definition of a base or derived SI unit, which is embodied by applying a primary (direct or ratio) reference measurement procedure, the operation of which is completely described by a measurement equation in terms of quantities using SI units. Accordingly, a quantity value and measurement uncertainty is assigned to a primary calibrator. This serves for calibration of a measuring system, operated according to a secondary reference measurement procedure, assigning a measurement result to a secondary calibrator. The hierarchy (Fig. 1) may continue sequentially through a manufacturer's selected measurement procedure, working calibrator, standing measurement procedure, and product calibrator, used in the medical laboratory as specified by the routine measurement procedure, to obtain the final measurement result and measurement uncertainty of a sample. A given hierarchy, for practical reasons, may omit consecutive pairs of the sequence from primary calibrator to standing measurement procedure.

The correct sequential transfer of values down the hierarchy requires measurement specificity and selectivity of each measurement procedure and commutability of each calibrator, i.e. it has the same behaviour towards the preceding and following measuring systems, working according to their respective measurement procedures, as have the routine samples. Lack of fulfilment of these requirements breaks the traceability chain. Unfortunately, the respective causes are often subsumed under the concept of "matrix effect".

Examples of a full hierarchy is the measurement of amount-of-substance concentration of cortisol in human plasma and catalytic activity concentration of γ-glutamyltransferase in human plasma.

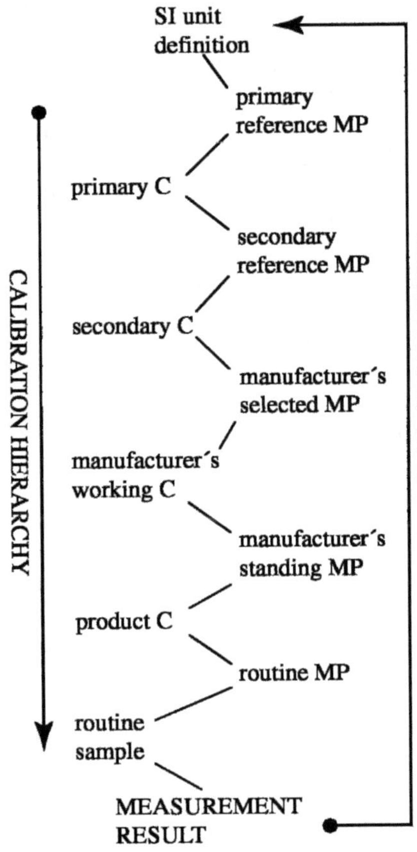

Fig. 1 Full calibration hierarchy providing metrological traceability from measurement results to SI unit definition as the stated reference. MP=measurement procedure, C=calibrator

Lack of a primary reference measurement procedure

If no measurement procedure has been devised with a completely describing measurement equation in terms of quantities using SI units, then empirical elements enter, and the unequivocal line from the defined SI unit is broken. The best substitute becomes an international conventional reference measurement procedure, which assigns a quantity value and measurement uncertainty to an international conventional calibrator.

An example is the amount-of-substance fraction of glycated haemoglobin A1c in total haemoglobin measured by high-pressure liquid chromatography-mass spectrometry [8] and a related BCR certified reference material 405 [9].

Any change in the international procedure or a shift to another measurement method means a new stated reference, usually giving new quantity values and measurement uncertainty as shown in the outcome of EQA. The international conventional reference measurement procedure is a part of the definition of the measurand.

Lack of primary reference measurement procedure and international conventional calibrator

An international conventional reference measurement procedure may indicate a measuring system allowing the assignment of quantity value and measurement uncertainty directly to the manufacturer's working calibrator. This is the case for number concentration of erythrocytes in blood by the ICSH measurement procedure [10].

Lack of internationally agreed measurement procedure

An international body may define an international protocol for value assignment allowing various measurement procedures or even measurement methods for the participating laboratories. This leads to an international conventional calibrator. Such is the case for many WHO International Standards [11] elaborated by the ECBS, e.g. bovine thyrotropin for 'bioassay'. This calibrator and the manufacturer's selected measurement procedure are then part of the definition of the quantity.

Lack of any international agreement

With no international agreement on measurement procedure or calibrator, the manufacturer's selected measurement procedure or even the individual laboratory's measurement procedure becomes the stated reference to enter into the definition of the quantity.

The sequence of five types of calibration hierarchy outlined above obviously constitutes decreasingly transparent and generalized traceability chains, and international collaboration is necessary to improve comparability.

Metrological traceability to an SI unit

When a calibration hierarchy starts with the definition of an SI unit followed by a primary reference measurement procedure and descends through commutable calibrators and specific and selective measurement procedures – 'full calibration hierarchy' above – the traceability to an SI unit is automatically ensured. The other four types of hierarchy do not preclude the use of a bona fide SI unit in the expression of a measurement result for a differential or rational quantity, but it is mandatory to specify the top measurement procedure and/or calibrator in the designation for the measurand. For example, 'amount-of-substance concentration of nitrogen(N) in human plasma by Kjeldahl procedure no. 3' (referring to the laboratory's list of procedures). The result in millimole per litre, however, is not unequivocally comparable with that of another Kjeldahl procedure, because the kinds-of-quantity are differently specified, but the unit is unchanged.

For other units, such as a WHO International Unit, an analogous rule applies; the top measurement procedure and/or calibrator must be specified.

International work on global reference examination systems

All of the elements in the reference examination (or measurement) systems listed earlier are continuously being improved by various means, and some recent outcomes should be mentioned here.

Written standards

The ISO and CEN have elaborated five relevant International and European Standards on:

- Presentation of reference measurement procedures
- EN 12286 (adopted as ISO 15193)

- Description of reference materials
- EN 12287 (adopted as ISO 15194)

- Requirements for reference measurement laboratories in laboratory medicine
- prEN ISO/FDIS 15195

- Metrological traceability of values assigned to calibrators and control materials
- EN ISO 17511

- Metrological traceability of values for catalytic concentration of enzymes assigned to calibrators and control materials
- EN ISO 18153

All five documents provide detailed instructions and obviously apply to various elements of calibration hierarchies.

The General Conference on Weights and Measures

At the metrological top, the CGPM defines SI units, and its executive, the CIPM, oversees their dissemination [12]. The rather newly established CCQM organizes Key Comparisons of measurements on reference materials among the national metrology institutes to ensure adequate measurement capabilities at the highest metrological level. Several of these materials are relevant to calibration hierarchies in laboratory medicine.

Recently, the CCQM has formed a Working Group on Bioanalysis which should be an important element especially in the infrastructure of laboratory medicine.

It goes without saying that the BIPM is the centre of activity for embodying the definitions of SI units, for organizing the work of the Consultative Committees, and for collaboration between national metrology institutes, regional intercomparisons, and international organizations.

International scientific organizations

Many scientific organizations are working on some elements of reference measurement systems in their respective disciplines. As an example, the current groups in the IFCC are listed in Table 2.

Table 2 Groups in the International Federation of Clinical Chemistry and Laboratory Medicine related to various elements of metrological traceability

IFCC Scientific Division	
C	Nomenclature, properties, and units
C	Molecular biology techniques
C	Plasma proteins
C	Standardisation of markers of cardiac damage
C	Standardisation of coagulation tests
C	Calibrators in clinical enzymology
WG	Selective electrodes
WG	Reference methods for apolipoproteins
WG	Standardisation of human chorionic gonadotropin
WG	Standardisation of Lp(a)
WG	Standardisation of HbA1c
WG	Standardisation of steroid hormone measurements
WG	Standardisation of osteocalcin measurements
WG	Intracellular and cell surface markers (flow cytometry)
WG	Standardisation of total plasma homocysteine measurement
IFCC Education and Management Division	
C	Analytical quality

Abbreviations: C=Committee, WG=Working Group

Infrastructure of metrology in laboratory medicine

To obtain reference examination systems for the many and very diverse types of property in the ever growing field of laboratory medicine is a complex and costly task which no single laboratory, institute or nation can achieve. The partially independent efforts in the scientific organizations, the large metrology institutes, and some funding by the EC have not proved sufficient to cover the needs.

It is, therefore, praiseworthy that internationally organized cooperation and collaboration within an agreed system is emerging from European and global consultations since 1999, not least prompted by the EU Directive on IVD MD's essential requirement of metrological traceability for values assigned to reference materials, which will become a challenge to industry.

Joint Committee on Traceability in Laboratory Medicine

A Symposium on Traceability in Laboratory Medicine, BIPM, 2002–06–09/11, called by Dr Terry J. Quinn, BIPM, and Professor Mathias M. Müller, IFCC, assembled some 60 delegates from Australia, Canada, Europe, Japan, South Africa, and United States. (The entities represented are indicated by an asterisk in Table 1.)

The general mission of JCTLM was agreed to be improvement in quality of healthcare with reduction in costs for governments and IVD industry through promotion of reference examination systems allowing traceability of examination results with improved comparability.

The JCTLM has been established as an interest group by an exchange of letters between the principal promoters and stakeholders CIPM/BIPM, IFCC, ILAC, and WHO. For two years the IFCC will chair with BIPM as Secretariat. Other important stakeholders in a general assembly are further scientific organizations (e.g. ICSH), CRM producers (IRMM, NIST), IVD MD industry (AdvaMed, EDMA, JACR), written standards developers (CEN, ISO, JCCLS, NCCLS), EQA/PT organizers (CAP, EQALM), regulatory bodies (EC, FDA), networks of reference measurement laboratories (DGKC, CDC). The JCTLM is not to be a legal entity and has no budget.

Two working groups have been created. WG 1 Reference materials and reference procedures, with IRMM and NIST in the chair, includes representations from AdvaMed, BIPM (CCQM), CAP, EDMA, EQALM, IFCC, JACR, NIBSC, and WHO; it will establish criteria for acceptance of materials and procedures and produce lists of such items.

WG 2 Reference laboratories, with IFCC in the chair, has representatives from AdvaMed, CDC, CIPM, EDMA, EQAS, ILAC, JACR, NMIs. This WG will set criteria for accreditation of reference laboratories at the calibration level, establish contacts to form networks, and promote parallel examinations. Already three networks could be identified, namely on enzymes (IFCC), glycohaemoglobin A1c (IFCC), and cholesterol (CDC).

The two WGs will report on progress around New Year 2003, and a second symposium should be planned for the summer of 2003.

Conclusions

Traceability of examination results is necessary to ensure reliability and the spatio-temporal comparability which is increasingly needed in the health services. The required global multielement reference examination system hitherto has been provided in a piecemeal fashion, but with the newly established Joint Committee on Traceability in Laboratory Medicine (JCTLM), coordinating all stakeholders with CIPM/BIPM, IFCC, ILAC, and WHO in the lead, the sparse resources should be distributed in a prioritised and structured way.

References

1. European Parliament and the Council of European Union (1998) Off J Eur Comm No L 331:1–37
2. ISO, IEC (2000) DS/EN ISO/IEC 17025:pp v, 26
3. ISO (2003) EN ISO 15189: v, 35
4. BIPM, IEC, IFCC, ISO, IUPAC, IUPAP, OIML (1993) International vocabulary of basic and general terms in metrology, English and French, 2nd edn. ISO, Geneva, pp 59
5. Dybkær R, Jørgensen K (1967) Quantities and units in clinical chemistry, including Recommendation 1966 of the Commission on Clinical Chemistry of the International Union of Pure and Applied Chemistry and of the International Federation for Clinical Chemistry. Munksgaard, Copenhagen, pp x, 102
6. IUPAC, IFCC (2002) Properties and units in the clinical laboratory sciences. Database accessed on http://dior.imt.liu.se/cnpu/
7. ISO (2002) ISO/DTR 15196 (in preparation)
8. Kobold U, Jeppsson J-O, Duelffer T, Finke A, Hoelzel W, Miedema K (1997) Clin Chem 43:1944–1951
9. IRMM (2001/2002) BCR reference materials. EU JRC, Geel, pp xi, 95
10. ICSH (1994) Clin Lab Haemat 16:131–138
11. WHO (1990) WHO Tech Rep Ser 800:181–214
12. Quinn TJ (1994/95) Metrologia 31:515–527

Accred Qual Assur (1998) 3:308–316

Freddy Adams

Traceability and analytical chemistry

Received: 31 March 1998
Accepted: 6 June 1998

F. C. Adams (✉)
Department of Chemistry, University of Antwerpen (UIA), Universiteitsplein 1, B-2610 Wilrijk, Belgium
Tel.: +32-3-8202010
Fax: +32-3-8202376
e-mail: adams@uia.ua.ac.be

Abstract The basic concepts of traceability as they are defined by the Comité Consultatif pour la Cluantité de Matière are contrasted with the practical exploitation in chemical analysis. The applicability of traceability concepts are tested for their practical applicability on four different analytical methodologies, neutron activation analysis, plasma mass spectrometry, beam microscopical analysis and speciation analysis of organometallic compounds.

Key words Traceability · Certified reference materials · Primary methods · Analytical chemistry

Introduction

The concepts of traceability are not always well accepted by the analytical chemistry community. There is a benign kind of neglect towards these ideas [1] or even straight hostility (e.g. the reactions of Alexandrov [2], and Ackermann et al. [3], on the papers of Hässelbarth [4]).

The reason for this reticent attitude stems from the specific culture in which analytical chemistry grew into a major scientific discipline and as a consequence of the development of another distinct activity, namely chemical analysis. In its diversity of approaches, analytical chemistry is complex, it contains a variety of different techniques, some of them more reliable and accurate than others. As a discipline, analytical chemistry cannot be treated as a collection of general, simple, absolute or dogmatic concepts. It is an immensely practical subject. Its driving force is power of detection, reliability (traceability could be a tool for achieving this) and efficiency and cost [5].

Analytical chemistry as a measurement science

Analytical chemistry is a science with its own theoretical underpinnings, its laws, axioms, corollaries and a guiding theory [6]. One could argue that it is a successful science, using the hyperbole that analytical chemistry is science's "Midas" turning everything it touches into gold, following Fabry's statement that "progress in analytics, driven by the demands of technology, is the key and flywheel of expanding engineering ..." [7].

Although related to metrology, analytical chemistry is a complex scientific discipline, measurement itself is only one, often a minor, aspect of the entire analytical process, among many other important parts. Many determinations are complex and multi-step procedures.

Analytical chemistry is complex because it draws on any available technique or on any information that is suitable for its specific purpose. As Murray [8] mentioned a good measurement scientist, in particular an analytical chemist, is an opportunistic scavenger always on the lookout for something lying around that can be adapted to a new purpose. Twentieth century analytical chemistry is similar to the revolution Picasso brought to twentieth century art when he declared "When there is something to steal, I steal it" [9].

Serious analytical chemists realise that too often analytical results are far from being accurate and that these inaccuracies do not only occur on the discipline's borderline, in the application of methods under development, but also during the application of well-established techniques. King [10] mentions a case on the determination of elemental lead in cabbage in an intercomparison exercise that led to errors of more than a factor of 10. Such errors cannot be tolerated, they can only be understood as sloppiness or incompetence (or both simultaneously) on the part of the analysts. Sloppiness will not be prevented by any formalised approaches, it is the result of people doing analysis, a technical act and not pursuing a science. Besides, most of these erroneous results were due to contamination, a source of error which cannot be corrected by the application of primary methods, not even by isotope dilution mass spectrometry. Incompetence is another matter, incompetent practitioners could be weeded out by laboratory accreditation procedures.

Why then would analytical chemistry be reluctant towards metrology in general and the traceability concepts in particular? According to its internal logic, this cannot be the case, because analytical chemistry has clearly demonstrated its readiness to adopt any ideas, methods or concepts that it can import. The concepts put forward in the traceability discussion are potentially valuable ones and could help in insuring analytical chemistry to keep track of its quantitative and accurate character as well as the comparability of its data.

However, it is clear for many practitioners of analytical chemistry that ideas put forward in publications on traceability seem rather oversimplified. These ideas tend to simplify analytical practice by assuming that every problem related to the comparability of data can be solved ultimately by adopting a simple chain of rigidly defined rules, and by the application of a few so-called primary or definite methods [11]. These methods (coulometry, isotope dilution mass spectrometry, and a few less generally applied methods) have a quite limited potential of application for the complex questions facing analytical chemistry as a problem solving discipline. For instance, one might wonder what kind of analytical problem can ever profit from the primary method relying on freezing point depression! The traceability advocates also tend to illustrate exceptions to the basic rule (traceability to stated references) in over-simple terms. They illustrate the case with a few (rather exceptional) "field" methods where any notion of accuracy fails for obvious reasons, e.g. because the analyte is not well defined (as in the determination of fat in meat products or the fiber content of corn flakes) [12]. Analytical chemists do not catalogue such measurements as analytical chemistry but as more or less useful measurement techniques somewhat related to the analytical enterprise.

The bulk of modern analytical chemistry, perhaps 99% of it, and most of its tools lie outside the limited scope and the simplified problems iterated in most recent papers on traceability. The application of the limited arsenal of primary methods in the remaining 1% often fails for other reasons: because the method is often too complex or cumbersome to apply in any real practice (as in isotope dilution mass spectrometry) or too labour intensive (as in most of volumetric analysis).

Maybe, however, there is a more profound reason for the antagonism: a debate between two cultures, between two sciences with entirely different paradigms, syntaxes and semantics. Fabry [7] demonstrated that analytical chemistry does not depend on the deductive approach for acquiring knowledge (Plato, Aristotle, Hegel), as I believe metrology does, but to the inductive method (Bacon, Hume, Mach). He claims that analytical work belongs to Francis Schaeffer's evidentialism that tries to answer the question "How can you be sure that?" with the answer "I am not sure and that does not disturb me. However, I can say that I am very sure due to the given data set and experiences". In short, analytical chemistry needs to develop in an unsupervised, entirely undogmatic climate.

Selected problems

A few pragmatically chosen examples will be used to show some of the problems that are appearing with the brainchild born from the marriage of metrology and analytical chemistry that has been named traceability. Four case studies selected from the direct laboratory experience of the author will be addressed which pertain to different areas in practical analytical chemistry:
(1) Neutron activation analysis (NAA). This method has been one of the most important methods for accurate trace elemental determinations. It has now (strangely) seemed to evolve in the light of traceability concepts;
(2) physical methods of analysis of solids by plasma techniques;
(3) microscopical instrumental analysis;
(4) speciation analysis of organometallic compounds in solid environmental materials, as an example of a much larger class of problems including the determination of organic compounds.

At present (2) and (3) lack any potential for SI-traceability as there are no primary standards and methods for analysis. As they cannot be assisted by any reliable basis for quality control, they have a tendency to show poor accuracy. This even leads to data that may be internally self-consistent but lack general comparability (e.g. the analysis of geological materials cited by Thomson [13]).

Example 1: NAA, a suitable method for trace element certification

NAA relies on the comparison of specific radioactivity, mostly gamma radiation, between a standard sample and an unknown sample after irradiation in the neutron flux of a nuclear reactor (or another neutron source). In principle the method is reliable and accurate, and more or less sensitive (dependent on the magnitude of physical constants involved in the reaction and the decay processes) for the determination of a number of elemental constituents in a sample. The physical background is completely understood and as nuclear phenomena are measured chemical effects are absent, except for (predictable) differences in neutron and radiation absorption in sample and standard. Hence, it does not come as a surprise that the method has a good reputation as far as its accuracy is concerned and was used in the past for a number of trace element certifications. The usual practice is to apply the method in a relative mode of standardisation using an elemental standard for every element to be determined and comparing the radiation produced.

Theoretically it is also possible to use absolute standardisation by calculating the unknown concentration from available microscopic nuclear data and the measurements of macroscopic quantities such as the neutron flux (with a flux monitor) and experimentally or calculated detection efficiencies. Is the method a primary method for the determination of quantity of substance when relying on relative standardisation? The answer is no as reference must be made to a standard of the quantity to be measured. Even when these standards are derived from pure elements, stoichiometric compounds or available certified reference materials (CRMs), according to the accepted definition, the method is considered indirect and not linked to the standard of quantity of substance. If, on the other hand, the absolute standardisation method is potentially traceable, could it be considered a primary method in the sense of the recommendations made by the Comité Consultatif pour la Quantité de Matière (CCQM)? Undoubtedly, according to the adopted definition the answer would be yes, although the method is not at present selected as such. Indeed, it can be assumed that published physical constants have been determined with the fundamental constants as a primary reference. Which one of the methods of standardisation is to be preferred for accurate measurement of the amount of the substance? From a strictly metrological point of view there can be no hesitation, it is the absolute method as it is directly linked to SI units and does not explicitly require a standard of the quantity to be measured. For the analytical chemist the answer belongs to the kind of questions with the reply: "the proof is in the eating of the pudding". Indeed, every practitioner knows that the application of the absolute standardisation method, despite its direct link with metrological concepts, is a risky undertaking and the results often show extremely poor accuracy. On the other hand, the relative method can provide excellent results if used professionally and if the sources of errors are properly accounted for. We clearly end up with a paradox, the most "traceable" method according to the CCQM definition is apparently not the preferred one in any real practice. But there is more connected to all this.

Nowadays there is a growing tendency to apply a third approach for quantification of NAA, an intermediate between the relative and the absolute standardisation methods, a formalised technique termed the k_0 method [14]. This method can be considered as an improved version of the absolute standardisation method in which (the often unreliable) nuclear data are replaced by accurately determined "compound" nuclear constants (k_0 factors, one for every element), which are used in a "protocol" type of environment. There are obvious practical advantages for the application of this method for routine multi-element determinations: one can get rid of the irradiation of a number of standards together with every unknown sample. It is clear that this method may considerably enhance the manageability of repeated analysis in NAA laboratories, it allows automation of procedures and hence, should lead to a higher throughput/manpower ratio in the routine analytical environment. One can certainly claim that the application of the method could upgrade quality of analysis in a given laboratory.

Nowadays, the k_0 concept tends to be promoted to the status of more than just a practical working tool. On the basis of ideas put forward on traceability (the removal of the standard of the quantity being measured), it has evolved into the preferred method for accurate NAA and a kind of a "near-traceable" method for trace elemental analysis. What is worse, it is advocated as the preferred method for element certification in trace element certifications.

Such an evolution is dangerous for several reasons. Despite its attractiveness as a practical tool in a given laboratory set-up there are potential dangers to rely on a method such as this for certification. The problems can be summarised as follows:

- Compared to the simplicity of the relative method, with its simple measurement equation, there is a hidden complexity in the k_0 method: complex algorithms, dedicated software for reactor neutron fluxes and gamma ray measurement efficiency and many problems associated with spectrum deconvolution. The method relies on a complex set of written standards which are not always fully understood by the average user. It uses non-transparent instrumentation and measurement processes. In short the method becomes, forgive the terminology, non-traceable to the user and this is, I believe, worse than non-traceable to SI units.
- With an effort to make the method work in an organised system of users in diverse laboratories, potential systematic error sources will tend to be diffused in a

large interlaboratory environment when the method is used with unlimited faith in its accuracy, as a black-box.

- In the relative NAA method accurate determinations boil down to irradiating sample and standard in the similar neutron spectrum of the same nuclear reactor in a reference irradiation position. In the k_0 approach a number of conventions must be addressed for insuring the "constancy" of the neutron flux and its energy spectrum within a particular reactor configuration [15]. The constants used are not fundamental (physical) constants. The neutron spectrum resulting from fission decay of U^{235} may be a well-known physical (but complex) entity, but one particular irradiation position of a given nuclear reactor is certainly not constant in time.

This leads to obvious conclusions:

1. The potential primary (traceable) method starting from the physical constants and a number of experimentally derived parameters is not able to provide accurate results.
2. The second best choice on the basis of traceability concepts, the k_0 method, contains a major flaw (in essence, there is no neutron spectrum for every irradiation position of every reactor available at BIPM in Sèvres!), neither is it transparent in its concepts and its error budget.
3. The naïve "prehistoric" (in the traceability sense) method devised long ago is quite appropriate for providing reliable quantitative results and should be selected as a suitable candidate for any certification exercise where it is applicable, on condition that the analytical chemists applying it realise the inherent limitations and pitfalls and make an adequate uncertainty budget.

The conclusion of this case study is that in certain conditions the traceability ideas could be misused to promote methods with the status of a "pseudo" primary method of analysis.

Example 2: physical methods of analysis – plasma mass spectrometry

Instrumental analytical methods are based on well-known physical laws concerned with the interaction of radiation with matter, and measurement of the resulting phenomena (radiation or particles). Often, the laws governing this interaction are reasonably well understood but were deduced from simple systems, usually one- or maximally two-component systems, not on complex samples. In practice they are often too general and too approximate for their straightforward use in analytical chemistry.

As demonstrated by Ramendik [16] the analytical signal in such methods depends on other factors than the physical and chemical characteristics of the analyte. In plasma methods of analysis, to take one example, the physical constants are the ionisation potential, atomisation energy, atomic or molecular mass. Experimental conditions, especially the sample matrix (the well-known, but often badly understood, matrix effect), play an important role and tend to degrade the accuracy of determinations. In essence, all these methods are non-traceable in the CCQM sense. Should this prevent their application in analytical chemistry? No, as even in their imperfect, unfinished state of development they are able to solve important technological and scientific problems. Their utility as analytical tools depends on the availability of reference materials (RMs) and on the application of one or more other reference methods with high accuracy.

Ramendik [16] pointed to the possibilities of the creation and development of theoretical foundations based on mathematical modelling in elemental mass spectrometry after the creation of a plasma. For laser plasma mass spectrometry of geological RMs and a quasi-equilibrium approach based on atomisation and ionisation temperatures without relying on reference RMs materials, he claims to be able to arrive at average uncertainties for 40 elements totalling 20% [17]. This may not be ideal but it is a suitable accuracy for solving many practical analytical problems.

Such approaches rely on so-called relative sensitivity coefficients (RSFs), ratios of the difference between the sensitivity of various elements, and these cannot be considered as fundamental constants. In fact, they provide no more than a quantitative measurement of the deviation of the method's result from the amount of substance, as issued from primary methods (if available). Other near-equilibrium plasma methods for the analysis of solids (glow discharge, sputtered neutrals secondary ion mass spectrometry) produce quite acceptable results for analytical practice.

The metrologists view is that these sensitivity factors are not traceable, although one can argue that this is only a formalistic point of view. Indeed, in the case of glow discharge mass spectrometry (GDMS) it is possible to transform these RSFs to quantities that are connected to both the unit of mass and to physically meaningful concepts. Referring to Bogaerts and Gijbels [18] the RSF in GDMS can be written as:

$$\frac{C_x}{C_s} = \mathrm{RSF}\left(\frac{x}{s}\right) \cdot \frac{I_x}{I_s} \qquad (1)$$

where I and C are the ion current and the concentration in mass units, respectively, and x and s represent the element x and the internal standard s. This RSF value is related to the relative ion yield (RIY) in the following way:

$$\mathrm{RSF}\left(\frac{x}{s}\right)=\frac{1}{\mathrm{RIY}\left(\frac{x}{s}\right)}\cdot\frac{M_x}{M_s} \tag{2}$$

where M_x and M_s are the atomic masses.

Following Vieth and Huneke [19], it can be stated that the RIY is determined only by transport and ionization/recombination effects:

$$\mathrm{RIY}\left(\frac{x}{s}\right)=S_\mathrm{T}\left(\frac{x}{s}\right)\cdot S_\mathrm{I}\left(\frac{x}{s}\right) \tag{3}$$

where S_T and S_I are real physical concepts describing the transport and the ionization/recombination, respectively.

Such methods, although not considered as primary methods of analysis because of a remaining "fudge factor", may be reasonable substitutes for exploitation in round-robin studies, for the production of "usable" RMs and for practical analysis.

As was shown recently by Bogaerts and Gijbels [20] state-of-the-art mathematical modelling of the plasma yields a very satisfactory agreement between calculations and experimental observations (sputtering rates and profiles, optical emission spectra, ion fluxes entering a mass spectrometer). They demonstrated that the models used present a quite realistic description of the glow discharge process.

In fact, plasma methods may belong to the most difficult methods to model completely: as many as 24 different collision processes are incorporated in Bogaert's model and non-equilibrium situations, and the presence of imperfect solid surfaces must be accounted for. Similar arguments could, if the processes are studied in detail, be formulated for atomic emission or atomic absorption analysis. Diagnostics of the inductively coupled plasma has resulted in a quite well-characterised sample environment.

Such methods are well on their way to achieving what Kolthoff and other analytical chemists achieved between the two world wars with titrimetric analysis. In this case, a thorough evaluation based on physical chemistry and the study of all relevant systematic errors led to a full account of all reasonable sources of error [21]. On the basis of this, volumetric analysis was given the status of a primary method by the CCQM, despite the fact that water remains one of the least understood solvents and that many interfering reactions can affect the accuracy of a titrimetric determination. If processes taking place in the glow discharge plasma are completely understood, will it then obtain the same status? I doubt it, or at the very least, I suspect it will take many years before this occurs. The reason being that there is no simple analytical equation linking the analytical signal and the result.

Example 3: instrumental elemental microanalysis

Most methods (electron probe microanalysis, micro-Auger, secondary ion mass spectrometry) cannot be considered as really accurate methods except when applied to quite simple systems. Their application relies on the use of CRMs but these are, with very few exceptions, not available. The reason for the lack of CRMs is the absence of any reliable methods of microanalysis. None of the limited range of primary methods is applicable for the analysis of a solid at a microscopical level. The world of microanalysis is badly in need of at least one method which is able to act as a reference for the other techniques and to link RMs or round-robin exercises to the SI units [22].

Although a primary method is not yet available, we argue that interesting developments are taking place that could bring x-ray fluorescence analysis (XRF) in line as a primary method.

According to its present reputation XRF is not a likely candidate for this purpose; it is considered as a poor method for any certification purposes because of its intense matrix effects. Using the wavelength-dispersive XRF technique reliable results can only be obtained through calibrations with a set of standards of closely similar composition. Energy-dispersive XRF also suffers from a number of drawbacks (spectral overlap, poor statistics). On the other hand, the physical basis of the interaction of x-rays with matter is fully understood and the physical constants governing the interaction and radiation absorption are known accurately from physical measurements conform to metrology measurements. In principle, it is possible to fully correct for deviations of linearity between measured intensities and elemental concentrations, provided that time and effort are spent to make the proper calculations. Certainly this is true in analytical conditions where the set-up is simplified as much as possible (by using monochromatic primary excitation with a solid state energy-dispersive detector on a sample with a well-characterised shape and surface condition).

Despite its bad reputation as an analytical tool, XRF is potentially a traceable method according to the CCQM definition and could be a primary method although it was not selected as such, and won't be for a long time. In fact, it is the only microanalytical method which can at present be considered as a candidate for accurate microscopic elemental analysis. Proof of this statement follows from Monte Carlo calculations in which experimental XRF spectra can be accurately modelled starting from first principles [23]. This is not an easy approach but with computing power now available it is feasible, though not worth the effort for bulk chemical analysis where other alternatives are available.

Synchrotron storage rings, for instance, are able to provide an extremely high flux of nearly monochromatic X-radiation on a small sample area. They could form the basis of XRF set-ups and enhance other microanalytical methods to provide accurate determinations. In the future they could serve as a reference method for elemental trace analysis on the microscopical level (with the quality of the random number generator, a non-SI concept, as the prime source of error).

Quantitative X-ray photon spectroscopy (XPS) and Auger emission spectrometry (AES) analysis are considerably more difficult to apply as quantitative tools for surface analysis as they require an understanding of the change in energy distribution of electrons as they move in solids. Compared with XRF, one has to deal with the outer electron cloud instead of the core electrons. Nevertheless as demonstrated in a recent review by Tougaard [24] principles and rules of general validity (universal cross sections) can be defined which give a reasonably accurate description of the inelastic scattering process and open the way for real (but again non-traceable) quantification.

The conclusion to be drawn from this case study is that, with the ongoing research and massive calculating power available nowadays, methods can evolve which are at least similar in their traceability status, if not better than "near-primary" k_0 NAA. Again, they seem too complicated in the overall analytical equation to win the confidence of the CCQM.

Example 4: speciation analysis of organotin compounds in solid samples

This example illustrates a range of problems that are symptomatic in quantitation and standardisation which appear in speciation analysis and in determination of organic compounds in solids. Organotin compounds (e.g. methyltin, butyltin, phenyltin) comprise one of the most thoroughly studied groups of organometallic compounds found in environmental samples.

The most sensitive methods of analysis are those involving conversion of ionic organotin compounds to volatile hydride or alkyl derivates for subsequent chromatographic separation and determination [25]. Hyphenated techniques have been used in environmental and biomedical studies based on the combination of gas chromatography with atomic absorption spectrometry (GC-AAS), microwave induced plasma atomic emission spectrometry (GC-AES) and inductively coupled mass spectrometry (GC-ICPMS). Advances over the past years have brought the instrumental detection limits to the sub-picogram level on a routine basis. The Community Bureau of Reference (BCR) organised a number of intercomparison exercises that convincingly demonstrated the reliability of the procedures developed. On the basis of these results a number of CRMs were developed for these compounds in various sediments and tissues.

However, this does not mean that at present reliable determinations of these compounds in solid samples are possible. The major difficulties in practical application are not related to the metrological aspects of the analytical procedures but to inherent difficulties of: (1) separating organometallic compounds from a solid matrix while (2) preserving information on their chemical identity throughout the entire analytical process. Up to now, the procedures have not been legally defensible because of the error-contributing part of the analytical procedure located at the immature front-end of the analytical process: the extraction step (recovery of the analytes) and the extent to which chemical integrity is maintained during analysis.

Variable recovery is a principal cause of non-equivalence of data and there is no straightforward solution to this problem [26]. Artificially made reference samples or pure compounds added to test material cannot be used for estimations of recovery of analytes. Direct speciation analysis from the solid sample [27] is not feasible at present, although analytical methods are appearing that could be useful in the future (X-ray absorption spectrometry, laser mass spectrometry, static secondary ion mass spectrometry).

Uncontrolled species transformations during analysis form another source of error. For methylmercury determinations in sediments it was demonstrated that errors of up to 80% resulted from the formation of the compound from inorganic mercury during separation and analysis [28, 29]. For the study of possible species transformations during analysis multiple isotope dilution could be used as a diagnostic tool for identifying the error and bias inherent in specific methods of storage, sample preparation and measurement [30, 31].

In such conditions what is needed are RMs containing an accurately known concentration of the analytes as they appear in a well-characterised solid material in its natural status (thus not as a doped substance). At present it is an open question as to what methods are applicable to round robins and certification exercises and what methods ultimately could serve as the primary reference to the SI. None of the so-called primary methods of analysis is appropriate for the purpose.

Until the situation becomes clear, further development should be based, as it is now, on reference samples that are utilised in conjunction with a "written standard" for the extraction process to generate a well-known concentration of the analytes in aqueous solution. Temporarily, these will be have to be used as "fit for purpose" references for the subsequent analytical determination process. However, the design of selective extraction schemes may not reflect the actual distribution within a test sample [32].

In conclusion traceability concepts as they are now defined are of little practical use. On the other hand, methods of analysis such as these are certainly needed in science and society and so are RMs and methods. Work must go on in the absence of a full uncertainty budget and lack of clarity concerning the traceability reference [9]. The BCR was right to develop a number of CRMs for speciation analysis as a diagnostic aid in the further development of this particular methodology.

To summarise:

1. Analysis is a lot more complex than the measurement process alone. The measurement step is often the best understood step in the overall analytical process. Error sources are largely situated outside the direct measurement step (Examples 2–4).
2. Many analytical methods (Examples 2 and 4) cannot be directly or indirectly linked to SI-units with any of the available primary methods of measurement. These can only be used in exceptional cases for a limited range of rather simple problems. Hence, many methods of analysis rely on other methods for the assessment of their accuracy. The instrumentation and measurement process used must be transparent and a full account of sources of error, relative and absolute, must be made. Unfortunately, as Examples 2 and 3 show, with the growing complexity of analytical methods this transparency becomes a major headache for both metrologists and analytical chemists. The relationship between signals measured and the derived concentration becomes a complex calculation [6] and the result of a measurement in many methods can only be traceable to the instrument, its electronics and integrated software [33].
3. Primary methods of measurement can, to some extent, be utilised for the preparation of "synthetic" RMs. In many situations these cannot be used in analytical chemistry as it is imperative that "real world" samples are used for standardisation purposes (Examples 3 and 4).
4. As follows from Example 1, complex methodologies tend to be promoted as a means of complying with traceabilty to the SI units; in this particular case it is attempted to derive quantitative results without relying directly on a standard of the quantity measured.

Traceability for measurements in (analytical) chemistry

We now move back to the meaning of traceability in analytical chemistry in its generalities and use the basic concepts as defined in the carefully worded article of Kaarls and Quinn [34] which summarises the conclusions of the CCQM [35]. The major items are as follows:

1. International comparability of chemical measurements is to be achieved by linking all measurements (analyses) to the SI. As analytical chemistry is aimed at the determination of the true composition of matter the goals are identical. Only the rationale and the path followed to arrive at the truth differ.

2. The expression of results is in basic SI units. Analytical chemistry is used to express results of the measurement process as concentrations expressed in different units according to the specific problem addressed. As long as these concentration units are properly defined there cannot be a problem in converting results. Neither can there be any real problem with the subtle difference between mass and amount of substance.

3. Measurement of the amount of substance: The definition of primary methods implies that a complete uncertainty statement can be expressed in terms of the SI units without reference to a standard of the quantity being measured. Only a very limited number of methods (isotope dilution with mass spectrometry, coulometry, gravimetry, titrimetric analysis) comply with these criteria and only these can be used for the determination of the purity of materials used in the preparation of (primary) RMs.

In the article of Kaarls and Quinn [34] primary methods are carefully defined as methods for the determination of the amount of substance in pure or simple compound systems, i.e. in samples which do not contain impurities acting as potential interferences. It is explicitly stated that it is a future task of the CCQM to investigate the applicability and robustness of these methods for complex mixtures encountered in practical analytical chemistry. Many other papers (e.g. [36]), however, tend to identify primary methods already as methods of analysis (to be used on complex samples of unknown overall composition). This over-optimistic (and unwarranted) enlargement of the definition implies that all titrimetric methods of analysis would be considered as primary methods putting aside any interference that occurs in complex samples. Considering all possible sources of error that may occur in both the stoichiometry of the reaction and with the determination of the equivalence point of a titration, this cannot be possible. Neither was this the intention of the CCQM.

4. RMs: Considering the limitations of available primary methods, emphasis is placed by the CCQM on the elaboration of synthetic RMs derived from pure materials. These would then be used for calibration of instruments and hence, could help in the metrological step of analytical procedures. As appears from the cited examples, such RMs are hardly suitable as reference samples in many applications of analytical chemistry.

All this implies that truly traceable CRMs relying on primary methods are only available in exceptional circumstances and it is highly desirable that methods are developed which provide greater access to them. Progress in analytical instrumentation and methodology might help us considerably. Isotope dilution GC-ICPMS might become a usable methodology for trace element determinations, at least for the poly-isotopic elements [37].

5. Stated uncertainty and levels of confidence of analytical measurements: There are two distinct points that need to be addressed here. Firstly, many papers discussing traceability concepts are heavily biased towards analytical determinations that require extremely high accuracy and, hence, full metrological orthodoxy. It is quite obvious that in specific situations strictly defined concepts must be enforced and adhered too, even though in some circumstances comparability of measurements are more important than the true analytical result. In nuclear safeguards, application of traceability concepts may prevent one, or more, plutonium bomb(s) remaining unaccounted for worldwide. One can clearly see the importance of a strict traceability of ozone measurements in monitoring global climatic change or in pollution abatement [38]. In both of these examples the methodology exists and there are very good reasons to enforce it through the strict implementation of traceability. Analytical chemistry is, however, a problem solving discipline and its results are providing answers to specific problems in science, society and industry. In defining the measurement process it is necessary to define the degree of accuracy necessary for "fitness for purpose". Economic considerations [3] come in here: it is not necessary that the accuracy of the determination is higher than necessary, provided that stated uncertainties include the true value, and that measurements are comparable [1]. In many applications measurement errors addressed in traceability concepts are insignificant compared to both fitness of purpose requirements and other error sources in the analysis or in sampling.

Secondly, there is the question of the estimation of errors. The approach of the International Standards Organisation (ISO) consists of estimating the individual error components in a total uncertainty budget [39]. Analytical chemists, on the other hand, are accustomed to report their measurement results by standard deviations of individual results (the repeatability). It has become clear from many interlaboratory comparisons that discrepancies (the reproducibility) are consistently larger than this repeatability estimate. Hence, it is necessary that more emphasis should be placed on the determination of the uncertainty of analytical methods [26]. Approaches for evaluating measurement uncertainty may rely on metrology laboratories and on analytical laboratories using standard operating procedures. CRMs and proficiency tests can be used for this purpose [40]. The total analytical error being larger than the measurement alone, it is necessary that other sources of measurement uncertainty are accounted for, e.g. those resulting from sampling [26, 41].

Conclusions

Accuracy is a central theme in any analytical process and its pursuit can no longer be the work of a single analytical chemist but requires comparative studies involving many methodologies and many experts. It is sensible to expect that pooling different kinds of data (methods and users) will lead to better estimates of the truth, even if some of the contributions are more reliable than others.

Traceability concepts are just some of the many means to foster quality assurance as a drive for analytical reliability. They are heavily centred on the measurement part of the overall analytical process. Their application is hampered by the adoption of a number of strict rules so that truly traceable CRMs relying on primary methods are only available in exceptional circumstances.

The pursuit of accuracy is a matter of quality assurance, in the laboratory and in the system, less one of using the concepts of traceability alone as a means for anchoring measurements in time or space. The Standards, Measurements and Testing Programme of the European Commission: 4th Framework Programme 1994–1998 includes measures for the maintenance and development of metrological systems. It places emphasis on both (1) the accuracy and traceability of measurements to the SI and on (2) the development of metrology in chemistry and routine analysis including the development, improvement and validation of analytical steps (sampling, digestion, preconcentration, separation and calibration).

Quality control of laboratories depends on the availability of CRMs, round-robin studies, intercomparisons and proficiency tests between methods and between laboratories. Of special importance is a full knowledge of the complex analytical process and the painstaking pursuit of the true value by defining all sources of errors and the application of an adequate error source budget. The application of Poisson and Bayesian statistics could have some advantage.

Organisations requiring a high throughput of analyses are now undergoing a paradigm shift in which performance is specified, not the methodologies used. The United States Environmental Protection Agency, for example, is presently shifting to a "performance based measurement system" for the environmental analyses of hazardous wastes. In this system performance needs (instead of specific measurement technologies) are specified for reliable, cost-effective analyses, thus avoiding costly measurement overkill[1]. Feedback on

[1] URL:http://www.epa.gov/epaoswer/hazwaste/test/pbms.htm

successes and failures are then used to expand knowledge on new or modified approaches in a flexible working environment and in "real world" conditions. In such an approach the generator of the data is responsible for demonstrating regulatory compliance by defining a sampling/analysis plan and strict record keeping of proofs of concepts and levels of validation.

Acknowledgements Thanks are due to B. Neidhart (GKSS, Geesthacht, Germany), J. Broekaert, University of Dormund (Germany) and P. De Bièvre (JRC, IRMM Geel, Belgium) for discussions and criticisms.

References

1. Thompson M (1996) Analyst 121:285–288
2. Alexandrov YI (1997) Fresenius J Anal Chem 357:563–571
3. Ackermann M, Breitschu W, Büchler H, Küppers G, Lührs C, Schlotmann BJ, Staats G, Zens R (1997) Fresenius J Anal Chem 357:562–564
4. Hässelbarth W (1995) Fresenius Z Anal Chem 352:400–402 and (1996) 354:263–265
5. Broekaert JAC, Tölg G (1997) Fresenius J Anal Chem 326:495–509
6. Booksh KS, Kowalski BR (1994) Anal Chem 66 15:782–7916
7. Fabry L (1997) Fresenius J Anal Chem 357:148–150
8. Murray RW (1995) Anal Chem 67:343 A
9. Gilot F, Lake C (1964) Life with Picasso. McGraw-Hill, New York
10. King B (1997) Analyst 122:197–204
11. Richter W (1997) Accred Qual Assur 2:354–359
12. De Bièvre P, Taylor PDP (1997) Metrologia 34:67–75
13. Thompson M (1997) Analyst 122:1201–1205
14. De Corte F, Bellemans F, De Neve P, Simonits A (1994) J Radioanal Chem 179:93–103
15. De Corte F, De Wispelaere A, van Sluys R, Bossus D, Simonits A, Kucera J, Frana J, Smodis B, Jacimovic R (1997) J Radioanal Chem 215:31–37
16. Ramendik GI (1997) Fresenius J Anal Chem 357:233–240
17. Oksenoid KG, Ramendik GI, Sil'nov SM (1990) J Anal Chem 45:613–624, translated from Zh Anal Khim (Russia) 45:858–871
18. Bogaerts A, Gijbels R (1996) J Anal At Spectrom 11:841–847
19. Vieth W, Huneke JC (1991) Spectrochim Acta Part B 46:137–153
20. Bogaerts A, Gijbels R (1997) Anal Chem 69:719–727
21. Beck CM II (1997) Metrologia 34:19–30
22. Richter W, Dube G (1997) Metrologia 34:13–18
23. (a) Vincze L, Janssens K, Adams F, Rivers ML, Jones KW (1995) Spectrochim Acta Part B 50:127–148; b) Vincze L, Janssens K, Adams F, Rivers ML, Jones KW (1995) Spectrochim Acta Part B 50:1481–1500
24. Tougaard S (1997) Surf Interfase Anal 25:137–153
25. Dirkx W, Lobinski R, Adams F (1994) Anal Chim Acta 286:309–318
26. IUPAC (1997) Harmonised guidelines for the use of recovery information in analytical chemistry. Protocol, 1997-02-14
27. Van Vaeck L, Adriaens A, Adams F (1997) Spectrochim Acta Part B (in press)
28. Bloom NS, Colman JA, Barber L (1997) Fresenius J Anal Chem 358:371–377
29. Hintelmann H, Falter R, Ilgen G, Evans RD (1997) Fresenius J Anal Chem 358:363–370
30. Kingston HM, Huo D, Lu Y (1998) Proceedings of the 1998 Winter Conference on Plasma Spectrochemistry, Scottsdale, Arizona, January 5–10, 1998. Also U.S. Patent 5414259, U.S. Patent Office, May 9, 1995
31. Intelmann H, Evans RD (1997) Fresenius Z Anal Chem 358:378–385
32. Cave MR, Wragg J (1997) Analyst 122:1211–1215
33. Buydens LMC, Schoenmakers PJ (eds) (1997) Intelligent software for chemical analysis. Elsevier, Amsterdam 1993
34. Kaarls R, Quinn TJ (1997) Metrologia 34:1–5
35. BIPM (Comité Consultatif pour la Quantité de Matière) (1995) 1, (1996) 2
36. Richter W (1997) Accred Qual Assur 2:354–359
37. Campbell MJ (1995) Quality assurance for environmental analysis. In: Quevauviller Ph, Maier E, Griepink B (eds.) Techniques and instrumentation in analytical chemistry, vol 17. Elsevier, Amsterdam, pp. 27–37
38. Konopelko L, Nekhlioudov I, Chelibanov V (1997) Metrologia 34:97–99
39. Guide to the Expression of Uncertainty in Measurements (1993) ISO, Geneva
40. Kane JS (1997) Analyst 122:1283–1288
41. Ramsey MH (1998) J Anal At Spectrom 13:97–104

Accred Qual Assur (1999) 4:342–346

Paul De Bièvre

Do interlaboratory comparisons provide traceability?

Received: 12 January 1999
Accepted: 26 January 1999

Presented at: CITAC International Symposium on Analytical Quality Assurance for the 21st Century, Sydney, Australia, 15–16 October 1998

P. De Bièvre (✉)
Institute for Reference Materials and Measurements, European Commission – JRC, B-2440 Geel, Belgium

Key words Traceability · Interlaboratory comparisons · Reproducibility · Traceability chain · Metrology in chemistry

Introduction

Observing that somebody else finds approximately the same measurement result when measuring the same measurand, has provided great comfort to many people: confirmation always gives a nice feeling. Thus, intercomparisons of measurement results obtained on the same measurand in the same material by different laboratories, are welcomed because they provide this feeling. They are also interesting because they provide a simple and clear means to estimate the actual degree of *reproducibility of results* attained by different laboratories. This is useful because a great many measurements on the same material are performed daily by pairs (or more) of laboratories and decisions must be made on whether observed differences are discrepancies or not. Knowing the usual degree of reproducibility of measurement results between laboratories is extremely valuable in such cases. Large compensations may have to be paid on the basis of perceived differences in measurement results when they determine (serious) differences in values of goods.

However, does an interlaboratory comparison provide "traceability" of the results? To answer this question, we first look at the definition of traceability and its scientific characteristics. Without a clear, unequivocal understanding of "traceability", we cannot answer the question raised.

Traceability

When different laboratories obtain different values this can point to a real difference in value of the measurands, or it can be caused by variations in the measurement parameters of the method during the measurement process (for homogeneous samples). But differences, observed by different parties, do not, in themselves, provide proof of "correctness". This requires an independent, *external* indicator or criterion, or correction, which enables one to convert the demonstrated interlaboratory reproducibility (a form of "precision") into "accuracy". Failing to show proof of this makes it impossible to present a statement of "accuracy".

This is where traceability comes in. Traceability of a measurement result means that the value of that result (each result of each laboratory) is demonstrated to be linked to an independent common "stated reference" [1], through an uninterrupted chain of comparisons (see Fig. 1 for the most popular traceability chain). That requires a priori knowledge of the measurement process and of the "stated reference", because the

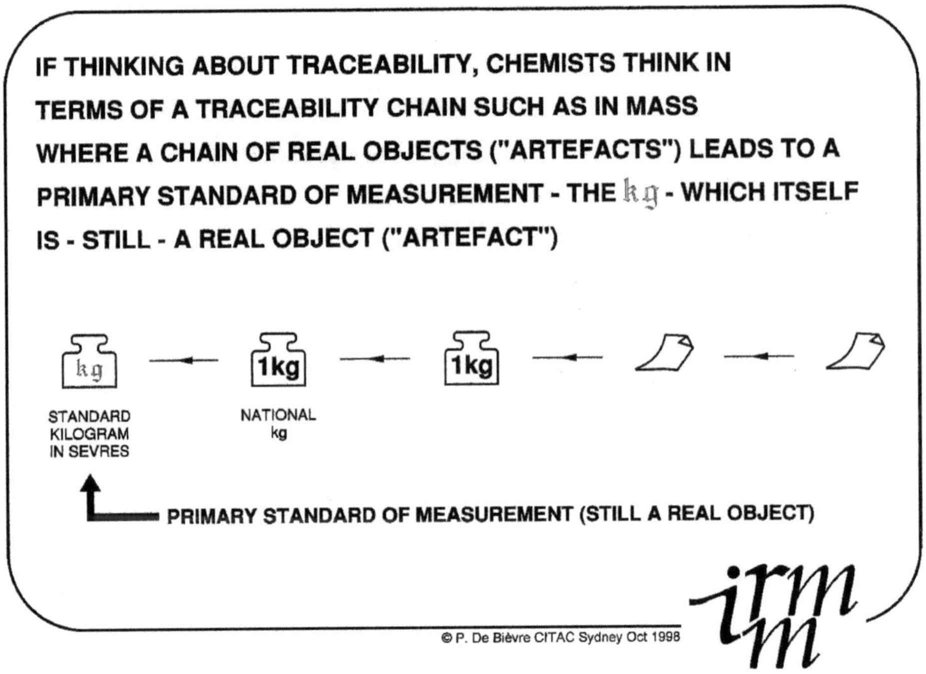

Fig. 1. The best known traceability chain, that of weights

measurement procedure to be followed in the measurement *must be planned in advance*. Traceability of values makes them *comparable* if the "stated reference" is the same for all. Establishing traceability of the value of a measurement result means each participant has to plan for him/herself. It requires a priori knowledge. Establishing traceability is a process that underpins, and is perpendicular ("vertical") to, the "horizontal" process of establishing the degree of reproducibility of results of several laboratories in an intercomparison (Fig. 2) (as a consequence, the expression "horizontal traceability" should be banned). There is no need for an intercomparison in order to establish traceability. It is worth noting that use of the results of interlaboratory comparisons can only be made *after* the measurement has been achieved (the result must be available); hence this is, of necessity, a posteriori knowledge, and therefore, of necessity, unfit to be used a priori.

The comfort of being confirmed by the measurement result of another laboratory is justified and ...nice. But it should not act as a drug, preventing the analyst from distinguishing between "accuracy" and "traceability" which are different concepts. A few examples will make this clear.

The first one is the traceability of the value resulting from the "core" measurement of isotope dilution (Fig. 3): a comparison of an unknown amount $n({}^{i}E,X)$ of isotope ${}^{i}E$ of element E in material X to a known

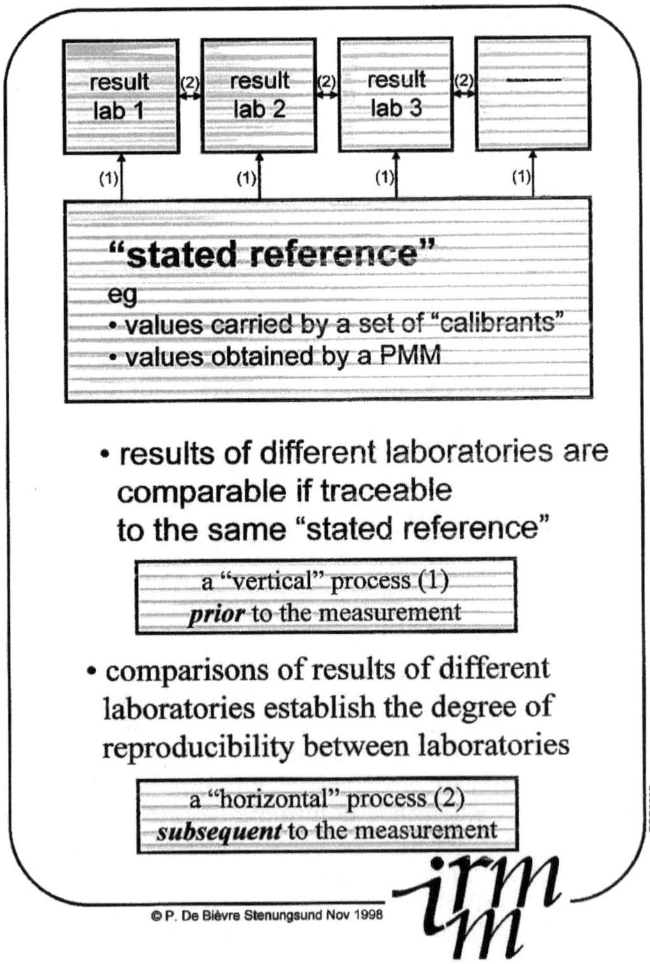

Fig. 2. Establishing traceability is a "vertical" process

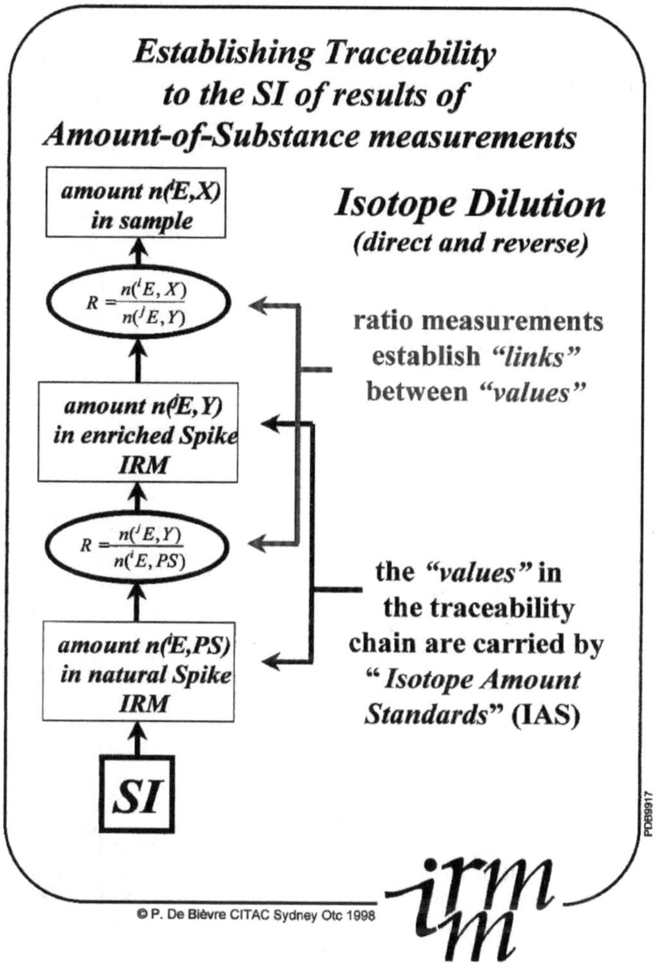

Fig. 3. Core part of a traceability chain of the measurement of an isotope amount ratio in an "isotope dilution" measurement procedure

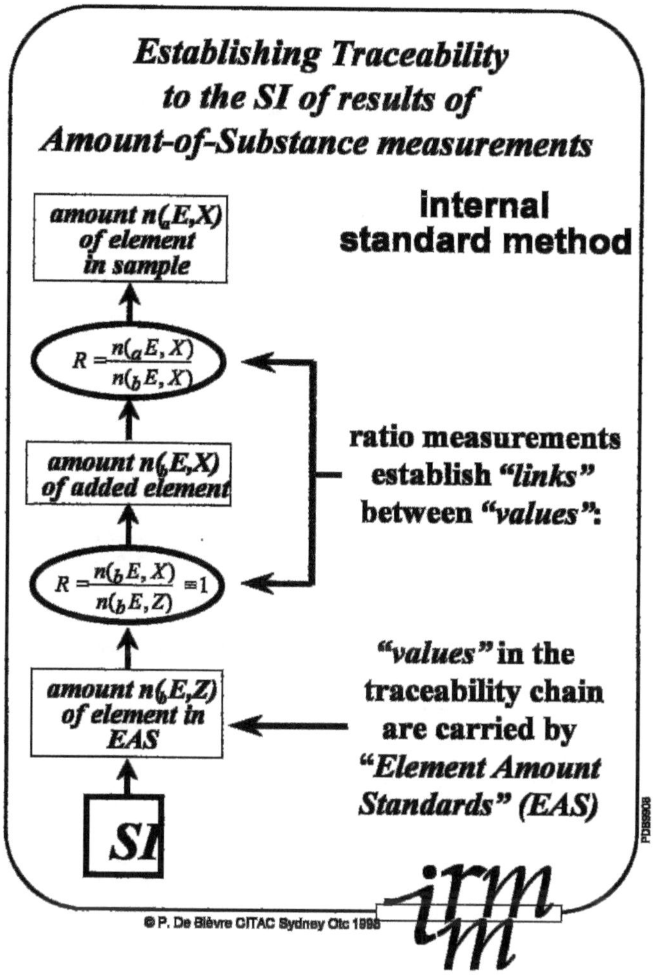

Fig. 4. Core part of a traceability chain in the measurement of an element amount ratio in an "internal standard" measurement procedure

amount of another isotope ^{j}E of the same element in (an isotopically enriched) material Y. The (part of the) traceability "chain" shown in Fig. 3 is simple and transparent. The link of $n(^{i}\text{E,PS})$ in a pure substance to the SI is established by a primary method of measurement (PMM): gravimetry. The entire purpose of establishing the chain is to demonstrate that the value of the measurement result is a (sub)multiple of the mole, the internationally agreed unit for amount of substance. We conclude from Fig. 3 that no interlaboratory comparison is needed to provide traceability.

The second example is establishing traceability in the case of the well known "internal standard method" as used in the measurement of an amount of element (Fig. 4). The "core measurement" is again the comparison of an unknown amount, now of an element $_{a}$E in material X to a known amount of another element $_{b}$E in material Z, in or added to the material X before the measurement. As in the previous example, $n(_{b}\text{E,Z})$ is linked to the SI by means of a measurement by a PMM: gravimetry. Again, the purpose of establishing the chain is to demonstrate that the value of the measurement result is a (sub)multiple of the unit for amount of substance.

A third and very important example is one where the unknown amount is not measured by a PMM directly in SI- or other units, but against commonly accepted references, usually values carried by reference materials (in whatever units). This is especially important in cases where the "substance" cannot (yet) be unequivocally identified as a unique substance (e.g. protein in beef or fibre in corn flakes, both very important in trade). Since an amount-of-substance measurement is not (yet) possible because the "substance" to be measured is not (yet) uniquely defined, we go back to the traditional definition of "calibration" as illustrated in Fig. 5 [2]. Several "reference" samples of different contents are prepared (or agreed) and their contents ex-

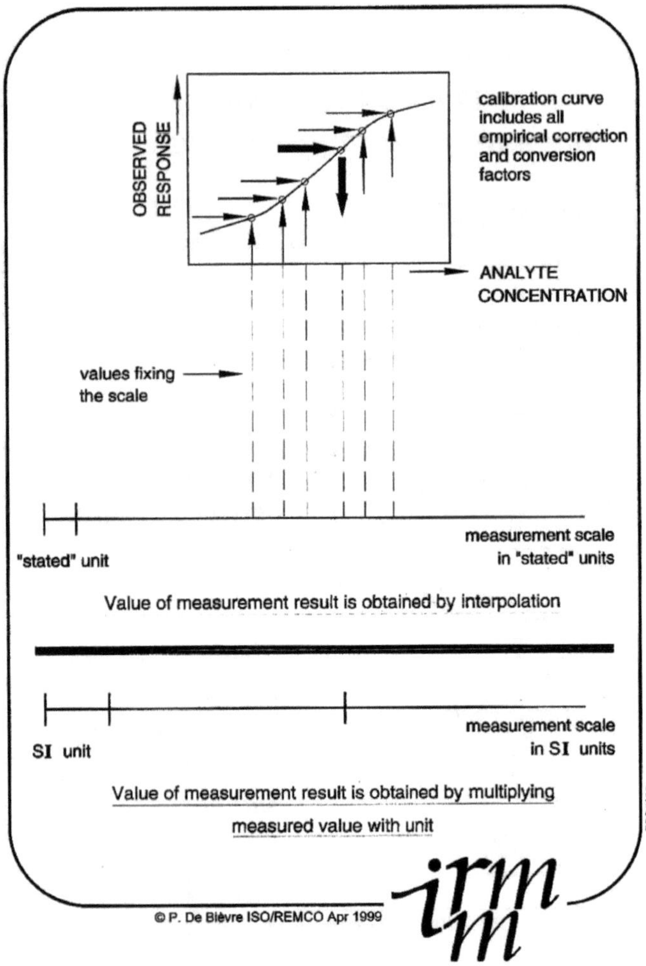

Fig. 5. Establishing traceability to a common scale through a "calibration" process

pressed (in whatever unit) in a "reference" laboratory. In the field laboratory, they can then be measured to make up a calibration curve using the observed responses given by the local instrument and the values carried by the "reference" samples. This curve enables the analyst to measure an unknown content of a sample and obtain a "calibrated" measurement value. It is measured on the same scale which was previously established by means of the values carried by the set of reference materials and is therefore automatically expressed in the same units. Again no interlaboratory comparison is needed to establish traceability. When the traceability of values (to a stated reference, here a common measurement scale) has been established, comparability is generated (Fig. 2). By means of an interlaboratory comparison the degree of reproducibility of traceable results from laboratories can be determined (Fig. 2). Thus interlaboratory comparisons are useful.

Fig. 6. Results obtained when measuringthe same measurand by different methods: estimating the degree of reproducibility, or the degree of equivalence, is achieved for the participating laboratories, not the establishment of the traceability of each result

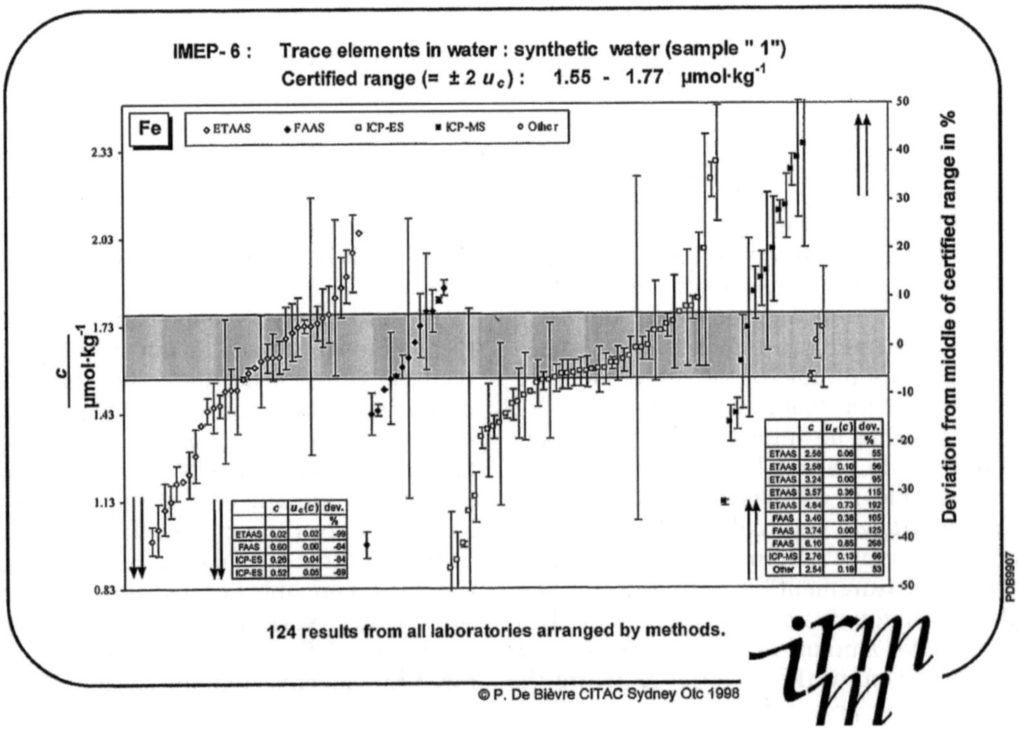

The same conclusion also follows from a totally different observation. Figure 6 displays the results of an interlaboratory comparison (in this case the international measurement evaluation programme: IMEP-6) according to methods as used by the participants. It is clear that this picture only reflects the degree of reproducibility of results obtained by various methods, but does not yield traceability of the measured values to an a priori stated reference.

Conclusions

Interlaboratory comparisons do not provide traceability of values of measurement results because they only deliver information a posteriori. Therefore, they cannot establish traceability. The establishment of traceability of a measurement result is a task for every single measurement laboratory on its own and does require knowledge a priori.

Interlaboratory comparisons of the results of different laboratories are an a posteriori process. They yield another useful product: the establishment of the degree of reproducibility of results of different laboratories, or degree of equivalence between the measurement capability of the participating laboratories.

It seems that these conclusions are also valid for the key comparisons organised by CCQM.

References

1. BIPM, IEC, IFCC, ISO, IUPAC, IUPAP, OIML (1993) International vocabulary of basic and general terms in metrology, 2nd edn. ISO, Geneve
2. De Bièvre P: Are we clear on the function of Matrix Reference Materials on the Measurement Process? An ISO/REMCO – IUPAC – BAM – EUROLAB-D Workshop on the Proper Use of Environmental Matrix Reference Materials, 23–24 April 1999, BERLIN. Proceedings to be edited by the Royal Society of Chemistry, UK

Accred Qual Assur (1999) 4:401–405

René Dybkaer

From total allowable error via metrological traceability to uncertainty of measurement of the unbiased result

Received: 5 March 1999
Accepted: 21 April 1999

Presented at: 4th Conference on Quality [R]evolution in Clinical Laboratories, Antwerp, Belgium 29–30 October 1998

R. Dybkaer
Copenhagen Hospital Corporation, Department of Standardization in Laboratory Medicine, H:S Kommunehospitalet, Øster Farimagsgade 5, DK-1399 Copenhagen K, Denmark
Tel.: +45-33-38-37 85/86
Fax: +45-33-38-37-89

Abstract The concept of "total allowable error", investigated by Westgard and co-workers over a quarter of a century for use in laboratory medicine, comprises bias as well as random elements. Yet, to minimize diagnostic misclassifications, it is necessary to have spatio-temporal comparability of results. This requires trueness obtained through metrological traceability based on a calibration hierarchy. Hereby, the result is associated with a final uncertainty of measurement purged of known biases of procedure and laboratory. The sources of bias are discussed and the importance of commutability of calibrators and analytical specificity of the measurement procedure is stressed. The practicability of traceability to various levels and the advantages of the GUM approach for estimating uncertainty are shown.

Key words Metrological traceability · Total allowable error · Trueness · Unbiased result · Uncertainty of measurement

Introduction

The important contributions of Prof. James O. Westgard to quality assurance in laboratory medicine have spanned a quarter of a century. His initial interest in statistical comparison of measurement procedures [1] soon led to criteria for judging precision and accuracy in a procedure [2]. Based on the concept "total analytic error", comprising constant systematic error, proportional systematic error, and random error, the concept "allowable total error" (originally called total allowable error) was defined with respect to clinical requirements, usually as a 95% limit. This measure has been maintained during all later developments by Westgard and his co-workers and has been recently applied to the "analytical model" used in a paper on the Validator® 2.0 which is a computer programme for automatic selection of statistical quality control procedures [3]. In worded form, the following equation is said to apply (where QC = internal quality control rules):

allowable total error
= constant inaccuracy of procedure
+ varying inaccuracy due to sample matrix
+ unstable inaccuracy detectable by QC
+ z (unstable imprecision detectable by QC)

where $z = 1.65$ yields a maximum allowable number fraction of defects of 5%.

The present discussion is about the constant bias of the measurement procedure (the first term, called constant inaccuracy, in the equation above). This component of overall bias is, in principle, a known detriment to trueness of measurement (defined as average closeness to a reference value).

Trueness and consequences of procedure-dependent bias

It is relevant to ask whether trueness is important or whether the sometimes heard pronouncement "precision is better than accuracy [meaning trueness]" relegates trueness to a lower priority. The reliance on precision is repeatedly seen in the results from external quality assessment (or proficiency testing) schemes all over the world, where method-dependent groupings of results for a given measurand are abundant.

Bias always impairs the comparability over space and time of the results for a given type of quantity and distorts the relationships between different types of quantity. Biological reference intervals are changed in comparison with a true distribution [e.g. 4, 5]. Harris even suggested a new term for such intervals, "medical indifference ranges" [6]. Whereas serial monitoring for change can sometimes live with a constant bias, this is not the case with screening, initial diagnosis, and movement towards a fixed discriminatory true limit, where diagnostic misclassifications are the outcome [e.g. 6–10]. A positive or negative bias of, say, 1 mmol/l in the amount-of-substance concentration of cholesterol or glucose in blood plasma has enormous effects on population health and economy.

Reduction of bias

Several approaches to the elimination of known bias should be considered when selecting, describing and operating a measurement procedure for a given type of quantity:

1. The type of quantity that is to be measured must be defined sufficiently well. This is particularly demanding when analyte isomorphs or speciation are involved.
2. The principle and method of measurement must be carefully selected for analytical specificity.
3. A practicable measurement procedure including sampling must be exhaustively described.
4. A calibration hierarchy must be defined to allow metrological traceability, preferably to a unit of the International System of Units (SI). Traceability involves plugging into a reference measurement system of reference procedures and commutable calibration materials.
5. An internal quality control system must be devised to reveal increases in bias.
6. Any correction procedures must be defined and validated.
7. Where possible, there should be participation in external quality assessment ("proficiency testing") using material with reference measurement values.

Metrological traceability

The necessary anchor for the trueness of a measurement procedure is obtained by strict metrological traceability of result, based on a calibration hierarchy. The official definition of traceability in metrology is: "property of the result of a measurement or the value of a measurement standard whereby it can be related to stated references, usually national or international measurement standards, through an unbroken chain of comparisons all having stated uncertainty" [11]. As stressed in the first resolution of the 20th General Conference on Weights and Measures (CGPM) in 1995 [12], the top of the calibration hierarchy, when possible, should be the definition of an SI unit.

The physical calibration hierarchy

In physics, the use of calibration hierarchies is well established and is used in any laboratory, e.g. for balances, volumetric equipment, spectrometer wavelengths, cuvette light path lengths, thermometers, barometers and clocks.

The chemical calibration hierarchy

For chemical quantities, involving the SI base unit for amount of substance, the "mole", its definition demands specification of the elementary entities of the component under consideration. According to the physical calibration hierarchy, a primary standard would be needed for each of the huge number of different compounds that are defined in the measurements. To circumvent this obstacle, the Consultative Committee for Amount of Substance of the International Committee on Weights and Measures (CIPM-CCQM) defines a primary reference method, which is claimed directly to give amount of substance in moles without prior calibration by a primary standard [13, 14]. Current examples of primary reference methods are isotope dilution-mass spectrometry and gravimetry. It should be realized, however, that establishing the more complicated measurement procedures based on such primary methods is by no means simple [15] and may require the expertise of the International Bureau of Weights and Measures (BIPM) or a national metrology institute (NMI). A primary reference measurement procedure (prim. RMP) assigns a value with uncertainty of measurement to a primary reference material [13], usually purified and stable, used as a primary calibrator (prim. C). The steps of the calibration series may be as follows with the responsible bodies in parentheses (accr. CL = accredited calibration laboratory; mf. = manufacturer).

SI unit (definition)	(CGPM)
prim. RMP	(BIPM, NMI)
prim. C	(BIPM, NMI)
sec. RMP	(NMI, accr. CL)

sec. C	(NMI→accr. CL→mf.'s lab.)
mf.'s selected MP	(mf.'s lab.)
mf.'s working C	(mf.'s lab.)
mf.'s standing MP	(mf.'s lab.)
mf.'s product C	(mf.→user)
routine MP	(mf., user)
routine sample	(user)
result	(user)

The length of the hierarchy can be reduced by eliminating pairs of consecutive steps, thereby reducing uncertainty.

Commutability and analytical specificity

There are two major reasons why a traceability chain may be broken and trueness lost due to the introduction of bias: insufficient commutability of a calibration material and non-specificity of a measurement procedure. The effect of these separate properties are often indiscriminately lumped together as "matrix effect". Commutability refers to the ability of a material, here a calibrator, to show the same relationships between results from a set of procedures as given by routine samples [16, 17]. Analytical specificity refers to the ability of a measurement procedure to measure solely that quantity which it purports to examine [16, 18]. Discrepancies between results of a reference procedure and a routine procedure applied to routine samples are often caused by non-specificity of the routine procedure. The use of a set of human samples as a manufacturer's calibrator to eliminate so-called matrix effects should only be accepted if the relationship between the results from reference and routine procedures is sufficiently constant to allow explicit correction with consequent increased uncertainty of assigned values.

Traceability in practice

It is relevant to ask how often the routine measurement procedures currently used in laboratory medicine provide results that are traceable to high-level calibrators and reference measurement procedures (Lequin: personal communication). It turns out that primary reference measurement procedures and primary calibrators are only available for about 30 types of quantity such as blood plasma concentration of bilirubins, cholesterols and sodium ion. International reference measurement procedures from the International Federation of Clinical Chemistry and Laboratory Medicine (IFCC) and corresponding certified reference material from BCR are available for the catalytic activity concentration of a few enzymes such as alkaline phosphatase and creatine kinase in plasma. For another 25 types of quantity, such as the catalytic activity concentration of aspartate aminotransferase in plasma and number concentration of erythrocytes in blood, no high-level calibrators exist. International calibrators, e.g. from WHO, but no high-level in vitro procedures characterize a couple of hundred types of quantity involving, for example, choriogonadotropin. An overwhelming number of types of quantity have no high-level ending of the traceability chain, but rely on the internal best-measurement procedure and calibrator of the reagent set manufacturer or individual laboratory. The end-user, as a rule, cannot be expected to establish the entire traceability chain if that goes above an in-house procedure. The laboratorian usually has to rely on the manufacturer which, in turn, may claim traceability of its product calibrators to the highest available level, preferably provided by a national metrology institute, an accredited calibration laboratory, or a reference measurement laboratory. In fact, this responsibility of the manufacturer is now enshrined in the EU Directive on in vitro diagnostic medical devices [19], which will be supported by four EN/ISO standards under development. The laboratorian should, however, bolster his or her belief in trueness and comparability – especially if the traceability chain does not reach high – by recovery experiments [20], comparison with a selected procedure [21], and interlaboratory parallel measurements [22], including external quality assessment [23], preferably on material with reference measurement procedure assigned values [24]. The internal quality control system finally checks, with a given probability, whether the current measurements are in statistical control with no sign of change in the assumed zero bias.

Uncertainty of measurement

The definition of metrological traceability (see above) stipulates that each link in the chain has a known uncertainty. Nowadays, this concept and its application have been reformulated by the BIPM and recently detailed in the "Guide to the expression of uncertainty in measurement" (GUM) [26]: "parameter, associated with the result of a measurement, that characterizes the dispersion of the values that could reasonably be attributed to the measurand". Useful explanations are provided in several other guides [26–30] as well as commentaries [e.g. 31–33]. The philosophy is to apply a bottom-up approach by formulating a function of all input quantities giving the measurand as output. An uncertainty budget of all sources of uncertainty is established. Important items to consider are:
- definition of the measurand
- realization of the measurand
- sampling
- speciation and matrix

- instability
- environment and contamination
- measuring system
- published reference data
- calibrator values
- commutability
- algorithms and software
- corrections and correction factor.

Each contribution is assessed as a standard uncertainty, either by statistical procedure on experimental data in the form of an a posteriori distribution, the so-called Type A evaluation, or by scientific judgement based on an a priori chosen distribution, Type B evaluation. The few standard uncertainties of important magnitude are combined quadratically, including any covariances, and the combined uncertainty, u_c, is obtained as the positive square root.

The advantages of this approach are important:

- The transparent budget invites improvement where major contributions are identified in the total sequence from definition onwards.
- There is no known significant bias allowing one, usually symmetric, measure of uncertainty.
- The combined uncertainty is comparable with that of other results.
- The combined uncertainty can be quadratically added to those of other results as demanded for traceability.
- The combined uncertainty can be compared with the classical top-down approach of calculating an uncertainty directly from replicate final results to reveal any discrepancy requiring further investigation.

The role of certified reference materials (with assigned value and uncertainty) in obtaining traceability and avoiding bias is obvious.

The GUM approach to uncertainty is rapidly gaining acceptance in metrological institutes and industry, and must be applied in ISO and CEN standards. It should be used in accredited laboratory work but chemists often find the implementation difficult and therefore hesitate [34]. Additionally, sometimes, there is a fear that honest GUM uncertainty intervals, which may be wider than classical precision intervals, are bad for business. Also, the perceived psychological effect on the customer of the term "uncertainty" seems to have led the food industry – naturally concerned about palatability – to propose the substitute term "reliability". Although it would be possible to define a concept with a "comforting" term inversely related to the measures of uncertainty – analogously to accuracy, trueness, and precision – the term reliability is already used for a more comprehensive concept covering several analytical performance criteria. There should be no doubt, however, that, as the GUM says, "The evaluation of uncertainty is neither a routine task nor a purely mathematical one; it depends on detailed knowledge of the nature of the measurand and of the measurement" [25]. To alleviate the calculations involved, commercial EDP programmes are being offered.

Conclusions

The upshot of these considerations is that one should cease to define a so-called allowable total error of result, with assessable biases of procedure and laboratory included. Instead, it is necessary to provide corrected results with a defined allowable maximum uncertainty at an agreed level of confidence. Likewise, a manufacturer may be asked to specify an expected uncertainty for a measuring system performing according to a measurement procedure under statistical control. Finally, the laboratorian can provide the customer with a corrected result and an accompanying uncertainty interval comprising a stated proportion of values that could reasonably be attributed to the measurand. This view is not in conflict with the 25-year-old statement by Westgard and co-workers – using classical terminology – that "In principle, only random error need be tolerated. Systematic errors can be eliminated by appropriate improvements in methodology" [1].

Acknowledgements Ms Inger Danielsen is gratefully thanked for her excellent secretarial assistance.

References

1. Westgard JO, Hunt MR (1973) Clin Chem 19:49–57
2. Westgard JO, Carey RN, Wold S (1974) Clin Chem 20:825–833
3. Westgard JO, Stein B, Westgard SA, Kennedy R (1997) Comput Method Programs Biomed 53:175–186
4. Gowans EMS, Hyltoft Petersen P, Blaabjerg O, Hørder M (1988) Scand J Clin Lab Invest 48:757–764
5. Hyltoft Petersen P, Gowans EMS, Blaabjerg O, Hørder M (1989) Scand J Clin Lab Invest 49:727–737
6. Harris EK (1988) Arch Pathol Lab Med 112:416–420
7. Ehrmeyer SS, Laessig RH (1988) Am J Clin Path 89:14–18
8. Hyltoft Petersen P, Lytken Larsen M, Hörder M, Blaabjerg O (1990) Scand J Clin Lab Invest 50 (Suppl 198):66–72
9. Hyltoft Petersen P, Hørder M (1992) Scand J Clin Lab Invest 52 (Suppl 208):65–87
10. Hyltoft Petersen P, de Verdier C-H, Groth T, Fraser CG, Blaabjerg O, Hørder M (1997) Clin Chim Acta 260:189–206
11. BIPM, IEC, IFCC, ISO, IUPAC, IUPAP, OIML (1993) International vocabulary of basic and general terms in metrology. ISO, Geneva

12. Comité International des Poids et Mesures (1998) National and international needs relating to metrology. Bureau International des Poids et Mesures, Sèvres
13. Kaarls R, Quinn TJ (1997) Metrologia 34:1–5
14. Quinn TJ (1997) Metrologia 34:61–65
15. Adams F (1998) Accred Qual Assur 3:308–316
16. Dybkaer R (1997) Eur J Clin Chem Clin Biochem 35:141–173
17. Fasce CF, Rej R, Copeland WH, Vanderlinde RE (1973) Clin Chem 19:5–9
18. Kaiser H (1972) Z Anal Chem 260:252–260
19. EU Directive 98/79/EC (1998) Off J Eur Comm L 331:1–37
20. Willets P, Wood R (1998) Accred Qual Assur 3:231–236
21. Hyltoft Petersen P, Stöckl D, Blaabjerg O, Pedersen B, Birkemose E, Thienpont L, Flensted Lassen J, Kjeldsen J (1997) Clin Chem 43:2039–2046
22. Groth T, de Verdier C-H (1993) Upsala J Med Sci 98:259–274
23. Hirst AD (1998) Ann Clin Biochem 35:12–18
24. Stamm D (1982) J Clin Chem Clin Biochem 20:817–824
25. BIPM, IEC, IFCC, ISO, IUPAC, IUPAP, OIML (1993) Guide to the expression of uncertainty in measurement. ISO, Geneva
26. Taylor BN, Kuyatt CE (1994) NIST Technical Note 1297. National Institute of Standards and Technology, Washington
27. Eurachem (1995) Quantifying uncertainty in analytical measurement.
28. EAL-G23 (1996) The expression of uncertainty in quantitative testing.
29. EAL-R2 (1997) Expression of the uncertainty of measurement in calibration.
30. ISO TR 14253–2 (1998) Geometrical product specifications (GPS) – Inspection by measurement of workpieces and measuring equipment – Part 2: Guide to the estimation of uncertainty in GPS measurement, in calibration of measuring equipment and in product verification. ISO, Geneva
31. Kadis R (1998) Accred Qual Assur 3:237–241
32. Bremser W (1998) Accred Qual Assur 3:398–402
33. Hässelbarth W (1998) Accred Qual Assur 3:418–422
34. Golze M (1998) Accred Qual Assur 3:227–230

Accred Qual Assur (2000) 5:142–150

Mirella Buzoianu

Practical considerations on the traceability to conventional scales

Received: 8 June 1999
Accepted: 13 December 1999

M. Buzoianu (✉)
National Institute of Metrology,
Sos. Vitan-Bârzesti No. 11,
75669 Bucharest, Romania
e-mail: office@inm.ro
Tel.: + 40-1-334 50 60
Fax: + 40-1-334 53 45

Abstract The basic concepts of traceability as they are defined by the Comité Consultatif pour la Quantité de Matière (CCQM) are difficult to apply to some chemical results. For instance, for some environments or chemical analyses measurement results are expressed in conventional units. Such units are realized on conventional scales relying on two fundamental pillars: reference materials and standard specification. The octane number of fuel or water turbidity measurements are typical examples of such units. Traceability concepts are discussed in terms of their practical applicability for turbidimetric analysis. Some outcomes on the validation of the metrological performance of turbidimeters and the comparability of turbidity measurement results are also presented.

Key words Traceability · Conventional scales · Primary reference materials · Turbidimetry

Introduction

The expression of results in analytical chemistry is mostly in SI units (all base units except the candela and many derived units). The principles to be followed to achieve the comparability and traceability of measurements to the SI have been clearly stated [1, 2]. However, certain types of measurements are expressed in conventional units. Turbidity evaluation in water quality analysis, determination of soluble content of fruit and vegetable products by the refractometric method, measurement of caking power of hard coal by the Roga test, determination of the octane number of fuel and seric protein analysis are some examples.

For such types of measurements it is necessary to create, sustain and use certain conventional units which are not within the scope of the SI. Such units are realized on conventional scales, relying on reference materials (RMs), realizing the fixed point(s), and the standard specification or similar document, giving the method of measurement. Therefore, both should be strictly defined to ensure the compatibility and traceability of measurements, since a result expressed in such a unit or a RM realizing a fixed point on a conventional scale should have the same quality all over the world. To what extent the above affirmation is true must be verified in each specific situation.

Within this framework, this paper attempts to structure some aspects of how to achieve traceability and comparability of turbidimetric results, as well as the validation of turbidimeters used in legal analysis.

The traceability of results in conventional units

As defined in the International Vocabulary of Basic and General Terms in Metrology [3], "traceability is the property of the result measurement or the value of a standard whereby it can be related to stated references, usually national or international standards, through an unbroken chain of comparisons all having stated uncertainties".

Thus a measurement result can be traceable either to a system of units (the SI for instance, but not necessarily) or to an agreed reference (RMs, or well-characterized reference standards based upon fundamental constants of nature, etc.)

According to the above definition, a result that is not expressed in SI units can meet the requirements for traceability. Note that in this case each traceability link is established for a stated chemical purpose. It is asserted that such a value is obtained by a measurement applicable to the measurand, in a particular laboratory using a specific procedure, over a time period during which relevant measurement operations are maintained under control with acceptable repeatabilities, and within a limited range of magnitude of the measurand.

In the case of turbidity measurement, the traceability chain should begin with the value of turbidity in a sample (the quantity being measured) then progress through higher levels of authority up to the unit of turbidity on an agreed and conventional measurement scale, based on the values assigned to RMs, stated in standard specifications, international recommendations or other reference documents.

Turbidity measurement usually directly relates the forementioned value in a sample to another value in a RM by a controlled comparison of values, or indirectly through an instrument calibration established for values for identical entities in similar RMs. This measurement is characterized, in part, by an observed repeatability and invariably by an estimated uncertainty – the sole indication of the quality for each link. The final link is made to a turbidity unit in an internationally recognized scale.

The turbidity measurement scale is a conventional scale based on RMs containing formazin, and standard specifications [4] which provide detailed information necessary to establish and use the scale, as well as experimental procedures available in turbidity measurements.

By means of appropriate RMs and relevant standard specifications, the end-user can realize the measurement scale against which he can measure his sample and/or calibrate his instrument within a specific uncertainty range. Consequently, to achieve traceability of a turbidity result, two aspects need to be considered: firstly that the RM is accurate and secondly that the instrument used for the measurement is calibrated in a traceable manner.

To estimate the uncertainty of a measurement on the scale, the user should consider the uncertainties due to the creation of the scale, and the uncertainty associated with the realization of its fixed points by the RMs. Also, selection of these RMs for realizing the fixed points of a scale depends on the required level of uncertainty of the end use.

To minimize the uncertainty of the measured value on the scale, the user should employ RMs and standards which have been certified in terms of the units of the scale. If this requirement cannot be meet, the correlation between the purity of the pure chemical compounds used for realizing the fixed points of the turbidity scale and the property on which this scale is based should be taken into consideration. Otherwise the uncertainty of measurement can be only roughly estimated.

An attempt to establish the traceablity scheme for turbidity measurements by the Romanian National Institute of Metrology (INM) is presented in Fig. 1.

Practical aspects of turbidity measurements

Turbidity as a measure of the relative clarity of a sample is a qualitative characteristic which is imparted by solids obstructing the transmittance of light through the sample. Thus, it is not a direct measure of suspended particles in a sample but, instead, a measure of the scattering effect such particles have on light.

Note that turbidity measurement plays an important role in many types of routine chemical analyses (e.g. nephelometric determination of water quality, the evaluation of the concentration of barium and sulphate, or potassium determination with sodium tetraphenylborate in water analysis, and determination of seric proteins). Also, the characterization of turbidity measurements is very important when evaluating some sources of uncertainty in gravimetry used in valid chemical metrology [5].

The quantity expressed as turbidity is defined as "the decrease of transparency of a liquid sample due to the presence of undissolved matters" [4]. As defined, this quantity cannot be quantified without reference to a method of measurement and it has no units independent of such a method. Unlike classical measurable

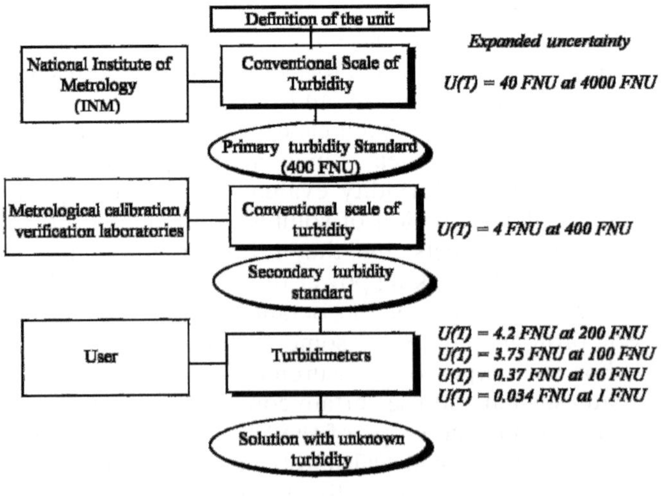

Fig. 1 An example of how the traceability of turbidity measurements at the field level can be established to a conventional scale

quantities, the quantity 'turbidity' cannot be entered into algebraic equations to define other measurable quantities. Its magnitude is determined by conventional measurement and it is made according to a written standard.

Accordingly, it is possible to quantify the presence of an amount of suspended particles (silt, clay, algae, organic matter, micro-organisms and other undissolved matters) by measuring its optical properties: either the amount of light scattered away from the direction of the incident light (nephelometry), or the amount of light absorbed from the incident beam. Consequently, turbidity units are stated with a qualifier that specifies the method of measurement: formazin nephelometric units (FNU) when the method of measurement of scattered light is used, or formazin attenuation units (FAU) for the measurement of the attenuation of the incident beam. Note that both of them are defined starting from the formazin primary reference standard.

Defining these units of measurement and establishing the formazin standard for turbidity measurement has not prevented the use of other units. For instance, nephelometric turbidity unit (NTU), identical to formazin turbidity unit (FTU) and equal to the FNU, are displayed by many types of commercial turbidimeters used in Romania. Some instruments also indicate Jackson turbidity units (JTU) or the silica unit (mg/l SiO_2) [6], that are still found as references today.

As defined in the Romanian national standard [6], a turbidity unit or mg SiO_2/dm^3 of water represents the dispersion of the incident beam when it passes through a suspension containing 1 mg of SiO_2 in 1 dm^3 of water.

Also, the FTU (as noted in [6]) represents the dispersion of the incident beam passing through a formazin suspension containing 0.5 mg formazin in 1 dm^3 of water.

As previously defined, turbidity is an optical property describing the interaction between light and suspended particles in a sample. A directed beam of light remains relatively undisturbed when transmitted through an absolutely pure liquid, but even the molecules in a pure water will scatter light to a certain degree. A particle interacts with incident light by absorbing the light energy and re-radiates the light energy in all directions, so that the relative spectral coefficient of attenuation ($\mu(\lambda)$) may be obtained as the sum of the spectral coefficient of diffusion ($s(\lambda)$) and spectral coefficient of absorption ($a(\lambda)$). Therefore, the intensity of light scatter depends on the wavelength of the incident light, the measurement angle and its optical configuration, as well as on the size, shape and composition of the particles in the solution [7]. No matter which optical property is being measured, a conventional scale is established to determine the turbidity of samples, relying on a direct correlation between the amount of light scattered or absorbed with the amount of suspended matter (several turbidity measurements are made in succession with the same (or same type of) instrument on samples with the same particulate make-up).

Today, many types of instruments are available to measure extremely low turbidity level over an extreme range of sample particulate sizes and composition. An instrument's capability to measure a wide turbidity range is dependent on its design. For instance, a nephelometer optical system typically comprises a light source, lens and apertures to focus the light, a 90° detector to monitor scattered light and, optionally, a forward-scatter light detector, a transmitted light detector and a back-scatter light detector, to minimize the impact of colour, stray light, and lamp and optical variabilities.

To measure turbidities according to the method of attenuation of the incident radiation, any photometer of minimum (400–900) nm wavelength range and maximum 60 nm bandwidth can be used.

Selection of a turbidimeter by its metrological characteristics

To function appropriately within the above described framework, a turbidimeter must satisfy certain criteria:

- The instrument specifications are adequate for standardization.
- The instrument meets the specifications.
- The instrument can be easily operated to verify attainable performance.
- Operators understand the behaviour of the instrument if measurement limits are exceeded.

To satisfy the above criteria, a thorough knowledge of the measurement characteristics of the instrument and their interaction with the chemical system are needed.

However, turbidity measurements are well known for the different technologies used in the apparatus configuration. After initial selection of an instrument focused on economic needs, the ultimate choice should be made taking into account technical criteria such as measurement range, accuracy and precision, and also the ergonomies. A comparison of several types of laboratory turbidimeters commonly used in Romania for turbidity measurements is illustrated in Table 1. Note that both measurement range, accuracy and precision were evaluated following the same procedure for an type of laboratory turbidimeters. The third criteria, software configuration was taken into account for its adaptability to routine work.

As recommend in [4], such comparisons, even for instruments using the same principle of measurement or being used under the standard's specification, are inap-

Table 1 Selection of a turbidimeter on its metrological performance

Specification	Instrument A	Instrument B	Instrument C	Instrument D	Instrument E	Instrument F	Instrument G
Principle of operation	Nephelometric signal (90°) ratioed	Nephelometric signal at 90°only	Nephelometric signal at 90°only	Nephelometric signal at 90°only	Nephelometric signal at 90°only	Nephelometric signal (90°) ratioed	Attenuation of incident light beam
Measurement mode	Select NTU, EBC, Nephlos	NTU	Select FNU, TE/EFTU, EBS, mg/l	FTU	UTF	NTU	Select T, E, C, NTU
Light source	W-filament lamp	W-filament lamp	LED (860 nm)	LED (890 nm)	W-filament lamp	W-filament lamp	W-filament lamp
Turbidity range	0–10,000 NTU	0.00–19.99 NTU0.0–199.9 NTU	0.00(1)–1000 FNU	0.00–50.00 FTU50–1000 FTU	0.1–10.0 UTF0.1–100.0 UTF0–300 UTF	0.001–2.000 NTU0.00–20.00 NTU0.0–200 NTU0–2000 NTU	0–999 NTU
Turbidity accuracy	±2% of reading plus 0.01 NTU from 0–1000 NTU;±5% of reading (1000...4000) NTU;±10% of reading (4000...10000) NTU	±5% of reading from 0.00–19..99 NTU;±7% of reading from 0.0–199.9 NTU	±4% of reading	±5% of reading or 0.5 FTUfrom 0–10 FTU±10% of reading (10...50) FTU±5% of reading (50...1000) FTU	±0.2 UTF from 0.0–10.0 TU±2% of reading from 0–100 UTF±6% of reading from 0–300 UTF	±2% of full scale	
Turbidity repeatability	±1% of reading or 0.01 NTU (whichever is greater)	No indication	±2% of reading	±2% of reading	±0.2% of reading from 0.0–10.0 TU±1% of reading from 0–100 UTF±3% of reading from 0–300 UTF	±0.1% of full scale	
Calibration	With step-by-step protocol buit-in	Two-point calibration 0 and 5 NTU	Built in instruments microprocessor	Two-point calibration 0 and 10 FTU	One point within every turbidity range	One point within every turbidity range	Multiple points within every turbidity range
Secondary standards	Gelex standards	Not available	Not available	Not available	Not available	Gelex Standards	Not available
Turbidity applications	Suitable for virtually any application	Water quality	Suitable for any application	Water quality mainly	Water quality	Water quality	Suitable for any turbidity application
Signal average	Adjustable and selectable	None	Adjustable	None	None	None	Selectable
Serial output	To a PC	To a printer	To PC or a printer	None	None	None	None
Printer	Build-in	None	None	None	None	None	None

propriate. Nevertheless, the lack of comparability in turbidity measurements is partly due to the instrumental uncertainties of the different types of turbidimeters routinely used in water quality laboratories.

An instrument's capability to measure a wide turbidity range is dependent mainly on the light source, the scattered light detector and optical geometry. The light sources most commonly used today in nephelometers are tungsten filament lamps and light emitting diodes (LED). When a tungsten filament lamp is used a monochromator providing at least a 60 nm bandwidth is recommended. Due to the statistical reproducibility of the nephelometric scattering of white light by the formazin polymer, instruments with the traditional tungsten filament white light optical design can be calibrated with a high degree of accuracy and reproducibility (e.g. Instruments A and F).

When the imposed light beam interacts with the sample, its response must then be detected by the instrument. Generally, for a given detector, when the incident light source is shorter in wavelength, the instrument is more sensitive to smaller particles. A combination of transmitted, forward-scatter and back-scatter detectors and black mirrors increase the accuracy and stability of the instrument and decrease the stray light (for instance in Instruments A and F). The source/detector combination defines the effective spectral characteristics of the instrument and the manner in which it responds to a sample.

Optical geometry incorporates the angle of scattered light, the path length traversed by the scattered light and ratio measurements, etc. Differences in the makeup of sample particles cause different angular scattering intensities. A 90 °C detection angle, recommended in [4], affords a simple optical system with low stray light, since interference and errors due to stray light or sample colour can reduce the accuracy of the instrument (Instruments B and D). In addition, low turbidity measurements require stability, low stray light and an excellent sensitivity.

The light reaching the detector (or detectors) of the turbidimeter comes from the sample – light scattered from the sample, and from the instrument – stray light. Stray light has a number of sources: sample cells with scratched or imperfect surfaces, reflections within the sample cell compartment, reflection within the optical system, lamps that emit diverging light and to a small extent the electronics. Unlike the case of spectrophotometry, stray light effects in turbidimitry measurements cannot be zeroed.

A quantified value for stray light within a turbidimeter is difficult to determine. One method uses a standard addition to a formazin suspension of a known low turbidity. But some instruments have a build-in system to determine the stray light and to correct the turbidity reading accordingly (i.e. Instrument C).

The path length traversed by scattered light is a design parameter affecting both instrument sensitivity and linearity. The sensitivity increases as the path length increases, but linearity is sacrificed at high particle concentration due to multiple scattering and absorbance. The use of a short path length can also increase the impact of stray light. A path length of less than 10 cm from the lamp filament to the detector is required in the instrument design [4].

Turbidity standards

The subject of standards dedicated to measurements on conventional scales partly depends on the variety of types in common use that are acceptable for reporting purposes, and partly on the terminology or definitions applied to them.

According to definition [3], a "primary standard is a standard that is designated or widely acknowledged as having the highest metrological qualities and whose value is accepted without reference to other standards of the same quantity". Thus, once defined, primary standards require no further reference. The essential quality of a primary standard is its intrinsic, long-term stability. Primary standards are mainly used to measure and determine the value of all other standards.

For chemical measurements, the CCQM also states thtat [8]: "A primary reference material (PRM) is one having the highest metrological qualities and whose value is determined by means of a primary method". For instance, in the practice of analytical chemistry, combining pure materials of well-determined purity according to an accurate gravimetric method, well-characterized PRMs can be produced repeatedly. These materials can then be used to calibrate instruments or to assign property values to secondary standards [1].

For turbidity measurements the only standard that has been demonstrated to fulfil the above mentioned requirements is the white polymer formazin – the single accepted traceable primary standard, used to realize the conventional scale of turbidity units. Formazin is synthesized through a condensation reaction involving hydrazine sulphate and hexamethylenetetramine dissolved in water. On accurately weighting and dissolving 5000 g of hydrazine sulphate (ACS Grade of >99% purity) and 50 g of hexamethylentetramine (of at least 98% purity) in 1 l of distilled water, the solution develops a white turbidity after 48 h. This is equal to 4000 FNU (equal to 4000 FAU). So, under ideal environmental conditions of temperature (25 °C ± 3 °C) and light, this formulation can be prepared repeatedly with an accuracy of ±1% from traceable raw materials. But its chemical stability is highly dependent on storage conditions, since exposure to heat or direct sunlight and prolonged exposure to ambient air can degrade its shelf-life significantly. Note that the ISO standard [4] recommends the preparation of a 400-FNU-formazin solution by mixing equal parts of an aqueous solution of hydrazine sulphate and analytical reagent quality hexamethylentetramine, which should be used for no more than 4 weeks if stored in dark place at (25 °C ± 3 °C).

Formazin has several desirable characteristics which render it appropriate as a turbidity standard. First, it can be reproducibly prepared from assayed raw materials. Second, its physical characteristics make it a desirable light scattering calibration standard, as the formazin polymer consists of chains of several different lengths, which fold into random configuration. This results in a wide array of particle shapes and sizes ranging from less than 0.1 to over 10 μm, yielding statistically reproducible scatter with all makes and models of turbidimeters. But formazin is not without its limitations. Besides the fact that the starting raw materials are listed as poisons and carcinogens, formazin is only stable at high concentrations. For instance, stability studies [7] indicated that formazin standards above 400 FNU are stable for longer than 1 year, standards between 20 and 400 FNU are stable for approximately 1 month, standards between 2 and 20 FNU are stable for approximately 12–24 h and standards below 2 FNU are stable for 1 h or less. On the other hand, turbidity standards of 1 FNU or less are rather difficult to prepare accurately. The necessary dilution ratio is so high that even small variation in the volumes measured leads to high uncertainties, and there is no sure way to obtain absolutely pure, turbidity-free water (multiple distillations, deionization and ultrafiltration can still leave residual particulate contamination in the water).

As defined, a primary standard or RM should have an indication of the value of uncertainty. Even the 4000 NTU commercial available turbidity standard gives no direct indication of its uncertainty. The lack of this piece of information is extremely important for metrological calibrations where uncertainties associated with each diluted turbidity solution have to be taken into account in the overall uncertainty of calibration. Assuming that a 4000 FNU turbidity standard may be obtained within ±40 FNU (considering the above mentioned accuracy of method of preparation [7]), to calculate the uncertainty of each turbidity solution establishing the points on the conventional scale of turbidity one should follow the steps described in the EURACHEM Guide [9]. Accordingly, using class A glassware, an expanded uncertainty (by applying a coverage factor of 2) as high as 1.17% from the nominal value was calculated within (50–400) FNU, 1.18% within (5–50) FNU, 1.21% within (1–5) FNU and 1.33% at 0.5 FNU. These calculated uncertainties should however be compared with turbidity measurement uncertainties that are estimated under real world conditions. For instance, a ±3 FNU confidence limit for a 100 FNU standard was recommended in a technical specification on the type C instrument.

The need for secondary standards in turbidity measurement is quite different from the application of secondary standards required for use in traditional metrology practice. Such standards are often referred to as transfer standards, which have been certified and can be traced to the original primary standard. By definition, anyone with the proper materials can directly synthesize a primary standard. The need for secondary standards is a matter of practicallity and convenience based on the instability of dilute formazin solutions. In addition, as apposed to the traditional metrological approach of secondary standards, secondary standards in turbidity measurements are mainly used to check a particular instrument calibration stability, since these standards are particulate suspensions formulated to match the turbidity of diluted formazin solutions. These are made to provide more stable reference standards and eliminate the need for preparing fresh formazin dilutions for routine calibration checks. For maximum accuracy, actual values of secondary standards are assigned at the first calibration of the specific turbidimeter against primary standards. Actual secondary standard values can be used for subsequent standardization checks up to the expiration limit provided by the manufacturer of such standards. Such turbidity standards (latex suspensions, styrene divinylbenzene beads, metal oxide gels or stabilized formazin) meet the requirements of stability but cannot be prepared repeatedly as stand-alone standards. Further, some of them have very narrow particle size and shape ranges, making them have very sensitive to wavelength and instrument model specific. Lastly, they must be traced to a formazin primary standard.

The uniformity of turbidity measurements

Since turbidimeters are used in legal measurements, these instruments are subject to compulsory metrological activities within pattern tests and metrological verification [10]. Also, INM has to ensure a valid scientific background for uniformity, consistency and accuracy for turbidity measurements regardless of their field of application. Thus, according to current Romanian legal requirements, the assurance of the legality of turbidimeters and the achievement of traceability in this field is the first prerequisite to accomplish uniform measurements.

Therefore, the first area of investigation is to determine the compliance of an instrument performance with the technical specification of the manufacturer and the standard requirements within pattern approval tests. For this purpose, a metrological procedure is issued establishing the technical and metrological characteristics to be tested, as well as valid methods to verify them. Mainly, instrumental accuracy and repeatability are determined and compared with technical specifications or with upper limits established by other standard requirements.

Some outcomes of the metrological evaluation of instrumental accuracy and repeatability of ten different types or model of turbidimeters frequently used for water quality analysis are discussed here. Considering eight points on the turbidity measurement range, uniformly distributed within 0.5–200 FNU, the result for each instrument was obtained in the form of five single values, from which the corresponding mean and standard deviation was calculated.

Note that some of the instruments tested were precalibrated by the manufacturer, others were calibrated on a standard procedure basis using fresh formazin calibration standards. The performances of the latter were consistently verified with fresh formazin standards. The instrumental mean and standard deviations of all turbidimeters tested compared with the nominal values of formazin standard solutions are plotted in Fig. 2.

Instrumental accuracy of measurement [3] was quantified as relative instrumental error, i.e. the mean of five results of measurement minus the nominal value of the formazin standard divided by the nominal value of the standard. Then, instrumental repeatability [3] was quantified as the relative standard deviation of the five repeated measurements. A relative deviation of the means from the nominal values of each standard to within −10% (at 5 FNU) and +5% (at 50 FNU) were determined. Instrumental repeatability did not deviated more than 3.5% from the nominal value of turbidity. Note that none of these errors exceeded the technical specifications supplied by the manufacturers of the instruments.

To verify the instrument accuracy in the low range of turbidity (0.5 FNU and 1 FNU), careful consideration was taken of the instrument design and particulate contamination. Special care was taken to make immediate measurements to prevent temperature influence.

Fig. 2 Comparative instrumental performances at a nominal value of turbidity

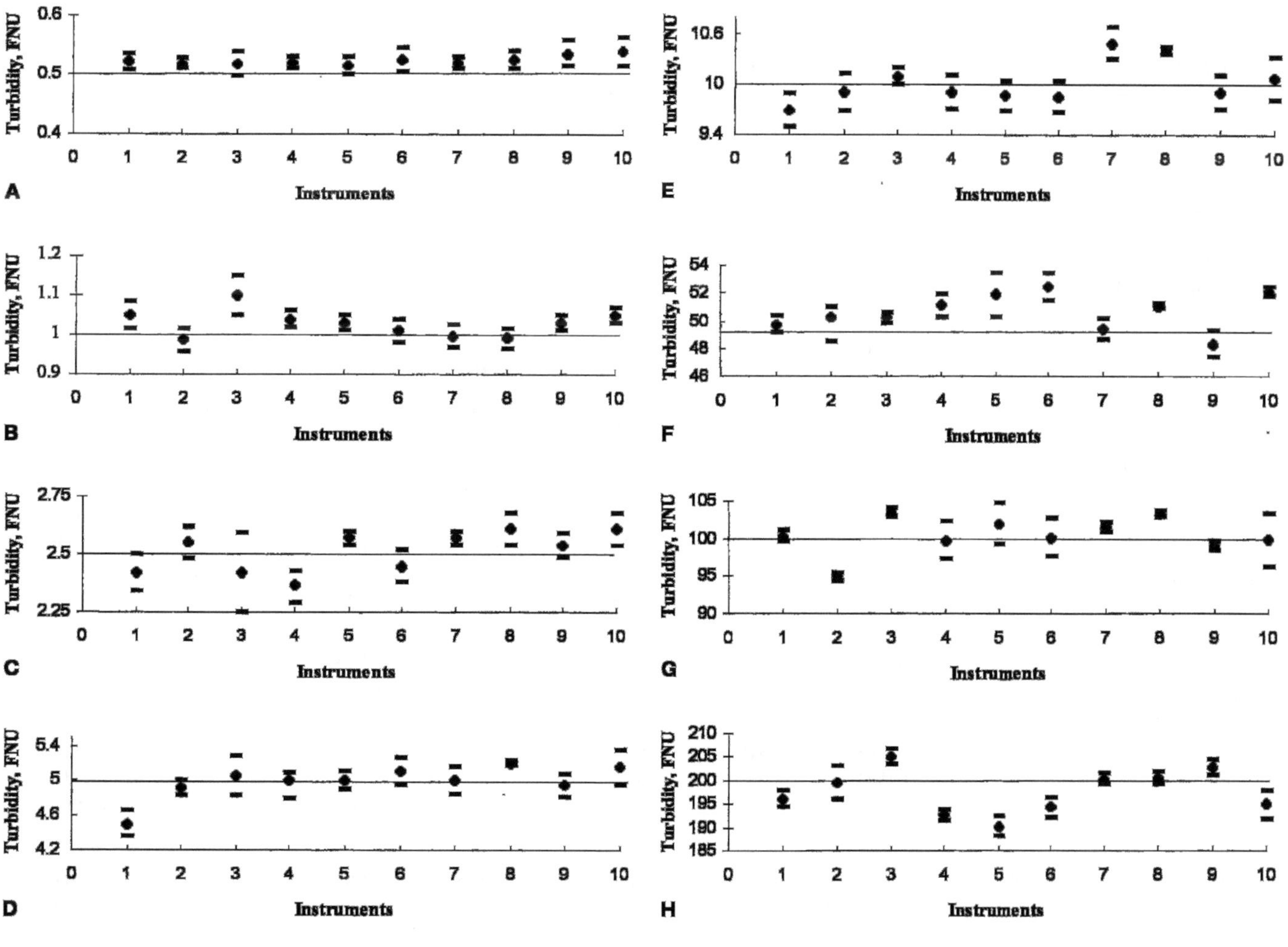

Where possible the stray-light was estimated, the instrument readings were corrected for the turbidity of the water, and sample cells and caps were carefully matched, orientated and cleaned (the outside of the sample cell was polished with silicon), air bubbles were removed, etc. Note that during these tests turbidity-free water (as low as 0.02 FNU by filtering distilled water through a 0.45-μm-membrane filter) was used. Even under these conditions the instrumental error at low turbidity values was within +3% and +8% at a nominal value of 0.5 FNU and –1.3% to +10% at a nominal value of 1 FNU.

Research was also conducted to determine the uniformity of the turbidity measurements. The comparability study involved the above-mentioned performance of several turbidimeters in the range of 0–200 FNU. Several makes and model of instrument (Table 1), some of them pre-calibrated by the manufacturer and others calibrated by the author using a formazin solution during the study, were used. Table 2 presents the instrument performance results.

The most significant practical problem that came to light was the difference in measured values among different instruments that had been calibrated with the same standard material. This was due to the difference in the spectral characteristics of the light source/detection combination. Even though standard specifications [4] aim to minimize variation by specifying the critical components of an instrument for turbidimetric measurement of water quality (such as the light source, the spectral peak response of the detector and the filter system, the distance traversed by incident light and scattered light within the sample tube, or the angle of light acceptance by the detector), by recommending some tolerance for these specifications, a substantial variability among instruments is obtained. However, successful correlation of measurements from different laboratories performing turbidity measurements was achieved when using the same instrument model.

To obtain some information on the magnitude of turbidity measurement uncertainty, the analysis of variance (ANOVA) method [11] was used to identify individual random effects in measurement so that they could be properly taken into account. The first estimate was the within-instrument component of variance (that is the variance of observations made on the same instrument) which was denoted as s_a^2. The second estimate, s_b^2, was the pooled estimate of variance obtained

Table 2 Turbidity measurement performances of the instruments tested

Instrument	Turbidity mean values measured (FNU) and instrumental standard deviation (FNU) at a nominal value of turbidity of							
	0.5	1	2.5	5	10	50	100	200
Instrument A mean	0.522	1.050	2.42	4.50	9.69	49.78	100.42	196.2
stand. dev.	0.015	0.035	0.08	0.148	0.197	0.632	0.638	1.823
Instrument A mean	0.519	0.987	2.55	4.92	9.91	50.28	94.98	199.6
stand. dev.	0.010	0.028	0.070	0.082	0.222	0.746	0.476	3.647
Instrument B mean	0.518	1.100	2.42	5.06	10.10	50.24	103.60	205.2
stand. dev.	0.020	0.050	0.170	0.228	0.100	0.362	0.548	1.674
Instrument B mean	0.520	1.040	2.37	5.00	9.91	51.10	99.80	192.7
stand. dev.	0.010	0.020	0.060	0.104	0.202	0.860	2.558	1.146
Instrument B mean	0.515	1.030	2.57	5.00	9.87	51.88	102.00	190.34
stand. dev.	0.015	0.020	0.030	0.105	0.170	1.564	2.722	2.026
Instrument B mean	0.525	1.010	2.45	5.12	9.86	52.42	100.22	194.26
stand. dev.	0.020	0.030	0.070	0.158	0.185	0.988	2.545	1.991
Instrument C mean	0.520	0.998	2.57	5.01	10.50	49.42	101.50	200.4
stand. dev.	0.010	0.028	0.030	0.158	0.184	0.736	0.620	1.168
Instrument D mean	0.525	0.992	2.61	5.21	10.41	51.08	103.34	200.66
stand. dev.	0.015	0.025	0.070	0.030	0.052	0.140	0.451	1.345
Instrument E mean	0.535	1.030	2.54	4.95	9.91	48.34	99.16	202.8
stand. dev.	0.025	0.020	0.050	0.129	0.200	0.929	0.568	1.674
Instrument E mean	0.540	1.050	2.61	5.16	10.08	52.18	99.88	194.96
stand. dev.	0.025	0.020	0.070	0.200	0.256	0.538	3.569	3.097
Overall mean	0.524	1.029	2.51	4.99	10.02	50.67	100.49	197.71
Estimate s_a	0.008	0.034	0.088	0.196	0.255	1.303	2.453	4.738
Estimate s_b ($\mathbf{u_T}$)	**0.017**	**0.029**	**0.079**	**0.145**	**0.185**	**0.823**	**1.875**	**2.107**

from the ten intruments tested. Since the estimate s_b^2 is based on the variability of the instrumental means and s_a^2 is based on the variability of the within-instrument observation, their difference indicates the possible presence of an effect that varies from instrument to instrument but that remains constant when observations are made on any single instrument.

The application of the F test with $\nu_a = 10–1 = 9$ degrees of freedom and $\nu_b = 10(5–1) = 40$ degrees of freedom, indicated the existence of an apparent between-instrument effect at higher turbidity values of 50 FNU and 200 FNU. Under these circumstances the best estimate of the uncertainty of the turbidity measurement was considered to be the pooled estimate of variance obtained from ten individual values of experimental variance of the observations made on each instrument. Consequently, relative uncertainties of measurement as high as 0.017 FNU at a nominal value of 0.5 FNU, and 1.87 FNU at a nominal value of 100 FNU were estimated. Note that an uncertainty of 3 FNU during the preparation of a 100 FNU trurbidity standard was previous mentioned.

Conclusions

This paper has examined a number of major practical problems arising from the request for traceability assurance to conventional scales in water turbidity measurement. To summarize, the need for accurate measurements of very low turbidities of samples containing fine solids demands good turbidimeter performance. Two major sources of error occurring from the variability in the calibration procedure and error in the instrument performance were detected. When combined, the propagation of these errors was evaluated to be as high as 3.4% of the reading.

References

1. Kaarls R, Quinn TJ (1997) Metrologia 34:2–5
2. De Bièvre P, Kaarls R, Peiser HS, Rasberry SD, Reed WP (1997) Accred Qual Assur 2/4:168–175
3. ISO (1993) International vocabulary of basic and general terms in metrology, 2nd edn. International Organization for Standardization. Geneva, Switzerland
4. EN 27027:94/ISO 7027:90 Qualité de l'eau – Détermination de la turbidité
5. Beck II CM (1997) Metrologia 34:19–27
6. STAS 6323 (1980) Drinking water. Turbidity determination
7. Sadar MJ Understanding turbidity science, Booklet No. 11. Hach
8. BIPM (1995) Comité Consultatif pour la Quantité de Matière 51
9. EURACHEM (1995) Guide to quantifying uncertainty in analytical measurement, 1st edn. Laboratory of the Governement Chemist, London
10. Buzoianu M (1998) Accred Qual Assur 3:328–334
11. ISO (1993) Guide to the expression of uncertainty in measurement, 1st edn. International Organization for Standardization, Geneva, Switzerland

Accred Qual Assur (2000) 5:224–230

Paul De Bièvre

Traceability of (values carried by) reference materials

Received: 11 November 1999
Accepted: 24 February 2000

Invited paper presented at the 2nd Central European Conference on Reference Materials (CERM.2), 9–10 September 1999, Prague, Czech Republic

P. De Bièvre (✉)
Adviser on "Metrology in Chemistry" to the Director of the Institute for Reference Materials and Measurements, European Commission – JRC, 2440 Geel, Belgium
e-mail: paul.de.bievre@skynet.be
Tel.: +32-14-851338
Fax: +32-14-853908

Abstract Traceability is a property of the result of a measurement. Since values carried by (reference) materials must also have been obtained, of necessity, by measurement, the definition of traceability also applies to reference materials. It is extremely helpful to give the traceability (of the origin) of a reference material a separate name, i.e. 'trackability'. An analysis of the function of values carried by reference materials, shows that they can fulfill different functions, depending on the intended use. One of the functions located outside the traceability chain – and hence not very relevant for establishing traceability – is evaluating the approximate size of the uncertainty of the measurement of an unknown sample by performing a similar measurement on a reference material, used as a 'simulated sample'. Another function is located inside the traceability chain, where the reference material is used as an added 'internal standard'. Then, the value carried by the reference material is essential for establishing the traceability of the measured value of an unknown sample. In the latter application, the reference material acts as an 'amount standard' (the certified value for amount is used).

Key words Traceability chain · Metrology in chemistry · Reference material · Amount standard · Validation

Introduction

Traceability is defined in the International Vocabulary of Basic and General Terms in Metrology, (VIM)[1] as the

- "property of the result of a measurement or the value of a standard whereby it can be related to stated references, usually national or international standards, through an unbroken chain of comparisons all having stated uncertainties".

Traceability is therefore clearly a property of the **value** of a result of a measurement. So what is a measurement? The same VIM defines it as a

- "set of operations having the object of determining the value of a quantity".

Again, the notion of **value** is very outspoken. Mostly, we obtain an unknown value by 'comparing' the unknown value to a known value (a simple definition of how a 'measurement' is carried out), i.e. measuring the ratio of an unknown value to a known value.

We have now moved the problem of establishing the traceability of one value to the establishment of the traceability of another 'known' value, which in turn must be compared to another 'known' value, etc. This process is quite acceptable, provided it stops somewhere. It stops when we arrive at a value which we *know* because we have *defined* it (consequently, it has an uncertainty of zero). It is the value of the unit in which we want to express the result of our measurement. Thus we can define a 'traceability chain' as follows:

– "a traceability chain is a chain of values linked by measurements which consist of comparisons of one value to another value, ending in the comparison with the value of the unit we have chosen to express the result of our measurement".

At this stage it is important to stress the prime function of a reference material, i.e. to *carry a **value***. It is also important to remember that such a **value**, including that carried by a reference material *per definition,* results from a measurement. The long – and still persisting – tradition of focusing on the **material**, should be reoriented to focus on the **value carried** by the material. All other characteristics of a reference material, essential as they may seem, draw their degree of importance from the prime consideration that they must ensure and safeguard the **value** carried, and that this **value** must be 'delivered' to the analyst in such a way that it is, and remains, meaningful for the intended purpose (stability, homogeneity, packaging, delivering sub-samples, etc). "Traceability of the **value carried** by a reference material" is therefore to be distinguished clearly from traceability of a **reference material**. The *traceability of a reference material* or *of a sample* has been called – quite appropriately – **trackability** (to the place or the producer it comes from) [2, 3].

The key elements of a traceability chain, *values* and *links* between values, have already been described [4]. More complete traceability chains are presented in Fig. 1. The symbols used are b = amount content [5, 6] in amount (mol) per mass (kg) of element (or compound) E in material X. Note that the chain is constituted by values 'linked' by operations called 'measurements', defined as above. The analyst could attempt the establishment of a complete traceability chain as shown (Fig. 1, left chain), but that would require a huge amount of work, or may not be possible, e.g. because the chain may be 'broken' (in the upper part under "chemical operations"). The same reasoning applies to the value carried by a reference material (central chain or right chain in Fig. 1). Every time we use a reference material, two traceability chains are involved as illustrated in Fig. 1: one for the measurement result obtained on the unknown sample (left chain), and one for the value carried by the reference material, (either the central or the right chain). They must, by their very nature, be similar. The first one must be demonstrated by the analyst. The second one must be demonstrated by the reference material producer. They sell the product and therefore must be accountable for the product.

The traceability chain of the value of a (certified) reference material

The traceability chain of a (certified) **value** is illustrated for two cases, i.e. for two different uses of a reference material, illustrated by the central and right chains in Fig. 1. The chains for both reference materials go via a purity investigation of the element/compound which is incorporated gravimetrically in the matrix. The resulting value takes into account the purity (with evaluated uncertainty) at point **2,** a statement (or studies) of stoichiometry (with evaluated uncertainty) at point **3** and, finally, the conversion from the mass of a pure substance (*PS*) into an amount at point **4.** A full traceability chain then ends in the value 1 of the mass (m) of the unit kg as shown.

(In the future, it will end in the value of the Avogadro constant and the value 1 of the atomic mass unit u defined as the mass of 1/12th of the mass of the ^{12}C atom. This will happen when the definition of the kg (now "the mass of the prototype of the kilogram") will have changed into the mass of a number of ^{12}C atoms, i.e. of "the mass of $\{N_A\}\cdot m(^{12}C)$. 1000/12)".

Alternatively, at point **2** in Fig. 1, the reference material producer could have chosen another traceability route, e.g. a coulometric measurement of the number of atoms of the element under investigation, which would have ended in the product $I\cdot t$ (electric current times time) and hence in amperes and seconds. Ultimately, the end of this traceability chain is then located in the value of the Avogadro constant and the value of the charge of the electron [7].

In many chemical measurements, however, we (must) use measurement scales which are based on values carried by commonly agreed reference materials. These enable the analyst to establish a 'calibration curve' through which observed signals from an instrument used in a measurement procedure, are converted to a concentration [8]. Any measurement result which is shown to be 'traceable' to such a common scale, is 'comparable' to any other measurement result which is 'traceable' to the same scale, i.e. established by the values of the same reference materials.

We now have a reference material at hand **carrying a traceable value**. How do we use it in an actual measurement process applied to an unknown sample?

Using the value, carried by a reference material, in the traceability chain of an unknown value in a sample:

1. 'Internal standard' method (value is located inside the traceability chain)

The **value** measured for the unknown sample is compared at some point to the **value** of an 'internal standard' added to the sample (left chain, point **1** in Fig. 1). That **value** comes from a reference material **carrying that value**, manufactured by a producer (right chain, point 1 in Fig. 1). Our analyst relies on the value supplied by the producer of the reference material from this point(identified by "HELP from (amount stand-

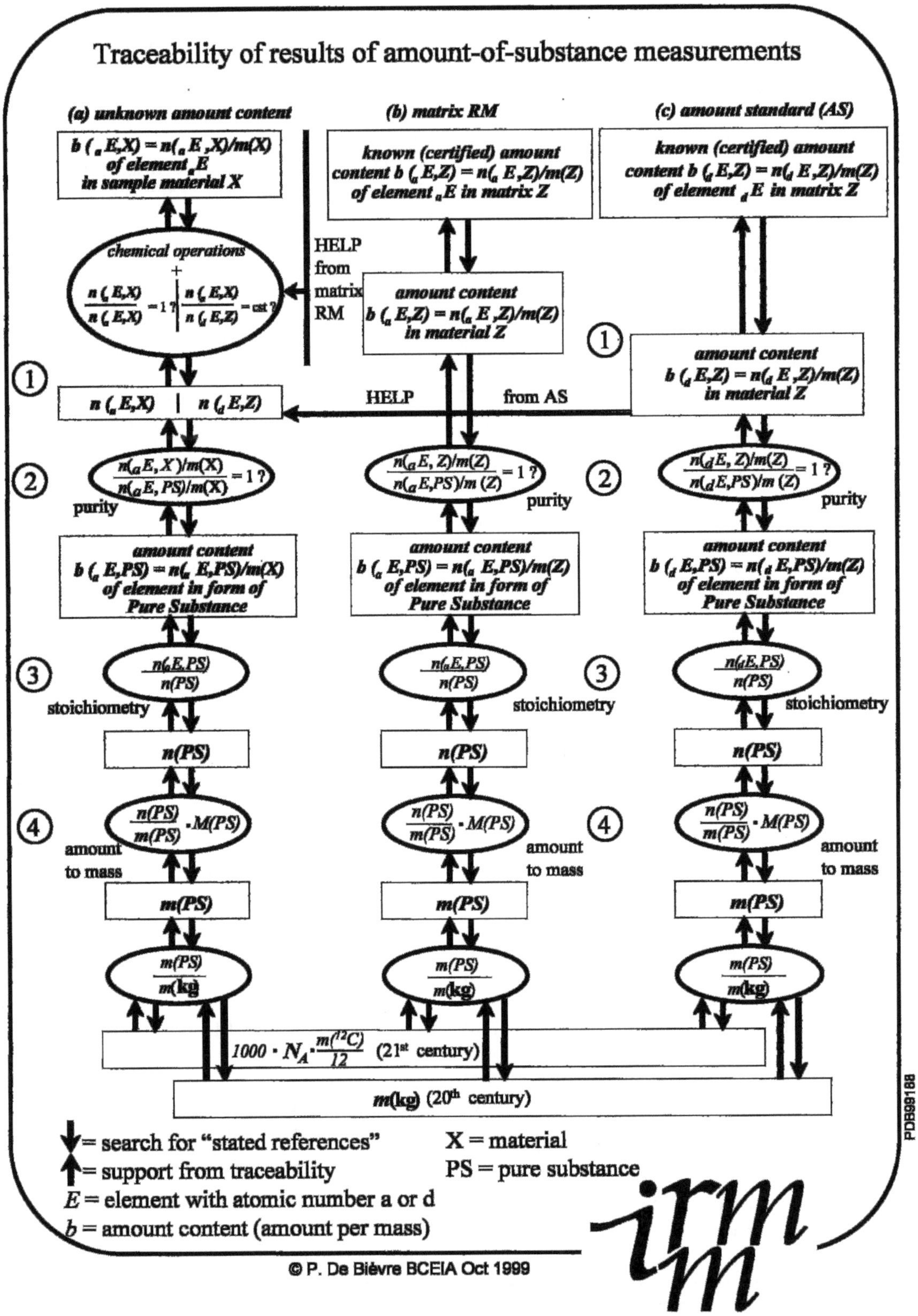

Fig. 1 Traceability chains of: **a)** the result of a measurement value obtained on an unknown sample; **b)** a value carried by a (matrix) reference material intended to reduce the uncertainty caused by "chemical operations" on the unknown sample; **c)** of a value carried by a reference material intended to be part of the traceability chain of the measured value in an unknown sample

ard) AS" in Fig. 1). It is the responsibility of the reference material producer to deliver or, at least, to 'state' to the analyst the **value** of the reference which the analyst wants/needs to 'state' as a 'reference'. This includes the establishment by the reference material producer of a traceability chain for the value offered to the analyst.

(Logically it follows that a traceability chain can equally be 'stated' to the ***value of a reference measurement*** *obtained by a reference measurement producer on the analyst's sample at point 1 in Fig. 1. This concept has the unknown sample travelling from the analyst to the (reference measurement) producer, rather than having the (reference) material travelling from the (reference material) producer to the analyst. This approach will be developed further on another occasion.)*

The role of the reference material producer is to reduce the burden of work for the analyst, by supplying the traceability chain of the **known value** of the **reference material** against which the analyst wants to measure the **unknown value** for an 'unknown sample'. This approach relieves the analyst from, e.g. measuring the purity of AS at point **2** in the right traceability chain of Fig. 1, or from doing the painstaking stoichiometric measurements on AS at point 3 in Fig. 1, right chain. It relieves the analyst from having to do the conversion from mass to amount at point 4 in Fig. 1 (which requires measurement of the isotopic composition in order to have a valid molar mass at hand). In short, it relieves the analyst from establishing the traceability chain of the measured **value** itself, because a "stated reference" can be bought from a *reference* ***value*** *producer*. The analyst can literally jump from the left chain on to the chain of the producer, buy the value and incorporate the purchased value into the traceability chain on the left by adding a portion of the reference material to the 'unknown sample'.

The name 'amount standard' (AS) is given to a reference material which has this essential function **in** the traceability chain. It is one of several possible functions of the **value** of a reference material.

We will now look at the traceability chain of the **value** carried by a 'matrix reference material', in order to understand its function in the measurement process. That is needed to produce reference materials which are adequate for the intended use ('fit-for-purpose').

2. Exploiting the value carried by a matrix reference material in the measurement of an unknown value in a sample (value is located outside the traceability chain)

This is illustrated in Fig. 2. Prior to the measurement, the **value** in the sample (left chain) may have had to undergo a "recovery" (extraction, distillation, dilution, or otherwise, summarized in the upper part of Fig. 2 under "chemical operations"). The question must therefore be asked: did this operation change the **value** in an unknown fashion thus putting the 'unbroken chain' under suspicion of having been 'broken'? Similarly, before the recovery operation, the **value** in that sample may have had to be digested. Again the question must be asked: did the digestion change the **value** in an unknown fashion thus putting the 'unbroken chain' under suspicion of having been 'broken'? And so on. A difficult point is where the analyst calculates the value of the claimed quantity at the end of the measurement, from the value of the quantity actually measured. Equation 1 shows that a 'conversion factor' $K({}_aE,X)$ is needed:

$$b({}_aE,X)_{\text{sample}} = K({}_aE,X) \cdot b({}_aE,X)_{\text{sample obs}} \qquad (1)$$

To this effect, in analytical chemistry it has been good practice that the analyst obtains a reference material with a matrix Z, similar – but of course not identical – to the one of the material X which needs to be measured. By performing the same operations on a sample of matrix reference material Z as performed on the sample with matrix X, an estimate can be obtained of the overall correction factor $K({}_a\text{E,Z})$. The **value** of the amount content $b({}_a\text{E,Z})_{\text{RM cert}}$ of the reference material as supplied by the reference material producer is known. The value $b({}_a\text{E,Z})_{\text{sample obs}}$ is observed by the analyst Hence, a 'correction factor' can be calculated $K({}_a\text{E,Z})$ for losses during digestion and recovery etc. (the "chemical operations" in Fig. 2) as determined with the help of the reference material. It can be applied to the measurement on the unknown sample. In short, this process can be described by Eq 2:

$$K({}_a\text{E,Z}) = b({}_a\text{E,Z})_{\text{RM cert}} \,/\, b({}_a\text{E,Z})_{\text{RM obs}} \qquad (2)$$

As shown in the central chain, upper part in Fig. 1, ("HELP from matrix RM"),the analyst can now use the correction factor from Eq 2 and substitute it for the unknown correction factor $K({}_a\text{E,X})$ in Eq 1. This enables a 'correction' to be made of the value observed in the unknown sample. In summary: the analyst simply substitutes $K({}_a\text{E,X})$ *for* $K({}_a\text{E,Z})$.

$K({}_a\text{E,X}) = K({}_a\text{E,Z})$?

However, the uncertainty of this substitution must be evaluated.

The evaluated uncertainty of this substitution, i.e. the uncertainty of *K*, will limit the uncertainty of the measurement result of the unknown sample and must be taken into account as a contribution to the total uncertainty budget of the result obtained on the unknown sample since:

a. the matrix of both samples is different.
b. the measurement conditions during the measurement of the reference material and of the unknown sample, are not identical.

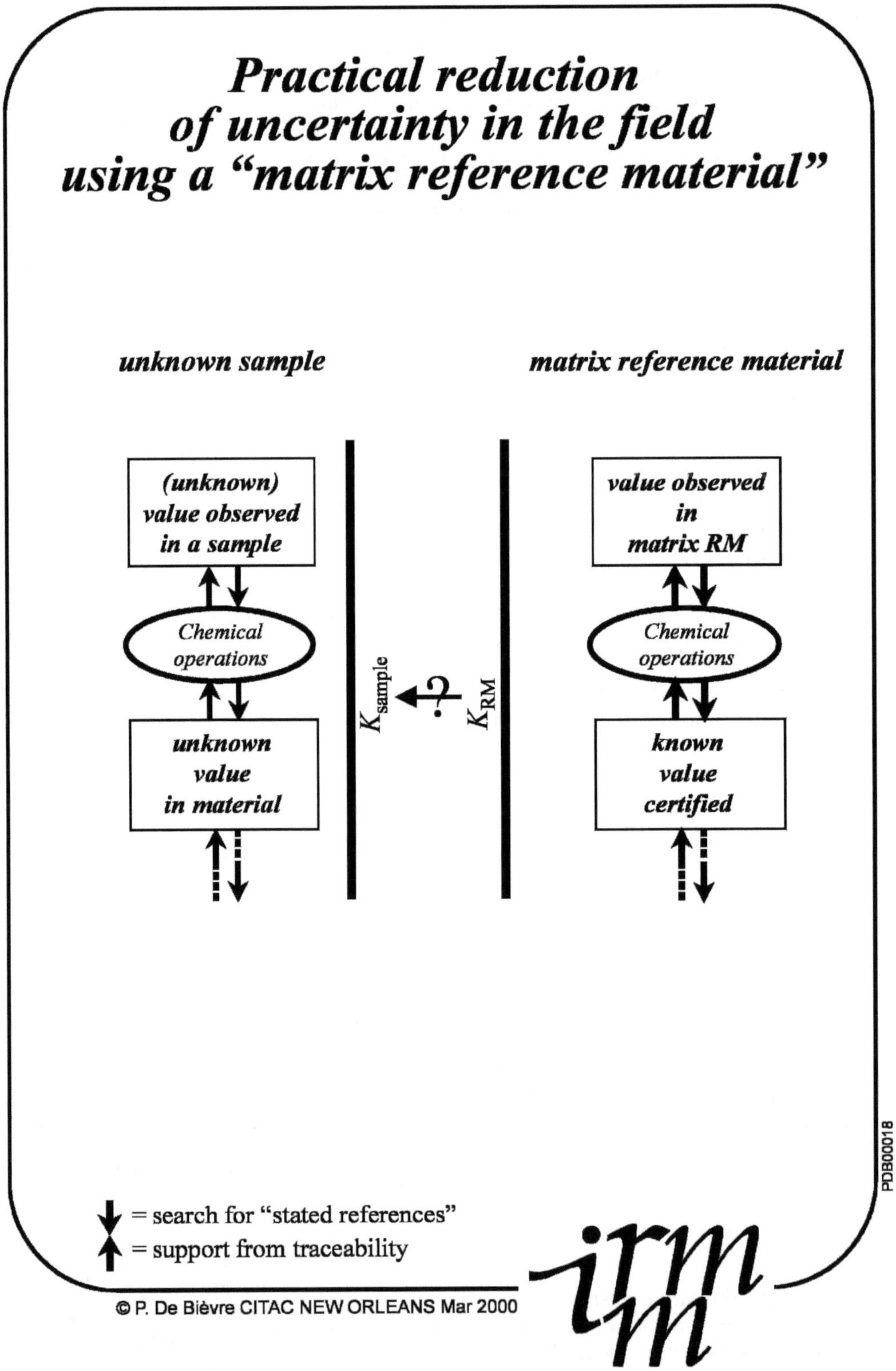

Fig. 2 A matrix reference material can be used to simulate chemical operations and estimate from that simulation a collection factor. The uncertainty of that collection factor must be evaluated and taken into account in the uncertainty ludget

This uncertainty is the quantitative result of the fact that the analyst had to ‘jump’ from one traceability chain to another.

Thus the matrix reference material has fulfilled another function than the ‘AS function’ for the analyst, a function which is not located **in** the traceability chain, but **outside** the chain: it enables the analyst to make an independent assessment of the possible magnitude of the conversion factor K, thus assessing – possibly reducing – the uncertainty of the measurement by carrying out a ‘correction’. But even this correction carries an uncertainty which must be evaluated. The problem of this ‘correction factor’ has been treated elsewhere in more technical detail under the name “recovery factor” [9].

From the above, we can now logically formulate requirements for traceability chains.

Requirements for traceability chains

For the value obtained by the measurement of an unknown sample:

- a transparent formulation of what is to be measured: identification of the measurand (related to the aim and intended use of the measurement result).
- a transparent formulation of the traceability chain wanted, including a clear and simple formulation of the ‘stated reference’ in the chain to which traceability of the measured value will be claimed. In other words, establishing a traceability chain is an a priori requirement, i.e. it must be planned *before* the measurement. It is not the *result* of a measurement.

1. For the value carried by a reference material, and used as an AS (pure elements, pure compounds):

- a transparent formulation of what will be claimed (related to the intended use of the AS).
- a transparent formulation of the measurand in the AS.
- a transparent formulation of the AS in the matrix reference material.
- establishment of the proper equation which relates what is actually measured to what is purported to be measured.
- a transparent formulation of the traceability chain to the values of a common measurement scale, i.e. to the value of the unit [possibly to the value of a SI unit or even to the value(s) of (a) fundamental constant(s)].

The base must be the same as for the measurement result if we want all our measurement results to be *coherent*.

2. For the value, carried by a reference material, and used to *simulate* the measurement of an unknown sample (most matrix reference materials):

- a transparent formulation of what will be claimed (related to the intended use of the matrix reference material).
- a transparent formulation of the measurand in the matrix reference material.
- a transparent formulation of the matrix of the reference material.
- a transparent formulation of the traceability chain to a common measurement scale, i.e. to the value of a common unit [possibly to the value of a SI unit or even to the value of (a) fundamental constant(s)].

The base must be the same as for the AS if we want all measurement results to be *coherent*.

Conclusions

1. There are different categories of reference materials according to the *function they carry out in the measurement process*:
   - those which perform a function **in the traceability chain** because the unknown value can be directly measured against their value; examples are: an amount $n(E,X)$ of another element, added to the unknown sample; the case of traceability of mass values described elsewhere [10].
   - those which perform a function **outside of the traceability chain** because they are used to estimate an *approximate* conection factor (with its uncertainty), for steps in the measurement of an unknown sample for which the degree of quantitativity must be ‘evaluated’, i.e. for which the size of the breakage of the traceability chain must be estimated (as it cannot be determined directly).
2. Establishing the traceability chain of a result of a measurement is too long and too expensive a process for an analyst to carry out; the analyst must be able to obtain ASs from reference material producers and/or National Measurement Institutes (NMIs), the values of which can be used as ‘stated references’; that puts the burden of such a programme (with justifications) on producers and NMIs.
3. Reference material producers and NMIs must provide **values**, which can perform two functions:
   - a function as part of the traceability chain because they provide **values in that chain** (AS function).
   - the function of estimating the correction factor (with uncertainties) for difficult or unknown steps in the traceability chain, by creating the possibility to simulate more or less the measurement process of an unknown sample on a known sample; this function lies **outside the traceability chain.**

Accred Qual Assur (2000) 5:414–417

A. Williams

What can we learn from traceability in physical measurements?

Received: 17 March 2000
Accepted: 20 June 2000

Presented at 3rd EURACHEM Workshop "Status of Traceability in Chemical Measurement", 6–8 September 1999, Bratislava, Slovak Republic

A. Williams
19 Hamesmoor Way, Mytchett, Camberley, Surrey GU16 6JG, UK
e-mail: aw@camberley.demon.co.uk
Tel.: +44-1252-542854
Fax: +44-8700-548926

Abstract An international system for providing traceability for the results from physical measurements has been under continuous development since the introduction of the Convention of the Metre over 100 years ago. Such a system has only been in existence for chemical measurements for about a decade and there is much that can be learnt from the way in which traceability has developed in physical measurements that will help its development for chemical measurements. First a number of myths that have grown up about the differences between these are examined. This is followed by a description of examples from physical measurements, which have important lessons for the ways in which traceability for the results of chemical measurements can be established.

Key words Traceability · Physical measurements · Chemical measurements

Introduction

The Convention of the Metre dates back to 1875, so there is a long history to the traceability of the results of physical and engineering measurements to international standards. Admittedly the mole was added to the base units in 1971 but there was little or no involvement of the International Committee of Weights and Measures (CIPM) in chemical measurements until a decade ago. It is therefore not surprising that traceability is not so well established for chemical measurements.

However the chemical community has not been dormant in the period since the Convention of the Meter came into being. A vast range of analytical methods has been developed with ever increasing sensitivity and selectivity. Comparability of measurement results has been achieved in many sectors by the use of collaboratively studied methods, reference materials, check samples and proficiency testing schemes. It is only recently that steps have been taken to develop traceability to international standards. In developing this traceability much can be learnt from the way traceability has developed for physical measurements.

Myths

First let me deal with some myths that have become prevalent in the chemical community at the moment.

Sample effects

It is widely stated that chemical measurements are so different from physical ones that nothing of value can be obtained from comparing them. Admittedly chemical measurements have their own problems, but this is true of the many types of physical measurements made. All measurements have their own problems and those arising in the many areas of physical measurements raise most if not all of those arising in chemical measurements.

It is often said that the main difference is due to the properties of the sample, which do not affect the results of physical measurements [1]. Unfortunately that is not true, the properties of the sample or of the object being subject to measurement, affect the results of all measurements. Some effects are well understood others are not.

A fundamental problem of all metrology is the difference in the response of the measuring system to the sample and the response of the measuring system to the standards used to calibrate it. A good measuring system minimises these differences, but all potential sources of difference have to be investigated and reduced during method development. In addition the uncertainty arising from these differences has to be evaluated.

Consider the simple case of weighing. The weights used on the balance, or in the case of a force balance those used to calibrate it, have their values traceable to the international kilogram and even in that case there are sample effects due to the difference in density between the balance weights and the international kilogram. An effect which is well understood and for which it is easy to apply a correction. However there are other effects arising from adsorption, over time, of dirt and moisture on the surfaces of the weights, for which it is much more difficult to apply a correction. For weighings in the laboratory, there may be large differences between the density of the sample and the weights, but the small correction needed can usually be applied with sufficient accuracy, since the effect is well understood. However it is much more difficult to apply a correction for the effect on the measured weight due to any difference between the temperature of the sample and that of the balance enclosure.

Another example often quoted is that there is no sample effect when measuring the length of a table. First, this is an incomplete definition of a measurand, since it is not clear what is meant by the length. But if, for example, the table forms a part of a measuring system to measure the precise shape of objects then the shape of the table can have an influence on the result. Then such factors as the material of the table, its temperature coefficient, the temperature gradients across the table, the loading of the table, the friction between its feet and the floor are sample effects that need to be taken into account.

It could be claimed that these are examples of sample effects on precise measurements that do not have their equivalents in chemical measurements, but they were introduced to show the universality of sample effects.

Measurements of radioactivity of a solution containing ^{35}S, a radioactive isotope that emits only beta particles, provide an example of sample effects closer to those occurring in chemical measurements. Sources for counting are prepared using small weighed aliquots of the solution, which are dispensed onto thin conducting films and dried. These sources are then placed into a $4\pi b$ counter, which has almost 100% detection efficiency to the beta particles that escape from the source. However a significant but unknown fraction of the beta particles is absorbed in the source. From the shape of beta spectrum of ^{35}S and an approximate value of the size of the crystals in the source it is possible to estimate that about 20% of the beta particles are absorbed in the source; a very close parallel to recovery in chemical analysis. There is not the time to go into how this problem was overcome, but it was not by utilising a standard method and ignoring the correction!

Just in case some are not convinced that sample effects in physical measurements can be as great in chemical ones, as a further example consider the measurement of absorbed neutron dose in a patient undergoing radiation therapy.

Error structure

Another claim is that the nature of the errors is different between chemical and physical measurements. It is claimed [2] that for physical systems, systematic errors predominate and that these are "corrected out of the result" whereas for chemical systems random errors predominate. I do not know the basis for these claims but they do not align with my own experience, e.g. the systematic error associated with recovery can easily be equal or greater than the random error. An important point about the error structures is that the ability to detect and correct for systematic errors is limited by the size of the random error, but this is true for all types of measurement. It requires 13 replicate measurements to have the sensitivity to detect an effect equal to the size of the standard deviation on 1 measurement. There is also the view that the uncertainties on physical measurements are of the order of one part per million, but again this is not true. Standards of radioactivity and neutron dose uncertainties are often in the range of 1–30%.

Traceability is not applicable to all measurements

It is sometimes implied that traceability is a concept that cannot be applied to all types of measurements. But all measurements are traceable to some base, without this it would not be possible to give a value. This base may not be satisfactory, it may not be stable, in fact, the base may not have been identified. The purpose of establishing the traceability is to identify this base and to ensure that it is satisfactory.

Comparison of traceability in physical and chemical measurements

Development of traceability

Results of measurements have to be traceable back to a common standard, but in many cases all that is necessary or all that may be possible is to provide traceability to some local standard. In many market towns throughout the United Kingdom it is still possible to see a local standard of length embedded in a wall, which was used for trading a few centuries ago. I have not found any records of comparisons of these local standards between the market towns, nor whether any enterprising tradesman took advantage of any differences.

However the use of local standards has persisted, for example, until the early 1960s the United Kingdom standard for radioactivity for ^{131}I was defined in terms of the current produced in an ionisation chamber held at the National Physics Laboratory (NPL), although the value was given in Curies. The ionisation chamber was of a simple design and copies of it were manufactured and sold commercially and this provided the transfer standard. This method of traceability was used because it was necessary to have a stable national reference, since ^{131}I was widely used for hospital treatments and it was difficult to make measurements directly in terms of Curies. When direct measurements became possible it was necessary to move the value of the United Kingdom standard by 3%. Such local standards have been common in many areas of measurement when it is more important, or only possible to ensure stable local comparability. For example for many years voltage measurements were related to national maintained standard cells.

Thus an important message is that traceability to international standards takes time to evolve and in many case traceability to a local standard may be all that is necessary or even all that is possible.

Traceability to SI

There is a general opinion that traceability to the base units of SI should be used whenever possible. However in many cases this not the best option. As was pointed out above, in some cases reference to a stable local standard may be preferable, if the measurements are used to check trends or control the values of certain parameters. Indeed traceability to more stable intermediate international standards is often used. For example, when the base unit of length was define in terms of the wavelength of light emitted by a particular atomic transition, traceability for length was provided using an iodine-stabilised laser. This was because the repeatability of wavelength of the light from the laser was very much smaller than the uncertainty on the realisation of the base unit. Such practices continue, measurements of resistance are referred to the quantised Hall effect using the conventional value $R_{K\text{-}90}=25812.807$ exactly and the uncertainty of the value of $R_{K\text{-}90}$ in the SI system is not included in the uncertainty assessments. The same approach is used for voltage measurements, which are related to a consensus value for the Josephson constant. Thus even in this most fundamental area of physical measurement traceability is not back to the SI base units because better intercomparability of measurements can be provided using a stable intermediate standard.

Thus the lesson from this is that even traceability to international standards does not necessarily have to include all the steps in the chain back to the base units. For as we have seen for many measurements it is better to use standards that do not include the final step or steps in the chain.

Other bases for establishing traceability

It is sometimes questioned whether traceability is a concept that can be applied to measurements of quantities such as pH, to the measurement of ratios and to the measurement of quantities for which there is no SI unit.

Again there are examples from physical measurements that demonstrate how traceability can be achieved. There is no SI unit for hardness but there is a well-established traceability system for the results of hardness measurements. Hardness scales, e.g. Brunel or Rockwell, are defined in terms of the properties of the machines used to measure the hardness. The properties defined include the shape of the indenter, the applied load and the rate of application of the load together with the tolerances on the properties. Thus utilising a machine that meets these defined properties provides the traceability for the results obtained using it. However it is a difficult task to keep monitoring the properties of the machines and reference materials are used in the form of hardness blocks whose hardness has been measured at a standards laboratory on a machine with very well defined and monitored properties.

This is very similar to the use of empirical methods for chemical measurements, the traceability is established by the use of a defined method.

Traceability has also been established for ratio measurements such as reflection coefficient, absorbance and angle measurements. Different techniques to establish traceability are used in different fields of measurement and it is a mistake to look for "a one size fits all" solution.

Conclusions

The differences between physical and chemical measurements should not be over emphasised. Many of the basic problems are very similar, e.g. sample effects are important in all types of measurement. Also similar error structures occur in both types of measurements and not all physical measurements have uncertainties less than those obtained in chemical measurements. Traceability applies to all measurement results.

Traceability for measurements in a particular area can take some time to develop. It is not always possible or even desirable to have traceability back to the SI base units.

It is possible to establish traceability for the results of measurements on all types of quantities, not just those for which there is an SI unit.

Various techniques have and can be used to establish traceability; it is not necessary to look for one technique that will apply to all measurements.

References

1. Ríos A, Valcércel M (1998) Accred Qual Assur 3:14–19
2. Horwitz W, Albert R (1997) Analyst 122:615–617

Accred Qual Assur (2000) 5:418–422

W. Richter

How to achieve international comparability for chemical measurements

Received: 16 March 2000
Accepted: 20 June 2000

Presented at 3rd EURACHEM Workshop "Status of Traceability in Chemical Measurement", 6–8 September 1999, Bratislava, Slovak Republic

W. Richter
Physikalisch-Technische Bundesanstalt, Bundesallee 100, 38116 Braunschweig, Germany
e-mail: Wolfgang.Richter@ptb.de
Tel.: +49-531-592 3200
Fax: +49-531-592 3015

Abstract It is the central aim of the current activities of metrology in chemistry to build confidence in the reliability of chemical measurement results so that they are accepted without costly duplication being necessary. An important prerequisite for such confidence is comparability based on traceability to recognised common references, ideally the SI units. Since metrology is organised within a national framework according to the national laws and regulations, a two-step procedure is to be followed to achieve international comparability for chemical measurements which is increasingly required as a result of the globalization of trade and economy: (1) establishment of national traceability structures for chemical measurements and (2) mutual recognition of the national traceability structures on the basis of equivalence criteria. The first step is at present being taken in many countries. Examples are presented for Germany. The second step has been initiated by the Mutual Recognition Arrangement (MRA) of the Meter Convention for national measurement standards and measurements and calibrations provided by national metrology institutes, which is based on international comparison measurements (key comparisons) carried out on the national standards level. Chemical analysis is included in this process through the Consultative Committee for Amount of Substance (CCQM).

Key words Chemical measurements · Comparability based on traceability · Traceability structures · International equivalence

Introduction

International comparability and hence acceptance of chemical measurement results is the common central aim of all international initiatives and organizations in the field of metrology in chemistry such as the Consultative Committee for Amount of Substance (CCQM), the Co-operation on International Traceability in Analytical Chemistry (CITAC), EURACHEM, EUROMET and others. This is a consequence of the globalization of trade and economy and other human activities, which requires that measurements made in one country be accepted in other countries without having to be repeated. The main goal is to build the necessary confidence in the reliability of the results so that these are accepted [1].

To achieve this goal, comparability of chemical measurements based on traceability to recognized standards and hence on thorough knowledge of uncertainty, must be established in analogy to the way in which the validity of measurement results is ensured in metrology in general. This task mainly consists of two parts:

1. Establishment of traceability on the national level
2. Mutual international recognition of the national traceability systems based on their demonstrated equivalence.

In the field of chemical measurements, part (1) is now underway in many countries. The basis for (2) has been laid for all kinds of measurements by the General Conference of the Meter Convention and includes high-level interlaboratory comparisons (key comparisons) as an important element [2, 3]. Metrology in chemistry is included in this comparison system through CCQM [4, 5]. In the following these two parts are dealt with in more detail.

Establishment of traceability of chemical measurements in a national framework

The ultimate reference point of traceability for any kind of measurement is the SI unit in which the result is expressed. Traceability is the prerequisite for complete evaluation of the uncertainty of a measurement, it is not a purpose in itself. Thorough knowledge of uncertainty in turn is the prerequisite for comparability. According to the way in which metrology is organized all over the world, traceability to the SI units is usually accomplished through national measurement standards. Every country has its own legal basis for its measurement infrastructure, an important part of which is formed by the national measurement standards.

In the field of chemical measurements, the question as to which are the national measurement standards is far from being completely answered. There is no doubt, however, that primary reference materials, e.g. high-purity substances which are ultimately necessary as reference points, will play a role as national (or even better international) measurement standards. But these alone will not be sufficient. As the task of chemical analysis is usually the determination of chemical composition, national reference points closer to this task, namely standard reference mixtures, are also required, and if the preparation of these is not feasible, e.g. due to instability problems, devices and procedures furnishing well-known compositions with small uncertainties must also be included as national measurement standards. All these kinds of national references or standards are currently in use or under development.

It is mainly due to the endeavours of the CCQM that the issue of national standards is being thoroughly discussed. A recent outcome has been a list of priority areas for which traceability of chemical measurements is most urgently required with a view to removing technical barriers to international trade after the World Trade Organization largely removed the tariff-based barriers [6], and in order to meet the requirements of accreditation and regulatory bodies. These priority areas to start with are:

- Health care
- Food
- Environment
- Advanced materials
- Commodities
- Forensics.

It is obvious that for every area a great number of references will be required to provide traceability chains for the working level. It is also quite obvious that national standards cannot be made available for every chemical measurement task. The central goal now is to develop a minimum set of national references which is large enough to provide traceability for the most important chemical measurement tasks in the above-mentioned priority areas, and to establish international equivalence for these national standards (cf. Mutual international recognition of national traceability systems for chemical measurements).

National standards are usually kept at the national institutes responsible for their development and maintenance and are used only for linking up secondary standards to them, which then are made available as transfer standards. In most countries this responsibility has been entrusted by law to the national metrology institutes (NMIs). In the field of chemical measurements for which traceability to the SI units and the development of national measurement standards have been the focal point of interest for only a few years, decentralized national responsibilities are now developing in such a way that high-level national chemistry institutes are entrusted in part with the task of maintaining national standards by agreements with NMIs. In this way the national reference level can be established on a broad basis, and this is underway now in many countries.

A system of national references is necessary but not sufficient for establishing international comparability of chemical measurements at the working level at which the cross-frontier exchange of goods and services usually takes place. For disseminating these references to the working level, an intermediate level is needed which can act as a multiplier between the national (mostly primary) and the working level. Such multipliers are well known for other parts of metrology (mass, length, electrical quantities, etc.) in the form of calibration laboratories and have proved to be an efficient element of the metrological infrastructure of many countries, in particular in Europe.

Such an intermediate level of secondary laboratories is also required in the field of metrology in chemistry. It is not so important whether these act as calibration or, more generally, as reference laboratories. What is important is that they are firmly linked to the national references and have the competence to carry these references to the working level. Competent reference material providers above all would be potential candidates for such laboratories. With such an intermediate or secondary level, realized in the form of accredited laboratories with dissemination tasks, complete and efficient traceability structures for chemical measurements can be established. It should not, however, be overlooked that worldwide harmonization of laboratory accreditation schemes is required also to finally achieve full comparability of chemical measurement results at the working level.

Practical realizations of national traceability structures for chemical measurements, as outlined here in general, already exist or are under development. A few examples are:

- A traceability structure for gas analysis in the United Kingdom
- Traceability structures for gas analysis, clinical chemistry, pH measurement and electrical conductivity of electrolyte solutions in Germany
- A traceability structure based on the National Institute of Standards and Technology (NIST)-traceable reference materials produced by commercial suppliers under the supervision of NIST in the United States.

Figures 1–3 show examples of national traceability structures for chemical measurements, which so far have been established and successfully used in Germany. It is a common feature of these structures that in the non-regulated area accredited calibration laborato-

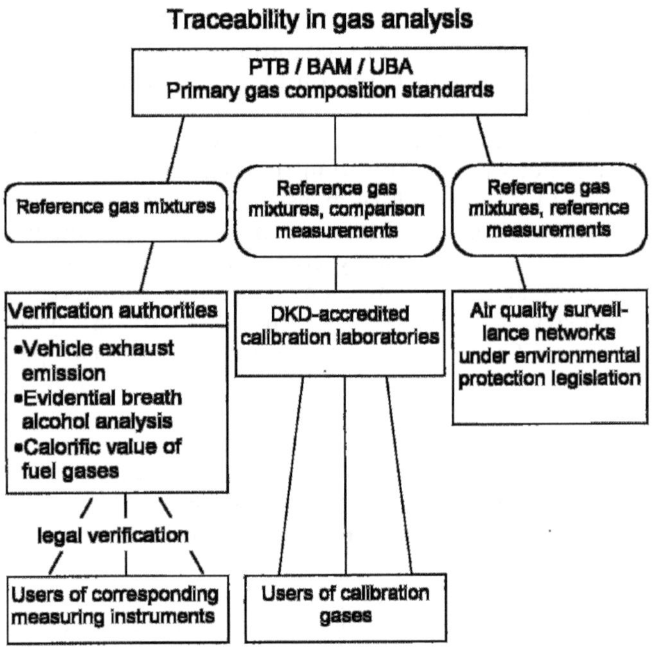

Fig. 2 The traceability system for gas analysis in Germany is composed of three chains, one for the legally regulated area via verification authorities, one for the non-regulated area via DKD-accredited calibration laboratories and one for the air pollution monitoring networks. Here the national reference level consists of PTB, BAM and UBA (cf. text) which share the responsibility for the national measurement standards needed, according to agreements with PTB

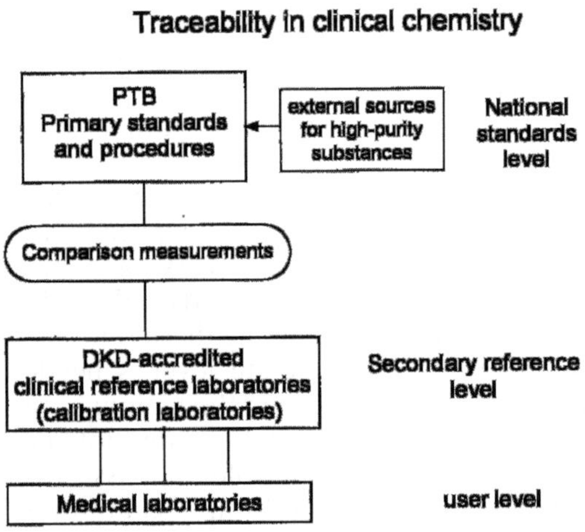

Fig. 1 Traceability system for the determination of the most important diagnostic markers in human body fluids in Germany. The clinical reference laboratories at the intermediate level providing calibration means to the routine medical laboratories are accredited as calibration laboratories in the framework of the German Calibration Service (DKD) and are firmly linked to the national metrology institute, PTB, by comparison measurements carried out on actual laboratory samples. Accreditation is in part required by the Federal Physicians' Council (BÄK) or is voluntary. The traceability system is still under development

ries act as multipliers in the dissemination chain (in the regulated area, verification authorities play this role). They are accredited within the framework of the German Calibration Service (Deutscher Kalibrierdienst – DKD) and their traceability is ensured by comparison measurements with, or transfer standards from, the national standards level which provides the primary standards and procedures. In the field of gas analysis (and in future probably also in other fields), the national standards level is provided by the NMI, the Physikalisch-Technische Bundesanstalt (PTB), the Bundesanstalt für Materialforschung und -prüfung (BAM) and the Umweltbundesamt (UBA) which, on the basis of agreements with PTB, share the responsibility for national standards for chemical measurements according to their capabilities. These three institutes, each in its field of responsibility, take part in the key comparisons of CCQM aiming at establishing international equivalence of the national measurement standards under the Mutual Recognition Arrangement (MRA) (cf. Mutual international recognition of national traceability systems for chemical measurements).

The extent to which technical barriers to international trade due to non-acceptance of chemical measurement results can be removed, largely depends on

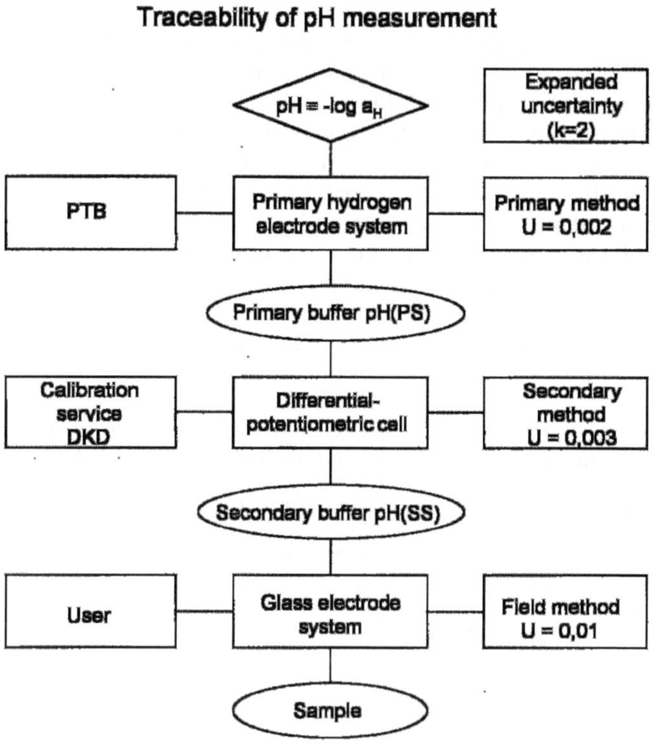

Fig. 3 Traceability system for pH measurements in Germany. This traceability chain is basically similar to those for clinical chemistry and gas analysis, as DKD-accredited calibration laboratories act as multipliers in dissemination. It is a special feature of this structure that traceability of pH measurements does not extend to the SI but to a conventional reference frame recognized worldwide. Traceability to the SI is possible but would imply a considerable increase in uncertainty

whether national traceability structures for metrology in chemistry can be successfully set up.

Mutual international recognition of national traceability systems for chemical measurements

While the establishment and implementation of national traceability systems is left to the countries and their legislation, the mutual recognition of these national systems calls for an international mechanism, at least if worldwide mutual recognition is aimed at. The MRA of the Meter Convention for national measurement standards and calibration and measurement certificates issued by NMIs [2] can be expected to at least provide for one important aspect of the task, namely the establishment of international equivalence at the national standards level. Together with the success of the work currently carried out by the International Laboratory Accreditation Cooperation (ILAC) for the harmonization of national laboratory accreditation schemes, the MRA will be the key to comparability and hence to acceptance of chemical measurement results at the working level.

The MRA is based on statements of equivalence which in turn are based on the results of so-called key comparisons, a well-selected set of high-level interlaboratory comparison measurements carried out on national measurement standards using methods which test the principal techniques in a given area of metrology.

Two kinds of key comparisons are distinguished: (1) Key comparisons carried out by the Consultative Committees (CC), in which the most experienced NMIs and other institutes entrusted with parts of the national references for metrology in chemistry take part and (2) key comparisons carried out by the regional metrology organizations (RMOs) like EUROMET, NORAMET, APMP, SIM in which one or more of the participants of the corresponding CC key comparison act as a link. In this way all national laboratories responsible for establishing traceability to national standards can take part in key comparisons and establish their degree of equivalence with others.

In the field of chemical measurements, the selection of key comparisons is not so straightforward as in the other metrological areas. Starting from the priority areas given in the second section, the following sub-areas have been identified by the CCQM for carrying out key comparisons.

- Automobile exhaust emission surveillance
- Breath alcohol analysis for drink-and-driving legislation
- Air quality surveillance
- Analysis of natural and drinking water with respect to toxic elements
- Calibration solutions for elemental analysis in general
- Diagnostic markers in clinical chemistry
- DDT metabolites in natural matrices
- pH measurement.

The first round of key comparisons in these fields was largely completed by the end of 1999 and showed that gas analysis, elemental analysis and pH measurement are already rather well developed and mature areas, whereas organic analysis, for example in the clinical and food areas, requires more attention, in particular with respect to sampling and sample pretreatment which are often the major sources of uncertainty. As regards sampling which is even more important when field measurements are linked up with national or international standards, comprehensive practical and theoretical knowledge is available, especially for particulate material sampling [7], which can be used where applicable to improve the comparability of chemical measurement results.

On the whole, good agreement based on known uncertainties has been achieved in these key comparisons, also in the more difficult areas. This shows that a group of laboratories exists at the national standards level capable of establishing, in a joint effort, a global reference system for chemical measurements.

The key comparisons carried out so far only define a starting point. The current list will be amended according to the increasing demand for traceability.

The results obtained by the participants in key comparisons, together with the CCQM reference values ultimately assigned to the key comparisons, are given in Appendix B of the MRA. Appendix C gives the calibration and certification capabilities of the national institutes based on the results of the key comparisons, after assessment by the RMOs and the Joint Committee (composed of RMO representatives and the Bureau International des Poids et Mesures – BIPM). All the data will be accessible via Internet so that finally a complete and transparent system will be available to the public, describing the capabilities of the institutes at the top of the national traceability systems and the degree of their equivalence. The key comparison database is already available at BIPM under www.bipm.fr

References

1. Inglis BD (1998) PTB-Texte 9:277–288
2. Göbel EO (1998) PTB-Mitteilungen 108:231–243
3. BIPM (1999) Mutual recognition of national measurement standards and of calibration and measurement certificates issued by national metrology institutes, Bureau International des Poids et Mesures (BIPM), Pavillon de Breteuil, 92312 Sèvres Cedex, France; Internet: www.bipm.fr
4. BIPM (1993) BIPM Proc.-Verb. Comité International des Poids et Mesures 82:122
5. BIPM (1995) BIPM Comité Consultatif pour la Quantité de Matiere 1
6. Liu V (1998) PTB-Texte 9:15–24
7. Gy P (1998) Sampling for analytical purposes, Wiley, Chichester, UK

Accred Qual Assur (2000) 5:423–428

Paul De Bièvre

The key elements of traceability in chemical measurement: agreed or still under debate?

Received: 17 March 2000
Accepted: 10 July 2000

Presented at 3rd EURACHEM Workshop "Status of Traceability in Chemical Measurement", 6–8 September 1999, Bratislava, Slovak Republic

P. De Bièvre
Adviser on "Metrology in Chemistry" to the Director of the Institute for Reference Materials and Measurements, European Commission – JRC, 2440 Geel, Belgium
e-mail: paul.de.bievre@skynet.be
Fax: +32-14-853908

Abstract Talking about "traceability" means talking about a "property of the **result** of a measurement", about "the **value** of a standard", about "stated references" and about an "unbroken chain of comparisons". It describes by which comparison, and to which other **value,** the result of a measurement has been obtained, i.e. is "traceable to". It is about the underlying structure of the measurement process of the **result of a measurement** and therefore about the *authority of the result.* Since values carried by (certified) reference materials have also been obtained by measurement, the definition of traceability equally applies. Traceability in the context of reference materials is also about the *authority of the* values *carried by the (certified) reference materials* and is, therefore, of key importance for the authority of the reference materials themselves. Hence, **values of results** of measurements constitute part of the traceability chain and their uncertainties are an intrinsic accompanying phenomenon. Uncertainties need a traceability chain against which they can be evaluated, and a traceability chain is an a priori requirement for evaluating the uncertainty budget of a measurement result. An attempt has been made to exemplify "traceability" chains in some types of chemical measurement and to identify the degree of international agreement on the key elements of "traceability". It is concluded that there is less than universal agreement on this issue. The debate should continue in order to arrive at the international understanding and agreement needed, as "traceability" is now being incorporated in the International Organization for Standardization (ISO), the International Laboratory Accreditation Co-operation (ILAC) and in other "guiding" or regulatory documents. It is also the reason why the Institute for Reference Materials and Measurements (IRMM) has taken up the study of the concept in its core programme on Metrology in Chemistry, and why it sponsored the Workshop in Bratislava.

Key words Traceability chain · Metrology in chemistry · Reference material · Comparisons values of standards

Introduction

Traceability is defined in the *International Vocabulary of Basic and General Terms in Metrology* (VIM) [1] as: "the property of the result of a measurement or the value of a standard whereby it can be related to stated references, usually national or international standards, through an unbroken chain of comparisons all having stated uncertainties". Traceability is therefore clearly a property of the **value** of a result of a measurement. So what is a measurement? The same VIM [1] defines it as: "a set of operations having the object of determining the value of a quantity". Again, the notion of **value** is very outspoken. But, then, what is meant by an operation? We use the following simple description: "the operation (of measuring the value of a quantity) is the performance of a comparison of an unknown value to a known value of the same quantity".

We have now moved the problem of establishing the traceability of an unknown value to establishing the traceability of a "known" value, which in turn must be compared to another "known" value, etc. This is quite acceptable provided this process stops somewhere. It stops when we arrive at a value which we *know* because we *defined* it. It is the value of the unit in which we want to express the result of our measurement. Thus a definition of a traceability chain follows naturally: "a traceability chain is a chain of successive comparisons (i.e. measurements) of one value to another value which ends in the value of the unit we have chosen to express the result of our measurement".

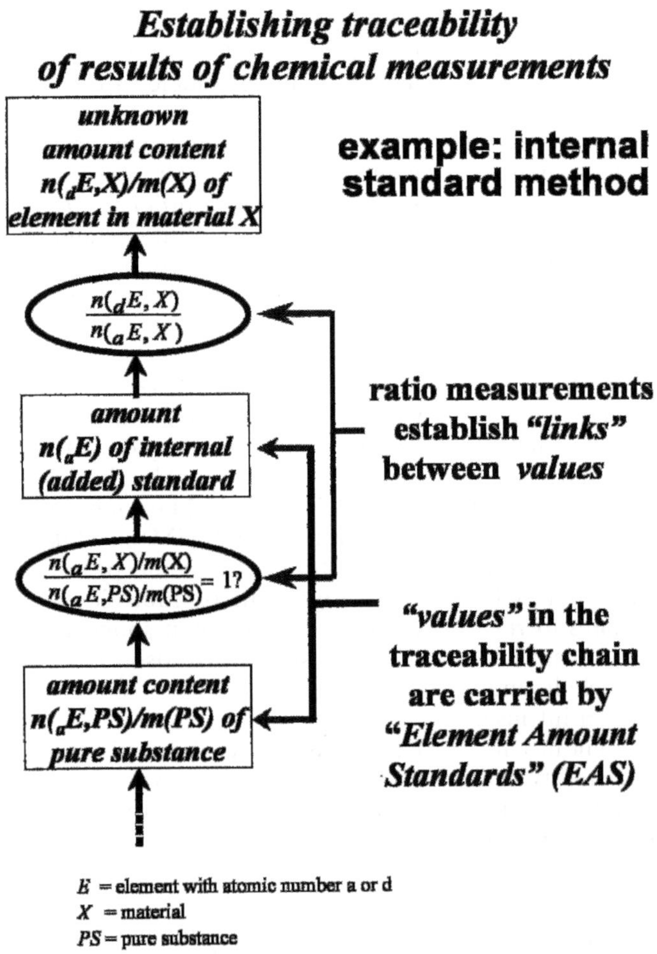

Fig. 1 The key elements of traceability chains: values and links (=measurements)

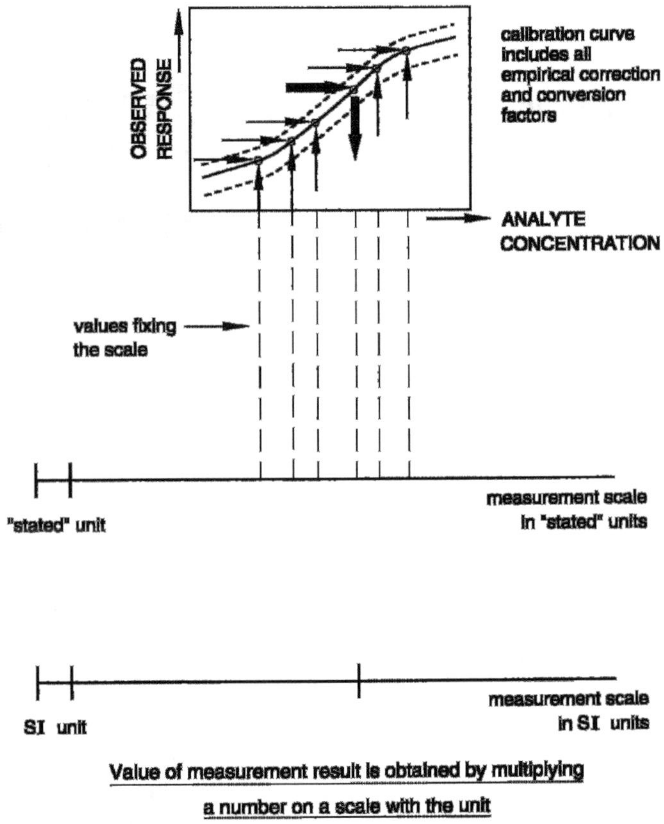

Fig. 2 Measurement scales can be constructed from observed responses which are related to values carried by commonly agreed reference materials

In Fig. 1, the essential elements of "traceability" are exemplified for the measurement of an element amount content using the "internal standard method": values linked by measurements to other values, thus constituting a chain (only a part of a chain is shown). The chain must ultimately end up in a value on a measurement scale (Fig. 2) [2] and therefore in a unit. All measurement results which are "traceable" to values on a common scale (or to the value of a unit) are *comparable* [3], meaning literally that they can be *compared* to each other. This does not imply that they must be *concordant* [3], i.e. result in the same value (within measurement uncertainty), but only that they can be validly *compared* for their magnitudes, even if these are (very) different. The measurement scale can be constructed from any sort of (internationally) agreed unit. If at all possible, this should be a base or derived SI unit because we have a solid international agreement on this. However, in chemistry, there are numerous measurement results which cannot be "traced" to (values of) SI units and we therefore must use the more empirical concept of measurement scales with empirical or even arbitrary units, provided they are agreed a priori between the people who want to *compare* their results.

At this stage it is important to observe the prime function of a reference material: it is *to carry a* value. This is further elaborated elsewhere [4]. Since any **value**, including that carried by a reference material, *per definition results from a measurement,* the long – and still persisting – tradition of focusing on the **material** should be reoriented towards focusing on the **value carried** by the material.

"Traceability of the **value carried** by a reference material" is to be distinguished clearly from *traceability of a reference* material which we describe as "trackability" (e.g. to the place it comes from) [4–6]. The basic idea that traceability chains, even in the case of the well-known "weights", are a matter of **traceability of values**, has been discussed before [7].

We now try to understand the **traceability chain of the value** obtained when performing a fully "traceable" measurement or when certifying a reference material for a "traceable" value.

The traceability chain of the result of a measurement, i.e. of a value

In Fig. 3a,b (right-hand chain), a full traceability chain of the **value** of a synthesized matrix reference material is shown. The left-hand chain is the traceability chain of the value resulting from the measurement of an unknown sample. The right-hand chain is the traceability chain of the value carried by the reference material and provided by the producer of the reference material. Both chains consist of successive comparisons from one value to another, all the way down to known values (of base units or of fundamental constants). The chain is constituted by **values** (in the rectangles) "linked" by operations called **"measurements"** (in the ovals), defined as above. Establishing the sequence value-measurement-value-measurement-value, etc, is establishing

Fig. 3a, b Traceability chains of values obtained from measurements, either on an unknown sample (left-hand chain), or on a reference material (right-hand chain)

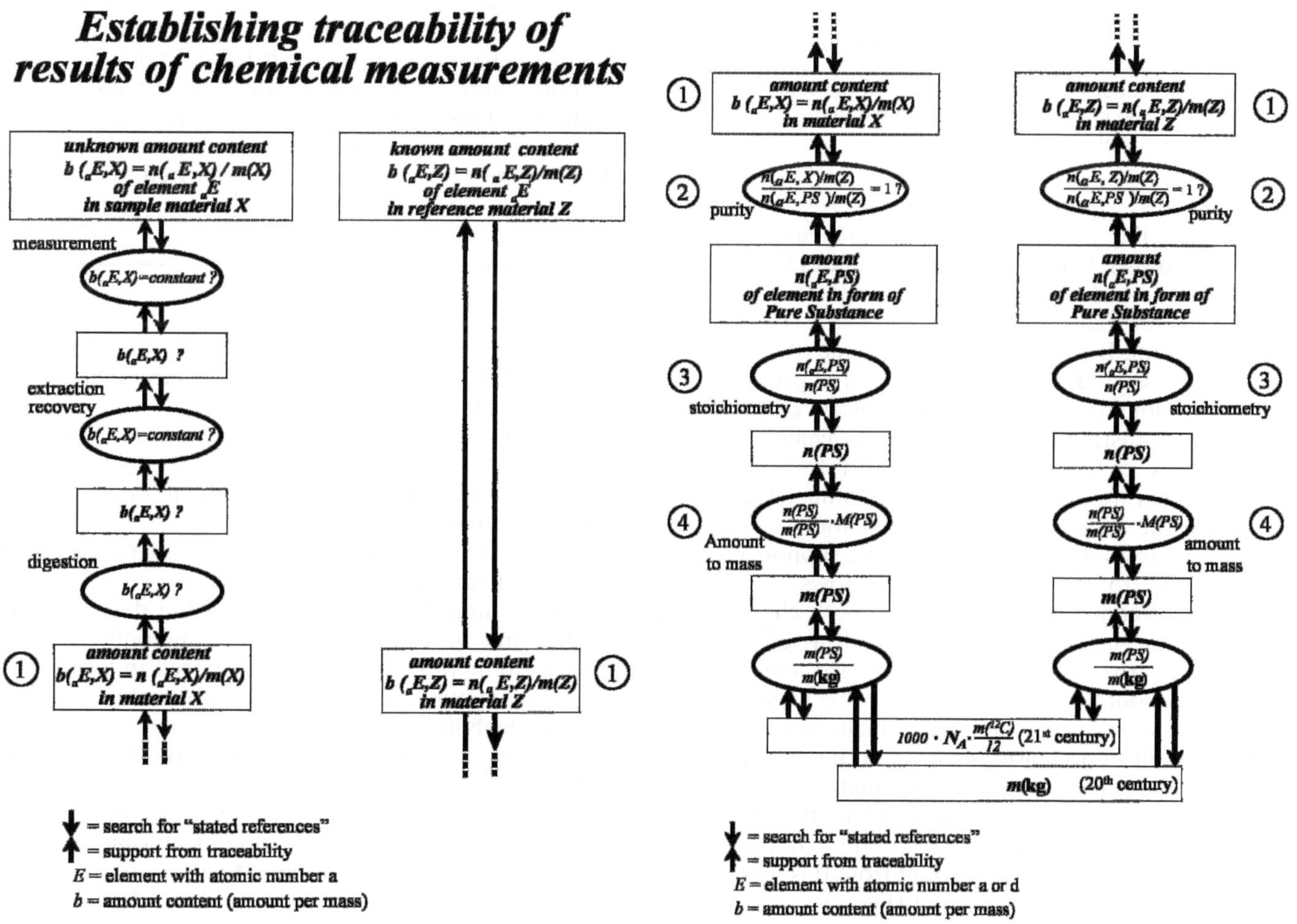

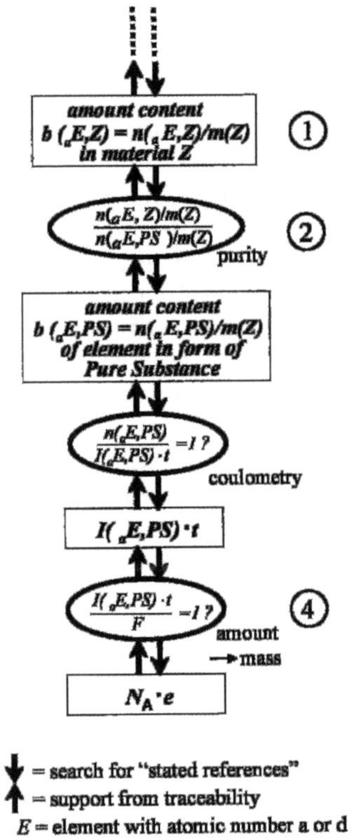

Fig. 4 A traceability chain need not end in the value of the kilogram. It can also end in the value of the Faraday, i.e. in the value of the ampere and of the second

the traceability chain of the value of a measurement result to a known value. This known value is the value we have given to the unit we have chosen to express the result of our measurement. In Fig. 3, it is the value of the kilogram. But in Fig. 4 it is the value of the ampere and of the second (the reference material producer may chose coulometry, i.e. measuring electric current times time – $I \cdot t$, instead of mass, m). Both chains are valid traceability chains. The chain constitutes the "trace" along which we arrive at this known value. The arrows indicate the direction of the search for "stated references" (arrows downward) and of the support given by traceability to a common base (arrows upward). It should not come as a surprise that both chains of the value of the unknown sample as well as of the value carried by the reference material, are similar, since both are about the traceability of a value which is measured. Thus a basis is established which makes the value of the result of a measurement of an unknown, as well as the value carried by a reference material, meaningful because the values are "traceable" to the value of a common measurement unit, or, better, to values of a common measurement scale built from this unit, whether an arbitrary unit or a SI unit (Fig. 2).

Traceability to a common unit, or to a common measurement scale, makes the results of measurements *comparable*, regardless of their magnitude. The very – and only – reason for the existence of the requirement "traceability" is to enable us to *compare* our measurement results (i.e. achieving *comparability*): measurements of the same measurand, carried out at different places and/or different times, yield *concordant* results (meaning that they fall within each other's stated uncertainty). It is emphasized again that traceability need not to be to (values of) SI units but can be to values of other (including arbitrary) units with which measurement scales can be built. As Fig. 2 shows, a measurement scale can be constructed from a "calibration" curve based on a number of values carried by commonly agreed reference materials (in this case: 5) [2].

We will see later [4] that it is this similarity which can be exploited by analysts to reduce their workload. Analysts could attempt to establish the complete traceability chain all by themselves in Fig. 3a,b (left-hand chain), but cannot do that in practice: too much work, too expensive. They can literally "buy" such values carried by reference materials. They add a known value from a reference material to their unknown sample and measures the ratio of these two values, – unknown to known – in the one and only measurement they must make. That was exemplified in Fig. 2 by means of the method wherein an internal standard is added to the unknown sample and measured under the same circumstances as the unknown sample. How the analyst can "buy" help by "purchasing" values carried by reference materials, is described elsewhere [4].

Conclusions

It has now become possible to identify the key elements in traceability of chemical measurement results, on which agreement is needed:

1. Traceability is about **values** of measurement results.
2. Values are linked by measurement operations to form a chain.
3. A traceability chain is a chain of comparisons of values to a commonly accepted value.
4. The commonly accepted value is the defined value of a unit or the reading on a commonly accepted measurement scale which is anchored in values carried by commonly agreed reference materials.
5. A traceability chain gives rise to uncertainties because it contains measurements (the links in the chain).
6. A traceability chain is therefore a prerequisite, i.e. an a priori requirement for the establishment of an uncertainty budget (the uncertainties are associated with the links in the chain); it follows that:

7. Traceability chains can exist with either large or small uncertainties in their links and values, and therefore with large or small "overall" uncertainty budgets; from which it follows that:
8. Traceability chains are the basic structures to attach uncertainties to, and uncertainties cannot be evaluated without traceability chains: uncertainty has to do with the strength of the traceability links and not with their existence.

It is hard to see that broad international agreement has been reached so far on these basic and logical elements of traceability chains in chemical measurement.

Fig. 5 A traceability structure can be set up locally, regionally or internationally, or against whatever commonly agreed references as required for the intended use

Traceability to a "stated reference"
of values obtained in field chemical measurements (FM)

establishing traceability is
comparing the value obtained by measuring an "unknown" to the value of a "known" (with its uncertainty)
which in turn is linked to the value of a "known" etc;
each "known" value has a smaller uncertainty than the previous "unknown" value

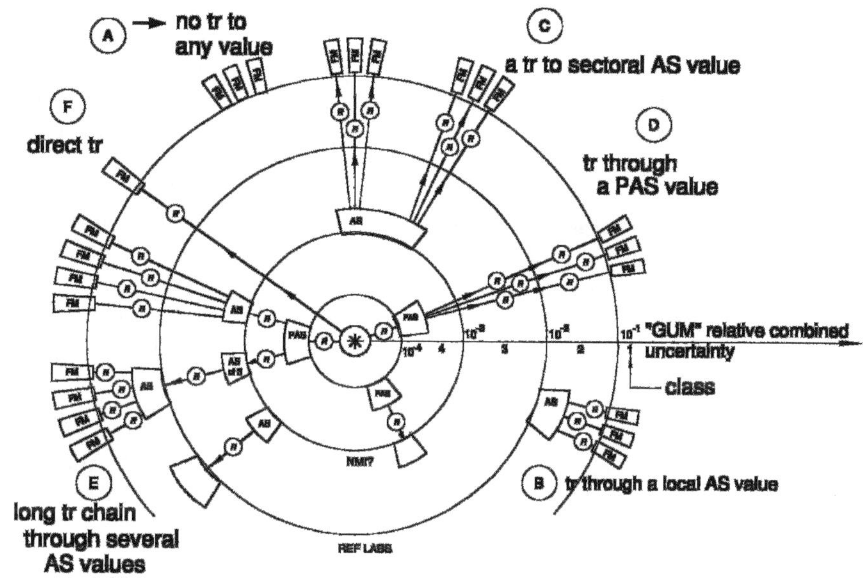

Notes:

1. **traceability (tr) implies a support with a direction (arrows) delivered by a higher measurement authority i.e. by a value with smaller uncertainty to the value of the unit chosen**
2. **the smaller the relative uncertainty, the greater the reliability and the stronger the link of the traceability chain**

Legend:

(P)AS = value carried by (Primary) Amount Standard
cl = class
✱ = "stated reference" = value of the measurement unit chosen
Ⓡ = value of a ratio measurement

Source: P De Bièvre, Chapter 7 in
"Accreditation and Quality Assurance"
Springer (1994 German ed. -1996 English ed.)

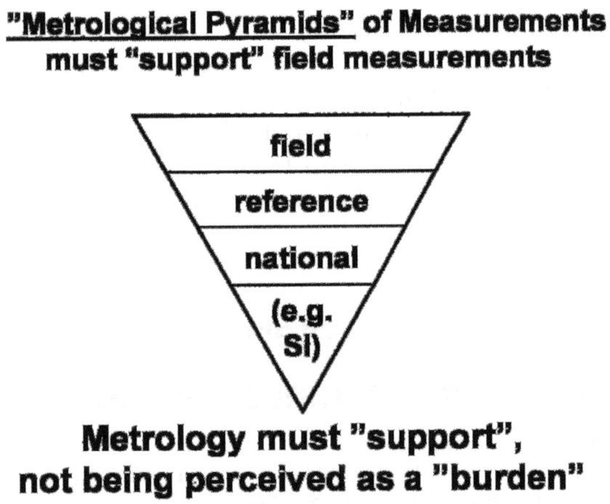

Fig. 6 "Reference" laboratories and traceability structures ought to support, not burden, the field laboratories

They are absent from the literature with the exception of publications by R Dybkaer [8]. Pictures of such traceability chains are almost completely missing. Yet such pictures provide simple, transparent insights. Probably international thinking and discussion ought to concentrate more on **chains of values** of the measurement results, as well as on **chains of values** carried by reference materials, rather than on materials, methods, instruments or institutes. This would also be closer to the spirit of the SI, which is to become independent of man-made artefacts and to anchor measurement results as far as possible in inalterable properties of nature and in units (with their values) derived from these. However, there are numerous chemical measurements where the latter is not (yet) possible. In those cases, it is useful to base traceability on the concept of commonly agreed measurement scales which are "calibrated" by values carried by commonly agreed reference materials. But also in the latter case, **traceability is a matter of values of measurement results.**

The task is to demonstrate the scientific/technical authority of the measurement result. This needs underpinning ("Untermauerung") of the measurement result in order to lead to the necessary credibility.

The concept of chains of values naturally leads to "ultimate" bases for traceability (to SI or other internationally agreed values) whenever possible. This is illustrated in Fig. 5 where different possibilities of traceability to "stated references" are visualized, depending on the level of uncertainty required. Realization of "cluster" or local traceabilities are perfectly possible at larger uncertainties if that is enough for the intended use (fitness-for-purpose). A segment from this circle leads naturally to the picture of an inverted pyramid (Fig. 6) which we prefer in this "upside down" version because it stresses the fact that metrology/traceability must *support* the (measurements in the) field laboratories and not *burden* them. It follows that any laboratory, institute or producer of (certified) measurements or materials which claims a "reference" role, has clearly the duty to *support* the field laboratories by providing them the underpinning shown.

From all of this, it is also clear that a better understanding of "traceability" requires a better understanding of the measurement process to which that concept is applied.

References

1. BIPM, IEC, IFCC, ISO, IUPAC, IUPAP, OIML, ISO (1993) International vocabulary of basic and general terms in metrology. ISO, Geneva
2. De Bièvre P, Kaarls R, Peiser HS, Rasberry SD, Reed WP (1997) Accred Qual Assur 2:168–179
3. Papadakis I, De Bièvre P (1997) Accred Qual Assur 2:347
4. De Bièvre P (2000) Accred Qual Assur 5:224–230
5. Fleming JW, Neidhart B, Wegscheider W (1996) Accred Qual Assur 1:35
6. De Bièvre P (1996) Accred Qual Assur 1:1095
7. De Bièvre P, Taylor PDP (1997) Metrologia 34:67–75
8. Dybkaer R (1998) Scand J Clin Lab Invest 58 (suppl 228): 38–39

Accred Qual Assur (2000) 5:429–436

Bernard King

The practical realization of the traceability of chemical measurements standards

Received: 17 March 2000
Accepted: 30 May 2000

Presented at 3rd EURACHEM Workshop "Status of Traceability in Chemical Measurement", 6–8 September 1999, Bratislava, Slovak Republic

B. King
National Analytical Reference Laboratory, 1 Suakin Street, Pymble, NSW 2073, Australia

Abstract Metrology is based on the concept of traceability. Traceability provides a means of relating measurement results to common standards thereby helping to ensure that measurements made in different laboratories are comparable. Good progress has been made in the application of metrological principles to chemical measurement, but there remains confusion about how you actually achieve traceability in a practical way.

This paper elaborates on the meaning and application of much used phrases such as 'the value of a standard', 'stated references', 'unbroken chain of comparisons', and 'stated uncertainties'. It also explains how traceability can be established in a practical way for different types of stated references, namely pure substance reference materials, matrix reference materials, and primary and reference methods. Finally, traceability chains for some typical examples of chemical measurement are described.

Introduction

Analytical chemistry has served society well over the past 100 years, but the concepts and systems that underpin chemical measurements are cracking under the stress of increasing requirements. The demands being placed on chemical measurements are increasing as society requires more complex, quicker, and cheaper measurements. A sustainable future depends on reliable, comparable, and traceable measurements and failure to provide such measurements will be costly in financial and human terms.

Although in its infancy, metrology in chemistry is already accepted by a number of governments as the means of ensuring that measurements made in different laboratories are comparable and traceable to common standards [1]. The ultimate aim is to develop a structured chemical measurement system which will lead to mutual recognition and trust in measurements made in laboratories all over the world. This in turn will facilitate international trade, wealth creation through innovation, and enforceable and trusted regulations.

Metrology is based on the concept of traceability, which is defined as follows [2]: Traceability is the property of a result of a measurement or the value of a standard whereby it can be related to stated references, usually national or international standards, through an unbroken chain of comparisons all having stated uncertainties.

Enormous progress has been made in applying this concept to chemical measurement and there is a growing amount of published work and associated conferences [3–5]. This includs work on concepts, terminology, and primary methods and has led to the new approach of 'metrology in chemistry', which is a synthesis based on input from physical measurement metrologists and analytical chemists.

The topic is complex with many interrelated strands. Despite successes, there remains some confusion in the debate and little clear guidance on how you actually achieve traceability in a practical way. Paradoxically, there is both 'loose talk' about what traceability means and a lack of recognition of the many elements of traceability that already exist. The definition of traceability succinctly describes the key issues and concepts and this paper will attempt to elaborate on their meaning and application.

Traceability; a deconstruction of the definition

The value of a standard

This phrase often means the purity or amount of major, minor, or trace component, or a physico-chemical property of a reference material (RM). It can also be the result obtained under carefully controlled conditions using a primary or reference method. It should be noted that it is not the RM that is traceable but the property value associated with the certification. By the same token, a primary method is not traceable, but the value(s) produced using the method is (are).

Related to stated references

For a value to be traceable it must be related to stated references. By definition and convention the stated references are taken to include SI [6] reference values (e.g., atomic mass values), reference materials (RMs), as well as primary, reference, and standard methods. It is sometimes stated that chemical measurements are traceable to the mole. This is an incomplete statement as chemical measurements are simultaneously traceable to a number of references, inter alia, the mole, kg, meter, etc. Whilst it is considered desirable to employ high level references, such as the SI, where feasible, this is not always necessary in terms of 'fit for purpose' criteria. Neither is it possible to relate all types of analyte (fat, fiber, protein, pH, etc.) to the SI. The key issue is that the references should be stated and fit for purpose.

The phrase 'related to' implies that the relationship is known and valid. This will only be realized if the relationship at every step of the process is clearly defined and valid. Hence the requirement for an 'unbroken chain of comparisons.' The parallel between these issues and those addressed by method validation is worth noting. Validation is the process of establishing that a method is capable of measuring the desired measurand (analyte), with appropriate performance characteristics, such as level of uncertainty, robustness, etc. It should also address systematic effects, such as incomplete recovery of the analyte, interferences, etc. These latter issues can be dealt with by designing a method to eliminate any bias, at a given level of uncertainty, or if that is not possible, to provide a means of correcting for the bias. This may be done at the method level, by applying a correction factor to all results, or at the individual measurement level.

Through an unbroken chain of comparisons

The 'comparisons' may take place every time a measurement is made (e.g., calibration of an analytical measurement using a standard solution), periodically (e.g., calibration of the balance), or infrequently (e.g., validation of a method). The reference value is used to either calibrate the process or to check its calibration or validity. The number of steps in the chain of comparisons should be kept to a minimum as each additional step introduces additional errors and increases the overall uncertainty. Interlaboratory comparisons provide evidence of comparability and provide confidence in traceability claims; they do not, however, provide traceability directly.

There is of course more than one chain of comparisons and all the component measurement processes associated with the chemical measurement need to be considered. These include physical measurements, such as mass, volume, etc., and chemical issues, such as identity and amount, which together constitute an 'amount of substance' (see later). The traceability of component measurements needs to be established at a level of uncertainty that is consistent with the required overall uncertainty of the final measurement. Components such as temperature, and even mass and volume measurements, often contribute little to the overall uncertainty and thus can be simply and easily addressed.

Dealing with the chemical issues is usually much more demanding. In a typical chemical measurement there are two main components to the chemical chain of comparisons, as illustrated in Fig. 1. One component is established by measuring the response from one or more chemical standard(s), often in the form of standard solutions, prepared from pure substance RMs. The identity, purity, and stability of the standards are important issues. The calibration factor F_c and its uncertainty, U_c, describe this part of the chain. Such methods are often called ratio methods and include most of the

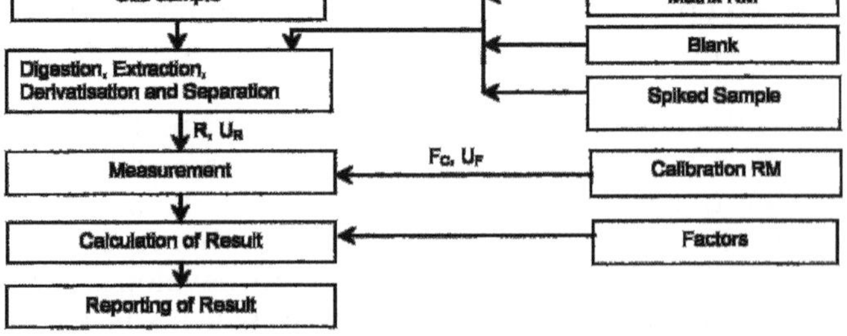

Fig. 1 Essential features of a typical chemical measurement. The concentration of analyte in the original sample (C) is given by $C = F_c \times C_s \times R$, where C_s is the concentration of the calibration standard

widely used techniques, such as GC, HPLC, MS, NMR, AAS, ICPMS, etc. The advantage of this approach is that it is not necessary to establish the traceability of intermediate parameters such as voltages or ion currents providing the sample and chemical standards behave in the same way within the detection system.

The second component of the chemical chain is the relationship between the original sample and the sample presented for measurement. Analyte loss due to decomposition, incomplete extraction, adsorption on containers or separation media can all affect the recovery factor, R. Also, species other than the analyte, that have not been fully separated, can contribute to the measurement signal, resulting in a positive bias. Finally, the matrix can enhance or suppress the measurement signal, further contributing to bias. The combination of all these factors can lead to a recovery factor, R, that is greater or less than unity. Ideally, the individual effects should be studied, understood, and minimized. This can be done by studying the inner workings of the method, comparing the value obtained from the method with that obtained using a reference or primary method and use of the other strategies illustrated in Fig. 1, such as the use of closely matched matrix RMs. Nevertheless, obtaining a good estimate of R can be difficult and thus the associated uncertainty, U_r, can be large. It is in these cases that consideration needs to be given to establishing traceability to lesser references than SI, for example, to the values obtained from a standard method or a reference material.

Stated uncertainties

This is an expression of doubt concerning the reliability of 'the value'. The process of deriving an estimate of the uncertainty is well described [7], even if it is difficult in some cases to produce a good estimate. Whilst traceability is a yes/no issue – you either have it or you don't – uncertainty is a matter of degree and describes the strength of the linkage.

The uncertainty associated with a traceable value must be related to a specified measurand (analyte) and be related to stated references. The following example illustrates the effect the choice of stated reference has on the stated uncertainty for the measurement of lead in milk. The uncertainty of a measurement of lead in milk, measured using a standard method, could be small, if stated relative to that standard method, where the measurand (analyte) is implicitly defined by the standard method. However, the method is likely to contain some additional errors and uncertainties if it were to be related to a primary method traceable to the SI, and these would need to be included in the estimate of uncertainty, if the SI was quoted as the stated reference. The interrelationship between uncertainty and stated references is illustrated in Fig. 2. The above approach has limitations in metrological terms if the results are expressed in SI units, e.g., mg/kg, etc. The use of SI units implies that a measurement is traceable to SI and, if this is not the case, then some form of qualifying statement is needed.

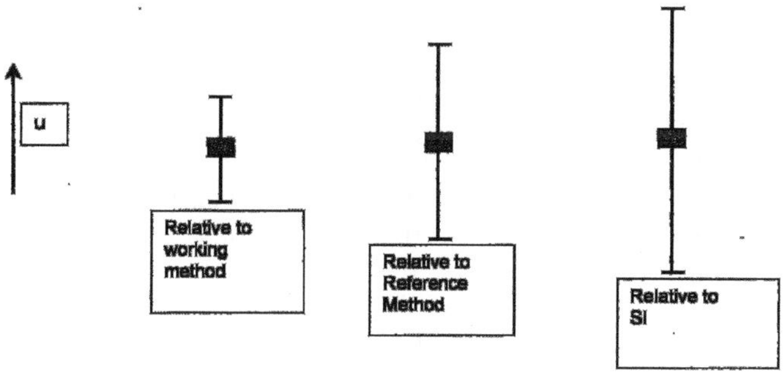

Fig. 2 Interrelationship between traceability and measurement uncertainty

The total combined uncertainty of a measurement is a function of the uncertainties associated with the component measurements, references, and processes. References such as RMs and primary methods are, therefore, important ways of providing uncertainty information for parts of the traceability chain.

The role of the mole in chemical measurement

Some chemists feel that the mole is an unnecessary SI unit as they make measurements in mass/mass or mass/volume units, using ratio methods. The definition and the importance of the mole has been discussed elsewhere [8], and the distinction has been made between its importance as a concept, the importance of the related atomic mass values, and the lesser role of the mole as a unit for actually reporting results. A distinctive feature of the mole is the need to define 'the entity'. This is an extra dimension compared with other SI units. For example, it is not necessary to ask, "is this a mass" when measuring the mass of an object, in the way that it is critical to ask, "is this lead" before attempting to measure the amount of lead. A mole measurement thus requires two issues to be addressed, namely identity and amount. It follows therefore that traceability claims must show unbroken chains covering both of these issues. It is because of the existence of a vast number of chemical species that it is necessary to clearly specify and separate the specified chemical entities from all other possible chemical entities prior to measurement. This leads to complex chemical measurement processes, with considerable attention to validation of the measurement method being required.

It is possible to have a mole of lead atoms, a mole of the d(+)-isomer of some organic molecule, or at a trivial level, a mole of bottles of red wine. Any problems are not in the concept but in the realization of a mole of poorly defined entities. As with the definition of the mole, it is necessary to specify the entity. The number may be absolute or a ratio, as in the definition of the mole. The definition of the mole specifies a relative number and relates the number to mass. It would be possible to redefine the mole in terms of an absolute number (Avogadro's number), but the linkage to mass would still exist. The relative atomic masses together with the related issue of chemical equivalence or stoichiometry provide the macro links between the different atomic and molecular entities and allow an amount of any entity to be expressed in terms of mass. It is worth noting, in passing, that the mole is a specific case of a more basic quantity, which is number, where the unit is one.

Measurement of the mole

For other SI units, the quantity is realized by either an artefact (e.g., the kg), or a measurement process leading to a value. The measurement process has to be capable of accurately measuring the quantity and where this is achieved by a number of laboratories independently at the highest metrological level, then the consensus value plus its uncertainty is taken as the primary standard. Others may use the primary standard to transfer traceability to their measurements, using a single or multi-step series of comparisons (methods/standards). A primary standard is established using a primary method, which by definition is carried out at the highest metrological level.

The mole can be realized in a similar way but, of course, there are millions of different types of mole. It is more appropriate to speak of realizing a mole and this can be done by measuring a specified entity and making use of chemical stoichiometry and atomic mass values to relate the measured property to mass, as defined in the definition of the mole, i.e., $M_x = N_x/A_n = m/M$, where M_x = number of moles of entity X; N_x = number of entities of X; A_n = Avogadro's number; m = mass of X; M = atomic mass of X.

Hence the measurement of a mole involves the measurement of a number of entities by, for example, mass spectrometry and/or measurement of the mass of the isolated entity X. For this reason, much of the science of chemical measurement is concerned with ensuring the identity of X and the isolation of X, so that its amount can be measured.

Other vital components of the traceability chain are atomic mass values, other fundamental constants, and associated physical measurements, such as mass, volume, etc. The results can be expressed in moles, moles per mole (mole fraction), or moles per kg, or if preferred in other related SI units such as mass/mass units. Whilst conversion between units will contribute to increased uncertainty at the highest metrological level, at more practical levels such considerations will be insignificant.

Primary methods

A primary method [2] is one that is capable of operation at the highest metrological level, which can be completely described and for which a complete uncertainty statement can be produced in SI units. The amount of substance can be measured either directly, without reference to any other chemical standard, or indirectly, by use of a ratio method which relates the amount of unknown entity X to a chemical standard. Primary direct methods, such as gravimetry and certain electrochemical and thermal methods are the exceptions in chemistry, as the majority of measurements are made indirectly by comparison with other pure substance RMs as discussed above and below. These ratio methods include isotope dilution mass spectrometry and chromatographic and classical methods. Hence the importance of pure substance RMs.

A method that has the potential to be primary can also be applied in a less rigorous way to provide a direct realization of the SI unit, but not at the highest metrological level. In chemistry, gravimetric analysis, which can be carried out at varying levels of uncertainty, is such an example. That is, it can be employed as a traceable routine method or applied at the highest metrological level, for establishing the purity of RMs.

Pure substance RMs

Following synthesis or isolation, separation and purification, materials are characterized for stability and homogeneity. The crucial characteristics are the identity and purity, which may be in the range 99% (or less) to 99.9999% depending on the material. To be of value there must be a high level of certainty concerning the structure or identity of the material. This is normally established from knowledge of the synthetic route and by using a number of independent characterization methods, which provide a mixture of analytical (e.g., elemental analysis, NMR, etc.) and fingerprint data (e.g., MS, retention times, etc.). The purity may be established by one or more of the following approaches: the direct measurement of the amount of the main ingredient; measurement of all possible impurities and their deduction from 100%; other methods, such as differential scanning calorimetry, which provide a direct measure of the level of certain impurities. The traceability of the purity certification will be established by describing the traceability of the contributory measurements.

The references, measurement processes, and the uncertainties associated with the traceability chain of an organic pure substance RM are illustrated in Fig. 3. For the amount measurement to be meaningful, the uncertainty of the identity must be close to zero. Although it is not yet possible to quantify this uncertainty, the use of a number of independent sources of information (synthetic route plus three or more independent analytical methods) helps ensure that the close to zero condition is met. The more subtle the structural characterization required, the more complex the task. Hence the difficulty in unambiguously identifying and measuring specific isomers of complex organic compounds such as steroids. Some metrologists may challenge the inclusion of identity within the traceability chain, but if it is not included then it is not possible to have an unbroken chain of comparisons to the SI unit, the mole. Some methods used for the characterization of purity have the potential to be primary, such as NMR, titrimetry, DSC. It is questionable, however, whether methods such as GC and HPLC, which are widely used for establishing the purity of organic RMs are primary methods. Nevertheless, providing the purity is high any errors will be small. Although there is still some debate about how best to assign the purity value and its uncertainty when a number of independent methods are used, it is clear that consideration needs to be given to all the information, whilst giving most weight to the 'best' estimates.

Pure substance RMs are used to derive the calibration factor F_c (see Fig. 1) and its uncertainty U_c for chemical measurements. Traceability of the amount of RM in a prepared standard solution is determined by the traceability of the purity certification, the mass measurement, and the volume measurements. For many trace chemical measurements the uncertainty (U_c) of the calibration derived from the pure substance RM is a small component of the final measurement uncertainty, because other factors, such as interferences, incomplete extraction. etc. (U_r) dominate the uncertainty budget. Thus, for many trace analysis methods, reagent grade materials are adequate for use as calibration standards. An exception is with complex organic substances, where purity of commercially available materials can be low and often many similar substances exist, leading to confusion concerning identity during measurement. In many cases the impurities are structurally similar to the main component making it difficult to purify and further confusing multicomponent measurements. In these cases RMs with a high level of certainty concerning identity (and purity) are essential.

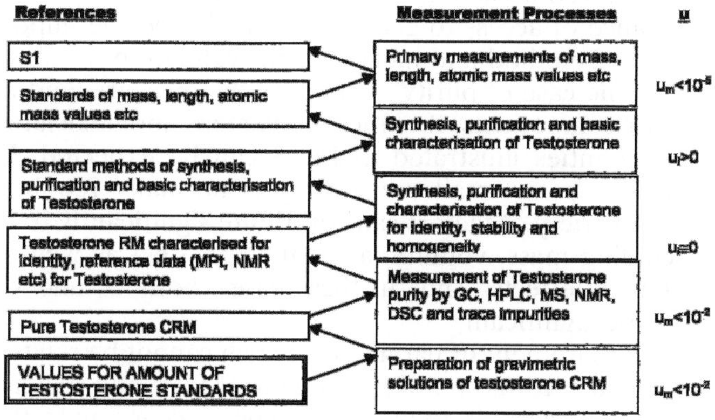

Fig. 3 Traceability of a pure substance RM of testosterone

Matrix RMs

Whilst pure substance RMs are mainly used for calibration purposes (determining F_c and U_c), matrix RMs are most often required to validate a measurement or method (determining R and U_R). To be of use matrix RMs must closely match the real sample in terms of analyte, matrix type, and concentration. In addition, the analyte must be incorporated into the matrix in the same way as in the real sample. RMs may be prepared by gravimetrically mixing the components or by characterizing the amount of the analyte(s) of interest in normal production or naturally occurring material. The former provides a more ready route to traceability, but in many cases such materials do not sufficiently closely match the real samples.

In sectors as diverse as metallurgy, food science, and environmental control, it is necessary to characterize complex materials and eliminate or control a wide range of potential interferences and incomplete recovery of the analyte during digestion, extraction, or separation. This presents a serious challenge to analytical science. The traditional strategy has been to take a pragmatic approach and to standardize the method as a way of achieving comparability. Interlaboratory precision statements are used to characterize the range of results that can be expected, but it is implicitly accepted that there are likely to be additional unidentified systematic effects and uncertainties.

Alternatively, interlaboratory consensus values based on a range of different methods are used to try to address systematic effects. Such approaches leave open the question concerning traceability, or how closely the certified value agrees with the true value. The establishment of traceability to SI requires the use of primary methods, such as isotope dilution mass spectrometry, or the use of other well understood and validated methods, where any systematic effects have been fully evaluated and corrected for. The uncertainty budget must include appropriate allowance for any suspected residual systematic effects. It is also worth remembering that if traceability to SI is not possible or appropriate, then traceability to some lesser reference should be specified and demonstrated.

Reference measurements

Although it is not common practice, another way of establishing traceability is by comparing measurements made using a primary or reference method with results obtained using a working level method. This can be done on an individual measurement basis or on a larger scale through interlaboratory studies, where the assigned value is based on a metrologically traceable value. Given the cost of such work it can be expected that such measurements will, when possible, be associated with more widely used and durable RMs.

Classification of reference materials and methods

Hierarchies, such as primary, secondary, and working level, or certified RMs and RMs are extensively used in describing traceability chains. Whilst such terms can be useful in explaining processes and links, they can also be confusing. For this reason their use has been limited in this paper. It is considered preferable to describe hierarchies in terms of the associated uncertainties. It can also be noted that, whereas in physical measurement it is common to have a hierarchy of references of the same basic type (e.g., a series of mass standards), this is rare in chemical measurement where the chain usually contains only one chemical RM, linked to a higher reference by a measurement process.

The practical realization of the traceability of routine chemical measurements

Providing the measurand can be defined in SI units, then in principle its measurement can be made traceable to SI. It would be a matter of convenience to stop the traceability chain at a lesser reference, such as a reference method. The associated uncertainty would be made relative to the stated reference, which would be assigned an uncertainty of zero. That is, relative to SI some systematic effects would be ignored. Increasingly, however, there is a drive to establish traceability to SI where feasible and to accept that this will result in a larger uncertainty.

It has been shown above how the traceability of RMs can be established. These RMs can be used to help establish the traceability of routine measurements as illustrated in Fig. 4. It will be noted that the uncertainties associated with the high level references are small compared to those associated with measurement of trace quantities of lead in a complex matrix such as blood serum. Where component uncertainties are less than one third of the combined uncertainty, then they will contribute little to that combined uncertainty. This does not mean that the higher level references are not necessary, only that they are not the major causes of difficulty. Also, although traceability of identity is always important it is not addressed in Fig. 4, as it is not a problem. This would not be the case with organic analysis, where confusion concerning identity can be a major problem, as discussed above and in Fig. 3. The major problem in trace analysis usually is the size of U_R, leading to a large combined uncertainty. The weak link in this chain is the validation of the routine method. Hence the importance of this in determining the overall quality of the measurement. This is typical of trace analysis involving complex matrices. The absence of the IDMS reference value for serum would reduce the traceability to just the routine method, with no control or knowledge of R and U_r. Estimates of R and U_R could be made by other means, for example by spiking and/or using a matrix RM. In trace analysis, U_R usually dominates the combined uncertainty. This underlines the importance of access to good matrix RMs and the importance of method and measurement validation.

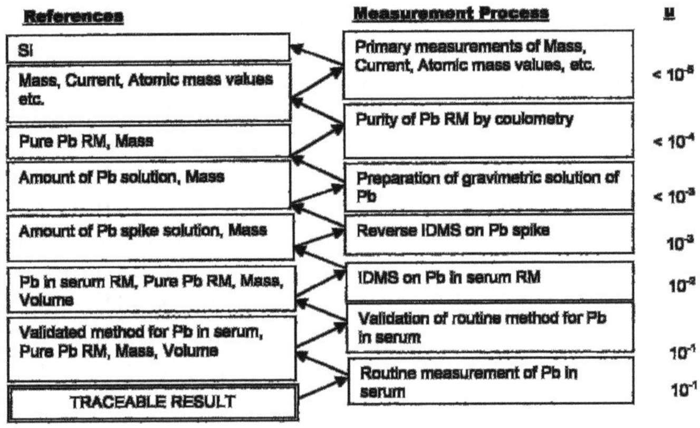

Fig. 4 Traceability of a trace of lead in blood serum measurement

In the case of purity analysis or the measurement of major components, the traceability steps and associated uncertainties illustrated in Fig. 3 can be expected. In these cases, a small uncertainty is normally required and a variety of factors can contribute significantly to it, including mass, volume, and other physical as well as chemical effects. Thus in these cases, both U_r and U_c may be significant.

Measurement of parameters which cannot be related to SI, such as fat and fiber content of food and pH can be made traceable to other references according to the same principles as discussed above.

Ensuring the equivalence of standards around the world

A number of organizations have contributed to the above developments and programs are being established at the national, regional, and international levels, to help provide the standards needed to facilitate traceability [1–9]. An important aspect of this work is the demonstration of the equivalence of the various standards used in different parts of the world.

A key organization is Comité Consultatif pour la Quantité de Matière (CCQM), which since its establishment in 1994 has made considerable progress on the agreement of definitions and the organization of interlaboratory studies to evaluate primary methods and to study the equivalence of national standards. Areas of work include gas analysis, trace element analysis, trace organic analysis, and the characterization of the purity of pure substance standards. The work is focused on the development of metrological tools and the demonstration of the feasibility of the metrological approach. CCQM aims to provide a framework for the demonstration of the equivalence of national standards through interlaboratory comparisons, known as 'Key Comparisons', which can be linked through regional and sectoral networks. It will only be possible to conduct a limited number of Key Comparisons, due to resource limitations. Each comparison will be carefully selected to cover an important measurement area and as far as possible address specific matrix, analyte, and measurement technique problems. For example, the measurement of trace levels of the pesticide metabolite pp' DDE in fish oil by IDMS is relevant to food safety and environmental concerns and represents the analysis of a complex organic material in a complex matrix, by IDMS.

It is envisaged that about 80 key comparisons will be needed to cover chemical measurements. Although it remains unclear how far 'the light will shine out' from a specific key comparison to other related areas of measurement, the demonstration of equivalence of national standards in selected areas will be of great importance for international trade. Despite the use of the term 'national standard' (which, perhaps, has more to do with history than with the future international vision of the world), it is not envisaged that every nation will have all the standards. The aim is more to do with demonstrating the equivalence of the different metrology in chemistry capabilities that are growing up around the world. Also, Key Comparisons will need to be underpinned by QA systems and accreditation to help transfer measurement traceability to the working level. The combined strategies will enhance international comparability of measurements and facilitate one stop testing.

Table 1 Examples of proposed topics for CCQM Key Comparisons

Health:	e.g., cholesterol in serum
Food:	e.g., arsenic in fish
Environment:	e.g., permanent gases in air
Advanced materials:	e.g., semiconductors
Commodities:	e.g., sulfur in fossil fuels
Forensics:	e.g., ethanol in air
Pharmaceuticals	to be decided
Biotechnology:	e.g., DNA profiling
General analytical applications:	e.g., pH

To date 3 key comparisons have been conducted, 6 are planned for 1999 and the areas listed in Table 1 are on the agenda.

Progress is illustrated by the following statistics. In 1998 CCQM organized 4 interlaboratory studies; in 1999 over 24 studies are planned. The progress made by laboratories participating in CCQM studies over time and their improved performance compared with working level laboratories is illustrated by the following performance data.

Only 2 out of 8 laboratories met the target accuracy by being within ±1% of the assigned value in a 1994 study of trace lead, compared with 9 out of 10 laboratories meeting the same target accuracy in a repeat exercise in 1997.

Also, in a 1998 trace element study involving much lower concentrations (1/1000) 9 out of 10 metrology laboratories established equivalence to within ±2.6%. This performance level can be compared with other laboratories (routine) where the range of results exceeded ±50%.

Developments at the national and regional levels have been described elsewhere [8].

Sectoral developments

Developments concerned with improving the validity, comparability, and traceability of chemical measurements are also taking place in specific sectors, such as food and agriculture, environment, clinical, pharmaceutical, forensic science and some areas of industry. For example, the clinical chemists [9] have adopted the metrological approach, expressing results in SI units and progressively developing measurement traceability at the working level. Another sector where the metrological approach is being pursued is gas analysis concerned with car exhaust pollution, environmental measurements, and drink-driving prosecutions. International trade and regulation are driving improvements in food and agriculture analysis but a different approach is often being taken in this particularly difficult sector. Some of the developments in this sector are the same or similar to the metrological approach, even if different terminology is used, but some of the champions have much less ambitious aspirations with regard to measurement accuracy and are content to establish comparability rather than traceability. Cross-fertilization between the groups will be important, in order to bring both approaches to a common focus. Failure to collaborate would result in parallel and unconnected measurement systems being developed with cost penalties and inferior measurement potential.

Conclusion

Traceability is a property of a measurement value whereby it meets certain criteria and in particular is related to stated references at a stated level of uncertainty. The nature of the stated references is open to choice on a fit for purpose basis and the level of uncertainty must be a reasonable estimate of the actual uncertainty and appropriate for the purpose. Where feasible, traceability to SI is, however, recommended as it provides stable references, unchanging in time or space. There is no requirement for traceability to imply high accuracy and for many purposes traceable measurements with a large uncertainty will be adequate. Primary and reference methods and reference materials provide the transfer standards that can help establish traceability for routine measurements. The measurement method must of course be well understood and described through a process of method validation and it must be recognized that this is often the most crucial and difficult part of establishing traceability. It is also clear that the traditional analytical chemistry strategies of calibration and validation are embraced by the metrological approach, but that the latter provides a fuller framework to describe measurement quality and provides a quality strategy that applies to all types of measurement.

In physical measurement, calibration standards are of prime importance, but in chemistry, standards such as mass standards and pure substance reference materials are necessary but not sufficient and often not the most problematic aspect of establishing traceability. As every analytical chemist knows, issues such as sampling, sample stability, contamination, interferences, and incomplete recovery of the analyte are usually the major contributors to measurement uncertainty. It is being increasingly recognized that if we wish to improve the traceability of chemical measurements, then we need to put the effort where the chemical problems are, and not where the problems are in physical measurement. It is a sign of maturity that this is now happening.

References

1. King B (1999) Metrology. In: Chemistry: Current Activities and Future Requirements in Europe, Report EUR N1907 EN, 1999, European Commission, Brussels
2. International Vocabulary of Basic and General Terms in Metrology, Second edition, 1993, ISO, Geneva
3. King B, et al (1999) Accred Qual Assur 4:3–13
4. De Bievre P, et al (1997) Accred Qual Assur 2:270–274
5. Valcarcel M, et al (1998) Metrology in Chemistry and Biology: a practical approach. EUR 18405 EN, European Commission, Brussels
6. Kaarls R, Quinn JJ (1997) Metrologia 34:7–11 [and other papers in the same issue]
7. EURACHEM Guide (1995) LGC, Teddington, UK
8. King B (1997) Analyst 122:197–204
9. Buzoianu M, Aboul-Enein HY (1998) Accred Qual Assur 3:270–277

Accred Qual Assur (2000) 5:437–440

M. Máriássy
L. Vyskočil
A. Mathiasová

Link to the SI via primary direct methods

Received: 17 March 2000
Accepted: 21 June 2000

Presented at 3rd EURACHEM Workshop "Status of Traceability in Chemical Measurement", 6–8 September 1999, Bratislava, Slovak Republic

M. Máriássy (✉) · L. Vyskočil
A. Mathiasová
Slovak Institute of Metrology, Karloveská 63, 842 55 Bratislava, Slovakia
e-mail: mariassy@smu.gov.sk
Tel.: +421-7-6029 4522
Fax: +421-7-6542 9592

Abstract The possible approaches to realising a link to the SI system and the status of primary direct methods in the traceability chain of chemical measurements are discussed. Some results obtained with the new coulometric standard system are presented.

Key words Traceability · Coulometry · Uncertainty · Primary methods

Introduction

There is general agreement that an artefact for the realisation of the unit mole as the top of the traceability chain is not needed/rational. However, there is a large variety of opinions on the nature of the link to the SI. They range from primary methods through pure elements to commercial substances. Often we hear objections that the uncertainty at this level is negligible compared to that in routine measurements, so that work at this level is unimportant. This is usually true for trace constituents, but for analysis of major or minor components standards may be a significant source of error.

Is there a difference between chemical and physical measurements? Most chemical analytical methods are relative, i.e. they compare the signal generated by a sample to the signal generated by the same quantity of a standard. In physical measurements, the standard is often incorporated into the measuring instrument, therefore the instrument may be calibrated for longer time periods. In contrast, in chemical measurements the instruments usually serve as comparators between the signals of unknown samples and external standards. In addition, chemical assay leads either to an intensive quantity or the sample (and standard) is used up.

The values of extensive quantities (in contrast to intensive quantities) depend on the system size (the amount of solute depends on the volume of the solution taken; its concentration does not).

Partial quantities are related to some part of the system; they have their opposite in integral quantities, whose value relates to the system as a whole(like length, volume, mass, voltage,...). As an example: having a mixture of two components, like sand and seeds – the mass of the mixture of both is an integral (and extensive) quantity, the masses of the individual components are partial in nature (extensive, too). Only integral extensive quantities are those that can be measured directly, partial or intensive ones are calculated from the results of other measurements.

In amount of substance measurements, we almost never face the need to determine the sum of all components. We try to determine specific substances that form a part of the whole. Amount of substance, which is of course an extensive quantity, can be considered as having partial nature – this is supported by its definition, too (entities must be stated). Any prepared standards cannot be dosed by amount of substance – for any dosing we need to measure an integral quantity like mass or volume. That means the certification of reference materials is best done in terms of concentration (mol/dm^3) or amount content (mol/kg).

The Consultative Committee for Amount of Substance (CCQM) has set up a definition of primary methods [1, 2] and has selected some methods with the potential of being "primary", from the viewpoint of the end user. From the point of view of metrology, methods used for linking the chemical measurements with the SI system at the highest level should not refer to other amount of substance standards. This requirement excludes methods which are relative in their principle. Some other methods identified as having the potential to be primary yield information expressed as amount fraction. This is essential for evaluation of purity, but in order to convert it to a value useful for transfer of the unit, additional information on the identity (molar mass) and content of the impurities is required. This additional information is needed to convert the result into amount content or similar quantities.

From practical considerations, for the area of inorganic analysis, there are two methods, the results of which are not dependent on a known amount of substance in some form of reference material (RM) (sometimes called absolute methods):
- Coulometry
- Gravimetry.

Coulometry is based on direct or indirect electrochemical transformation of the determined substance. For a complete electrochemical transformation of amount of substance n of the substance determined, we need electric charge Q quantitatively described by the Faraday law:

$$n = \frac{Q}{z \cdot F}$$

where n is amount of substance, Q is electric charge, z is charge number of the electrochemical reaction and F is the Faraday constant (96485,3415 ± 0,0039) C/mol (CODATA 1998).

Determination of the amount of substance is thus in direct relation to basic units of the SI system and does not need a RM for comparison. The Faraday constant is one of the fundamental constants (it can be expressed as the product of the electron charge and the Avogadro constant). It enables the attainment of high precision and accuracy and is independent of the atomic weights of the elements in the sample. Its drawback is lower selectivity, a feature common to titration methods. This makes coulometry especially suitable for determination of relatively pure substances used as standards by other (relative) methods. The Faraday constant has been proposed as an ultimate standard in chemistry [3].

Gravimetric analysis is one of classical analytical methods. It is based on chemical transformation of the sample using excess of a reagent to a substance, which is weighed after processing. The weight of the substance obtained serves as a base for calculation of amount of substance.

The advantage of the method is its feasibility with common laboratory equipment. The disadvantage lies in lower selectivity and in the integral character of mass measurements, i.e. they determine a property of the object as a whole. Moreover, for attaining the highest accuracy needed at this level, the need of determination of actual atomic weights cannot be overlooked in some instances.

Strictly speaking gravimetry belongs to the relative methods. The only difference is in the position of the RM in the measurement process: the weighed product itself serves as an RM – usually with the a priori assumption, that its purity is 100% and stoichiometry is correct. The mass fraction of the weighed substance should be taken into account in the equation used. Problems associated with the formation of solid phase [4], e.g. surface adsorption effects (ionic species and water [5]) are significant in analyses aiming at relative uncertainties at about 10^{-4}.

Another possible approach, which is broadly used, is to use high-purity substances (indirectly assayed) as standards. For use at the highest level, this approach requires the determination of *all important impurities* in the sample. This means not only metallic impurities, commonly stated in the manufacturers certificates, but also non-metals as oxygen, carbon, etc. The content of impurities is not always known in advance. If the total content of impurities is very low, the uncertainty of their determination does not affect the required uncertainty of the sample assay. Some other problems are discussed in Ref.[5]. The need of determining the molar weight may equally apply here.

A schematic view of the traceability and uncertainties in inorganic analysis is depicted in Fig. 1.

The Slovak Institute of Metrology (SMU) has recently rebuilt its high-precision coulometric equipment to be used as a national standard for amount of substance measurements. Its main purpose is certification of primary reference materials of composition with directly determined main component. These primary reference materials can be subsequently used for disseminating traceability into different chemical measurements.

Results and discussion

The new, computer-controlled equipment was constructed based on past experience with coulometry gained at SMU. Some parts were constructed specially

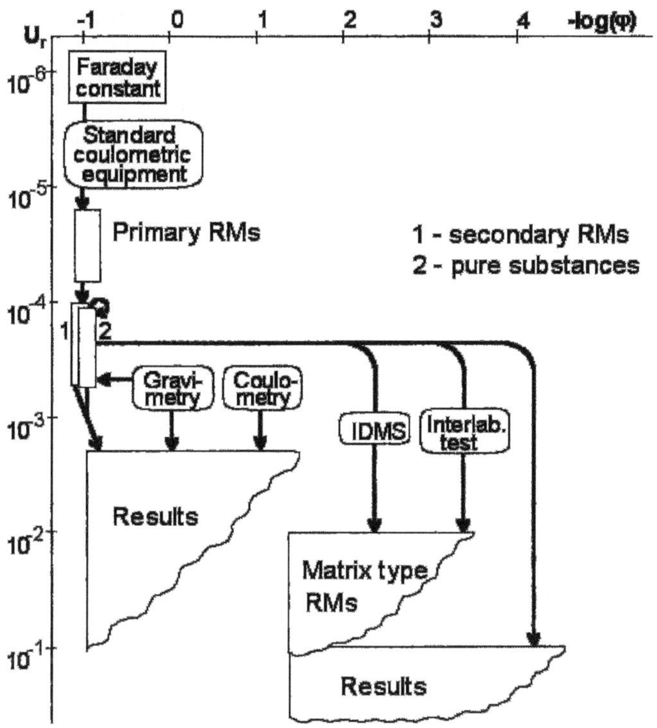

Fig. 1 A schematic view of the traceability and uncertainties in inorganic analysis (φ–amount content)

for this purpose. The control program was written in Turbo Pascal.

The entire constant current coulometric system consists of several instruments, completed by auxiliary equipment. The main parts of the coulometric system are given in the following list:

- Current source 7961 (I< 1 A) (Applied Precision)
- Indication unit 8971 (potentiometric, amperometric) (Applied Precision)
- Relay/valve unit (Applied Precision)
- Digital voltmeter Solartron 7071 (Schlumberger)
- Microbalance S4 (Sartorius)
- Analytical balance AE240S (Mettler)
- Standard weights
- Standard 1 Ω resistors (Metra)
- Piston burette 665 Dosimat (Metrohm)
- Coulometric cells
- PC Pentium, 266 MHz
- Auxiliary equipment

The main metrological parameters of the new system are as follows:

- Typical RSD · 0.002 %
- Typical Type A uncertainty (for $n=10$) 0.0007%
- Typical Type B uncertainty 0.002–0.006%
- Combined uncertainty 0.002–0.006%

Table 1 Selected assays made on the new system

Material	Analyte	Amount content φ/mol·kg^{-1}	$\varphi_{exp}/\varphi_{theor}$/%	RSD/%	u_C/%	Remark
$K_2Cr_2O_7$	Cr^{VI}	6.79732	99.9832	0.0010	0.0019	RM
		6.79694	99.9777	0.0024	0.0022	CCQM-P7
HSO_3NH_2	H^+	10.29878	99.9960	0.0032	0.0019	RM
		10.29652	99.9740	0.0015	0.0019	COOMET
KCl	Cl^-	13.4113	99.9722	0.0033	0.0062	CCQM-P7
NaCl	Cl-	17.1061	99.9829	0.0029	0.0066	CCQM-P7
		0.0250224	–	0.032	0.021	Solution

Table 2 Uncertainty budget for 0.5 g samples (relative contributions in 10^{-6})

	Amidosulfuric acid	Potassium Hydrogen Phthalate	Potassium Dichromate	Potassium Chloride	Arsenic Oxide
Incomplete rinsing	2	4	2	2	2
Spray losses	2	2	0	0	0
Sample introduction	0	0	0	0	5
Current efficiency	1	1	1	30	1
Electrolyte impurities	10	10	10	10	1
Inert gas impurities	5	15	5	5	2
Diffusion	1	10	10	5	10
Adsorption	–	–	–	50	–
Total:	**11.6**	**21.1**	**15.2**	**59.6**	**11.6**
Weighing	6.5	6.5	0.5	6.5	6.5
Air buoyancy correction	0.8	1.3	8.5	0.9	0.5
Voltage measurement	9.8	9.8	3.0	9.8	8.8
Electric resistance	3.0	3.0	3.0	3.0	3.0
Uncertainty of mass and charge:	**12.2**	**12.2**	**11.1**	**12.2**	**11.4**
Total Type B uncertainty	**17**	**24**	**19**	**61**	**16**

- Expected change per year (drift of components) –0.0015%
- Expected change per year (after drift correction in program) –0.0002%

Results of some measurements made on the new coulometric system from November 1998 are given in Table 1.

The main problem in evaluating the uncertainty of measurements in coulometry lies in identification of important uncertainty sources and estimation of their contribution (Table 2). With very low instrumental uncertainty, other factors become limiting to the achievable uncertainty, mainly those connected to the chemical processes in the cell and the homogeneity of the material.

The situation is seldom favourable enough to enable the use of a single method for establishing a link to the SI system. The information on the content of impurities (by means of relative methods) is needed in most cases. Except for coulometry, determination of molar weight is of importance for highest level work. Thus seeking for a "method that stands alone" seems to be over optimistic. In order to link an amount of substance measurement of a given species to the SI system at the highest level, the choice of a particular method will depend on the nature of the species and possibilities of the methods under consideration.

The "ultimate" link to the SI system is currently possible in some cases only by using pure substances (e.g. elements) and their corresponding atomic weights; however, a more straightforward way is to use direct methods like coulometry, which eliminate the problems associated with the dissolution step. Their use can fulfil both the role of a standard (they incorporate the unit mole into the measurements) and the role of measurement capability.

Acknowledgements This work has been supported by the U.S. – Slovak Science and Technology Joint Fund, Grant N° 94 086.

References

1. CCQM (1995) Report of the 1st meeting. BIPM, Sèvres
2. CCQM (1999) Report of the 4th meeting (February 1998). BIPM, Sèvres
3. IUPAC, Commission on Electroanalytical Chemistry (1976) Pure Appl Chem 45:127-130
4. Beck II CM, Watters RL (1996) Document CCQM/96-13
5. Yoshimori T (1975) Talanta 22:827-835
6. Moody JR, Greenberg RR, Pratt KW, Rains TR (1988) Anal Chem 60:1203A-1218A

Accured Qual Assur (2000) 5:441–445

Adolf Zschunke

The role of reference materials

Received: 17 March 2000
Accepted: 15 August 2000

Presented at 3rd EURACHEM Workshop "Status of Traceability in Chemical Measurement", 6–8 September 1999, Bratislava, Slovak Republic

A. Zschunke
Bundesanstalt für Materialforschung und -prüfung (BAM),
Richard-Willstätter-Strasse 11,
12489 Berlin, Germany
e-mail: adolf.zschunke@bam.de
Tel.: +49-30-81 04 58 00
Fax: +49-30-81 04 59 72

Abstract In this article the role of reference materials is confined to chemical measurements only. Recognized reference materials are one of the tools to obtain comparability of analytical results. Recognition demands confidence in the reference materials and in the reference material producers. A reference material producer is a technical competent body that is fully responsible for the certified or other property values of the reference material. The "analyte" has to be specified in relation to the selectivity of analytical procedure. The full range of reference materials can be presented as a three-dimensional space of the coordinates: analyte, matrix and application. If reference materials are used for calibration or correction of calibrations they establish the traceability of results of chemical measurements. The traceability is only valid within a stated range of uncertainty. Pure substances can represent the unit of amount of substance. A precondition is the microscale specification of the analyte and the accurate determination of the main component and/or the impurities.

Key words Reference materials · Traceability · Chemical identification · Amount of substance

Introduction

A role is an acting part in a play which consists of teamwork with other actors. Similarly, the role of reference materials in chemical measurements is important but should be described in context with uncertainty, traceability, comparability.

Chemical measurements are a special part of the scientific discipline "Analytical Chemistry". They are also applied in many other testing fields and other scientific disciplines, such as biology, physics and medicine. The results of chemical measurements become more and more important for decisions in economy, trade, science, medical care, environmental protection, consumer protection, sports, jurisdiction and politics. Comparability is needed on different levels, beginning at the laboratory and ending in a global exchange of analytical results.

Global comparability of analytical results

More and more decisions based on chemical measurement are having global effects. Reference materials are important tools to obtain global comparability of results of chemical measurement. However, the use of validated analytical methods and the proof of personal skills by proficiency testing are other tools of the same rank (Fig. 1) [1].

A prerequisite for global comparability is the mutual recognition of reference materials provided in different countries. Reference materials have to fulfill certain requirements to become accepted (Fig. 2).

Certification of reference materials according to the requirements of ISO Guides 34 and 35 [2] or Bureau Communautaire de Reference (BCR) guidelines [3] is an important mean to establish international acceptance. The certificate has to prove the traceability as an

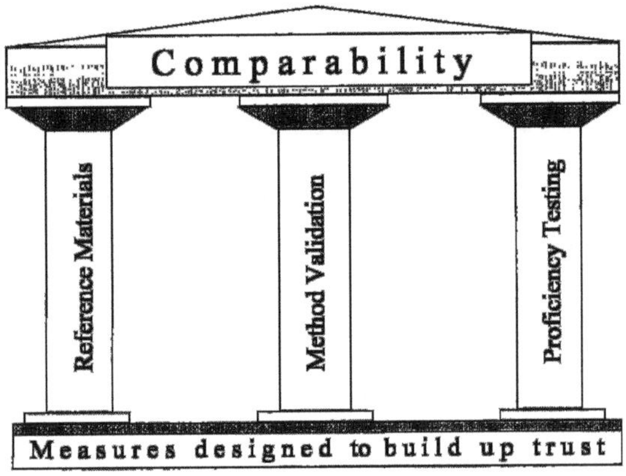

Fig. 1 Components of comparability

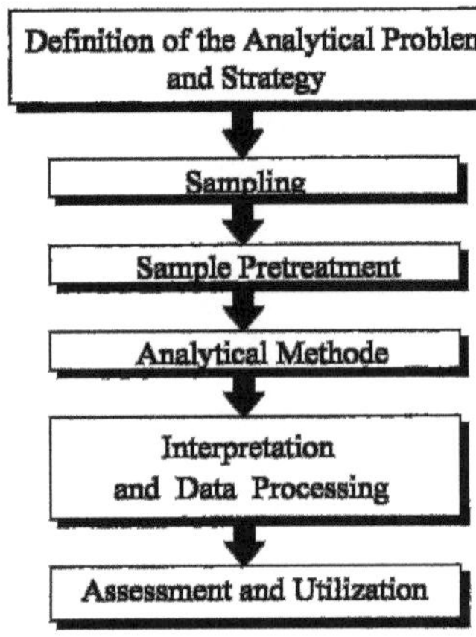

Fig. 3 Steps of analytical procedure

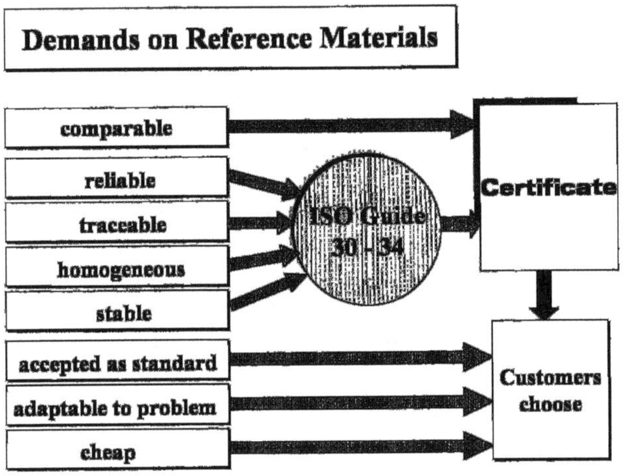

Fig. 2 Requirements of reference materials

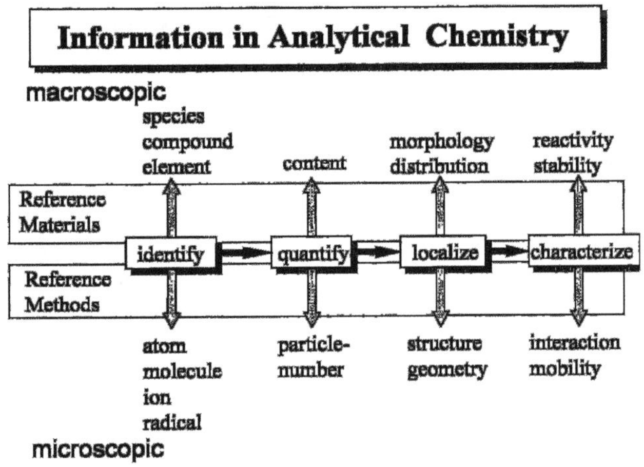

Fig. 4 Information gain in analytical chemistry

additional authorization. Another aspect of international recognition is the creation of confidence in reference materials by third-party assessment of the producer (bodies who are responsible for preparation, homogeneity and stability assessment, testing, assignment of property values and their uncertainties, packing, labelling and distribution of a reference material). The International Laboratory Accreditation Co-operation (ILAC) [4] is preparing a worldwide system of third-party assessment of reference materials producers in co-operation with ISO-REMCO, EURACHEM, EUROLAB and EA.

Range of reference materials

The term chemical measurement emphasizes the measuring aspect of chemical analysis. However, chemical measurements are always embedded in an analytical step by step procedure, which also has to consider many chemical problems (reactivity, chemical equilibrium, etc.) (Fig 3).

Reference materials often allow the assessment of the whole analytical procedure. Analytical chemistry can be defined as a scientific discipline which develops and applies methods, instruments and strategies to obtain information on the composition and nature of matter in space and time [5]. During the course of this information gain special entities are determined on a macroscopic or microscopic scale.

The term chemical measurement can cover all these determinations, including identification [6]. Identification defines the so called "analyte" by means of chemical, electrochemical, spectroscopical and other physical properties. In most cases identification is done by measurements. Identification is valid only in a reference system. The terms describing the analytical problem (see Fig. 4), the measuring system used, the reference methods and the reference materials, belong together as the reference system.

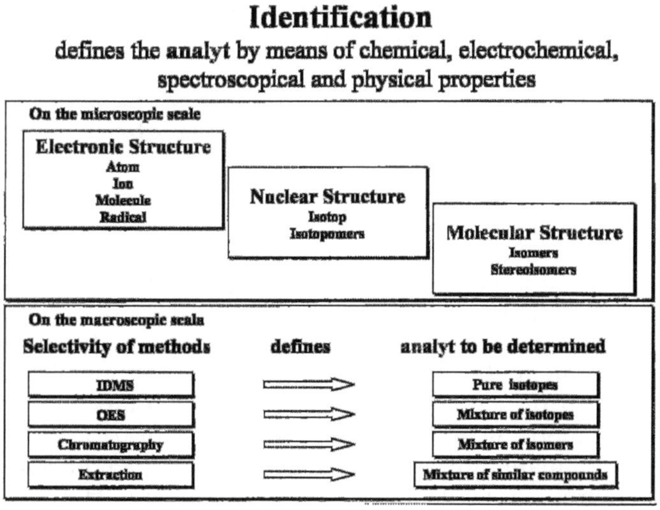

Fig. 5 Identification on the microscopic and macroscopic scale

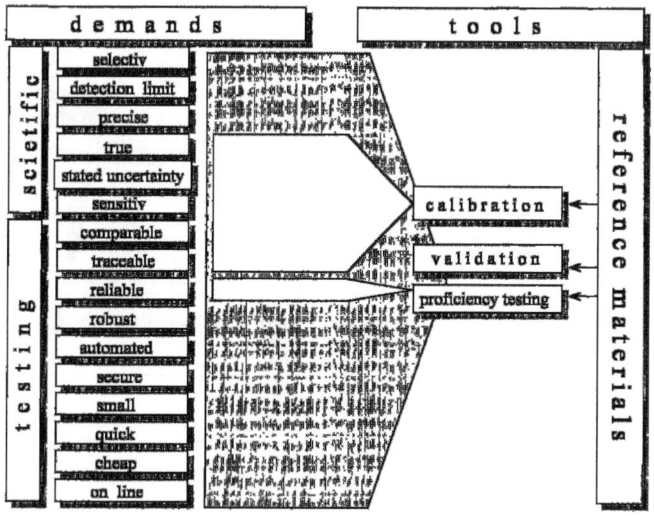

Fig. 6 Demand of customers on analytical chemistry

On the microscopic scale, the identity can be defined by considering the electronic, nuclear and molecular structures (Fig. 5).

On the macroscopic scale, in most cases, only groups of identities can be identified (isotope mixtures, isotopomer mixtures, stereoisomer mixtures). The definition of the "analyte" depends on the selectivity of the analytical method including sample pretreatment, e.g. extraction (see Fig. 3). In a complex composition sometimes only classes of compounds (e.g. fat, polycyclic aromatic hydrocarbons) are identified. In these cases the analyte is designated as a sum parameter.

All quantities and properties on the macroscopic scale of analytical chemistry can be represented by standards. In many cases the standards are divisible without changing the properties. Then they are called reference materials. Reference materials can be:

- Pure chemical substances
- Blends or synthetic mixtures
- Simulates or artifacts
- Spiked and unspiked real life samples

The range of reference materials covers a three-dimensional space of co-ordinates:

- Analytes
- Matrices
- Applications

Systems of classification very often follow the application fields, e.g. the catalogues of the Institute for Reference Materials and Measurements (IRMM), the National Institute for Standards and Technology (NIST), Laboratory of the Government Chemist (LGC), etc. or the database for certified reference materials COMAR. In all application fields like food and agriculture, environment, health and safety, industry and services, etc., reference materials are used for:

- Calibration of measuring systems
- Assessment of analytical procedures
- Performance test of instruments
- Definition of measurement scales
- Interlaboratory comparisons
- Qualitative analysis.

The selection of a suitable reference material is determined by the following viewpoints:

- The analytical task, considering the reference systems
- The definition of "analytes", considering the analytical procedure
- The intended use.

The demands on analytical chemistry differ to some extend depending on whether they come from purely scientific sources or arise in the field of testing. In the latter case the customer defines the demand. Fitness for purpose determines the quality of the reference material. The gathered demands (see Fig. 6) have partly overlapping meanings and are not independent of each other. Other reference materials have to meet all demands.

Reference materials as the base of traceability

If reference materials are used for calibration or correction of calibration they establish traceability of chemical measurements. Traceability is the link or the "vertical comparison" between an analytical result and a national or international accepted standard, preferable a realization of the SI unit (see Fig. 7).

The certificate of a reference material is only proof of traceability if the certified values are accompanied by uncertainties. The certified reference material

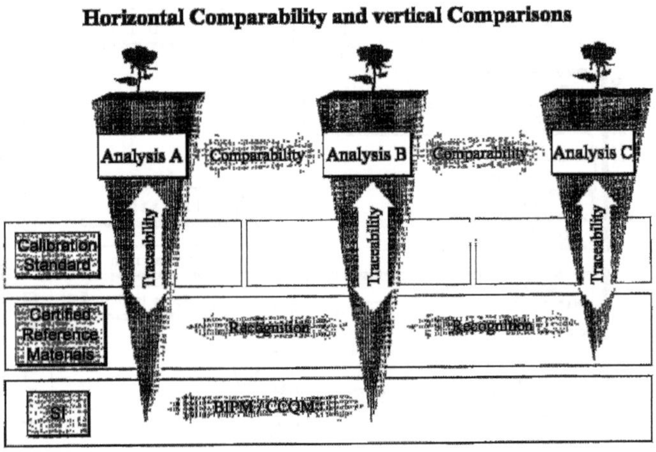

Fig. 7 Horizontal comparability and vertical comparisons

should be the end-point of the traceability chain for the user.

The producer (certifier) of certified reference materials is responsible for the link to the SI or, if this is not possible, for mutual recognition of the reference materials within the respective sphere of validity. For global comparability a global recognition is needed.

Traceability has an additional confidence building aspect: traceability is the proof of trueness and the proof of the reliability of an analytical result. Traceability is only valid in connection with an uncertainty range. Every reference material has to be traceable to a stated reference independent of the intended use. But use defines the needed uncertainty range.

Representation of the unit amount of substance

If the reference materials are pure substances and can be specified on the microscopic level, then they represent the unit amount of substance. Because there are no absolute pure substances the representation is in all cases an approximation. The degree of approximation is given by the accuracy of the contents of the main component. In case of pure elements, e.g. metals Fe, Cu, Zn the determination of the main component by coulometry is limited by an uncertainty of 0.01%. The determination of all impurities needs completeness and requires a great deal of analytical equipment. However, a combination of inductively coupled plasma-mass spectrometry (ICP-MS), atomic absorption spectrometry (AAS) and isotope dilution mass spectrometry (ID-MS) covering all elements of the periodic table allows a decrease of total uncertainty to 0.0032% (Cu, see Fig. 8).

Many elements are specified on the isotopic level because the natural isotopic ratio is well known. The best characterized pure element is the best approximation of the unit.

In the case of pure compounds, the specification can become a problem when isotopomers, stereoisomers and other species are possible. Because of its structural

Fig. 8 Multielement characterization of m4 N copper

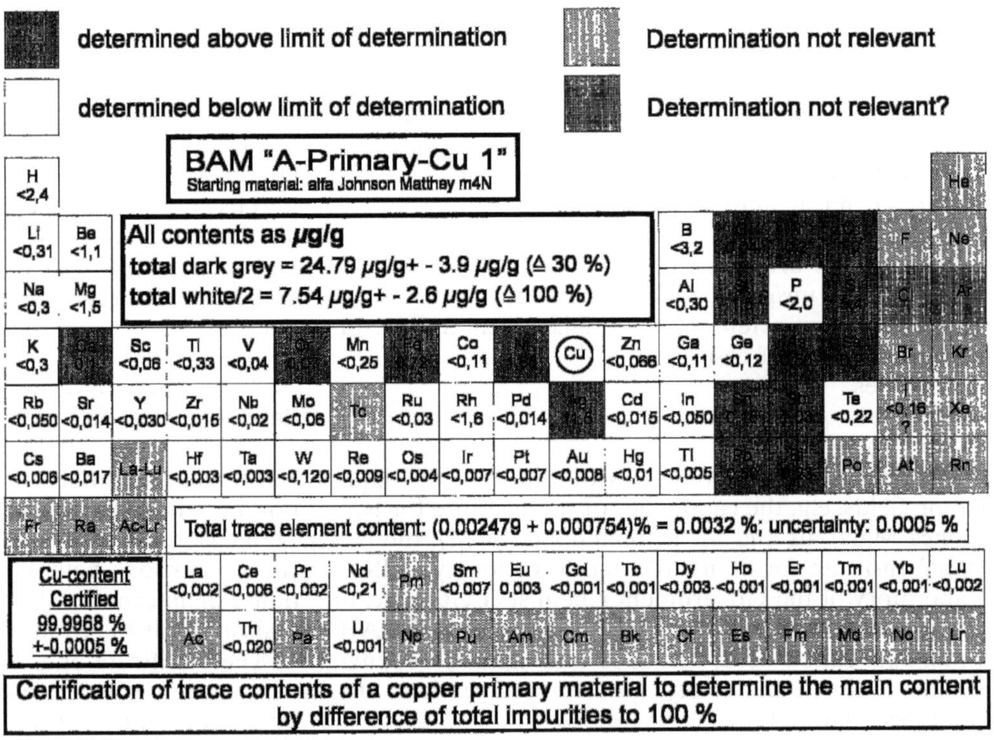

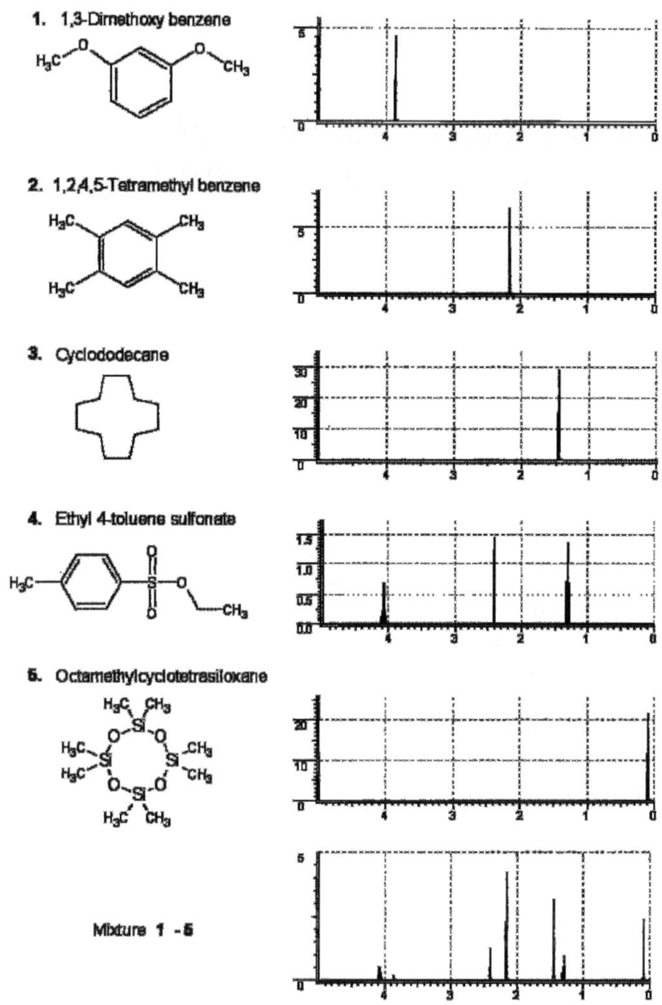

Fig. 9 Composition of organic compounds for a quantitative nuclear magnetic resonance comparison experiment

and isotopic selectivity nuclear magnetic resonance (NMR) spectroscopy is the most promising method to characterize representations of amount of substance. The high isotopic and structural selectivity of NMR spectroscopy is supplemented by the primary character of the quantification.

In the NMR spectrum integrated signals are exactly proportional to the number of contributing nuclei. The Comité Consultatif pour la Quantité de Matière (CCQM) has started international comparison of quantitative NMR experiments. In the first round the possible reproducibility should be established. The composition of a mixture of organic compounds has been determined by integration of the NMR signals. Already the first experiments (Fig. 9) have shown the problems arising by isomerization (ethyl-4-toluene sulphonate), decomposition (1,3-dimethoxybenzene), purity of standard compound and superimposition of isotopic satellites. Additional experiments with a new composition are necessary.

References

1. Zschunke A (1998) Accred Qual Assur 3:393 – 397
2. ISO Guide 34 (1996) Quality system guidelines for the production of reference materials; ISO Guide 35 (1989) Certification of reference materials – general and statistical principles. ISO, Geneva
3. European Commission Doc BCR /48/ 93 (1993) Guidelines for the production and certification of BCR reference materials. European Commission, Brussels
4. ILAC-G12:2000 (2000) Guidelines for the requirements for the competence of reference material producers. International Laboratory Accreditation Cooperation (ILAC see www.ilac.org)
5. IUPAC, Analytical Chemistry Division (1999) Division's aims; http://www.iupac.org/divisions/V/index.html
6. Karpov YA (1999) Accred Qual Assur 4:174 – 176
7. De Bièvre P, Taylor PDP (1997) Metrologia 34:67–75

Accred Qual Assur (2001) 6:100–102

Brian Belanger

The measurement assurance concept in calibration and traceability at NBS/NIST

Received: 23 October 2000
Accepted: 2 November 2000

B. Belanger
Executive Director,
Visiting Committee on Advanced Technology (Retired)
National Institute of Standards and Technology
Gaithersburg, MD 20899–1004, USA
e-mail: bcbelanger@aol.com
Tel.: +1–301–258 0708
Fax: +1–301–947 3338)

Abstract During the last quarter of the twentieth century, The United States National Bureau of Standards (NBS), later the National Institute of Standards and Technology (NIST), introduced a measurement quality control concept called "measurement assurance," and developed measurement assurance programs, or MAPs, for high-level calibration processes. The measurement assurance approach has, over time, become increasingly popular in the metrology community, and in recent years has become well accepted both inside and, to some extent, outside the United States as a rigorous way to ensure the quality of calibrations. The concept has also found application in defining traceability to national standards. This paper traces the history of the measurement assurance concept.

Keywords Traceability · Measurement assurance · Calibration · Statistics

The National Institute of Standards and Technology (NIST) and its predecessor, the National Bureau of Standards(NBS), prides itself on having the ability to make measurements characterized by state-of-the-art precision and accuracy. In industrial settings, measurements usually must be made quickly and cheaply. While it is important to find ways to reduce the cost of measurement and to make measurements more quickly, NIST's national standards mission is such that quality generally takes precedence over speed. NIST scientists and engineers take sufficient time to ensure that a valid statement of uncertainty accompanies data we provide.

In the first half of the twentieth century, the terms precision and accuracy were used, but there was a lack of agreement within the United States calibration community as to their precise meanings, and widespread naiveté regarding how to achieve adequate precision and accuracy in critical instrument calibrations. Measurements with insufficient accuracy for the application can result in erroneous decisions, unnecessary costs, and sometimes, serious health or safety consequences. For example, a radiation therapy dosage that is too high can harm the patient and one that is too low may fail to cure the disease. For that application and many others, the importance of achieving adequate measurement quality is readily apparent. By virtue of its mission, NBS has been sensitive to the need for measurement quality and strives to help others achieve adequate measurement quality.

Quality control experts, such as Shewhart and Deming, decades ago called attention to the importance of minimizing measurement errors in achieving quality control in manufacturing and the importance of statistical monitoring of process data. But tying these concepts to equipment calibration in a statistically rigorous way did not occur until the second half of the twentieth century.

In 1961 R. B. Murphy wrote about the importance of maintaining a measurement process in a state of control [1]. Then in 1963, the Chief of the NBS Statistical Engineering Laboratory, Churchill Eisenhart, wrote a seminal paper titled "Realistic Evaluation of the Precision and Accuracy of Instrument Calibration Systems" [2]. The underlying philosophy can be summarized in the following statement: "a measurement operation must have attained what is known in industrial quality control language as a state of statistical control...before it can be regarded in any logical sense as measuring anything at all." This observation was readily accepted at NBS, and no doubt by at least a few in other national measurement laboratories who appreciated the importance of quantified measurement uncertainty. But typically the people who performed routine calibrations in most industrial and government laboratories were not familiar with the need to establish and monitor the state of statistical control of a calibration process.

Eisenhart's classic paper was followed by additional publications by Eisenhart and his NBS colleagues. W. J. Youden contributed important ideas related to conducting interlaboratory testing and quantifying bias between laboratories [3, 4]. Another NBS statistician, Joe Cameron, developed calibration designs for using check standards to control NBS' calibration processes. Ultimately, the body of work by these and others at NBS led to a more systematic approach to ensuring the quality of instrument calibrations, which NBS called "measurement assurance programs or MAPs" [5–7]. A key feature of a MAP is producing data on a continuing basis to quantify and monitor the measurement uncertainty. When such data are available, one has the ability to prove that the uncertainty is sufficiently small to meet the requirements of the measurements or calibrations.

After first employing MAP techniques for its own calibrations, NBS launched efforts to teach the concept to others. Detailed descriptions of how to implement the MAP concept for particular kinds of calibrations were published during the late 1970s and the 1980s by Croarkin, Varner, and others. References [8] and [9] provide two examples of MAP techniques for dimensional calibrations. References [10] and [11] (taken together) describe MAP techniques for maintaining dc voltage.

Within a relatively short period, measurement assurance became NBS' preferred technique for controlling its calibration processes. And, once NBS published information on the MAP approach, top-echelon industrial and government calibration laboratories in the United States also began to use the technique to ensure the quality of their calibrations.

Beginning in the 1960s, a number of United States government agencies required their contractors and other organizations subject to government regulation to calibrate critical measurement equipment on a regular basis and to certify that such equipment was "traceable to NBS." The Department of Defense had the most extensive calibration requirements, but agencies such as the National Aeronautics and Space Administration (NASA), the Food and Drug Administration (FDA), and the Nuclear Regulatory Commission (NRC) also imposed calibration traceability requirements. Unfortunately, for those who had to comply, the agencies requiring traceability to NBS were often remarkably vague about specifying acceptable compliance procedures. During the 1960s and 1970s companies frequently contacted NBS to inquire, "What must I do to establish traceability to NBS?" (The question was frequently phrased, "What is the MINIMUM I must do to comply?) NBS' metrologists believed that to establish traceability to national standards for an instrument calibration process in a meaningful way, one must monitor precision (the type A random error of the process) on an ongoing basis, quantify bounds to the possible offsets relative to national standards (type B errors), and combine estimates of these two kinds of errors into an overall uncertainly statement that would stand up to scrutiny. But prior to the 1970s, few people responsible for calibration programs in industry and government agencies actually did so.

Recognition began to grow that the prevailing understanding and implementation of traceability prior to 1980 was inadequate. The following story was often repeated at NBS. (It is not clear whether this incident actually occurred or is just folklore. In either case, it illustrates the concern.)

A company sent a set of gage blocks to NBS at regular intervals to be calibrated. NBS dutifully calibrated the blocks and provided a calibration certificate and data report. The company kept their current NBS calibration certificate on file to prove to their auditors that they were "traceable to NBS" for dimensional measurements. The problem was that each time the blocks returned to NBS, our people found the seal unbroken. The gage blocks had never been used, yet the company satisfied the auditors that they were "traceable to NBS" as required.

Certainly any organization that requires traceability to national standards ought to focus on whether the measurements made by the organization subject to the traceability requirement are sufficiently accurate for their intended purpose and not simply on whether NBS calibration certificate(s) is on file. Without a valid uncertainty statement and evidence that the measurement process remains in a state of statistical control, no one can determine whether a given measurement or calibration is adequate for its purpose.

In a 1980 paper titled "Traceability: an Evolving Concept," [12] Brian Belanger argued that for traceability to be meaningful, it needs to be coupled with the concept of measurement assurance. In other words, unless a valid uncertainty statement (encompassing estimates of both random error and possible offsets relative to national standards) accompanies a given measurement, it should not be considered "traceable to national standards."

With continuing publicity from NBS about the measurement assurance concept, these ideas gradually caught on. At about the same time that the paper referenced above appeared, the American Society for Quality Control (ASQC) formed a standards writing group to develop standards for calibrations and measurements. ANSI/ASQC Standards M1 and Q4 resulted, which embraced the measurement assurance philosophy.

During the 1980s, presentations on measurement assurance programs were given frequently at major United States metrology conferences, such as the National Conference of Standards Laboratories annual conference and the Measurement Science Conference. From time to time NBS offered training courses at various locations around the country on the MAP approach to calibrations. This wide exposure helped to gain acceptance and understanding of the measurement assurance concept in the United States By 1990, most top-notch calibration laboratories in the United States were using measurement assurance programs in some form to control the quality of their critical calibration processes.

The fact that the uncertainty of a calibration process may change with time (may improve or may deteriorate with time) is an essential concept in measurement assurance. This was mentioned in Belanger's 1980 paper and strongly reinforced in a recent (1998) discussion of this topic by Ehrlich and Rasberry [13]. Because of this fact, traceability is "a moving target," and metrologists today generally appreciate that continuing scrutiny of a calibration or measurement process is needed.

The 1993 edition of the International Vocabulary of Basic and General Terms in Metrology (VIM) defines traceability as a property of a measurement, not a process, and that traceability should involve quantified uncertainties. This definition is consistent with the measurement assurance approach and shows that the concept of quantified uncertainty as a requirement for "good" traceability is becoming accepted globally.

So, today the measurement assurance approach to calibration is widely used in the United States and in many other places around the world. Increasingly, people appreciate the fact that traceability to national standards is not rigorous unless measurement uncertainty is quantified. While other laboratories and individual metrologists outside NBS also advanced the concept, NBS/NIST is proud of its contributions to putting measurement assurance on a solid technical foundation, and to gaining widespread acceptance for it in the international metrology community.

References

1. Murphy RB (1961) On the meaning of precision and accuracy. Materials Research and Standards, ASTM, Philadelphia, Pa., pp 264–267
2. Eisenhart C (1963) J Res Nat Bur Stand, Sect C: Engineering and Instrumentation 67:161–187
3. Youden WJ (1962) "The collaborative test," presented at the Referees' Seventy-sixth Annual Meeting of the Association of Official Agricultural Chemists, Washington, D.C., 16.October
4. Youden WJ (1961) How to evaluate accuracy, Materials research and standards. ASTM, Philadelphia, Pa. pp 268–271
5. Cameron JM (1977) Measurement assurance, NBS Internal Report, pp 77–1240
6. Belanger BC (1983) Measurement assurance programs Part I: General introduction. NBS Special Publication 676-I
7. Croarkin C (1983) Measurement assurance programs Part II: Development and implementation. NBS Special Publication 676-II
8. Croarkin MC, Beers JS, Tucker C (1979) Measurement assurance for gage blocks. NBS Monograph 103
9. Croarkin MC, Varner RN (1982) Measurement assurance for dimensional measurements on integrated circuit photomasks. NBS Technical Note 1164
10. Field BF, Finnegan TF, Toots J (1973) Metrologia 9:155–166
11. Eicke WG, Auxier LM (1974) IEEE Trans Instrum Meas 23, No. 4, December
12. Belanger BC (1980) ASTM Stand News 8:22
13. Ehrlich CD, Rasberry SD (1998) J Res NatInst Stand Tech 103:93

Accred Qual Assur (2001) 6:107–110

Ilya Kuselman
Boris Anisimov
Irena Goldfeld

Lifetime of the traceability chain in chemical measurement

Received: 5 October 2000
Accepted: 3 December 2000

Presented at the EUROANALYSIS XI, 3–9 September 2000, Lisbon, Portugal

I. Kuselman (✉) · B. Anisimov
I. Goldfeld
The National Physical Laboratory of Israel (INPL), Givat Ram, Jerusalem 91904, Israel
e-mail: kuselman@netvision.net.il
Tel.: +972–2–6536534
Fax: +972–2–6520797

Abstract Since the uncertainty of each link in the traceability chain (measuring analytical instrument, reference material or other measurement standard) changes over the course of time, the chain lifetime is limited. The lifetime in chemical analysis is dependent on the calibration intervals of the measuring equipment and the shelf-life of the certified reference materials (CRMs) used for the calibration of the equipment. It is shown that the ordinary least squares technique, used for treatment of the calibration data, is correct only when uncertainties in the certified values of the measurement standards or CRMs are negligible. If these uncertainties increase (for example, close to the end of the calibration interval or shelf-life), they are able to influence significantly the calibration and measurement results. In such cases regression analysis of the calibration data should take into account that not only the response values are subjects to errors, but also the certified values. As an end-point criterion of the traceability chain destruction, the requirement that the uncertainty of a measurement standard should be a source of less then one-third of the uncertainty in the measurement result is applicable. An example from analytical practice based on the data of interlaboratory comparisons of ethanol determination in beer is discussed.

Keywords Traceability · Uncertainty · Chemical analysis · Regression analysis · Ethanol · Beer

Introduction

According to the definition [1] the *traceability chain* is the unbroken chain of comparisons or calibrations from the result of a measurement or the value of a measurement standard to the national or international standards, all having stated uncertainties. The uncertainty of each link in this chain (measuring analytical instrument, reference material or other measurement standard) changes over the course of time. Therefore, the calibration intervals of measuring equipment used in testing (analytical) laboratories [2, 3] and of measurement standards used for their calibration are very important. The same applies to the shelf-life of a certified reference material (CRM) as a measurement standard. So, taking into account these intervals, one can say that *the traceability chain on the whole is stable only for a limited time.*

If the dates of the last calibration and dependence of the uncertainty changes versus time are known for each link of the traceability chain, the stability (i.e. lifetime) of the taceability chain is predictable. As an end-point criterion of the chain destruction, the requirement that the uncertainty of a measurement standard should be a source of less than one-third of the uncertainty in the measurement result is applicable. After the chain destruction, a new one, planned in advance (for example, with a new reference material) should be used.

An analysis of the traceability chain lifetime in the field of chemical measurement is discussed in the present paper.

Statistical aspect

For calibration of measuring equipment by measurement standards or CRMs, a straight-line response function is usually postulated

$$Y = \beta_0 + \beta_1 X + \varepsilon, \tag{1}$$

where Y is a response (for example, a peak area in chromatography); X is a predictor or independent variable (for example, a certified value of the CRM – a concentration of the analyte); β_0 and β_1 are the intercept and the slope; ε represents the "error", i.e. deviation of Y from the straight line. The ordinary least squares technique (regression analysis) is used for β_0 and β_1 estimation from the calibration data (Y_i, X_i). Unfortunately, many of us forget that regression analysis assumes Y as being subject to error and not X [4,5]. However, without doubt, X is also subject to error in analytical practice. Moreover, as mentioned above, uncertainty in X of a standard increases during its life. The question is, when is the uncertainty negligible, and when is the lifetime of the standard at an end and therefore it should be substituted by a new one?

Assuming that X is also subject to error, lets write η_i for the true value of Y_i, and ξ_i for the true value of X_i, i=1,2,..., n:

$$Y_i=\eta_i+\varepsilon_i, \tag{2}$$

$$X_i=\xi_i+\delta_i, \tag{3}$$

where ε_i and δ_i are random errors of η_i and ξ_i, respectively; ε_i is independent of $\xi_i + \delta_i$; and δ_i is independent of η_i and ε_i. The postulated model in this case is

$$\eta_i=\beta_0+\beta_1\xi_i, \tag{4}$$

or in other words (combining Eqs. (2–4) in the following)

$$Y_i=\beta_0+\beta_1 X_i+\varepsilon_i^*, \tag{5}$$

where

$$\varepsilon_i^*=(\varepsilon_i-\beta_1\delta_i) \tag{6}$$

Lets assume that errors ε_i and δ_i are independent values which have normal distributions with mean values equal to zero, and variances σ_y^2 and σ_x^2, correspondingly. In this case, if σ_x is known as the standard uncertainty of the measurement standard (CRM), the slope (β_1) and the intercept (β_0) of the calibration curve can be estimated by the following equations [6]:

$$b_1=S_{xy}/(S_x^2-\sigma_x^2) \text{ and } b_0=Y_{av}-b_1X_{av}, \tag{7}$$

where

$$S_{xy} = [\sum_{i=1}^{n}(X_i - X_{av})(Y_i - Y_{av})]/n, \tag{8}$$

$$S_x^2 = [\sum_{i=1}^{n}(X_i - X_{av})^2]/n, \tag{9}$$

$$X_{av} = \sum_{i=1}^{n} X_i / n \text{ and } Y_{av} = \sum_{i=1}^{n} Y_i / n. \tag{10}$$

When the uncertainty σ_x^2 is negligible in comparison with S_x^2, the estimations by Eq. (7) do not differ from the estimation used in the ordinary least squares technique. However, the uncertainty increase will influence the result of such estimation. Theoretically it may even happen that $\sigma_x^2 \geq S_x^2$, and Eq. (7) lead to an absurd result. In such cases $\beta_1=\infty$ is accepted and no absurd results will be obtained [6]. This influence is very important not only for determination of the analyte concentration X_0 corresponding to the response Y_0 by the calibration curve, but also for the correct uncertainty evaluation in the determination result.

If Y_0 is obtained from *m* replicate readings, the standard uncertainty in X_0 is evaluated by the following equation [7]:

$$S_{X0} = \frac{S_{y/x}}{b_1}\left\{\frac{1}{m}+\frac{1}{n}+\frac{(Y_0 - Y_{av})^2}{b_1^2\sum_{i=1}^{n}(X_i - X_{av})^2}\right\}^{1/2}, \tag{11}$$

where

$$S_{y/x} = \left\{\sum_{i=1}^{n}(Y_i - \hat{Y}_i)^2 / (n-2)\right\}^{1/2} \tag{12}$$

is the overall uncertainty of the fit of the calibration curve by the regression line; and $\hat{Y}_i$ is the i-th point on the calculated (regression) line corresponding to the measured Y_i.

Substituting b_1 from Eq. (7) into Eq. (11) one can obtain the formula which shows how the uncertainty in the measurement standard σ_x influences the uncertainty in the determination S_{X0}:

$$S_{X0} = \frac{S_{y/x}(S_x^2+\sigma_x^2)}{S_{xy}} \times \left\{\frac{1}{m}+\frac{1}{n}+\frac{(S_x^2+\sigma_x^2)^2(Y_0 - Y_{av})^2}{S_{xy}^2\sum_{i=1}^{n}(X_i - X_{av})^2}\right\}^{1/2}. \tag{13}$$

According to the rules of combined uncertainty evaluation [8, 9], σ_x can be considered as negligible, if it leads to increase of S_{X0} for less then one-third of its initial value (calculated by ordinary least squares technique). An example of the σ_x influence on the calibration parameters b_1, b_0 and S_{X0}, and corresponding lifetime of the traceability chain are analysed below.

Ethanol determination in beer

INPL participates in the Brewing Analytes Proficiency Testing Scheme (BAPS) organized by the Laboratory of the Government Chemist (LGC) in the United Kingdom. So, once a month INPL analysts (co-authors of the present article – B. Anisimov and I. Goldfeld) perform etha-

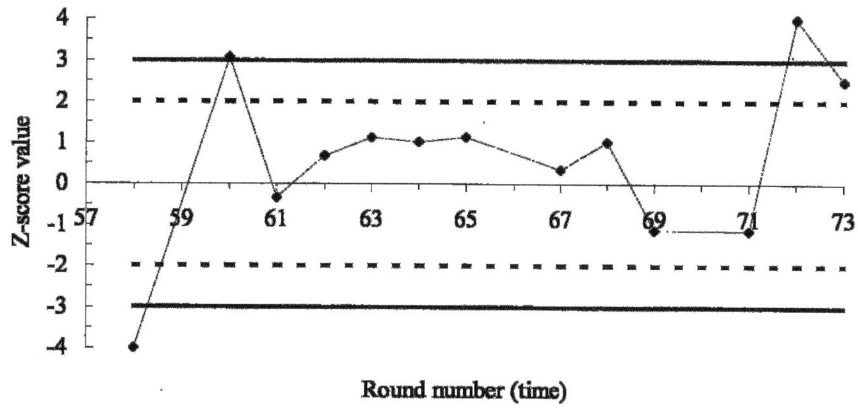

Fig. 1 Z-score control chart. The INPL results are shown by *dots*, satisfactory limits (Z=±2) by *dotted lines*, and unsatisfactory limits (Z=±3) by *solid wavy lines*

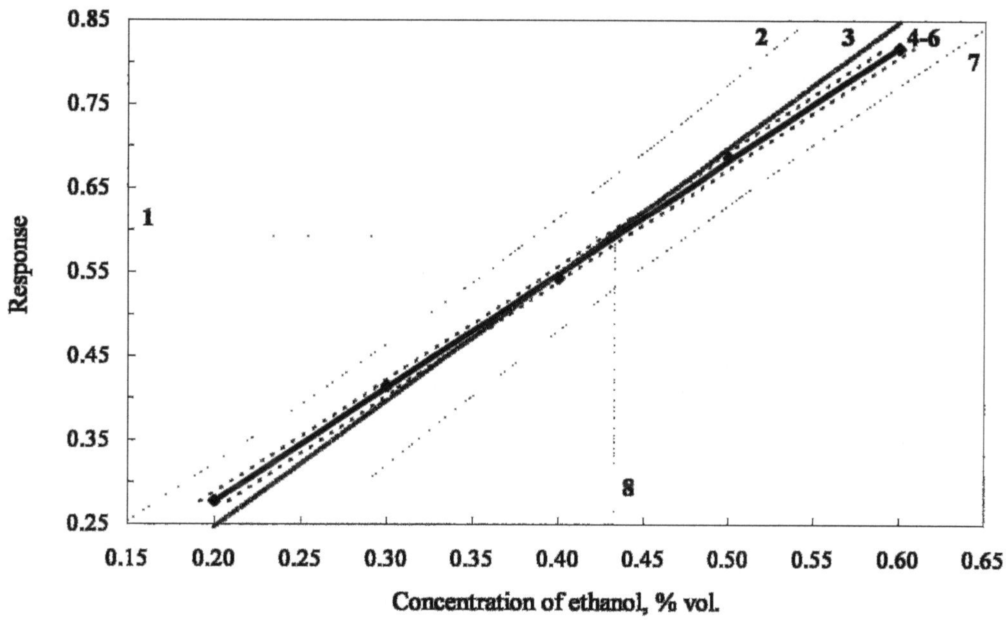

Fig. 2 Calibration curve in Round No. 73. Calibration data are shown by *dots*; the calibration curve obtained by the ordinary least squares technique, by the *solid* regression line 5; the limits of the regression corridor by *dotted lines* 4 and 6; the result of the ethanol determination in the sample by the *dotted lines* 1 and 8; the calibration curve for 3-times increased uncertainty of the certified value by the *solid* regression line 3, and the limits of its corridor by *thin lines* 2 and 7

nol determinations in beer samples obtained from LGC using a standard gas-chromatographic method. In Fig. 1 one can see a Z-score control chart for our results (X_0) during 16 rounds from November 1998 up to January 2000. The Z-score is calculated as the difference of X_0 and the ethanol assigned value in the beer sample (in the units of the standard deviation established by LGC), satisfactory limits being $|Z|\leq 2$ (shown by dotted lines), and unsatisfactory limits $|Z|\leq 3$ (shown by solid wavy lines). The analyte determinations in the first two rounds (Nos. 58 and 60) were performed by using in-house reference materials (RMs) of ethanol in water for calibration of the measurement instrument (GC HP 5880A with a capillary column DB-FFAP). The results were unsatisfactory. Then another traceability chain was built, based on the calibration using CRM LGC5404 purchased from LGC, with the certified value of the ethanol concentration in water of 5.00% by volume. The CRM expanded uncertainty is 0.03% vol at the 95% level of confidence and, correspondingly, σ_x=0.015% vol. Two beer CRMs LGC5004 and LGC5005 were used for the measurement quality control. The shelf-life of CRM LGC5404 is 12 months from the day of shipping, and really 12 months all the INPL results in BAPS were satisfactory including Round No.71. Then the result of Round No.72 was unsatisfactory. Many factors could have led to the unsatisfactory result, but the CRM shelf-life was found to be the most important. To correct the situation a new sample of the CRM LGC5404 (at the beginning of its shelf-life) was opened in Round No. 73 and the round was performed successfully.

To demonstrate the influence of the CRM uncertainty σ_x on the determination, the calibration curve obtained in Round No. 73 by the ordinary least squares technique is shown in Fig. 2 by regression solid line 5. The limits of the expanded uncertainty "corridor" formed by $t_{0.95,\,3}{*}S_{X_0}$ values are shown in Fig. 2 by dotted lines 4 and 6, where $t_{0.95,\,3}$=3.18 (see tables of the Student distribution at level of confidence 0.95 and 3 degrees of freedom [7]), n =5, m=5. Response in the calibration is the ratio of peak area for analyte (alcohol) and internal standard (1-butanol). The result of the determination X_0=4.33% vol is shown by the dotted lines 1 and 8. The corresponding standard deviation S_{X_0} without correction for the uncertainty in the measurement standard is 0.0020% vol. If σ_x=0.015% vol is taken into account, the recalculated S_{X_0} equals 0.0025% vol (the combined uncertainty increase caused by the CRM dilution is neglected here to simplify the calculation). So, there is no practical change in the S_{X_0} value. This value will increase to one-third, if σ_x is 0.017% vol, i.e. to 13% rel. larger, than in the CRM certificate. However, when the CRM shelf-life is at an end and σ_x is increased 3 times (to 0.045% vol.), for example, the changes are dramatic. Corresponding calibration curve (regression solid line 3) and the "corridor" limits (thin lines 2 and 7) by Eqs. (7) and (13) are also illustrated in Fig. 2.

Conclusions

1. Stability of the traceability chain in chemical analysis is dependent on the calibration intervals of measuring equipment and shelf-life of CRMs used for the calibration of the equipment.
2. Ordinary least squares technique, used for treatment of the calibration data, is correct only when uncertainties in the certified value of the measurement standards or CRMs are negligible. If these uncertainties increase (for example, close to the end of the calibration interval or the shelf-life), they are able to influence significantly the calibration and measurement results. In such cases, regression analysis of the calibration data should take into account that not only the response values are subjects to errors, but also the certified values.
3. As an end-point criterion of the traceability chain destruction, the requirement that the uncertainty of a measurement standard should be a source of less then one-third of the uncertainty in the measurement result, is applicable.

References

1. ISO (1993) International vocabulary of basic and general terms in metrology. ISO, Geneva, Switzerland
2. OIML International Document No.10 (1984). Guidelines for the determination of recalibration intervals of measuring equipment used in testing laboratories. OIML, Paris, France
3. NCSL RP-1. Recommended Practice (1996) Establishment and adjustment of calibration intervals, 3rd edn. Boulder, USA
4. Draper NR, Smith H (1981) Applied regression analysis, 2nd edn. Wiley, New York, p 122
5. Padilla GJ (2000) Proceedings of the International Conference on Metrology – Trends and Applications in Calibration and Testing Laboratories. Jerusalem, Israel, p. 321
6. Madansky A (1959) J Am Stat Assoc 54: 173–205
7. Miller JC, Miller JN (1993) Statistics for analytical chemistry, 3rd edn. Ellis Horwood PTR Prentice Hall, New York, p 113
8. EURACHEM/CITAC Guide (2000) Quantifying uncertainty in analytical measurement, 2nd edn. Teddington, UK
9. ISO (1993) Guide to the expression of uncertainty in measurement, 1st edn. ISO, Geneva, Switzerland

Accred Qual Assur (2001) 6:252–256

Stanley Rasberry

Proficiency evaluation as a traceability link in chemical metrology

Received: 14 November 2000
Accepted: 11 December 2000

S. Rasberry
National Institute of Standards and Technology (retired)
Current address:
31 Longmeadow Drive,
Gaithersburg, MD 20878
e-mail: rasberr@attglobal.net

Abstract Traceability implies comparison of the results of measurements, or comparison to national or international measurement standards. One of several approaches that have been used in chemistry to provide for such comparisons is distribution of proficiency evaluation materials which have been measured by a reference laboratory. A newer approach is based on receipt and measurement at a reference laboratory of materials that have been produced and analyzed by other laboratories. Traceability concepts and approaches to realization will be described together with discussion of the relative merits of various approaches. Extension into metrological fields other than chemistry will also be explored.

Keywords Chemical metrology · Traceability · Proficiency evaluation · Accuracy of chemical analysis · Quality assurance in chemistry

Introduction

Analytical chemistry spans a wide range of applications when compared to many other fields of measurement. This breadth of applications means that different aspects of sound metrological practice may have more or less importance depending on particular circumstances. Traceability [1, 2], and ultimately accuracy, are two such considerations that have quite variable importance to analytical chemistry. Examples can be cited where traceability has no importance. Counterexamples can also be given where it is useless to perform an analytical measurement without establishing a firm link of traceability.

Examples where traceability has no importance arise in the production of most commodity materials. Here the role of analytical chemistry may be merely to confirm that material is being produced with an acceptable maximum or minimum chemical characteristic. Assuring that specific properties of each lot of material produced compare well with previously-produced, acceptable batches implies that the new lots of material will also be suitable for use. What is required here is essentially not accuracy; rather, it is a precise ratio of characteristics with earlier, acceptable product. Such a situation arises in an aluminum plant where the objective is to produce a specific alloy (e.g., type 6061) by sampling the molten material and adding elements until they are found, by spectrometric analysis, to match the levels present in earlier, acceptable batches.

In this example, internal consistency, over time, can suffice to assure production, with minimal analytical cost, of acceptable alloy. Indeed, portions of previous batches which have yielded high-quality materials and products can serve as measurement standards without the need for traceability to SI or national standards. Measurement traceability to a higher laboratory may be useful in attaining certain objectives, such as inter-plant comparisons and reaching agreement between producer and buyer laboratories, but, it is not essential to immediate quality assurance for production in a single plant.

The purpose of this paper is to consider the other end of the spectrum, namely, to look at cases where traceability links are essential to the success of the application. From a societal point of view, some of the most pressing applications have to do with food safety, clinical laboratory results, and protecting the quality of the environment. Traceability of measurements to those of a national laboratory is of great importance to these latter applications.

This can be illustrated by an example dealing with environmental measurement. Numerous environmental studies depend on accurate analyses being performed at widely separated locations, perhaps groundwater from every county in the U.S. being analyzed by dispersed local laboratories. The overall objective is to have analytical results which can be compared without regard to which laboratory took the data. A practical way of assuring comparability is to insist that each laboratory produce accurate results, as evidenced by the quality of their traceability to a common basis of measurement. Economical approaches to demonstrating traceability of measurements are usually based on comparing measurements with a reference laboratory which has facilities to relate measurements to suitable national or international standards.

The environmental example given leads to the observation that analytical chemistry, when accuracy is important, is not different from other fields of metrology. Namely:

- To be judged as to accuracy, a result must be accompanied by an uncertainty
- Uncertainties also have uncertainty that can be assessed by demonstrating traceability
- Traceability requires comparisons of results of measurements, or comparison with appropriate standards
- To attain a given level of accuracy, it is necessary to exchange measurements with a reference laboratory that can measure more accurately than the given level, or make a comparison to a standard that is more accurate than the required given level

In analytical chemistry, one of the ways that results of measurements are compared has been given the special designation "proficiency evaluation." The remainder of this paper will treat the role of proficiency evaluation in accreditation programs, including how it serves as a traceability link, and how looking at this link from both ends has suggested a new approach to performing proficiency evaluation.

The need for proficiency evaluation

The earliest requirements for laboratories to undertake formal demonstration of their proficiency are probably based on direct contractual obligations. Several circumstances could produce the need, but two account for most. In the two-party situation a buyer might require that the laboratory of a supplier provide evidence that its laboratory is tested by an independent agency. If the laboratory gives evidence of producing accurate analyses of materials it supplies, the buyer can minimize additional testing of incoming goods.

In the second circumstance, three parties are involved: buyer, seller, and independent laboratory. Here, both the buyer and seller have a keen interest in the adequacy of results of the independent laboratory. In some cases neither the buyer nor the seller have a laboratory, in others the results of an independent "third-party" laboratory are sought to arbitrate disputes or corroborate claims.

More recently, especially since 1970, formal laboratory accreditation programs have been established to provide independent assessment of laboratory capabilities. These continue to gain the acceptance of producers, buyers, service providers, independent laboratories, and some, but not all, regulators. Developers of programs include the National Voluntary Laboratory Accreditation Program (NVLAP) located at NIST, the American Association for Laboratory Accreditation (A^2LA), and the College of American Pathologists (CAP). The first two organizations develop and provide a wide range of programs that cover a large number of the nation's testing, physical, chemical, and calibration laboratories. The CAP accredits more than 10,000 clinical testing laboratories to help assure the quality of healthcare provided in the U.S.

Usual requirements for a laboratory to be accredited in one of the programs mentioned above include on-site assessments by trained assessors and periodic participation by the laboratory in rounds of proficiency testing that are recognized by (but not necessarily provided by) the accrediting organization. Often a national laboratory, such as NIST, is sought out to serve as a reference laboratory for proficiency tests. Other possible sources are usually laboratories that are accepted by all parties as operating at a high level of measurement capability and may be referred to as reference laboratories. When a laboratory is assessed for accreditation, it is asked to give evidence of the calibration of its instruments and to demonstrate how accuracies of measurements are determined. An independent proficiency evaluation provides an excellent confirmation or painfully obvious denial of the laboratory's claims.

In the U.S., proficiency testing has been held to be of utmost importance in accreditation programs and is typically mandated wherever it is possible to conduct. European systems of accreditation have been slower to include proficiency testing as an integral part of accreditation, but, in recent years, have begun to make greater use of proficiency testing.

Traditional approaches to proficiency evaluation

Most situations where proficiency evaluation is applied are rather narrowly defined as to scope. For example, clinical laboratories may be asked to demonstrate that they can determine certain constituents occurring in human serum with uncertainties not to exceed specific limits. Accrediting bodies will require successful participation in periodic proficiency tests conducted by a reference laboratory which they recognize.

To conduct a proficiency test, a reference laboratory will prepare a quantity of material appropriate for distribution to all the laboratories under test. Requirements for the material will include that it is well characterized, and that it is sufficiently stable and homogeneous that no tested laboratory will be put at disadvantage by receiving a sample not representative of the lot. Further, the material should be typical of that of interest, and have constituents with concentrations within the ranges of interest. Additionally, it is important that concentration values be kept confidential during the period of the test, so that no participant will have an unfair advantage.

The proficiency test provider and reference laboratory may be the same organizational entity. If not, they must closely coordinate and document their quality assurance procedures and division of responsibilities. For a test round to be successful, each laboratory under test must receive a sample which is the same as every other sample from the lot, within the limits specified for the test. This means that careful attention must be given to material packaging, stability, and handling during distribution.

A set period of time is established for laboratories under test to carry out the prescribed chemical analyses and any other required measurements. A deadline is also given for return of results to the test provider, who will develop the statistics necessary to confirm the validity of the test round and to provide a figure of merit for each participant. The latter score may be as simple as pass/fail.

A new NVLAP accreditation program

In 1994, NVLAP began accreditation of calibration laboratories. Initially the program began with coverage of physical metrology laboratories and subsequently has included analytical chemistry as a category called "chemical calibration." A factor prompting the inclusion of analytical chemistry was the recognition that chemical analysis has a basis in metrology not fundamentally different from that of physical measurement.

The U.S. Environmental Protection Agency (USEPA) requested, in 1997, that NIST develop what has become the first program within chemical calibration. The NVLAP accreditation program is called "Chemical Calibration: Providers of Proficiency Testing." It is initially limited to organizations that provide chemical metrology services by conducting proficiency tests of laboratories which support drinking water and wastewater compliance monitoring, and ground and surface water quality monitoring. It is important to note that the program will extend only to questions of quality assurance at the top end of a metrological chain and not to the working level of the numerous environmental laboratories. The new program is described in NIST Handbook 150–19, Chemical Calibration: Providers of Proficiency Testing [3].

New approaches to proficiency evaluation

One of the first issues faced in development of the new program was how best to test the proficiency of the providers. The test needs to be quite comprehensive because the providers' technical activity spans a wide scope including material procurement, preparation, packaging, characterization, and distribution, and collection, processing, and reporting of proficiency test data. The desire to test the full scope of activity raised questions about the practicability of using traditional methods of proficiency evaluation. An additional consideration is that the number of initial applicants for the program was small, thus presenting only a small base for prorating the costs of preparing proficiency test materials for use in the traditional way.

A new approach was developed and presented in Handbook 150–19, using the newly coined term "indirect proficiency testing." In the new approach, each provider of proficiency testing submits to NIST a portion of each lot of material distributed in the proficiency test (PT) rounds that it operates. NIST can then evaluate the proficiency of the provider by checking the validity of the provider's claims regarding assigned value. Further, some appraisal is possible regarding the material's suitability, homogeneity, and stability based on examination of the results of the completed PT round, which will also be submitted to NIST. Where the approach can be used, it offers the possibility of 100% surveillance of materials distributed, in addition to allowing NIST to evaluate the provider's proficiency.

In lieu of analyzing every PT material received, NIST will have the option of developing proficiency information in alternative ways. One of these is a ratio approach; that is, the relative comparison of several similar PT materials submitted by different providers. If ratio comparisons for the materials do not agree with the ratios of their assigned values, it is necessary to analyze some of the materials, possibly selecting them from among providers on a rotational basis. An advantage found in this approach is the lower costs usually associated with ratio methods as opposed to absolute analyses.

Another alternative could be especially useful after a profile of success had been demonstrated by a specific provider. Taking such background information into account, NIST might analyze PT materials from that provider only on a random basis. PT materials received and not selected for immediate analysis would be set aside until the conclusion of the PT round and the provider delivers the results of the round to NIST. In the event that anomalous results arise for the test round, the set-aside material would be available for NIST analysis and examination.

The new approaches described in this section are seen as being the most suitable for application to organic and inorganic constituents found in the environmental matrices which comprise the initial program. The reason for this is that the cost of conducting a traditional proficiency test for a limited number of providers is estimated to exceed that attached to the costs for NIST surveillance of their materials. Even at equal costs, the indirect approach has added value in that it provides more information than constituent values alone, and it provides the possibility for review of all the materials issued by each producer. The indirect approach tests the value assignment process rather than the operations for verification analysis that would be tested by the direct approach. Further, it tests the specific concentration levels at which the provider is producing and distributing proficiency test materials, a situation that would occur only randomly in the direct approach.

Outside the organic and inorganic constituents, the traditional approach may find favor for tests of other constituents. Currently, radiological testing appears to be such a case. Here the economics are reversed from those presented above and it is more feasible to issue traditional PT materials from NIST on a regular, periodic schedule.

Extension to other fields of metrology

The new, alternative approaches described in this paper have been developed in support of quality assurance systems for environmental chemistry, particularly for water including drinking water, ground and surface water, and waste water. It is possible that the approaches could find use in other applications. The approaches will most likely be useful to the extent that the following conditions are met:

- Material produced in lots – that is, the laboratories under test produce defined lots of materials as part of the tests they conduct. This condition is necessary so that a defined "population" of material is represented by that which is sent to NIST or another oversight body for monitoring proficiency of the laboratories.
- Small number of participants – this condition is not a required one in any strict sense. However, it is one that bears on economic considerations. Simply stated, if there are a large number of participants, it may be economically advantageous for the oversight body to prepare and distribute proficiency test materials using traditional approaches.
- Process in reasonable control – the indirect proficiency test will be most economical of resources when the participating laboratories are consistently producing PT material lots having good quality. This condition facilitates use of the ratio methods mentioned in the previous section of this paper, and thus reduces the number of accurate, traceable measurements that are required.

The most obvious extensions from the field of water chemistry remain within the general field of chemistry. There, approaches could be used to test providers of proficiency evaluation for any type of material.

Extension into physical metrology is less obvious, but has no conceptual barrier. The direct approach could be applied if proficiency tests were conducted by a reference laboratory distributing to device producers, preparing lots of material measures and calibrants such as gage blocks, masses, or thermometers. Distribution of such material measures is already a feature of most measurement assurance programs in physical metrology. Use of the indirect approach would have units randomly selected from production and sent to a reference laboratory. This could prove, in some cases, more economical than direct proficiency testing of the producers of the devices.

Conclusions

Historically, there have been several ways of demonstrating metrological linkages of traceability. One of the earliest was for a reference laboratory to "calibrate" an artifact by measuring it and then provide it, along with certified results, to one or more laboratories so that they could also measure it and then compare their results with those of the reference laboratory. Later, distribution of certified reference materials, produced in lots by reference laboratories, provided an exactly analogous linkage for those kinds of materials amenable to production and certification in lots.

More recently, proficiency evaluation programs have produced similar traceability linkages, but with a novel twist. The proficiency test material (transfer standard) is sent without attachment of results. The laboratory under test receives both an independent confirmation of measurement ability and, indirectly, a traceability linkage when the test is concluded and the assigned value of the measurand is disclosed. It is unlikely that proficiency evaluation will supplant traditional forms of providing traceability links. However, the concept does provide metrologists with an additional tool.

This paper has described new approaches to providing quality assurance within the operation of proficiency evaluation programs. The approaches are used in a new NIST program to monitor providers of proficiency testing in an environmental field. A part of that monitoring is conducted by means of indirect proficiency evaluation of the providers. The new program demonstrates that both direct and indirect proficiency evaluation can provide traceability links, in chemical metrology, to measurements carried out at NIST.

References

1. International Organization for Standardization (ISO) (1993) International vocabulary of basic and general terms in metrology, 2nd edn. ISO, Geneva
2. De Bièvre P, Kaarls R, Rasberry S, Reed W (1995) Metrology and the role of reference materials in validation and calibration for traceability of chemical measurements. Proceedings of the 1995 NCSL Workshop and Symposium, pp 479–494
3. Faison CD, Parris RM, Rasberry SD (1999) NIST Handbook 150–19 Chemical calibration: providers of proficiency testing. NIST, Gaithersburg, MD

Accred Qual Assur (2003) 8:184–190
DOI 10.1007/s00769-003-0625-9

Paul Armishaw
Bernard King
Roderick G. Millar

Achieving traceability in chemical measurement – a metrological approach to proficiency testing

Received: 14 October 2002
Accepted: 11 March 2003
Published online: 23 April 2003

Presented at the Interact 2002 Conference, July 21–25, 2002, Sydney, Australia

P. Armishaw · R. G. Millar (✉)
National Analytical Reference Laboratory, AGAL, Clive Road 3, Cottesloe, 6011 Western Australia, Australia
e-mail: rod.millar@agal.gov.au

B. King
Bolton Gardens 25, Teddington, Middlesex, TW11 9AX, UK

Abstract ISO/IEC 17025 requires that testing laboratories establish the traceability of their measurements, preferably to the SI units of measurement. The responsibility for establishing traceability lies with each individual laboratory and must be achieved by following a metrological approach.

The results of measurements made in such a way are traceable to the standards used in method validation and to the calibration standards used during the measurement process. If these standards are traceable to SI then the measurements will also be traceable to SI.

Participation in appropriate proficiency studies (an ISO/IEC 17025 requirement) enables laboratories to demonstrate the comparability of their measurements. If the materials used for the studies have traceable assigned values, then proficiency testing also provides information about measurement accuracy and confirms, or otherwise, that appropriate traceability has been established. This paper will report on a new approach for the establishment of traceable assigned values for chemical testing proficiency studies. The work is conducted at a "fit for purpose" level of measurement uncertainty, with costs contained at a level similar to previous "consensus" based proficiency studies. By establishing traceable assigned values in a cost effective way, NARL aims to demonstrate the added value of the metrological approach to participant laboratories.

Keywords Traceability · Metrological · Chemical · Proficiency · Testing

Introduction

Since 1993, Australia's National Analytical Reference Laboratory (NARL) has been conducting inter laboratory (proficiency testing) studies according to international guidelines [1, 2, 3] for a range of chemical tests. Our laboratory is certified to ISO 9001 [4] and holds technical accreditations to ISO/IEC 17025 [5] for a range of its test methods. In 2000, NARL gained independent recognition to ILAC-G13:2000, "Requirements for the competence of providers of proficiency testing schemes" [6]. More recently, NARL has been exploring ways of establishing a metrological approach to its proficiency testing program involving the establishment of traceable assigned values accompanied by uncertainty estimates at a "fit for purpose" and affordable level. Our work is based on the assumption that for measurements to be valid they need to be made using an approach that requires the laboratory to:

- Define clearly what is to be measured (e.g. total analyte or extractable analyte defined by a specific method etc)
- Use a test method that has been validated to ensure correct identification and selective, accurate measurement of the analyte of interest
- Establish the bias associated with the mass transfer steps (digestion, extraction, evaporation, derivatisation etc.) by measuring the recovery of spikes and by

using control samples and appropriate matrix reference materials
- Calibrate the chemical measuring instrument using standards prepared from pure substance reference materials of known identity and purity
- Calibrate physical measurement processes with traceable physical standards
- Estimate associated measurement uncertainties

The results of measurements made in this way are traceable to the standards used in method validation and to the calibration standards used during the measurement process. If these standards are traceable to SI then the measurements will also be traceable to SI.

Only a few chemical proficiency study organisers in the world provide metrologically assigned values as part of their schemes. Such an approach, which will help participating labs to themselves adopt a more metrological approach to their work, is traditionally considered to be desirable, but too costly.

The NARL approach, reported here, has been established, without increasing costs, by re-engineering the process to improve efficiency and focus effort on critical issues. An important part of the process is the quality assurance and quality management systems that have been established.

The NARL proficiency study program

The NARL inter laboratory study program is a proficiency testing scheme currently covering pesticide residues (in animal fats, soil, fruit and vegetables), pollution by petroleum products (in soil and water) and illicit drug analysis (heroin, cocaine, amphetamines and LSD). The objectives of the NARL studies are as follows:

- To provide testing laboratories with a tool to improve the accuracy and traceability of their chemical measurements
- To provide Australian chemical testing laboratories and the national laboratory accreditation body, NATA, with information on the current "state of the practice" in each area of analysis
- To evaluate and encourage improvements in laboratory methods and performance from "state of the practice" to "fit for purpose"
- To enable participating laboratories to assess their performance relative to domestic and international peer laboratories and hence to improve the comparability of results between laboratories and between countries
- To develop and promote a fit for purpose and affordable metrological approach to proficiency testing.

In keeping with these objectives, NARL aims to establish traceability with an associated estimate of measurement uncertainty for all its study assigned values (the values which are the best available estimate of the true concentration of the analyte). These uncertainties are expressed as expanded uncertainties, approximating a 95% confidence interval.

Test material preparation, stability, homogeneity

The material from an appropriate bulk supply of the matrix material is weighed into a suitable container and homogenised. Where appropriate, portions of the matrix material are tested for the analytes of interest and for possible interferences. Pesticide and environmental test samples are usually prepared by spiking known amounts of substances of specified chemical purity into the bulk sample matrix. The spiking process is witnessed and cross-checked by a second scientist and full records of the process and of reference standard solutions used are maintained so that they can be verified if required.

Use of test samples containing naturally incurred substances has the advantage of testing the ability of study participants to extract complexed or "bound" contaminants. However, the concentrations of naturally incurred substances in the sample matrix cannot usually be known to the same level of accuracy as spiked substances. One compromise that is sometimes useful is to prepare test samples by spiking, but to "age" the samples before dispatch to allow time for the added substances to interact with the matrix and therefore better simulate "real world" situations.

After preparation and final homogenisation, the prepared samples are sub-divided and packaged into labelled, individual test samples.

Materials are prepared with a sound knowledge or experimental evidence of the stability of both the matrix and the analyte chemical in that matrix. This information may be obtained from published storage stability data or from results of a storage stability trial.

Once prepared, the study samples are stored and shipped to participant laboratories under conditions that ensure the continued stability of both the sample matrix and analyte. Prior to dispatch, the homogeneity of the packaged samples is assessed by statistical batch testing for all analytes of interest according to published procedures [1, 3, 7]. This involves duplicate analysis of ten separate packaged units selected at random from the entire batch or at regular intervals through the fill sequence if fill trend effects are anticipated. The analyses are performed in as short a time as is practical and in a random order using an accurate test method that is sufficiently precise for the purposes of the study. Each individual test sample is mixed thoroughly in its container prior to taking the test portion which must be sufficiently large, particularly for solid samples, so as not to influence the precision of the test results. Test results are subjected to analysis of variance (ANOVA) to assess the variance between test sample units. This variance is then compared to the predetermined target standard deviation (σ) for the study to establish whether the material is sufficiently homogeneous for the purposes of the study.

Sometimes for convenience, and to enable homogeneity testing to be completed in a practical timeframe, one or two representative analytes are selected as markers for homogeneity and undergo full testing as described above. Spiking concentrations of the remaining analytes are checked by duplicate analysis.

If the above approach creates practical problems, an alternative is to perform *single* analyses on a minimum of five test portions of the study sample. The standard deviation of replicate analysis results is an indicator of sample homogeneity and method precision. The disadvantage of this approach is that it does not provide a simultaneous measure of the analytical variance under the homogeneity test conditions. Analytical variance must be estimated from historical data (e.g. method validation) or spiked recoveries run with the homogeneity test samples.

Communication with participants and confidentiality considerations

To maintain confidentiality, participating laboratories are randomly allocated a code letter or code number, which is used for reporting the results of the study. Participants are instructed to analyse the test materials using the test method of their choice and to report results according to their usual procedures. It is recognised that sometimes laboratories are measuring different things and in these cases care is needed in assessing the agreement between results. As results are received from participants, they are transcribed from faxed result sheets into the study-specific results spreadsheet. Transcription is checked and signed by a second NARL officer.

Establishing traceable assigned values

The assigned value is the value to which participants' results are compared, and must be the best available estimate of the true concentration of analyte. It is important to clearly define the measurand such that the assigned value only relates to that measurement. For example, if the measurand is the amount of analyte extracted by a specific method, then it is important that this is clearly understood.

Traceability of the assigned value is achievable provided there are direct links to stated references, together with sound estimates of the uncertainty of the links. NARL aims to establish and maintain traceability to SI, where this is technically feasible, but not necessarily at the highest metrological level. This is achieved by establishing and maintaining the following types of links.

Traceability of physical measurements

For physical quantities such as mass, volume etc. measurements are made with equipment calibrated using standards traceable to SI, such that their total contribution to the overall uncertainty is less than one-fifth of the desired overall uncertainty.

Traceability of NARL homogeneity test results

NARL test methods include details of how traceability was established at the time the method was validated. Provided that the documented test method is followed, and all critical control points are addressed, measurements made using the test method will correctly identify and selectively and accurately measure the analyte of interest. Such measurements are traceable to the standards used in method validation and the calibration processes. For the test methods used in homogeneity testing of study samples, traceability of chemical measurements is maintained by:

1. Calibration using simple solutions of well-characterised pure substance standards or matrix matched standard solutions. Calibration solutions are prepared from materials whose identity and purity have been established to an appropriate level of uncertainty and where the effects of any impurities have been evaluated. Where appropriate and where available, standards provided by metrology institutes with demonstrated capability are used. In other cases, materials from other reputable suppliers or prepared in-house are used after appropriate characterisation. Where necessary, professional judgement is used to estimate the uncertainty associated with chemical standards. The target uncertainty of the identity is "for practical purposes zero" and for purity less than one-fifth of the desired overall uncertainty.
2. The established precision and reproducibility of the method.
3. Sound knowledge of method bias and recovery from thorough investigation of interferences and matrix effects and by use of matrix CRMs, and/or spiking studies. Consideration is given to any mismatch between standards and samples, such as concentration and matrix differences and differences in the way the analyte is incorporated into the matrix.
4. Reference to uncertainty budgets developed according to ISO/GUM principles [8, 9] which are included in each test method. These budgets are used to estimate the uncertainty associated with measurements made using the method. The effects of sample homogeneity and analyte stability are included in the overall estimate of measurement uncertainty.

By addressing all of the above, traceability to the mole will normally be established. Exceptions include measurements of poorly defined analytes, such as fat, and where it is not feasible to establish reliable bias data. In such cases, it may be appropriate to define the measurand (analyte) in terms of the method of measurement, but the measurements can still be made traceable to the SI. Uncertainty associated with bias and recovery are often the largest components of the overall measurement uncertainty and the most difficult to address.

Traceability of formulated (spiked) concentrations

For study samples prepared by direct formulation (spiking), the concentration is potentially traceable, through calibrated balances and volumetric glassware, to the pure standards used for spiking. If these pure substances are well characterised, the traceability chain extends to SI. However, this traceability chain can be broken if the process used for spiking is not well understood and controlled (e.g. unquantified losses due to evaporation). The process of preparing a test material by formulation is quite simple in concept; a known amount of analyte is added to an analyte-free matrix. However, the situation may be more complex in the real world. For example, the analyte may be volatile, or strongly bound to the sample matrix, or there may be some uncertainty as to whether the matrix is "analyte-free". Traceability may also be compromised if the analyte is not well defined (e.g. "gasoline range organics").

All of these contribute to the uncertainty of the spiking process, and to the extent that they are unknown or uncontrolled, may break the traceability chain. When planning proficiency studies, strategies for maintaining the links in the traceability chain must be considered.

Traceability of reported participant results

Without an accompanying statement of traceability, individual participant results (and the consensus derived therefrom) may or may not relate to the same measurand and may or may not be traceable to common references. However, provided that the consensus (median) agrees with at least one (traceable) independent measure of the assigned value, it can be used, in conjunction with the other independent measure(s), to set the (traceable) assigned value. The (reproducibility) uncertainty of the (consensus) median of participants' results can be calculated from the median of absolute deviations, MAD [3].

Setting the assigned value from independent measures

Guidelines for setting the Assigned Value are available [1, 2, 3]. There are several independent measures of concentration that may be used to set the assigned value:

- Formulation (e.g. spiking concentration)
- Direct comparison with certified reference materials
- Homogeneity testing results
- A primary method of chemical measurement
- Consensus value from expert laboratories
- Consensus of participants

In a typical NARL study there are two or more independent measurements of the analyte concentration, each with an associated statement of traceability and estimate of uncertainty. In setting the assigned value, any significant differences between the independent measures must be considered and where possible the causes identified. If these differences are too great, it may not be possible to determine an assigned value for an analyte, in which case a consensus value or indicative value may be used. The rationale for determining the assigned value is always described in the study report.

Where the different measures of the analyte concentration agree within their estimated uncertainties then one of the following procedures is used:

1. If one particular independent measure is considered to be significantly more reliable, that measure is used as the best estimate of the assigned value and its uncertainty.
2. If there are significant technical deficiencies with an independent measure (e.g. no consensus for participant results) then unreliable data are excluded when setting the assigned value.
3. If there is no reason to choose one specific measure, the separate (reliable) measures are combined [10] into a single assigned value, being the inverse-variance weighted (for uncertainty) average of the independent measures of analyte concentration.

An example drawn from a recent NARL pesticide residue proficiency study [11] is shown in Fig. 1.

For this analyte, the assigned value was set from the inverse-weighted (for uncertainty) mean of two independent measures of component concentration:

1. **S**, the formulated (spiked) concentration (accompanied by measurement uncertainty estimate) – traceable to SI through gravimetric preparation using pure substance chemicals of specified purity
2. **R**, the NARL reference analytical value (accompanied by measurement uncertainty estimate) – traceable to SI through the validated test method used and the pure substance reference standards used to calibrate the measuring instrument

A third measure **M** (the median of participants' results) was not used in this case because, although in agreement

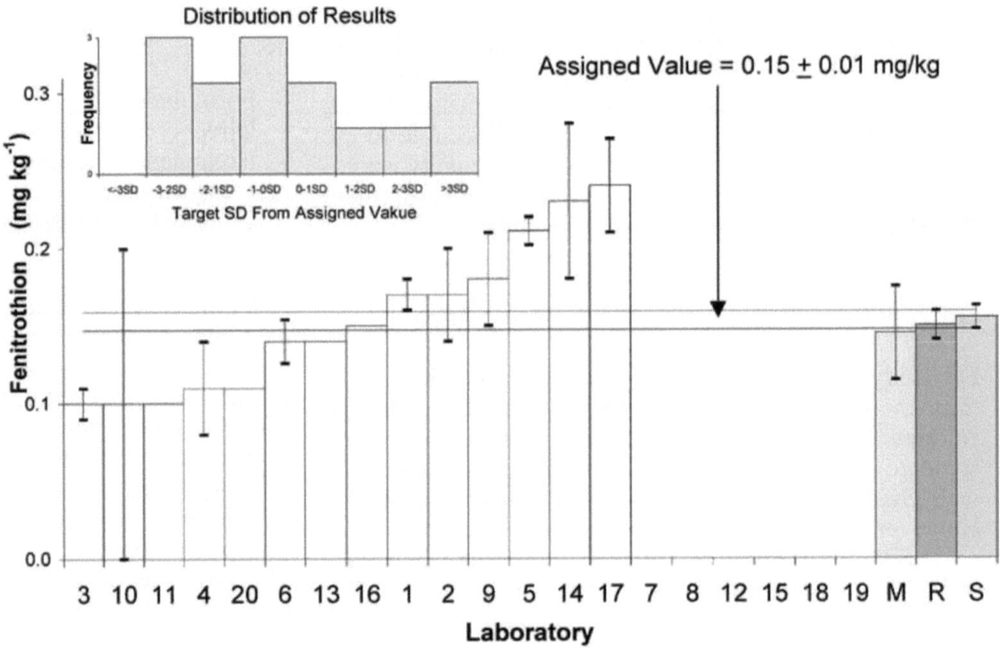

Fig. 1 Graphical presentation of proficiency study results fenitrothion in homogenised tomato

with the other two values, there is no real consensus (as evidenced by the distribution of results in the histogram at top left of Fig. 1).

Some interesting observations and conclusions are evident from Fig. 1:

- Of the 14 participants, only labs 2, 6, 9, 10 and 16 have reported a result that agrees with the assigned value (within the respective measurement uncertainties).
- Lab. 10, although in agreement with the assigned value, has a large uncertainty estimate and the reported result may not be fit for its intended purpose.
- Labs 5, 14 and 17 are in agreement with each other but not with the assigned value indicating the presence of individual laboratory biases which have not been accounted for in their uncertainty estimates.
- Labs 1, 3, 5 and 6 have almost certainly underestimated their measurement uncertainty.
- Labs 11, 13, 16 and 20 have not reported an uncertainty estimate.

Statistical treatment and performance assessment

NARL uses the assigned value to assess and report on laboratory performance. The statistical treatment used by NARL is based on accepted, international, standard published procedures [1, 2, 3, 10].

A central aspect is the calculation of z-scores (Eq. 1):

$$z=\frac{\chi-X}{\sigma} \tag{1}$$

where z=z score, χ=participant's result, X=assigned value and σ=target standard deviation.

The assigned value ($\boldsymbol{X}$) and the target standard deviation ($\boldsymbol{\sigma}$) have a critical influence on the calculation of z-scores and must be selected with care if they are to provide a realistic assessment of laboratory performance.

The target standard deviation ($\boldsymbol{\sigma}$) is the between-laboratory coefficient of variation that, in the judgment of the study coordinator, would be expected from a group of laboratories given the concentration of analyte and degree of homogeneity of the test materials. Published data, fitness for purpose and generalised models such as the Horwitz equation [12] are taken into account to determine $\boldsymbol{\sigma}$. It is important to note that the target standard deviation is selected by the study coordinator and is *not the same standard deviation calculated from results returned by participant laboratories*. This approach allows z-scores to be used as a fixed reference point for assessment of laboratory performance, independent of variations in group results from one study to the next. Calculated z-scores are indicative of the current state of the practice among the group of participant laboratories. By setting a realistic target standard deviation, laboratory performance can be compared to achievable performance providing a benchmark for progressive improvement.

E_n-scores (sometimes called E_n numbers) are an alternative to z-scores [2]. They provide a measure of how closely an individual laboratory result agrees with the assigned value, taking account of uncertainties in both the result and assigned value. The E_n-score (Eq. 2) is an objective measure of whether or not an individual result is consistent with the assigned value:

$$E_n = \frac{\chi - X}{\sqrt{U_\chi^2 + U_X^2}} \quad (2)$$

where En=En-score, χ=participant's result, $U\chi$=expanded uncertainty of participant's result, X=assigned value and U_X=expanded uncertainty of the assigned value.

Unlike z-scores, E_n-scores do not require the setting of a target standard deviation. An E_n-score of ≤ 1 indicates that the result and assigned value are in agreement within their respective uncertainties.

An E_n-score of >1 indicates that the result is different from the assigned value, and therefore that the uncertainty associated with the result has been understated.

A property of the E_n-score is that if the uncertainty reported with the result is large enough, E_n will always be <1. However in such cases, the fitness for purpose of the test result should be questioned.

Presentation of data and reporting of participant performance

Within two weeks of the study closing date, we issue an interim report, the purpose of which is to provide rapid feedback to participants. At the conclusion of each study, a detailed final report is prepared and issued to participants. This report contains a full description of the study together with statistical analysis and graphical presentation of the results. The report is prepared in a standardised format consistent with ISO and ILAC guidelines for proficiency test reports.

In presentation and interpretation of results, NARL aims for objectivity, clear presentation, and statistical data treatment that is transparent to participants, internationally accepted and metrologically sound. Sources of chemical standards, statements concerning traceability and estimates of measurement uncertainty are included in the study report.

The report also provides a way of promoting the metrological approach and helps provide training material for laboratory staff.

Fitness for purpose and cost effectiveness

NARL studies are designed to test the proficiency of laboratories by using test samples that resemble "real world" samples. Homogeneity of the test samples is established at a level that is sufficient for the purpose of the study taking into consideration the degree of interlaboratory variability appropriate for the particular analysis.

For any proficiency study, test samples must be properly prepared and the quality of the prepared samples (sufficient homogeneity, stability) must be controlled and demonstrated. This usually means statistically based chemical testing to demonstrate the (sufficiently) homogeneous distribution of all test analytes throughout the sample matrix. If the homogeneity testing method, as well as being sufficiently precise, is accurate (i.e. produces traceable results with an estimate of uncertainty) then the homogeneity test results can be used to set a traceable assigned value.

In this way, proficiency studies can be conducted that incorporate traceable assigned values, but require little extra effort over the consensus based approach [13, 14].

Use of study test samples as reference materials

At the conclusion of the study, surplus test materials are offered for sale as reference materials for quality control and method development purposes. These materials have been studied for homogeneity and stability and have been assigned values for specific analytes. The more recently produced materials (since July 1999) have traceable assigned values with rigorously evaluated uncertainties. Older materials have consensus-based assigned values.

Conclusion

Laboratories accredited to ISO/IEC 17025 must establish the traceability of their test results, preferably to SI units of measurement. Participation in appropriate proficiency studies enables laboratories to demonstrate the comparability of their measurements. If the materials used for the studies have traceable assigned values, then proficiency testing also provides information about measurement accuracy and confirms, or otherwise, that appropriate traceability has been established.

Laboratory accreditation bodies also offer formal recognition (accreditation) of proficiency testing providers to ILAC-G13:2000 (International requirements for competence of providers of proficiency testing schemes). This process is intended to provide confidence that proficiency testing schemes are designed and conducted to internationally acceptable standards.

However, confidence can only be fully established if the proficiency study provider adopts a metrological approach, involving the establishment of traceable assigned values accompanied by full uncertainty estimates.

Acknowledgements The authors would like to thank the Australian Government Analyst for permission to publish and wish to acknowledge the contribution of NARL staff to the successful operation of the NARL proficiency program.

References

1. Thompson M, Wood R (1993) JAOAC Intl 76:926–940
2. ISO (1997) Proficiency testing by interlaboratory comparisons. Part 1. Development and operation of proficiency testing schemes. ISO Guide 43-1, ISO, Geneva
3. Lawn RE, Thompson M, Walker RF (1997) Proficiency testing in analytical chemistry. RSC, Cambridge, UK
4. ISO (1994) Quality systems model for quality assurance in design, development, production, installation and servicing. ISO 9001, ISO, Geneva
5. ISO (1999) General requirements for the competence of testing and calibration laboratories. ISO/IEC 17025, ISO, Geneva
6. ILAC (2000) Guidelines for the requirements of the competence of providers of proficiency testing schemes. ILAC-G13:2000, International Laboratory Accreditation Co-operation (ILAC)
7. Fearn T, Thompson M (2001) Analyst 126:1414–1417
8. ISO (1993) Guide to the expression of uncertainty in measurement. ISO, Geneva
9. EURACHEM/CITAC (2000) Quantifying uncertainty in analytical measurement, 2nd edn. LGC, UK
10. Hasselbarth W, Bremser W, Predel R (1998) Fresenius J Anal Chem 360:317–321
11. Millar RG (2001) NARL Analytical Quality Assurance Study No. 01-04, Australian Govt. Analytical Laboratories, Cottesloe, Western Australia
12. Thompson M, Lowthian PJ (1995) Analyst 120:271–272
13. van der Veen MH, Horvat M, Milačič R, Bučar T, Repinc U, Ščančar J, Jaćimović R (2001) Accred Qual Assur 6:264–268
14. Baldan A, van der Veen MH, Prauß D, Recknagel A, Boley N, Evans S, Woods D (2000) Accred Qual Assur 6:164–167

Accred Qual Assur (2003) 8:319–322
DOI 10.1007/s00769-002-0567-7

Andrew Wallard

Traceability issues in measurement

Received: 2 October 2002
Accepted: 2 November 2002
Published online: 8 April 2003

Presented at the International ILAC/IAF Conference on Accreditation in Global Trade, September 23–25, 2002, Berlin, Germany

A. Wallard (✉)
Bureau International des Poids et Mesures, Pavillon de Breteuil, 92312 Sèvres, France
e-mail: awallard@bipm.org

Abstract This paper reviews the current state of play of the Mutual Recognition Arrangement created by the International Committee for Weights and Measures in 1999. The aim of the MRA is to provide a framework within which National Metrology Institutes can demonstrate the equivalence of their realisations of the units and quantities of the SI system to which accredited laboratories are traceable. The article offers some views on the need for traceable measurements, their relevance to technical barriers to trade, and the use that is being made of the MRA framework by national and international bodies.

Keywords Traceability · Mutual recognition · CIPM

Introduction

All companies today have to think big: national markets are no longer large enough to ensure that all the economies of scale can be gained; international standards must be complied with if exports are to grow and off-shore manufacture is, in many cases, essential if companies are to take advantage of differing labour costs. For the instrumentation industry, new markets inevitably imply servicing and replacement of critical sub-system elements that require calibration. Faced with these challenges, it is a brave – or perhaps foolhardy – company that thinks only of national calibration and traceability facilities and services when markets operate globally. In parallel with this market-led dynamic there is an additional trend, driven by governments and international trade bodies that want to reduce and eliminate trade barriers. Indeed, successive international economic and trade summits have committed governments to the reduction of trade barriers and to open markets so as to help alleviate poverty, challenge monopolies, reduce protectionism and open markets so as to benefit consumers. Up to now, much of the concentration economically and by bodies such as the World Trade Organisation has concentrated on price barriers and to the introduction of global standards. Now, attention is turning to non-price barriers, the foremost of which are often technical. How often do we hear of bespoke national twists to the internationally negotiated and agreed standards? How often do we hear of problems encountered by companies that manufacture in imperial units when international markets demand metric? How often do we see the reluctance to accept tests and measurements made in one country when exports are to another? As metrologists our task is not only to ensure technical consistency at a national level, but more and more to address the international context of equivalence and traceability. This is really nothing new in the history of world metrology but it takes on a new significance when driven by the above issues of market access and fair trade. This article tackles some of the current issues and brings readers up to date with some of the potential solutions.

Traceability: the use of ISO/IEC 17025

There are, essentially, two elements to international confidence in measurement. The first is evidence of traceability to national realisations of the SI units and quantities, the only "stated reference" that is commonly agreed world-wide. The second is the international equivalence of the national standards themselves. The first is normally provided by the use of test facilities and laboratories that are accredited to the international ISO/IEC 17025 standard that assesses the technical competence of calibration and testing staff as well as the conditions in which the measurements take place. Accreditation, incidentally, is not to be confused with certification, which is normally to the ISO 9001 suite of standards and which do not incorporate the same requirements for rigorous technical competence as does ISO/IEC 17025. Accreditation has grown rapidly in the last 25 to 30 years and now is the norm amongst quality conscious companies and purchasers. The core approach in the standard is accepted world-wide although certain sectors of industry occasionally add sector-specific requirements. In our experience, however, many of the bespoke requirements can be catered for within the framework of the standard itself. However we appreciate that some sectors that are unfamiliar with the application of the standard and which may be major standard component purchasers may feel more comfortable with addenda which can be acceptable, provided they are not designed to act as a technical barrier to world trade. Suppliers are increasingly aware of the benefits of accreditation and are – at last – seeing assessment by a recognised third party body as aiding market competitiveness and customer confidence rather than as a cost that hits the bottom line. Internationally recognised accreditation bodies are themselves assessed to the international guide for national accreditation organisations – ISO Guide 58 – and those that do comply and are members of the International Laboratory Accreditation Co-operation – ILAC – enjoy the benefits of increasing world-wide acceptance of their calibrations and tests. This has served markets well and the ILAC system is recognised as conferring credibility with legislators, regulators and users in general.

However, we now need to turn our attention to the second of our two criteria for full acceptance – and one that is increasingly important as industry best practice and requirements for accurate measurement approach the capability of national standards laboratories (normally called National Metrology Institutes or NMIs). Put simply, ISO/IEC 17025 requires evidence of traceability to national realisations of the standards of the SI but says nothing about how well these NMI standards agree world-wide. If such matters were not generally well ordered we could be faced with entirely consistent national systems but ones which differed substantially and significantly because the SI realisations at different NMIs turned out to be different. The Mutual Recognition Arrangement (MRA) of the International Committee for Weights and Measures (CIPM, from its French title) deals with just this situation.

The mutual recognition arrangement of the CIPM

Briefly, because it is not my intention in this article to go into the full details, the aim of the MRA is that participating NMIs agree on a framework within which they can recognise each others' realisations of the SI standards as well as the calibration certificates that they issue. The MRA has been signed by over 50 countries, economies and international institutes but the coverage actually extends to well over 100 NMIs as well as national institutes "designated" to hold certain national standards in countries where the NMI does not have full technical coverage. The "output" is the Key Comparison Data Base (KCDB) maintained by the BIPM and within which one can find details of the comparisons being undertaken within the MRA framework as well as a listing of the "calibration and measurement capabilities" of participating organisations.

In order to comply with the MRA requirements a participating NMI or institute must:

- State the uncertainty with which it expects routinely to provide calibration services to customers;
- Have this calibration and measurement capability (CMC) reviewed and agreed by its neighbours within its local Regional Metrology Organisation (RMO) and then between RMOs (RMOs are networks of NMIs based roughly on five economic and trading blocs in the Americas, Europe, Euro-Asian, Asia Pacific, and Africa);
- Participate in a world-wide series of comparisons of standards which provide the ultimate technical validation of the CMC claims. At present there are some 400 key comparisons registered into the BIPM's database (the KCDB), conducted by the CIPM and the Regional Metrology Organisations (RMOs);
- Maintain a quality system that gives confidence in their procedures and which provides assurance that the NMI continues to offer the claimed and validated uncertainties until such time as there is another comparison.

Key comparisons provide the ultimate technical basis for validating CMCs and are carried out under special procedures that ensure that no participant knows its results before all the measurements are completed and also that uncertainties are stated before the comparison begins. In this way, the key comparisons are a true test of NMI performance and reveal any differences in the day-to-day realisation of the units and quantities of the SI by the NMIs involved.

The end result of the MRA is a database of agreed and technically validated and reviewed CMCs – currently there are over 14,000, generally in the fields of physical and chemical measurements – that can be used as references by a multitude of users. Access to the database is through www.bipm.org.

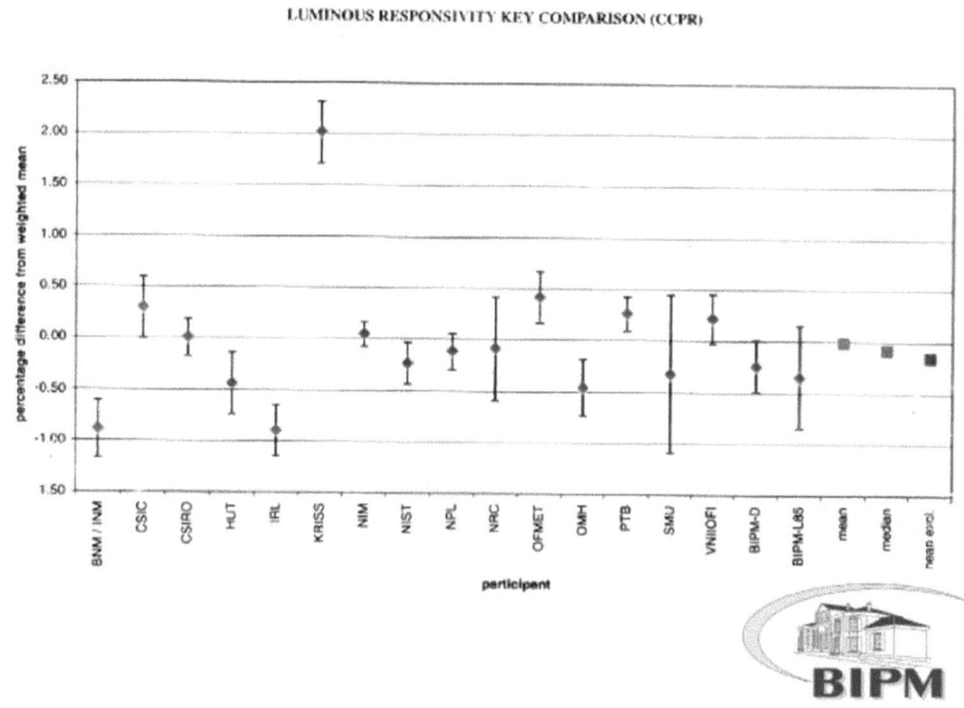

Fig. 1 Typical graphical representation of the results of a key comparison

The facilities offered by the KCDB meet a variety of user needs through its four appendices. Appendix A lists the signatories to the MRA; Appendix B contains details of the results of the technical comparisons undertaken by the signatories and details any differences. Appendix C lists the CMCs by subject area or by country and Appendix D lists the comparisons completed or in train. By searching Appendix B, the user can call up graphical representations of the results of comparisons; see Fig. 1. These show the differences in the realisation of the standards by the participants together with the uncertainties that they claimed. Often it also shows the reference value that can be derived from the comparison together with **its** uncertainty. These pieces of information tell the user whether the NMIs in question are state of the art and whether any differences between individual realisations realisation are statistically and significantly different from the state of the art.

The search engines within Appendix C of the KCDB also enable a user to log into a particular subject area – electrical, dimensional, chemical, etc. and then search for NMIs that offer a particular service. Searching is made easy and internationally consistent because all entries are in a standard format and are also described in terms of agreed detailed service categories and ranges over which the CMC applies. Checking back to see if results from the supporting comparisons are available in Appendix B enables information on equivalence to be viewed. However, in the absence of results from the comparisons, the user can be assured that the CMC claims in Appendix C have already been reviewed intensively by the relevant experts and that their judgement is that the claim is valid. Of course, the acid test is the results from the comparisons but not all of these will be completed for some time. By and large, the comparisons have so far validated the CMC claims but there have been some surprises that have led to re-appraisals of the performance of some of the participants. With a few exceptions these are of more interest to the NMI metrologists than to day to day users because the results give metrologists new information about the effects of some of the type B – systematic – uncertainties and help their understanding of the factors that limit performance. This area, though, requires more work and greater understanding of when NMI differences may begin to have an effect on uncertainties at working levels that may impact on the requirements of legislation or on conformity with technical specifications.

This arrangement has been in place since October 1999 and has attracted attention from accreditors, regulators and Government trade negotiators. The ILAC community, for example, intend to use it to check that the uncertainty claims of accredited laboratories are not "better" than the CMCs of the NMI to which they claim traceability through the ISO/IEC 17025 requirements. Regulators are increasingly aware that the MRA, particularly when combined with similar agreements within the accreditation community, can help them in their job of checking whether imported products meet the relevant national or international requirements. In addition, trade negotiators are seeing its relevance to the reduction of technical barriers to trade: for example, the CIPM's

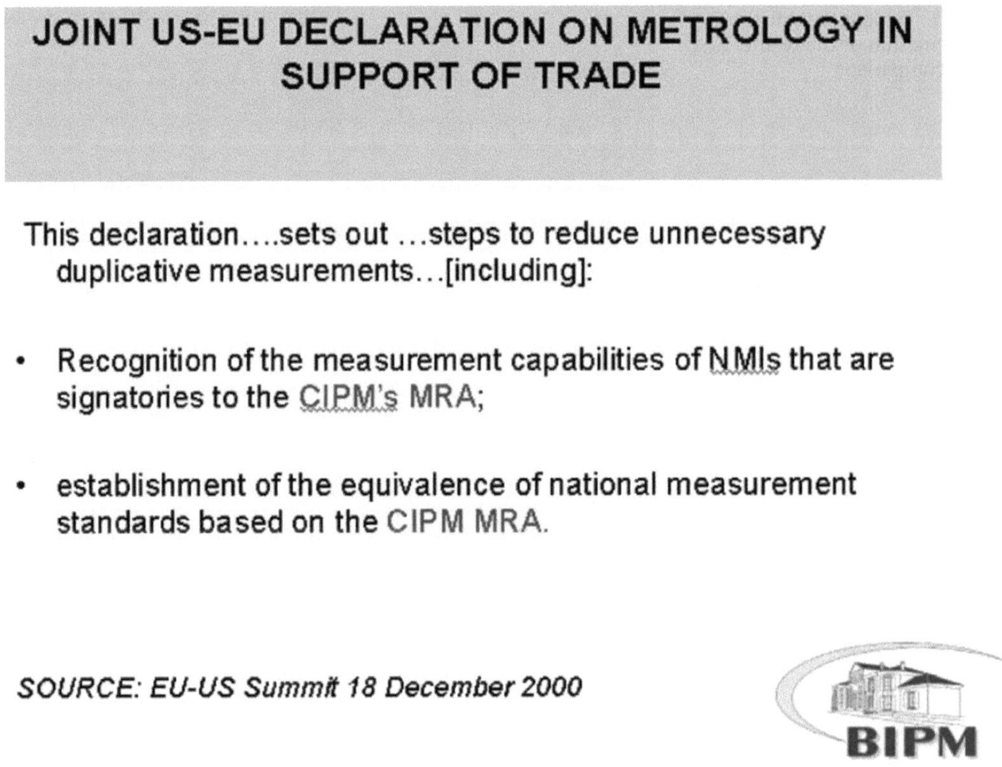

Fig. 2 Text accompanying the EU-US declaration on metrology in support of trade

MRA is cited in the metrological documents that underpin the December 2000 European Union/United States trade agreements (Fig. 2). It has even more power in the market place because calibration customers can use it to "shop around" and find NMIs anywhere in the world that can offer the services they want. They can check out the accuracy offered and with a few e-mails or telephone calls can find details of their quality of service such as price and turn-round. Many of the NMIs are now experiencing competition for their services and some are seeing 10% or more of their customers coming from outside their home base.

The MRA therefore helps market acceptance of certificates and helps reduce technical barriers to trade in two specific ways. First, it provides an agreed framework within which the calibration certificates from NMIs are formally accepted by the NMI signatories. Second, it provides a way of identifying international traceability and equivalence links between NMIs. This information is being used constructively as a way of demonstrating compliance with legislative or other national requirements for traceability to a named NMI.

The KCDB and MRA are still in their early days but independent economic analysis studies have already shown its huge impact on the efficiency with which multilateral equivalence arrangements can be put in place between NMIs. It has also shown that they could reduce the costs of trade compliance by approximately € 4 thousand million. It is also clear to those involved that it is now time to promote the benefits and the use of the database to potential user communities and invite feedback and comment on its value and ease of use.

Accred Qual Assur (2004) 9:2–4
DOI 10.1007/s00769-003-0716-7

Wolfgang Richter

Comparative study of the presentations at the CCQM workshop on traceability

Received: 24 October 2003
Accepted: 24 October 2003
Published online: 27 November 2003

W. Richter (✉)
Stadeweg 1,
38106 Braunschweig,
Germany
e-mail: wolfgang.k.richter@t-online.de

The articles in this and the previous issue of ACQUAL (Accred Qual Assur 8:10) originate from the Consultative Committee for Amount of Substance (CCQM) Workshop on Traceability held at the Bureau International des Poids et Mesures (BIPM) in April, 2002. Not all of the workshop presentations have been submitted for inclusion in these special issues. However, some articles which were not presented at the workshop but which focus on the same topic have been included.

As described by BIPM in the introduction to the workshop programme, the aim of the workshop was to provide a clear description of how metrological traceability can be disseminated to "field" laboratories at the routine working level. The current situation was described from the viewpoint of metrologists, and society, regulators and industry.

The contributors to the workshop were requested to tackle such questions as:

- What does traceability mean to you?
- How do you ensure traceability?
- How do you disseminate traceability?
- How do you ensure traceability of your own results?

All in all, these questions have been tackled in the contributions but not in a systematic way and not by every author. Traceability is treated quite differently depending on the individual viewpoint. National Metrology Institutes, for example, as providers of metrological traceability (more precisely: providers of reference points or end-points of traceability), are more concerned with ensuring the traceability of these reference points and their dissemination down to the working level. For field laboratories which "receive" traceability, the most important question is how to establish and demonstrate metrological traceability of their analytical results, when asked by customers, regulatory bodies and accreditation bodies. This diversity of views and kinds of treatment turn out to be of great advantage, because a complete picture of the present state of traceability of chemical measurements results emerges.

A very broad spectrum of applications is encompassed in the papers, ranging from the determination of element concentrations in water to biotechnology-created substance concentrations in food. This range is particularly broad in terms of the degree of difficulty of establishing traceability. The reports clearly show that whereas traceability in the former case has been largely implemented, its establishment in the latter case is still outside the available possibilities.

The presentations published can be roughly grouped into five categories:

1. Establishment of traceability infrastructures (six contributions)
   The largest group describes existing or planned infrastructures to carry out the establishment of traceability, mostly within national frameworks, but also pursuing international aims. Building up practical traceability infrastructures is obviously regarded as the most important task in metrology in chemistry, after a common understanding has largely been reached that the metrological concepts of traceability and measurement uncertainty are necessary and useful tools to promote comparability of measurement results, and

hence the mutual acceptance of these in a global framework.
It is described in more or less detail how traceability can be disseminated through such infrastructures down to the working level where chemical testing laboratories are increasingly required to demonstrate traceability of their measurement results to reliable references. How these references are provided by (mostly) national institutes is also explained. Descriptions of historical developments are also included, not without some positive appraisal of what has been achieved already in the individual institutes.

2. Guidance on how to establish traceability in practice (one contribution)
Reference is made to the new Eurachem/CITAC guidance document on traceability for chemical laboratories, which was still in a draft state at the time of the workshop. The basic principles and some important implications are discussed. The document is a very useful guide for practitioners in the field of chemical analysis, facilitating the understanding of the necessity and benefit of traceability of chemical measurement results, and giving detailed guidance on how to establish it at the laboratory level. Such guidance is very important for extending and completing the traceability chain down to the working level, so that full advantage can be taken of structural elements for traceability existing above the field level.
There is, however, a tendency to regard traceability as an end in itself in this connection and not merely as a tool. This may be justified here, because from the viewpoint of field laboratories the establishment of the metrological traceability of their results is in fact a central issue which is worth regarding as an important aim.
3. Basic considerations on traceability (two contributions)
Here issues of clarity of understanding and of properly defining traceability are in the foreground of the discussion. The question "Traceability to what?" is discussed in detail. It is proposed that traceability be regarded as the ability to demonstrate that measurement results are what they are purported to be. Since measurement results are always expressed by a product of a numerical value and a unit of measurement, this view implies that it is the relevant unit which forms the end-point of a traceability chain, or, formulated in the reverse way, traceability provides the units in which results are expressed.
Taking into account the complexities of practical chemical and biological measurements, the proposal is made that, in practice, a shortening of the distance between units and their expressions should be contemplated. Reference laboratories are regarded as useful for this purpose because they are able to perform matrix-independent reference measurements on samples submitted by field laboratories, the samples then being returned with reference values attached. This idea is also presented in the next category of contributions.
4. Traceability in clinical chemistry/laboratory medicine (three contributions)
Clinical chemistry rightly claims to have the longest history of applying the concept of metrological traceability in chemical measurement. Nevertheless, it belongs to the areas where establishing and disseminating traceability is particularly difficult, due to the great variety and complexity of analytes and matrices. Driven by the enormous importance of clinical measurement results and the severe consequences of errors, reference systems have been developed and tailored to the different requirements in different fields of application. The EU directive on in-vitro diagnostic medical devices acts as an additional driving force for such developments. Measurement results obtained by means of reference methods form the end-points of a traceability chain. Isotope dilution mass spectrometry (IDMS), a potentially primary method, is often used to establish traceability to SI units, at least for well-defined measurands/analytes. Many of the physiologically important metabolites and substrates, electrolytes, hormones and drugs belong to this group. Dissemination to the numerous medical laboratories in the field usually takes place via proficiency testing systems in the framework of external quality assurance. "Real-life" samples with traceable reference values attached to them are used for proficiency testing rounds. The majority of the clinical measurands are not yet sufficiently well defined. For example, important substances or groups of substances such as proteins, enzymes, proteohormones, tumor markers and cardiac markers belong to this group. Here the next step is the definition of the measurands before reference systems can be established. Conventional reference systems will be the utmost that can be achieved in these cases in the near future. Essential progress with respect to global solutions is expected from the Joint Committee on Traceability in Laboratory Medicine (JCTLM) recently founded under the leadership of the Comite International des Poids et Mesures (CIPM)/BIPM.
5. Traceability in testing food derived from modern biotechnology (one contribution)
The current methods available for testing food derived from modern biotechnology, e. g. for the content of genetically modified food, provide results in internal, not easily convertible units. This makes compliance with legal limits, which are usually given as mass fractions, difficult or even impossible. In order to improve this unsatisfactory situation, a definition of the measurands and standardization of all details of the evaluation procedure is proposed. Such measures

will at least create comparability within a system in which the standardization requirements are strictly followed. This is probably the best that can be achieved in the near future. Traceability to SI units requires further research and development.

In summary, the contents of the contributions can be used to make the following statements with respect to the present situation of traceability of chemical measurement results:

- It is now widely accepted that metrological traceability, i.e. traceability to internationally recognized references like SI units, is an indispensable prerequisite for achieving comparability and hence confidence in, and acceptance of, chemical measurement results in a worldwide framework.
- National traceability or dissemination structures, internationally linked via comparison measurements under the CIPM Mutual Recognition Arrangement, are beginning to play a central role in establishing traceability of chemical measurements in practice.
- Reference materials are the most important transfer standards for disseminating traceability (more precisely: the units of measurement) to the user level. Availability of certified reference materials with evidence of their metrological quality is, however, very limited. Great efforts are necessary to improve this situation.
- In addition, reference measurements carried out by competent laboratories on customer samples are increasingly required to eliminate or, at least, reduce the matrix mismatch problem which is almost ubiquitous in chemical measurement and cannot be solved with off-the-shelf reference materials due to the great variety of different matrices occurring in practice. This is particularly important in "difficult" areas like clinical chemistry and food analysis.
- In proficiency testing, as a means of linking laboratories to existing traceability structures, the concept of traceable reference values assigned to the test samples is increasingly adopted instead of using the laboratory average as the reference value. The leading example of proficiency testing on the basis of traceable reference values is the International Measurement Evaluation Programme (IMEP) of the Institute for Reference Materials and Measurements (IRMM) of the European Union (in existence since 1986, within the nuclear measurement community since 1976). Proficiency testing is a very useful tool in quality assurance; it can propagate traceability, but cannot establish it.
- The metrological concept of traceability has the longest tradition in clinical chemistry as compared with other fields of chemical measurement. The severity of consequences of erroneous measurement results for patients, clinicians and health care systems was the driving force behind introducing it over nearly two decades ago. A clear distinction is made here between homogeneous or well-defined measurands/analytes for which traceability to SI units can be established, and heterogeneous or not well-defined measurands. Reference networks are urgently required for the latter category. Progress is expected from JCTLM.
- In clinical chemistry there are "sectorial" and "local" traceability structures, depending on whether comparability (as a result of traceability) within such structures is restricted to distinct sectors of application or to geographical regions, respectively. Although traceability to SI units is the ultimate metrological goal, at least for the time being, sectorial or local traceability structures are the best that can be provided to establish some kind of traceability. This is the case in such fields where the result critically depends on how the measurand is defined. The global network of reference laboratories of IFCC, which provide primary reference procedures for enzyme activity measurements, is an example of a sectorial traceability system providing worldwide comparability. On the other hand, the Cholesterol Reference Method Laboratory Network (CRMLN) of the Center for Disease Control and Prevention (CDC) is an example of a sectorial traceability system which is mainly restricted to one region, the United States.

As a whole the contributions show that the discussion about traceability is more intense and fundamental in the field of chemical measurement than in metrology in general. A simple explanation could be that in "classical" metrology it can usually be taken for granted that traceability is in place whereas in chemistry it still has to be built up. The problems which have still to be solved are impressively described in these two special issues of ACQUAL, particularly in the contributions on clinical chemistry and food testing. As a result, a greater insight and better understanding are achieved for the benefit of the whole of metrology.

Accred Qual Assur (2003) 8:340–345
DOI 10.1007/s00769-003-0666-0

Mathias M. Müller

Traceability in laboratory medicine

Received: 7 January 2003
Accepted: 5 May 2003
Published online: 28 June 2003

Presented at the International ILAC/IAF Conference on Accreditation in Global Trade, 23–25 September 2002, Berlin, Germany

M. M. Müller (✉)
Institute of Laboratory Diagnostics,
Kaiser Franz Josef Hospital,
Kundratstrasse 3, 1100 Vienna, Austria
e-mail: mathias.mueller@kfj.magwien.gv.at
Tel.: +43-1-60191-3301
Fax: +43-1-60191-3309

Abstract In laboratory medicine meaningful measurements are essential for diagnosis, risk assessment, treatment and monitoring of patients. Thus methods applied in diagnostic measurements must be accurate, precise, specific and comparable among laboratories. Inadequate or incorrect analytical performance has consequences for the patients, the clinicians, and the health care system. One key element of metrology is the traceability of a measurement result to the SI system ensuring comparable results. This principle is described in the ISO/TC 212/WG2 N65 prEN 17511 Standard. In addition to the principles of metrology, the clinical usefulness, the diagnostic needs, and the biological and disease associated variations in patients' specimens have to be considered when the analytical biases for diagnostic purposes are defined. It must be the general goal of diagnostic laboratories to produce results that are true and comparable worldwide. The recent European in vitro diagnostic (IVD) Directive 98/79 EC follows the above mentioned standard of the International Organization for Standardization (ISO) and the European Committee for Standardization (CEN) requesting its application for all IVD reagents used within the European Union. This new European legislation will have a worldwide impact on manufacturers and clinical laboratories and will be implemented in 2003. It states that "traceability of values assigned to calibrators and/or control materials must be assured through available reference measurement procedures and/or available reference materials of a higher order". Thus a worldwide reference system needs to be established by collaboration and mutual recognition between the United States National Institute of Standards and Technology (NIST), European Metrology Institutes (EUROMET), regulatory bodies (e.g. United States Food and Drug Administration, FDA) the IVD industry and professional organizations (e.g. International Federation of Clinical Chemistry and Laboratory Medicine, IFCC). In June 2002, in Paris, representatives of international and regional organizations and institutions decided to form the "Joint Committee on Traceability for Laboratory Medicine" (JCTLM), which will support industry in registration and licensing of the "CE" label to test systems conforming to the IVD Directive.

Keywords Traceability · Laboratory medicine · Biological variation · Analytical bias · Quality assurance · Standardization · Reference measurement system · Joint Committee on Traceability for Laboratory Medicine

Introduction

In laboratory medicine meaningful, accurate and precise routine measurements are essential for diagnosis, risk assessment, treatment and follow-up of patients using physical, chemical, biochemical, immunological and molecular biology techniques for measurement and detection of compounds in body fluids, tissues and cells. In order to achieve these goals within the integrated health care system and to improve the quality of test results diagnostic laboratories implement quality systems. Many guidelines for quality management have been published and it is generally assumed that by implementing them the pre-analytical, analytical and post-analytical steps in the overall diagnostic process will be improved and become more efficient. The overall quality of the laboratory report depends on the following steps:

- Rationale, disease-oriented test selection, diagnostic algorithms
- Preparation of the patient, sampling of specimens
- Pre-analytical handling of specimens
- Accurate and precise analytical performance
- Selection of appropriate, sensitive and specific methods
- Calibration of analytical systems
- Internal and external quality assurance
- Post-analytical handling of test results
- Clinical interpretation of test results
- Reporting process
- Patient- and disease-oriented consultation with clinicians.

Standardization of all these important steps will improve the overall diagnostic quality and will have an enormous economic impact. One key element in this complex process is related to the analytical measurement in biological samples. Incorrect and non-comparable analytical performance of laboratory tests have severe consequences for clinical medicine and the patient due to wrong diagnosis, wrong treatment, psychological stress and additional costs for diagnostic procedures. Major difficulties in accurate measurement in clinical laboratories are related to the complexity of the measurands (Table 1), their biological variations due to age, sex, diet, time and posture, as demonstrated for short- and long-term intra-individual variations of measurands performed in clinical laboratories. It was concluded that the use of laboratory data for clinical diagnosis is considerably improved when intra-individual variations and critical differences including biological and analytical variations are used instead of differences compared to group reference ranges (Table 2) [1, 2, 3]. In addition, the complexity of the matrix has an impact on the analytical performance when investigating body fluids (blood serum or plasma, urine, cerebrospinal fluid, ascites, tissue, cells).

Table 1 Classification of measurands in laboratory medicine

Homogeneous	Heterogeneous
Glucose	Proteins peptide-bond (biuret reaction) epitopes (immunoassays—antibodies)
Creatinine	Enzymes activities (defined conditions) mass (immunoassays–antibodies)
Cholesterol total–fractions	Glycoproteins isoforms—glycoforms
Electrolytes total–activity–free	
Steroids–Thyroxine free bound to proteins	

Table 2 Intra-individual variations of clinical chemistry measurands in serum [1]

Measurand	Analytical CV (%)	Daily CV (%)	Weekly CV (%)	Monthly CV (%)
Na	0.6	1.4	0.8	1.3
K	1.0	7.8	6.7	7.3
Glucose	1.5	25.8	16.8	20.8
Uric acid	1.0	9.8	12.4	14.3
Alanine aminotransferase	0.9	10.3	32.2	47.5

Quality assurance—the present state of the analytical performance

Proficiency testing, and internal and external quality assurance are nowadays mandatory and integral to running a diagnostic laboratory [4]. The results of proficiency testing programmes not only give insight into the analytical performance of individual laboratories but also allow comparison of commonly used analytical techniques. In most of these programmes fitness-for-purpose criteria based on so-called "peer-group" target values are used for grading the participating laboratories. These fixed limits of acceptability may be based on biological variation and the potential of routine methods used. Thus these limits might change with time, methodology and the analytical performance of participating laboratories; they are not based on medical needs but they reflect the common state. For assigning the values to the specimens (usually lyophilized human-based serum) either reference methods are used or the values transferred with a reliable method from a certified reference material. A kind of traceability is established. In the International Measurement Evaluation Programme (IMEP) conducted by the Institute for Reference Materials and Measurement (IRMM), Geel, Belgium, target values are assigned

Table 3 Acceptance limits set by external quality assessment (EQA) organizers for inorganic components in human serum based on biological variations and the potential of routine methods and observed ranges in the International Measurement Evaluation Programme (IMEP)-7 Evaluation Programme [5]

Measurand	Target range ± % EQA accepted limits	Observed range ± % of majority of IMEP-7 participants
Ca	3.0–4.7	5
Cl	2–3	4
Cu	10–12	15
Fe	11	7
K	5.5	3
Mg	6–15	10
Na	2.2	2
Se	9–10	20
Zn	10–12	15

by reference methods, thus a traceability chain and the analytical bias of routine methods can be established (Tables 3 and 4) [5]. However, in most proficiency programmes the analytical performances reported may not reflect the true state of the art of an individual laboratory, since the control specimens may not behave like a patient's sample with the routine method applied due to matrix effects and a lack of commutability of the control specimen.

With the introduction of quality assurance in the diagnostic laboratory 56 years ago [6], a kind of educational and benchmarking process started forcing laboratories, national and international organizations, and the IVD industry to improve the methods applied in clinical laboratories. Comparison of the measurements of enzyme activity demonstrate that the analytical performance of the methods applied 30 years ago were far beyond the biological variation and most probably insufficient for medical needs. Interlaboratory comparisons show that with the new routine methods based on recommendations of the International Federation of Clinical Chemistry and Laboratory Medicine (IFCC) (Table 5) comparable results can be obtained irrespective of time and space and thus small individual variations can now be detected. Similar improvements in the analytical process in clinical laboratories can be reported generally for homogenous measurands.

However, the results obtained with immunoassays for proteins are quite different owing to their biological variation, the existence of various isoforms in health or disease and the lack of standardization so far. A variety of heterogeneous measurands like tumour markers and proteo-hormones have not been characterized properly, thus no reference measurements procedures and certified reference materials are available. Therefore no conformity of measurements can be obtained. The target values assigned to human serum matrixed reference materials used in proficiency programmes are method or reagent dependent, with analytical bias sometimes far beyond the clinical decision criteria. This means that clinicians might misclassify patients when non-method-specific reference ranges are applied for the interpretation of test results.

Table 4 Interlaboratory comparison of enzyme activity measurements. Source: Austrian Proficiency Testing Programme

Year	1970	1982	1992	2002
Participants	36	269	603	1358
AP	33.9	11.5	5.6	5.9
ALAT	67.8	17.2	5.0	5.9
ASAT	36.3	20.8	5.3	6.0
LDH	32.0	10.1	5.9	3.7

Table 5 International Federation of Clinical Chemistry (IFCC) reference methods

Reference methods	
PH	Maas A. et al. [7, 8, 9, 10]
Tonometry	Burnett AW et al. [11, 12, 13, 14, 15, 16]
Na, K	Burnett AW et al. [17]
Ca^{++}	Burnett AW et al. [18]
Apo A1	Barr JR et al. [19]
HBA1c	Jeppsson J-O et al. [20]
ALAT	30 °C: Bergmeyer H-U et al. [21]; 37 °C: Schumann G et al. [22]
ASAT	30 °C: Bergmeyer H-U et al. [23]; 37 °C: Schumann G et al. [24]
Amylase	30 °C: Lorentz K [25]
CK	30 °C: Horder at al. [26, 27, 28, 29]; 37 °C: Schumann G et al. [30]
GGT	30 °C: Shaw LM et al. [31]; 37 °C: Schumann G et al. [32]
LDH	30 °C: Bais R et al. [33, 34, 35, 36]; 37 °C:Schumann G et al. [37]

Standardization—reference systems—traceability in laboratory medicine

This analytical dilemma and the non-conformity of test results obtained for complex, heterogeneous measurands stimulated a movement towards standardization implemented by various international organizations such as the IFCC (Table 6). All these efforts tried to follow metrologically established rules according to the *International Vocabulary of Basic and General Terms in Metrology* (VIM) [38] and ISO-CEN standards (Table 7)

Table 6 IFCC reference materials

Reference materials	
Apo A1, B	WHO: SP1, SP3
Albumin	WHO: 74/1
Plasma Proteins	IRMM: CRM 470
PSA free, complexed	WHO: 96/668, 96/700
α-Amylase	IRMM: 456
ALAT	IRMM: 454
ASAT	IRMM: in preparation
CK-MB	IRMM: 455
GGT	IRMM: 452
LDH-1	IRMM: 453
Cortisol reference panel in fresh frozen human sera (1–17)	IRMM: 451
HCG (6 primary standards):	WHO:
Intact	99/688
Alpha subunit (hCG-alpha)	99/720
Beta core fragment (hCG beta-cf)	99/708
beta subunit (hCG-beta)	99/650
nicked (hCG-n)	99/642
nicked beta subunit (hCG-beta-n)	99/692
Lp(a), HbA1c, myoglobin	In preparation

Table 7 ISO/CEN Standards essential for reference systems in laboratory medicine

Standard	Title
ISO/EN 15195	Requirements for reference measurement laboratories in laboratory medicine
EN 12286	Measurement of quantities in samples of biological origins—Presentation of reference measurement procedures
EN 12287	Description of reference materials
ISO 17025	General requirements for the competence of testing and calibration laboratories
PrEN ISO 15189	Competence requirements for medical laboratories
ISO/EN 17511	Measurement of quantities in samples of biological origin—Metrological traceability of values assigned to calibrators and control materials

by establishing a **reference system** in laboratory medicine [39, 40]. Based on the characterization of the measurands, definitive reference measurement procedures, or at least commonly approved consensus methods were developed by expert laboratories and used for assigning target values to certified reference materials. By using these kinds of reference systems (reference measurement procedure, reference material, reference laboratories) target values for field calibrators can be assigned and the analytical bias of field methods can be established. Thus traceability by an unbroken chain of comparisons each having a stated uncertainty can be implemented [38]. In most of these standardization projects, field studies are conducted in order to link the new "standardized" measurement results to the clinical needs. With this kind of approach a better comparability and trueness of test results worldwide is envisaged.

One key element of metrology is traceability of a measurement result to the SI-system ensuring comparable results. **Traceability** is defined as the property of the result related to national or international standards through an unbroken chain of comparisons all having stated uncertainties. The overall objective in the hierarchy of measurement procedures is dedicated to the highest available analytical quality. This principle is described in the ISO/TC 212/WG2 N65 prEN 17511 Standard [41]. Following these rules, results of diagnostic measurements have to be comparable and true wherever they are performed in the world. The recent European IVD directive 98/79 EC follows the ISO/CEN Standard, requesting its application to all IVD reagents used within the European Union [42]. This new European legislation will have a worldwide impact on manufacturers and clinical laboratories and will be implemented in 2003. It states that "traceability of values assigned to calibrators and/or control materials must be assured through available reference measurement procedures and/or available reference materials of a higher order". Thus a worldwide reference system needs to be established by collaboration and mutual recognition between the United States National Institute of Standards and Technology (NIST), European Metrology Institutes (EUROMET), health authorities, regulatory bodies, EQAs organizations, theIVD industry and professional organizations (e.g. IFCC).

In laboratory medicine more than 500 measurands are used for diagnosis and follow-up of patients [43, 44]. At presence only some 100 are traceable to the SI as described in Table 1. The vast majority of measurement results are not traceable to the SI, but to arbitrary units, for example World Health Organization (WHO) International Units or manufacturer's mass units. Virtually all proteins, lipoproteins and glycoproteins examined belong to this group: most of them are measured or determined by means of an immunochemical reaction, i.e. nephelometric, turbidimetric, immunoassay, saturation analysis. These proteins/glycoproteins are most important parameters in the medical field, such as in oncology, endocrinology/fertility and virology. For all these measurands the establishment of a reference measurement system by interdisciplinary research is urgently needed in order to fulfill the requirements of the written ISO/CEN standards and European legislation.

In order to achieve this large task, all relevant organizations and institutions (Bureau International des Poids et Mesures—BIPM, IFCC, International Laboratory Accreditation Cooperation—ILAC, IRMM, NIST, Advanced Medical Technology Association—AdvaMed, European Diagnostic Manufacturers Association—EDMA) formed the Joint Committee on Traceability for Laboratory Medicine (JCTLM) as a joint venture of professionals in 2002 [45]. The main goal of JCTLM will be: "to achieve international equivalence in laboratory medicine by development of international conventional reference systems comprising reference materials, reference measurement procedures, implementation of reference measurement laboratories for selected and prioritized analytes in relation to medical needs". The aim of the JCTLM following the concept of a global reference system in laboratory medicine was defined as follows:

- Promotion of the traceability concept
- Dissemination of information on reference measurement procedures and certified reference materials
- Coordination/guidance in the establishment of reference measurement systems
- Establish links between reference laboratories and metrology institutes
- Identification of clinically relevant projects conducted by appropriate professional organizations, encouraging interdisciplinary collaboration
- Support and encourage application of reference measurement systems by the IVD industry.

So far two Working Groups (WGs) have been established:

- WG on Reference Materials and Reference Measurement Procedures
- WG on Reference Measurement Laboratories.

Networks of expert laboratories for quantification of well-defined measurands will be competent in using the best internationally recognized analytical procedures, and their main responsibility will be to assign values to reference materials. In addition to the principles of metrology, the clinical usefulness, the diagnostic needs, and the biological and disease associated variations in patients' specimens will be considered when the analytical biases of field methods for diagnostic purposes are defined. JCTLM will enable registration and licensing of the "CE" label to test systems conforming with the IVD Directive. It is envisaged that the JCTLM initiative will result in harmonization and/or standardization of procedures used in medical laboratories, achieve true and worldwide comparable results and have an impact on clinical decision criteria.

References

1. Costongs GM, Janson PC, Bas BM, Hermans J, van Wersch JW, Brombacher PJ (1985) J Clin Chem Clin Biochem 23:7–16
2. Costongs GM, Janson PC, Bas BM, Hermans J, Brombacher PJ, van Wersch JW (1985) J Clin Chem Clin Biochem 23:69–76
3. Costongs GM, Bas BM, Janson PC, Hermans J, Brombacher PJ, van Wersch JW (1985) J Clin Chem Clin Biochem 23:405–10
4. Rej B (2002) Accred Qual Assur 7:335–340
5. Örnemark U, Van Nevel L, Taylor PDP, Robouch P, De Bièvе P, et al. (1999) Accred Qual Assur 4:463–472
6. Belk WP, Sunderman FW (1947) Am J Clin Pathol 17:853–861
7. Maas AHJ, Weisberg HF, Burnett RW, Mueller-Plathe O, Wimberley PD, Zijstra WG, Durst RA, Siggaard Andersen O (1987) J Clin Chem Clin Biochem 25:281–289
8. Maas AHJ, Weisberg HF, Burnett RW, Mueller-Plathe O, Wimberley PD, Zijstra WG, Durst RA, Siggaard Andersen O (1987) Clin Chim Acta 165:97–109
9. Maas AHJ, Weisberg HF, Burnett RW, Mueller-Plathe O, Wimberley PD, Zijstra WG, Durst RA, Siggaard Andersen O (1987) Ann Biol Clin 45:229–236
10. Maas AHJ, Weisberg HF, Burnett RW, Mueller-Plathe O, Wimberley PD, Zijstra WG, Durst RA, Siggaard Andersen O (1987) Biochim Clin 3:241–249
11. Burnett RW, Covington AK, Maas AHJ, Müller-Plathe O, Weisberg HB, Wimberley PD, Zijlstra WG, Sigaard Andersen O, Durst RA (1989) JIFCC 1:78–81
12. Burnett RW, Covington AK, Maas AHJ, Müller-Plathe O, Weisberg HB, Wimberley PD, Zijlstra WG, Sigaard Andersen O, Durst RA (1989) Eur J Clin Chem Clin Biochem 27:403–408
13. Burnett RW, Covington AK, Maas AHJ, Müller-Plathe O, Weisberg HB, Wimberley PD, Zijlstra WG, Sigaard Andersen O, Durst RA (1989) Ann Biol Clin 47:373–376

14. Burnett RW, Covington AK, Maas AHJ, Müller-Plathe O, Weisberg HB, Wimberley PD, Zijlstra WG, Sigaard Andersen O, Durst RA (1989) Biochimica Clinica 13:945–949
15. Burnett RW, Covington AK, Maas AHJ, Müller-Plathe O, Weisberg HB, Wimberley PD, Zijlstra WG, Sigaard Andersen O, Durst RA (1989) Clin Chim Acta 185:S17-S24
16. Burnett RW, Covington AK, Maas AHJ, Müller-Plathe O, Weisberg HB, Wimberley PD, Zijlstra WG, Sigaard Andersen O, Durst RA (1989) J Biomed Lab Sci 2:185–192
17. Burnett RW, Covington AK, Fogh Andersen N, Kulpmann WR, Lewenstam A, Maas AHJ, Muller-Plathe O, Sachs C, Siggaard Andersen O, VanKessel AL, Zijlstra WG. (2000) Clin Chem Lab Med 38:1065–1071
18. Burnett RW, Christiansen TF, Covington AK, Fogh Andersen N, Kulpmann WR, Lewenstam A, Maas AHJ, Muller Plathe O, Sachs C, Siggaard Andersen O, VanKessel AL, Zijlstra WG (2000) Clin Chem Lab Med 38:1301–1314
19. Barr JR, Maggio VL, Patterson, DG Jr., Cooper GR, Henderson LO, Turner WE, Smith SJ, Hannon H, Needham LL, Sampson EJ (1996) Clin Chem 42:1676–1682
20. Jeppsson JO, Kobold U, Barr J, Finke A, Hoelzel W, Hoshino T, Miedama K, Mosca A, Mauri P, Paroni R, Thienpont L, Umemoto M, Weykamp C (2002) Clin Chem Lab Med 40:78–89
21. Bergmeyer HU, Horder M, Rej R (1986) J Clin Chem Clin Biochem 24:481–495
22. Schumann G, Bonora R, Ceriotti F, Ferard G, Ferrero CA, Franck PFH, Gella FJ, Hoelzel W, Jorgensen PJ, Kanno T, Kessner A, Klauke R, Kristiansen N, Lessinger JM, Linsinger TPJ, Misaki H, Panteghini M, Pauwels J, Schiele F, Schimmel HG, Weidemann G, Siekmann L (2002) Clin Chem Lab Med 40:718–724
23. Bergmeyer HU, Horder M, Rej R (1986) J Clin Chem Clin Biochem 24:497–510
24. Schumann G, Bonora R, Ceriotti F, Ferard G, Ferrero CA, Franck PFH, Gella FJ, Hoelzel W, Jorgensen PJ, Kanno T, Kessner A, Klauke R, Kristiansen N, Lessinger JM, Linsinger TPJ, Misaki H, Panteghini M, Pauwels J, Schiele F, Schimmel HG, Weidemann G, Siekmann L (2002) Clin Chem Lab Med 40:725–733
25. Lorentz K (1998) Clin Chem Lab Med 36:185–203
26. Horder M, Elser RC, Gerhardt W, Mathieu M, Sampson EJ (1989) JIFCC 1:130–139
27. Horder M, Elser RC, Gerhardt W, Mathieu M, Sampson EJ (1990) JIFCC 2:26–35
28. Horder M, Elser RC, Gerhardt W, Mathieu M, Sampson EJ (1990) JIFCC 2:80–83
29. Horder M, Elser RC, Gerhardt W, Mathieu M, Sampson EJ (1991) Eur J Clin Chem Clin Biochem 29:435–456
30. Schumann G, Bonora R, Ceriotti F, Clerc-Renaud P, Ferard G, Ferrero CA, Franck PFH, Gella FJ, Hoelzel W, Jorgensen PJ, Kanno T, Kessner A, Klauke R, Kristiansen N, Lessinger JM, Linsinger TPJ, Misaki H, Panteghini M, Pauwels J, Schimmel HG, Vialle A, Weidemann G, Siekmann L (2002) Clin Chem Lab Med 40:635–642
31. Shaw LM, Stromme JH, London JL, Theodorson L (1983) J Clin Chem Clin Biochem 21:633–646
32. Schumann G, Bonora R, Ceriotti F, Ferard G, Ferrero CA, Franck PFH, Gella FJ, Hoelzel W, Jorgensen PJ, Kanno T, Kessner A, Klauke R, Kristiansen N, Lessinger JM, Linsinger TPJ, Misaki H, Panteghini M, Pauwels J, Schiele F, Schimmel HG, Weidemann G, Siekmann L (2002) Clin Chem Lab Med 40:734–738
33. Bais R, Philcox M (1994) Eur J Clin Chem Clin Biochem 32:639–655
34. Bais R, Philcox M (1994) Ann Biol Clin 50:475–492
35. Bais R, Philcox M (1994) Biochim Clin 18:180–212
36. Bais R, Philcox M (1994) J Automat Chem 16:167–182
37. Schumann G, Bonora R, Ceriotti F, Clerc-Renaud P, Ferard G, Ferrero CA, Franck PFH, Gella FJ, Hoelzel W, Jorgensen PJ, Kanno T, Kessner A, Klauke R, Kristiansen N, Lessinger JM, Linsinger TPJ, Misaki H, Panteghini M, Pauwels J, Schimmel HG, Vialle A, Weidemann G, Siekmann L (2002) Clin Chem Lab Med 40:643–648
38. BIPM, IEC, ISO, IFCC, IUPAC, OIML (1993) International Vocabulary of Basic and General Terms in Metrology, 2nd edn. ISO, Geneva, Switzerland
39. Müller MM (2000) Clin Chem 46:1907–1909
40. Müller MM. (1998) Clin Biochem 31:433–436
41. ISO/TC 212/WG2 N65 prEN 17511 (2000) In vitro diagnostic medical devices—Measurement of quantities in samples of biological origin—Metrological traceability of values assigned to calibrators and control material. ISO/CEN, Geneva, Switzerland
42. EU Lex: Directive 98/79 EC (1998) On in vitro diagnostic medical devices. Official J L 331:1–37
43. Thomas L (1998) Labor und Diagnose. TH-Books, Frankfurt, Germany
44. Jacobs DS, De Mott WR, Grady HJ, Horvat RT, Huestis DW, Kasten BL (1996) Laboratory test handbook. Lexi-Comp, Hudson, USA
45. http://www.bipm.org/enus/2_Committees/JCTLM.shtml

Accred Qual Assur (2003) 8:454–460
DOI 10.1007/s00769-003-0647-3

Markus Lipp

Testing for foods derived from modern biotechnology: opportunities and limitations for metrology

Received: 17 January 2003
Revised: 27 March 2003
Published online: 3 July 2003

M. Lipp (✉)
Monsanto Company,
800 North Lindbergh Boulevard,
St. Louis, Missouri, 63167, USA
e-mail: markus.lipp@monsanto.com
Tel.: +13-14-694-8644
Fax: +13-14-694-8562

Abstract Various countries have established labeling schemes for food derived from modern biotechnology. As a consequence, test methods need to be available to industry and regulators. The three test options, bioassays, protein-based and DNA-based test methods, are discussed. None of these methods is able to directly measure the percentage of foods derived from modern biotechnology by weight (weight-%), the unit in which most of the thresholds for food labeling in the different countries (if any) have been established. The conversion of the measurement units to weight-% is difficult to achieve and influenced by a number of biological factors. Metrology can aid the standardization of methods enormously by defining clearly the relationships between measurement units and other units of interest (e.g. legal thresholds).

Keywords biotechnology · Genetically modified organisms · Polymerase chain reaction · Reference materials

Introduction

Since the introduction of biotechnology in agriculture, there has been an ongoing debate regarding the need to label foods containing the products of modern biotechnology. Historically, foods are labeled for the following reasons:

1. Safety concerns (e.g., allergenic components)
2. Nutritional interest (e.g., fat content)
3. Ethical concerns (e.g., vegetarian, kosher)
4. Right of consumer to be informed (e.g., preservatives)

Today, many countries practice a rigid review system in order to guarantee food safety with respect to foods derived from modern biotechnology. Safety studies for each product are reviewed by regulatory agencies of these countries to ensure that neither its population nor its environment suffers from the consumption and placement on the market of foods derived from modern biotechnology. The overwhelming scientific conclusions from these agencies as well as from most scientific studies are that:

1. Foods derived from modern biotechnology are as safe as current food products and do not pose a hazard to the environment.
2. Current foods derived from modern biotechnology are nutritionally equivalent to their conventional counterparts.
3. Current foods derived from modern biotechnology do not pose additional ethical concerns.
4. However, it is felt that the consumer has the right to be informed about the presence of foods derived from modern biotechnology.

The right of the consumer to be informed has been introduced into the legislation of several countries for the labeling of final foods derived from modern biotechnology. The specific regulations across countries for labeling of final foods derived from modern biotechnology show considerable differences but two distinct regimes can be differentiated: mandatory labeling regimes and voluntary labeling regimes.

Once labeling guidelines or regulations are in place, means of enforcement are needed. In other words, test systems for the presence of foods derived from modern biotechnology are needed for the final food as well as throughout the food production chain. The following will discuss some of the details with respect to labeling and testing of foods derived from modern biotechnology.

Labeling

Voluntary labeling. The USA and a few other countries have imposed a voluntary labeling scheme. For foods derived from modern biotechnology that are considered to be substantially equivalent to their conventional counterpart, i.e., equivalent with respect to nutritional aspects and consumption patterns, no labeling is mandated by those countries. However, with respect to foods that do NOT contain any products of modern biotechnology, the producer has the right to inform the consumer accordingly. In order to avoid a misuse of this 'Does not contain products derived from biotechnology' label, several prerequisites have to be fulfilled, which will not be discussed here in detail.

However, for all products derived from modern biotechnology that would be considered NOT to be substantially equivalent to current commercial products, all these countries would require an appropriate label.

Mandatory labeling. The European Union and other European countries as well as several of the Asian and South American countries have imposed a mandatory labeling scheme. That is to say, the presence of foods derived from modern biotechnology must be indicated on all food labels. However, most countries do grant exemptions from their mandatory labeling schemes. In general, the exemptions from labeling are related to the adventitious (inadvertent) presence of foods derived from modern biotechnology and the associated thresholds vary from country to country, ranging between 1% and 5%, typically by weight. Additional exemptions may be related to specific food matrices, e.g., refined oils or to its function in the final food, e.g., process aids, food additives, flavorings and colorings. All thresholds refer to the amount of foods derived from modern biotechnology in relation to the total weight of food (weight-%). However, this is not necessarily explicitly stated in all of the corresponding regulations, e.g., the Australian labeling provisions do state explicitly weight-% as the unit for a threshold, while the European provisions do not give any indication of to what its 1% threshold refers.

Mandatory labeling schemes that do not provide a threshold for exemption of labeling are very difficult to enforce and are virtually impossible to comply with. Under such regulations, labeling will be driven by the sensitivity of testing methods, which are rapidly improving. This leads to a very unstable situation for enforcement laboratories and the food producing industry. Mandatory labeling schemes need to be accompanied by provisions for a threshold in order to be predictable and manageable for enforcement and industry.

Testing for food derived from modern biotechnology

In general there are two categories for testing for the presence of food derived from modern biotechnology. Testing for the phenotype (herbicide tolerance or proteins) includes:

1. Testing for the phenotype (herbicide tolerance or proteins)
   1. Bioassays, with respect to detecting a specific phenotype which confers herbicide resistance
   2. Detection of proteins that are responsible for the altered phenotype, using (a) enzyme-linked immunosorbent assays (ELISA), and (b) lateral flow devices.

2. Testing for the genotype (DNA)
   1. Qualitative polymerase chain reaction (PCR)

Bioassays

This reliable but simple test of seeds (and some grain) is designed to determine the number of seeds/seedlings that show herbicide tolerance. The test consists of a medium in which the seeds/seedlings are germinated or grown in the presence of an herbicide, e.g., [1, 2]. The germination or survival rate indicates the amount of biotechnologically enhanced seeds/grains that are tolerant for a selected herbicide. This test is only applicable if a high germination rate of the seed/grain can be assumed and by its very nature relies on herbicide tolerance as the trait under investigation. This test is not applicable for varieties that do not exhibit any tolerance against a selected herbicide or for processed or heat-treated seeds.

Detection of proteins

Most of the biotechnologically enhanced plant varieties currently on the market express a new protein, which confers the desired phenotype. For example, Roundup Ready varieties express an introduced form of the protein 5-enolpyruvyl-shikimate-3-phosphate synthase (EPSPS), which is slightly different to the endogenous EPSPS protein produced by the plant. This slightly different form is less susceptible to glyphosate (being the active ingredient in the family of Roundup herbicides) and thus will allow these plants to continue to grow after an application of glyphosate.

Analytical tools to detect this particular protein rely on the use of antibodies, which specifically recognize the newly introduced protein. Two major applications are available: lateral flow devices (also known as strip test or dipstick), and ELISA. An overview and an in-depth discussion of protein-based methods for the detection of grain derived from modern biotechnology can be found in the literature e.g., [3, 4].

Lateral flow test devices are very easy to use, reliable, fast and specific. Several manufacturers supply these tests and their performance is verified and published (http://www.usda.gov/gipsa/biotech/rapidtest.htm) with respect to the manufacturers specifications by the US Department of Agriculture-Grain Inspection, Packers and Stockyard Administration (USDA-GIPSA) [5]. Although this test is qualitative by nature, it is possible to derive an answer in respect of compliance to a given threshold through the application of appropriate sampling techniques. USDA-GIPSA published details on this sampling scheme on the internet as well (http://www.usda.gov/gipsa/biotech/sampling_grains_for_biotechnolog.htm).

For example, calculating the size of a single sample that ensures that in 95% of all cases the lot under investigation does contain less than 1% of biotechnologically enhanced seeds of interest, it can be calculated that a sample size of 299 kernels is needed. The underlying assumption is however, that the test applied is sensitive enough to detect the presence of at least one single biotechnologically enhanced kernel and that the sample is representative for the bulk of material.

Another format to test for newly expressed proteins is provided through different ELISA assays. Typically, one antibody is coated on a microtiter plate and serves as a capture antibody while a second antibody (added later in the process) is labeled with a reporter molecule allowing the read-out with optical devices. These ELISAs can be operated in a quantitative manner, but need to be calibrated. The measurement unit can be traced back to the amount of protein present in the calibrator, independent of whether the calibrator consists of purified proteins or other biological materials (e.g., seeds, leaves). The amount of proteins within a plant-derived matrix (leaves, seeds, grains), however, depends on several factors, including environmental conditions and can thus not directly be related to thresholds expressed in weight-%.

As antibodies may cross-react with other proteins both methods could lead to false positive results. Further, the alteration of the structure of the protein could render the protein unrecognizable for the antibodies causing false negative results. Changes in the structure of proteins can be induced through any kind of processing, notably, heat treatment, and changes in pH. Additionally, for processed food any separation steps, e.g., separating proteins from cornstarches, can influence the sensitivity of the method. Consequently, protein-based methods are best suited for the analysis of unprocessed grain and/or seed. Finally, the accuracy of these methods is critically dependent on the ability to extract efficiently the introduced protein from the plant material or food. Therefore, extraction conditions and buffers must be validated for each matrix individually.

Testing for the genotype

All current varieties of plants enhanced by biotechnology have new pieces of DNA introduced in their genome. These pieces of DNA are unique and well characterized and thus can be tested for. The most commonly used analytical technique is PCR, which consists of a set of repetitive enzymatic amplification steps performed on a small and specific part of the genome in order to obtain a high enough number of copies of specific DNA that it can be detected by the assay.

PCR can be performed as a qualitative or a quantitative method. PCR techniques are typically more sensitive than protein-based tests and the utmost care has to be taken to avoid false-positive results due to contamination of samples, equipment, working areas, etc. The specificity of a PCR reaction depends only on the specificity of the stretch of DNA the analyst is focusing on. The analyst has to be very careful in selecting the appropriate method. For example, a method detecting the 35S promoter of the cauliflower mosaic virus could be an indicator for several biotechnological enhanced varieties at the same time (but not all of them). However, it will be a poor analytical target for quantitative analysis as the number of promoters present within the different biotechnological enhanced varieties is not constant. Furthermore, the cauliflower mosaic virus infects many vegetables of the Cruciferae family and can thus easily cause false positive results in processed foods. The main feature of the analytical techniques are summarized and compared in Table 1.

Method validation

As for all other analytical methods, methods for the analysis of products derived from modern biotechnology need to be validated. This validation should follow international standards (e.g. ISO 5725) and comply with, 'The IUPAC/AOAC/ISO harmonized protocol of method validation' [6] or the standards cited above. Validation of testing methods is greatly facilitated by the use and availability of the appropriate reference materials. The following paragraphs specify some important

Table 1 Comparison of three test options for foods and seeds derived from modern biotechnology. *IP* Identity preservation

	Bioassay	Protein methods	DNA methods
Scope	Whole kernel	Unprocessed material	All
Traits	Herbicide tolerance	Most herbicide tolerant and insect protected plants	All
Complexity/price	Low/cheap	Low/cheap	High/expensive
Speed	Slow	Fast	Medium
Robustness	High	High	Medium
Reliability	High	High	High-low
Identification	Not possible	Not possible	Possible
Typical application	Seed and some grain testing	Grain, leaves and some seed testing	Food or food ingredients, feed or feed ingredients
Typical scope	Seed purity	Grain movements, IP-systems or channeling	Processed food or feed

issues for standardization of methods and reference materials.

Anklam et al. [7] as well as Ahmed [8] recently published a comprehensive overview of different PCR assays that have been published in the literature. The authors tried to include performance data adding to the value of the review articles. The validation of PCR methods and thus the establishment of such performance criteria is still the subject of much debate. Hübner et al. [9] suggested an approach for the validation of PCR assays. In general, it is currently the view of most researchers that validation of a PCR assay should not differ essentially from the validation of other analytical methods. Thus, all principles outlined in the ISO standard 17025 'General requirements for the competence of testing and calibration laboratories', ISO standard 5725 'Accuracy (trueness and precision) of measurement methods and results' as well as the principles as laid down by Codex Alimentarius (http://www.codexalimentarius.net), are applicable to PCR.

Method standardization

It is important that any attempts to standardize analytical methods comprise all necessary steps for performing the analysis:

1. Sampling (statistically valid sampling plans)
2. Extraction (recovery of extraction procedure, quantity of DNA, quality of DNA)
3. Determination/detection (specificity, sensitivity)
4. Expression of results: measurement units, absolute quantities, and relative quantities

While correct and appropriate sampling procedures are of ultimate importance for obtaining representative results, this document will not further discuss aspects of sampling and the reader is referred to other resources e.g., [10, 11].

The next operation typically consists of some sort of extraction step, where the analyte is brought into solution and, most of the time, purified simultaneously. Different extraction procedures may result in different yields and different qualities of the extracted analyte. As this can affect the subsequent detection a careful assessment of the extraction methods used is necessary.

For proteins, most applications rely on immunoassays and alternatives are not readily available; total recovery and possible denaturation that renders the extracted proteins undetectable by antibodies are the most critical factors. Depending on the tissue type analyzed (leaves, seeds, roots) validation needs to be performed separately on all different tissue types. For each tissue type the extraction efficiency as well as sensitivity, linearity and range of the method (amongst other parameters) need to be assessed.

Protein based assays are influenced by the relative amount of protein in the cells, the so-called expression level. Expression of a specific protein varies according to different parameters, such as climate, soil, drought conditions, etc. If the expression level in a given sample is different from the expression level of the materials used as calibrators, additional uncertainty will impact the result expressed in mass fractions. While again, for individual and intact kernels other analytical approaches (pooled sampling procedures from GIPSA, see above) are possible that would not be affected by this phenomenon, this would not be applicable for processed samples.

For DNA-based test methods, the extraction procedure cannot be validated independently from the determination system used. There is little or no alternative but to amplify the target DNA within the extracted material in order to determine the quantity and quality of the DNA-sequence actually used for the amplification reaction. Of course, there are several methods suitable for the determination of the total amount of DNA, but in the subsequent amplification reaction only a small fragment of the DNA is amplified; the abundance of this small fragment can only be measured after appropriate amplification. Moreover, the total quantity of DNA in some samples is so small that the sensitivity of methods for the determination of total DNA is not sufficient.

For DNA analysis, the amount and quality of the extracted DNA are of importance. Due to the fact that different methods for the quantification of DNA are influenced in different ways by the physical and chemical state of the DNA (double-stranded vs. single-stranded, length of DNA fragment) and that the state cannot be accurately described, all data on the quantity of DNA will have a relatively high uncertainty. A typical, but misleading, concept from molecular biology is to express the amount of DNA in a given reaction vial in terms of copy numbers. In order to derive this value, the measured absolute amount of DNA is divided by the weight of the haploid genome of the plant. There are few references in the literature on genome size and none of them has any uncertainties determined for the value of the genome size they report, e.g., [12, 13, 14, 15]. More importantly the values of the genome size can differ with respect to variety, seed treatment and other factors [12, 13, 14, 15]. It is impossible to determine the genome size in complex foods. If the limit of quantification or the limit of detection is subsequently expressed in absolute figures of copy numbers, misleading or inaccurate statements will result.

Certified reference materials offer the chance to limit the impact of most of the factors described above. They are certified to the mass fractions of the plant materials (typically seeds) used [16]. This makes these standards fully traceable to SI units. The materials are produced under highly standardized and well-characterized conditions and their homogeneity is confirmed. Every analyst calibrating his/her method using these standards will have an analytical system that is traceable to SI units.

However, it is important to note some of the disadvantages of this approach. The principle of measurement, almost invariably, will not be based on detecting mass fractions, but instead related to parameters that only correlate with the mass fraction, i.e., the amount of protein or DNA extracted from the sample. Even if the system is calibrated against mass fractions of a seed mixture using appropriate calibrators (e.g., the certified reference materials from the Joint Research Center of the European Commission (the JRC, and here, in particular, the Institute of Reference Materials and Measurements of the JRC in Geel, Belgium), the units of measurement will remain the amount of DNA. The analyte (DNA or protein) to mass ratio from samples to be analyzed may be influenced by additional factors. For DNA analysis the zygosity (homozygous or heterozygous) and hybrid status (hybrids may have more than the two basic sets of chromosomes) of a plant may be different. In homozygous plants both alleles at a specific locus are the same, i.e., they have two identical genes with respect to a specific feature or trait. In heterozygous plants the two alleles at a specific locus are different, i.e., they have two different genes with respect to that feature or trait. If, now, reference materials used for calibration differ in their zygosity from the sample under investigation a relative correction factor of 2 is needed (homozygous cells contain twice the number of genes of interest than heterozygous cells) to calculate the exact mass ratios. While in the individual intact grain or seed kernel, further investigation by single-kernel analysis would potentially allow the determination of zygosity, this is not possible for any processed materials. Moreover, in some crops, especially in corn, different tissues demonstrate different patterns of heritage. For example in corn the embryo is diploid, the endosperm (the part of the seeds that contains the nutritive part, but not the embryo) triploid, and the pericarp (the outer wall of the seed) haploid in its chromosomes. During processing of corn, the embryo of a kernel is separated and used in oil production, whereas the endosperm is used for starch production. While the embryo of a corn kernel receives one set of chromosomes from the paternal plant and one set of chromosomes from the maternal plant to form a diploid genome, the cells differentiating into the endosperm received one set of chromosomes from the paternal plant and two from the maternal plant. Typically, both parents are not from the same genetic make-up, as most modern corn varieties are F1 hybrids. In this case, either parent can be the conventional or the biotechnological enhanced plant. While the embryo maintains the 1:1 ratio of diploid cells with respect to its Mendelian pattern, the genetic make-up of the triploid endosperm cells would result in a 2:1 (or a 1:2) pattern. Consequently, the observed frequency of genes will depend on the tissue of a corn seed that is analyzed. It is of importance to realize that processed samples derived from commodity corn always consist of more than one variety, and it will not be possible to differentiate what the original genetic make-up of these varieties was as the structure of cells and kernels is disrupted.

If polyploid hybrids (plants that have three or more sets of chromosomes) are considered, it is obvious that the discussion above is much more complicated, especially, as the trait of interest may be present in only one of the several chromosomes.

These biological factors also affect the reference materials. It is easy to envisage that different reference materials, all certified with respect to their mass fractions, could be derived from different cultivars or derived from different tissues (processed fractions) and would differ in their genetic make-up. This would result in different relative amounts of DNA for an identical mass fraction. Due to these biological factors these uncertainties can only be avoided if all producers of certified reference materials establish a system of comparable reference materials, accounting for the underlying biological factors.

Several research groups propose the use of plasmids as a calibrator. Plasmids are relatively short strains of DNA, which can easily be produced through bacteria. They can be obtained in abundant quantities and are typ-

Table 2 Influence of the efficiency of PCR reactions on quantitative results. The explanation of the symbols used can be found in the text. *Conc.* Concentration

X_{ct}	X_0	E_X	c_t	Δc_t	$\Delta\Delta c_t$	Conc. (%)
3.00E+11	200	0.98	30.93	6.74		1.00
3.00E+11	20,000	0.98	24.19			
3.00E+11	200	0.99	30.70	6.52	–0.23	1.17
3.00E+11	20,000	0.98	24.19			
3.00E+11	200	0.95	31.64	7.45	0.71	0.61
3.00E+11	20,000	0.98	24.19			
3.00E+11	200	0.98	30.93	6.19	–0.55	1.47
3.00E+11	20,000	0.95	24.74			
3.00E+11	200	0.95	31.64	6.90	0.15	0.90
3.00E+11	20,000	0.95	24.74			

ically of very high purity. Plasmids can be readily characterized with respect to their size and thus a value for the number of copies of a specific plasmid solution can be established with a relatively high degree of confidence. Moreover, they are stable and easy to handle. All these features make them very attractive for use as a calibrator. The disadvantages of using plasmids, in addition to their being a very potent source of possible contamination, is the fact that PCR reactions will differ in their efficiency of replicating the target site of the DNA, with respect to purity and length of the DNA as well as other less defined factors. It is well recognized that the efficiency of a PCR reaction is higher if a plasmid is used compared to genomic DNA. Due to the exponential amplification of the target DNA sites during the PCR reaction even small differences in efficiencies can have a big influence on the final result. The equation that describes the exponential amplification of a PCR reaction is given as (User's guide of ABI7700, Applied Biosystems, Foster City, Calif., USA):

$$X_n = X_0 \times (1+E_X)^n$$

where X_n is the number of target molecules at cycle n, X_0 the initial number of target molecules, E_X the efficiency of target amplification, and n the number of cycles. This formula can be re-arranged to $n=\log(X_N/X_0)/\log(1+E_x)$, and rewritten to calculate the threshold cycle c_t at which a pre-set number, X_{ct}, of molecules, sufficient to give a reliable instrument reading is produced: $c_t=\log(X_{ct}/X_0)/\log(1+E_x)$

In order to demonstrate the impact of efficiency on quantitative results, the following example is calculated. Two PCR reactions are performed for each sample, one on an endogenous gene and the other specific for the trait introduced by modern biotechnology. The calibration using plasmids is assumed to show near perfect (98%) amplification efficiency. The other two PCR reactions are assumed to be inhibited at various degrees, which can easily happen for food samples. For illustrative purposes, it was assumed that total of 3×10^{11} target molecules are necessary to elicit a significant reading. Starting amounts were 200 target molecules for the target specific for the biotechnological enhancement and 20,000 target molecules for the endogenous gene (simulating a sample containing1% DNA from modern biotechnology). For each row in Table 2, the number of cycles, c_t, to reach the 3×10^{11} molecules necessary for a significant reading was calculated and is reported in Table 2. The difference in c_t between the first and second simulated PCR reaction was calculated and reported as Δc_t and is directly related to the ratio of the amount of molecules at the start of the reaction. To reflect the impact of calibration by plasmids with high amplification efficiency and the analysis of a sample with somewhat lower efficiency, the difference in Δc_t with respect to the first set of calculation is reported as $\Delta\Delta c_t$. It is worth noting that c_ts (and the differences thereof) relate in a logarithmic manner to concentrations; an increase (decrease) of three c_ts results in approximately a tenfold increase (decrease) of the relative concentrations. For illustrative purposes $\Delta\Delta c_t$ was converted into DNA-% through the simple formula $1/2^{\Delta\Delta ct}$ and the results are given in the right row of Table 2. For example in rows 3 and 4 of Table 2, a PCR reaction is simulated with an efficiency of 97% for the target from the biotechnologically enhanced variety and an efficiency of 98% for the crop specific target. The corresponding c_ts are calculated to be 31.16 and 24.19 resulting in a difference Δc_t of 6.97, which differs by 0.23 ($\Delta\Delta c_t$) from the standard reaction where both targets were amplified with 98% efficiency. The result of the second reaction would be 0.85% instead of 1%.

It is obvious from this table that differences in PCR efficiency will affect the reported results. While plasmids typically have a high efficiency, other samples derived from grain or food may result in a lower efficiency through the presence of inhibiting substances. It remains very difficult to assess and determine small differences in PCR efficiency and thus it will be very difficult for the analyst to determine if and to what extent his/her results are affected by differences in PCR efficiency adding to the overall uncertainty of the results.

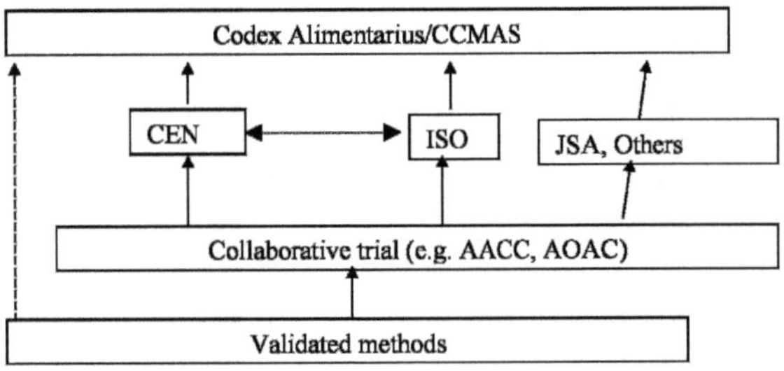

Fig. 1 Exemplified overview and interconnectivity of standardization of analytical methods for food. *CEN* The European Committee for Standardization, *ISO* International Organization for Standardization, *JSA* Japanese Standardization Agency, *AACC* American Association of Cereal Chemists, *AOAC* Association of Official Analytical Chemists, *CCMAS* Codex Committee of Methods of Analysis and Sampling

International standardization of methods for detection of foods derived from modern biotechnology

Several methods and procedures have been developed aiming at the detection of the same food derived from modern biotechnology. This has repeatedly caused problems in trade, as analytical results generated in different laboratories did not concur. It is of high importance to achieve a level of standardization of analytical methods, prevents any future trade disputes due to incompatibility of analytical methods.

There are several agreements of the World Trade Organization that explicitly refer to problems in trade with respect to foods derived by modern biotechnology. Within the Sanitary and Phytosanitary Agreement and within the agreement of Technical Barriers to Trade, reference is explicitly made to ISO standard and standards established by Codex Alimentarius to solve trade disputes on food. Codex Alimentarius established standards for analytical methods through its Codex Committee on Methods of Analysis and Sampling (CCMAS). CCMAS and ISO require methods to be validated before they will be considered for adoption or for being included in a standard. The validation requirements themselves are detailed in e.g., ISO standard 17025. Figure 1 gives an overview of the process.

In conclusion, many countries are in the process of or have already established provisions for labeling of foods derived from modern biotechnology on either a voluntary or a mandatory basis to ensure the right of the consumer to information. Analytical tools are needed for industry to ensure compliance with labeling provisions and for regulatory authorities to enforce these provisions. However, there seems to be an important disparity between the units in which a threshold is expressed in (e.g., 1% weight/weight) and the measurement units of the three test options available: bioassays, protein-based tests and DNA-based tests. None of these tests is able to measure weight-% directly and thus measurements such as concentration of protein or DNA, need to be converted into weight-%. Numerous biological factors affect the exact conversion between these units and the interpretation of analytical results need to take these factors into account. Metrology can aid the standardization of methods enormously by defining clearly the relationships between measurement units and other units of interest (e.g., legal thresholds). Biometrologists should point out the advantages and disadvantages of having SI-traceable standards compared to non SI-traceable standards, thus aiding the current discussions on the use of bioassays, protein-based tests or DNA-based tests for enforcement.

References

1. Rashid A, O'Donovan JT, Khan AA, Sharma MP, van Nguyen H (1997) Weed Technol 11:527–531
2. Murray BG, Friesen LF, Beaulieu KJ, Morrison IN (1996) Weed Technol 10:85–89
3. Stave J (2002) J AOAC Int 85:780–786
4. Lipton CR, Dautlick JX, Grothaus GD, Hunst PL, Magin KM, MihaliakCA, Rubio FM, Stave J (2000) Food Agric Immunol 12:153–164
5. United States Department of Agriculture (2002) Grain Inspection Packers and Stockyards Administration Federal Grain Inspection Service Directive 91812 2-7-02 (http://wwwusdagov/gipsa/reference-library/directives/9181–2pdf)
6. Horwitz (1995) Pure Appl Chem 67:331–343
7. Anklam E, Gadani F, Heinze P, Pijnenburg H, Van Den Eede G (2002) Eur Food Res Technol 214:3–26
8. Ahmed F (2002) Trends Biotechnol 20:215–223
9. Hübner P, Waiblinger HU, Pietsch K, Brodmann P (2001) J AOAC Int 84:1855–1864
10. Lischer P (2002) Mitt Lebensm Hyg 92:290
11. Remund KM ,Dixon DA, Wright DL, Holden LR (2001) Seed Sci Res 11:101
12. Arumuganathan K, Earle ED (1991) Plant Mol Biol Rep 9:211–215
13. Ho I, Rayburn AL (1991) Plant Sci 74:255–260
14. Rayburn AL, Pedersen WL, McMurphy LM (1993) Plant Sci 37:79–82
15. Rayburn AL, McMurphy LM (1991) Crop Sci 31:63–67
16. Trapmann S, Schimmel H, Kramer GN, Van den Eede G, Pauwels J (2002) J AOAC Int 85:775–779

Accred Qual Assur (2003) 8:448–453
DOI 10.1007/s00769-003-0682-0

Wolfgang Richter
Bernd Güttler

A national traceability system for chemical measurements

Received: 21 March 2003
Accepted: 26 July 2003
Published online: 16 September 2003

W. Richter · B. Güttler (✉)
Physikalisch-Technische
Bundesanstalt (PTB),
Bundesallee 100,
38116 Braunschweig,
Germany
e-mail: bernd.guettler@ptb.de
Tel.: +49 531 592 3200
Fax: +49 531 592 3015

W. Richter
Stadeweg 1,
38106 Braunschweig,
Germany

Abstract Current developments in Germany for establishing a traceability system for chemical measurements are reported. The focus is on a dissemination mechanism which employs chemical calibration laboratories accredited within the framework of the German Calibration Service (DKD) and acting as "multipliers" between the national standards level and the user level by providing the user with calibration means which are traceable to the SI via national standards. At the national standards level, a network of high-level chemistry institutes coordinated by the national metrology institute, PTB, provides the primary references for chemical measurements.

The use of the metrological dissemination system provided by the DKD also for chemical measurements is a logical extension of a traceability mechanism, successful for more than two decades in general metrology, to metrology in chemistry. In detail, traceability structures in clinical chemistry, electrochemistry, elemental analysis and gas analysis are described. This system has become an important part of the efforts made in Germany to support chemical laboratories in meeting the traceability requirements of the market and of legal regulations.

Keywords National traceability system · Chemical calibration laboratories · Clinical chemistry · Electrochemistry · Elemental analysis · Gas analysis

Introduction

The continuing globalization of trade and economy requires confidence in measurement results of any kind, including chemical measurements. Chemical measurement results in particular are often the basis for decisions and agreements, for example in health care, environmental protection and international trade and must therefore be reliable and trustworthy.

An important prerequisite for confidence in measurement results is knowledge of the measurement uncertainty, based on traceability to recognized references, ideally to the SI units. Traceability of chemical measurement results has therefore become a key issue in the last decade, and its establishment in an organized and structured way is an important goal in all industrialized countries. This increasingly also holds for emerging economies.

Due to the great variety and complexity of chemical measurement tasks, the establishment of traceability in the field of chemical analysis is more difficult than in other areas of metrology and therefore requires concentration of the efforts on the most urgent demands for traceability. It is the central aim of the CIPM Consultative Committee for Amount of Substance (CCQM), which today is the leading organization for traceability issues of chemical measurements, to promote and harmonize an international primary reference framework for the most important chemical measurement tasks. This is an ongoing process which recently gained additional impetus from the Mutual Recognition Arrangement for national measurement standards and for calibration and measurement certificates issued by national metrology institutes (CIPM-MRA) [1], drawn up by the CIPM under the Metre Convention in 1999 in order to raise the confidence in measurement results of any kind and hence their acceptance. The CIPM-MRA is the response of metrology to the globalization of the markets.

In order to disseminate the units of measurement to the field laboratories in an efficient way, traceability infrastructures are necessary, first within national frameworks. In the field of chemistry the examples of traceability chains available in general metrology, for example, in length measurement, which consist of a considerable number of intermediate steps in the form of artefacts arranged in a hierarchy, are not optimally suited because chemical analysis is largely method oriented. It is, however, very important that there is at least one intermediate level in a chemical traceability chain which acts as a multiplier to the user level since it is impossible for the small number of institutes at the primary level to meet directly the ever increasing demand for traceability of chemical measurements. Germany started about 10 years ago to set up a traceability system for chemical measurements including calibration laboratories accredited within the framework of the German Calibration Service (DKD) as such multipliers. This is described in the following.

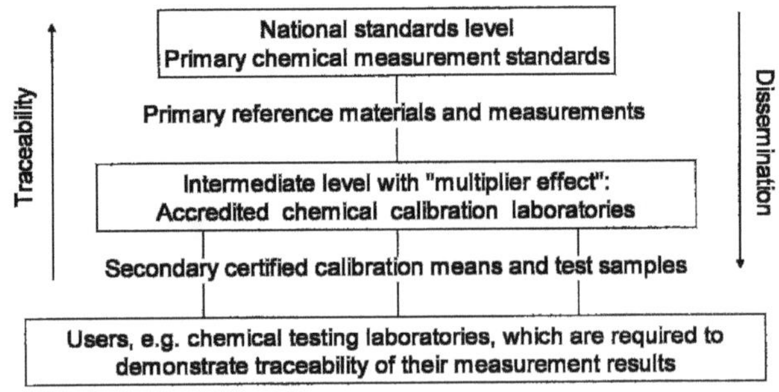

Fig. 1 General structure of the traceability system for chemical measurements

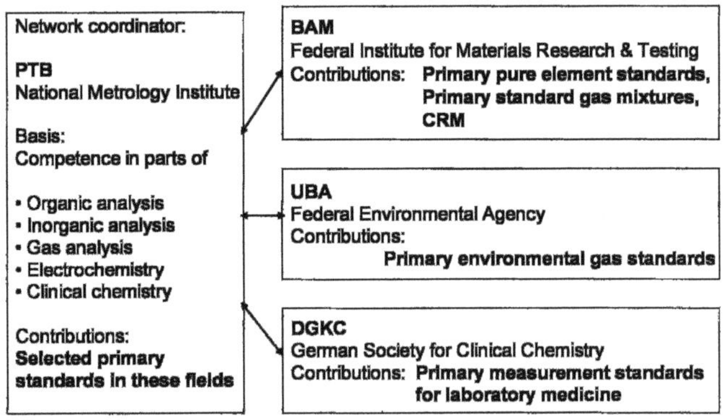

Fig. 2 Structure of the national standards network for chemical measurements. The double arrows indicate formal agreements between PTB and the three network partners, in which the division of labour and the contributions to the network are defined

Structural principle of the traceability system

At present the traceability system consists of structures (traceability chains) in the fields of clinical chemistry, electrochemistry and gas analysis. A traceability structure for elemental analysis is under development. Figure 1 shows the structural principle of the traceability system, which is applied to all the fields mentioned.

It consists of three levels. At the top of the structure a network of national laboratories provides the primary chemical measurement standards and ensures that these are linked up with the international reference framework for chemical measurements. Via primary reference materials and reference measurements, a secondary level consisting of accredited chemical calibration laboratories, including verification authorities in the regulated area, is connected to the national standards level.

This secondary or intermediate level has an important "multiplier" function. It is firmly linked to the national standards and provides traceable calibration means (mainly certified reference materials) and test samples to the workshop level, which consists essentially of chemical testing laboratories (including medical laboratories) which are required to give evidence to their customers that their measurement results are traceable to recognized references. In the case of medical laboratories, the traceability requirement also has a legal background.

This multiplier function is of growing importance because it will not be possible in future for the national laboratories at the top of the traceability chain to serve the users directly, due to the growing demand for traceability in the field of chemistry. It was the growing demand for traceability in metrology in general that led to the establishment of the calibration services some 20 years ago after the calibration workload had become unbearable for the national metrology institutes. Now we are facing a similar situation in metrology in chemistry, although here other paths of dissemination exist as well.

National standards network for chemical measurements

The network at the top of the traceability system providing the primary standards for chemical measurements consists at present of four institutes as shown in Fig. 2.

The national metrology institute, PTB, coordinates the network on the basis of its legal mandate and its competence in those parts of chemical analysis which are relevant to the major demands for traceability. The contributions to the network as listed in the boxes are based on agreements between PTB and the three network partners, the Federal Institute for Materials Research and Testing (BAM), the Federal Environmental Agency (UBA) and the German Society for Clinical Chemistry (DGKC), represented by the Reference Institute for Bioanalysis. In these agreements parts of PTB's responsibility for the national measurement standards, for which PTB does not have the necessary resources, are transferred to partner institutes, at which these resources are available. The division of labour is permanently under review.

With the joint capabilities of this network, the following sectors corresponding to the CCQM priority list of areas where traceability of chemical measurements is particularly important, can be addressed:

- Health care
- Environmental protection
- Advanced materials
- Commodities
- Forensics

The sector "Food" which is also on the CCQM priority list is, however, not yet covered by the network. One reason is that this sector is strongly regulated by legislation which restricts the possibilities of adapting to new developments, even if these are regarded as useful improvements.

An important task of the network is also to ensure that the national references are firmly linked up with the international reference framework for chemical measurements, represented by the results of the CCQM key comparisons listed in the BIPM key comparison database (KCDB). For this purpose, every network member takes part in CCQM key comparisons on its own responsibility and submits its own calibration and measurement capabilities (CMC) to the international evaluation procedure for entry into the KCDB. The key comparisons and the CMCs form the technical basis of the CIPM-MRA.

To have a network of laboratories at the top of a traceability system for chemical measurements instead of just the national metrology institute seems to be a requirement typical of metrology in chemistry and is under consideration in many industrialized countries, because the competence for chemical analysis in most countries (except U.S.A.) largely lies outside the domain of the metrology institutes. Another example that underpins this view is the development of metrology in chemistry in Switzerland, where the Swiss Federal Office of Metrology and Accreditation (METAS) and the Swiss Federal Laboratories for Materials Testing and Research (EMPA) jointly provide the national references for chemical measurements [2].

Accredited chemical calibration laboratories as "multipliers"

The purpose of the national standards network is to provide the primary measurement standards as references for the measurements carried out on the workshop floor. In order to reach the working level in an efficient way, a dissemination mechanism is required.

It is obvious to think of accredited calibration laboratories as the most important link to the working level also in the field of chemical measurements, after the concept of the dissemination of the national measurement standards via accredited calibration laboratories has been successfully applied for more than two decades in metrology in general.

The calibration laboratories in question are accredited within the framework of the German Calibration Service (DKD). The competence of the DKD calibration laboratories for their dissemination tasks is above all based on two central requirements for which the laboratories are thoroughly assessed before accreditation:

1. A firm link must exist to the national institute, in metrology in chemistry to the national standards network, at the primary level via transfer standards (e.g. reference measurements and/or reference materials) or other means by which traceability to the national standards is ensured.
2. The laboratory must demonstrate its capabilities in comparison measurements with the national institute, here again the network, on real laboratory samples and provide a complete uncertainty budget according to the ISO Guide on the Expression of Uncertainty in Measurement (GUM) for its calibrations, i.e. value assignments to the reference materials or other calibrators which, in its capacity as a calibration laboratory, it is going to supply to the field laboratories (e.g. testing laboratories).

Meeting these stringent requirements enables the chemical calibration laboratories to act as providers of calibration means at the secondary level. This task requires that the value assignment to the reference materials and other calibration means provided to the user is more accurate than the measurement results at the user level need to be. Chemical calibration laboratories also exist in other countries, for example in the Netherlands and in the U.K.

At present the state of accreditation of chemical calibration laboratories in Germany within DKD in accordance with EN 45 001, now ISO/EC 17025, is as follows:

- Two calibration laboratories for pH measurement.
- Two calibration laboratories for electrolytic conductivity measurement.
- Two calibration laboratories for measurands of clinical chemistry. Further accreditations are under way.
- One accreditation for gas analysis is under way.

The experience so far gained with this approach to an efficient traceability system for chemical measurements is very positive, although the system, and particularly the dissemination mechanism, are still in an early stage of development. In the following the traceability structures already available are described.

Clinical chemistry (laboratory medicine)

The quality assurance guidelines issued by the Federal Physicians Council (BÄK) and based on the medical products legislation is the main driving force behind metrology in clinical chemistry. The aim is to increase the reliability and recognition of the chemical measurements carried out for diagnostic and therapeutic purposes by the numerous medical laboratories as part of the German health care system. The central goal is to minimize repeat measurements and hence costs and physical strain on patients. The key to higher reliability and recognition is demonstrated traceability to recognized standards, as far as possible to the SI units, and a structured system ensuring this traceability, in addition to a fully implemented quality assurance system [3]. The traceability requirement is further supported by a new EU directive on in vitro diagnostics, which requires traceability of the values assigned to clinical calibrators and control materials to higher-order references.

As a consequence of these driving forces, a traceability structure for clinical chemistry was set up. At the national standards level PTB and DGKC, the latter represented by the Reference Institute for Bioanalysis, are providing the primary standards and procedures to which the measurements in the medical laboratories on the working level are ultimately referred. At present, national references are provided for the following groups of analytes (concentrations in human blood serum), which are subject to the quality assurance measures of the BÄK guidelines. Only the most important analytes are given in brackets as examples.

- PTB
  Metabolites and substrates (cholesterol, creatinine, glucose, uric acid), hormones (cortisol, progesterone), electrolytes (Li, Na, K, Mg, Ca, Cl)
- DGKC
  Metabolites and substrates (urea, triglycerides, bilirubine, lactate), enzymes (the measurands are the enzyme activities), hormones (aldosterone, estradiol, estriol, testosterone, thyroxin), drugs (theophylline, digoxin, digitoxin), total proteinAs far as possible, isotope dilution mass spectrometry is used for the primary measurements in both institutes (e.g. [4]).

The primary references maintained by PTB and DGKC are disseminated to the medical laboratories at the working level mainly via ring tests on well characterized samples (undisclosed to the participants), which are traceable to the primary references, within the framework of the so-called external quality assurance as required by the BÄK guidelines. In the case of measurands for which accredited calibration laboratories exist at the intermediate level according to Fig. 1, a two-step procedure is used:

1. The calibration laboratory is connected to the national standards level via comparison measurements on laboratory samples taken from the calibration laboratory, which are analysed by the national standards laboratory (PTB or DGKC) and the calibration laboratory to be accredited or re-evaluated. Agreement within predefined limits is required as a proof of the competence of the calibration laboratory. The sample with the known value (the national laboratory's value) is then used by the calibration laboratory as measurement standard for its work. It is the advantage of this kind of transferring standards over the transfer of reference materials from the shelf that these standards perfectly match the matrices occurring in the calibration laboratory.
2. The calibration laboratory in turn provides the calibrated test samples for the ring tests to the medical laboratories.

A total of about 30,000 ring test measurements are performed every year by approximately 4,000 medical laboratories. For several of the measurands for which external quality assurance is required by the BÄK guidelines, accredited calibration laboratories do not yet exist.

Here the ring test samples are directly provided by DGKC. It can be expected that the number of accredited calibration laboratories operating as multipliers between the national standards and the user level will increase in future. As already mentioned, further accreditations are underway.

Electrochemistry

Traceability structures for pH and electrolytic conductivity measurement have been built up in this field. The measurement of these quantities is of high relevance to metrology in chemistry and of great economic and scientific importance. This is demonstrated by the high and still growing demand for traceability of measurement results for these quantities to national measurement standards. Furthermore, a new series of European standards on these topics is in preparation within CEN/TC 332. While traceability of pH measurement has been established as one of the PTB's first actions in metrology in chemistry [5], the necessary building blocks for a traceability structure for electrolytic conductivity have been installed only recently [6].

pH measurement

The national standard maintained at PTB is a primary electrochemical measuring system in which the definition of the pH value is very closely realized by the Bates–Guggenheim approximation, the conventional procedure adopted by the national metrology institutes leading in this field, and also adopted by IUPAC. At the secondary level accredited calibration laboratories use the buffer solutions measured at the primary level for the multiplication process. A special differential measuring set-up is used for this purpose which increases the uncertainty only slightly. The calibration laboratories provide certified secondary buffer materials to the working level at which the materials are used for calibration purposes in a great variety of fields. Glass electrode measuring systems, which require frequent re-calibration, are mostly applied at the working level. It is a special feature of this structure that traceability does not extend to the SI at the uncertainty level provided by PTB but to a conventional reference framework which is recognized worldwide. Traceability to the SI can be established if needed, but with increased uncertainty.

Electrolytic conductivity

Providing traceability for electrolytic conductivity measurements is a new activity of PTB. It is a consequence of the growing demand for reliable calibrations of electrolytic conductivity measuring cells. The measurement of electrolytic conductivity is a useful analytical tool often applied in various fields of science and technology, in particular in the case of aqueous media, for which electrolytic conductivity is a measure of the concentration of ionized substances. Although it is a non-specific sum parameter, it can, under given conditions, be used as an easily accessible quantitative measure of the water quality, replacing cumbersome and expensive chemical analyses.

Accurate electrolytic conductivity measurements are required, for example, in water purity assessment which is needed by the pharmaceutical and semiconductor industries and in power plants, for the evaluation of the water quality under regulatory requirements and for water analysis in environmental monitoring.

The national standard for electrolytic conductivity measurement is a primary measuring set-up developed and maintained at PTB. Its central element is a measuring cell of exactly known geometry in which the distance of the electrodes can be changed and exactly measured. Resistance measurements are carried out with at least two different electrode spacings with exactly known shift, with all other conditions kept constant. The measured electrode shift, the cross section of the cell and the two resistance values allow the electrolytic conductivity to be determined in absolute terms with an uncertainty comparable to that of leading institutes in the field.

The dissemination of the unit to the users takes place via DKD-accredited calibration laboratories as described for pH measurement.

Elemental analysis

Element solutions with mass concentrations of elements of nominally 1 g/l, either as single or multi-element solutions are among the most frequently used calibrators in chemical analysis, and traceability to the SI units of the concentrations stated by the manufacturers is increasingly required. In response to this growing demand a traceability structure for elemental analysis is at present being set up.

At the national standards level, BAM and PTB jointly provide the primary references. BAM provides the high-purity elements with known uncertainty for the purity as primary chemical standards and PTB uses these materials to prepare primary element solutions.

These primary solutions will be used as transfer standards to link up accredited calibration laboratories which in turn will provide element solutions as CRMs in the required amounts to the chemical testing laboratories. So far, accredited calibration laboratories do not exist in this field, but a first accreditation is in preparation.

In a joint project started in 2001, PTB and BAM are developing the basis for providing primary reference solutions for the 60 most frequently required elements. At present, primary solutions are available for ten elements at PTB.

Gas analysis

The national standards for the various fields in which gas analysis is of importance are provided by BAM and UBA. PTB's contribution is the type approval of gas analytical measuring instruments whose metrological control is required by the legal regulations.

The whole traceability structure for gas analysis can be subdivided into three parts.

1. Gas analysis within legal metrology

Gas analytical instruments for vehicle exhaust emission surveillance, evidential breath alcohol analysis in road traffic and calorific value determination of fuel gases are subject to legal control and require type approval and initial and subsequent verification. The national standards required in this part of gas analysis are provided by BAM. PTB uses in-house standards prepared by dynamic blending to ensure traceability of its type approval measurements. For the type approval of the breath alcohol analysers as well as for the tests of calibrators used for their initial and subsequent verification, thermodynamic air–alcohol mixture generators are used at PTB for which ethanol–water solutions are provided by BAM as certified reference materials.

The multiplier function at the intermediate level is fulfilled by the verification authorities of the federal states. The result is the deployment of a large number of verified gas analysers whose measurements are traceable to the SI units through the structure described.

2. Gas analysis under environmental protection legislation

For the gas analytical measurements performed by the air quality monitoring networks, UBA provides the primary standards. In most cases these are low-concentration mixtures of pollutants in air prepared by static or dynamic blending. A completely different approach is used for ozone measurement where so-called standard reference photometers (SRP) operating in the UV spectral range are applied as primary references in several countries. These SRPs, and also that operated by UBA, are linked within an international ozone reference network which is coordinated by the BIPM.

The primary standards are disseminated to the air quality monitoring networks by calibration of their gas analytical equipment at the UBA Pilot Laboratory at Langen. Within the networks air quality monitoring laboratories are appointed and provided with the necessary calibration gases.

3. Gas analysis in the unregulated area

Traceability requirements in this area have so far mostly been fulfilled with calibration gas mixtures directly provided by BAM to the user level. A traceability structure of the kind shown in Fig. 1 does not yet exist but can be expected within a short time when the first calibration laboratory for gas mixtures as used in the automobile sector has finalized its accreditation process. There is now considerable interest in DKD accreditation for gas analysis which will increase the importance of establishing traceability via accredited calibration laboratories.

Conclusions

It is now generally accepted that traceability of chemical measurements to recognized standards is an indispensable prerequisite for achieving comparable and trustworthy analytical results. After it has been largely clarified how traceability can be established for chemical measurements, structured systems are now called for to realize traceability in practice.

A tested example of a practical traceability structure has been described which has already proved useful in several fields of chemical analysis. It makes use of a national calibration service as a successful and efficient dissemination mechanism. An essential part of the structure is a network of high-level chemistry institutes at the national standards level providing the end points of traceability and coordinated by the national metrology institute. The need for a network of competence seems to be typical for metrology in chemistry for which in most countries the resources are largely to be found outside of the national metrology institute.

An important feature of the traceability structure described above is an improved reliability of data as a result of reference measurements rather than the use of reference materials alone. Reference measurements allow exact matching of the sample matrix. This is very important in clinical chemistry where "difficult" matrices are quite common.

References

1. BIPM (1999) Mutual recognition of national measurement standards and of calibration and measurement certificates issued by national metrology institutes. Bureau International des Poids et Mesures (BIPM), Pavillon de Breteuil, 92312 Sèvres Cedex, France. Internet:www.bipm.org
2. Haerri H (2001) metINFO 2:9-14, Internet:www.metas.ch
3. Dybkaer R, Örnemark U, Uldall A, Richter W (1999) Accred Qual Assur 4:349–351
4. Henrion A, Dube G, Richter W (1997) Fresenius J Anal Chem 358:506–508
5. Spitzer P, Eberhardt R, Schmidt I, Sudmeier U (1996) Fresenius J Anal Chem 356:178–181
6. Spitzer P (2000) DKD aktuell 2:4

Accred Qual Assur (2004) 9:5–17
DOI 10.1007/s00769-003-0718-5

L. Siekmann

Establishing measurement traceability in clinical chemistry

Received: 28 April 2003
Accepted: 25 August 2003
Published online: 23 December 2003

Presented at the CCQM Workshop on Traceability 16–17 April 2002 at the BIPM, Sèvres, France

L. Siekmann (✉)
Institut für Klinische Biochemie,
Universitätsklinikum der Universität Bonn,
Sigmund-Freud-Str. 25, 53127 Bonn,
Germany
e-mail: lothar.siekmann@ukb.uni-bonn.de
Tel.: +49-228-2875911
Fax: +49-228-2875033

Abstract Measurement traceability is probably the most important tool for achievement of comparability in clinical chemistry. As stipulated by the In Vitro Diagnostica Directive of the European Union and several ISO standards, values assigned to calibrators and control materials must be traceable to reference materials and/or reference procedures of higher order. In the German proficiency testing system, statutory use of reference measurement procedures for several measurands has been in force since 1988. As a result, reference procedures are now regularly applied for the setting up of target values in the control samples of internal and external quality assessment and for assigning values to the manufacturer's calibrator and control materials. Noticeably, the comparability of results obtained by different diagnostic tests has greatly improved for the measurement of many metabolites and substrates, e.g. creatinine, cholesterol, uric acid, total glycerol and urea. For many measurands in laboratory medicine the implementation of the concept of traceability proves to be much more difficult; this mainly concerns the measurement of proteins, in particular enzymes, proteo-hormones, tumour markers and cardiac markers. For such measurands the analyte must first of all be distinctively defined before a reference system can be established which comprises reference procedures, reference materials and networks of reference laboratories.

Keywords Traceability · Clinical chemistry · Reference measurement procedure · Isotope dilution mass spectrometry (IDMS) · External quality assessment

Introduction

The concept of measurement traceability provides probably the most important strategy to achieve standardisation in laboratory medicine aimed at comparable measurement results regardless of the method, the measurement procedure (test kit) and of the laboratory where the analyses are carried out.

Consequently the In Vitro Diagnostica Directive of the European Union stipulates [1] that values assigned to calibrators and control materials must be traceable to reference materials and/or reference methods of higher order.

This requirement should be adhered to not only by the diagnostic kit manufacturers but—most importantly—also by the organisers of external quality assessment schemes when they assign target values to the control materials, which are distributed to the laboratories participating in external quality assessment. This will contribute to standardisation and comparability of test results.

According to the Vocabulary in Metrology (VIM) [2] and the Guide to the Expression of Uncertainty in Metrology (GUM) [3] measurement *traceability* is defined as

> *property of the result of a measurement or the value of a standard whereby it can be related to stated references, usually national or international standards, through an unbroken chain of comparisons all having stated uncertainties.*

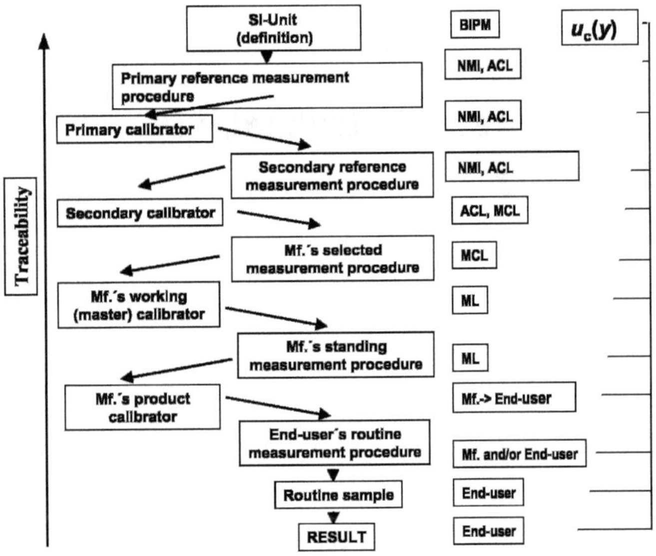

Fig. 1 Metrological traceability and hierarchy of procedures and materials (according to ISO/IEC 17511). $u_c(y)$: Uncertainty; BIPM: International Bureau of Weights and Measures; NMI: National Metrology Institute; ACL: Accredited Calibration Laboratory; MCL: Manufacturer's Calibration Laboratory; ML: Manufacturer's Laboratory; Mf: Manufacturer

According to EN/ISO 17511 [4] traceability of a value attributed to a routine sample, a calibrator or a control material is established by a series of comparative measurements using measurement procedures and reference materials in a chain of decreasing hierarchical order (Fig. 1). The figure demonstrates the calibration hierarchy from top to bottom and the traceability chain from bottom to top. Since each link in the chain contributes to the uncertainty of the result it is advisable to omit as many steps as possible. In terms of metrology it would be ideal to omit all in-between steps of the traceability chain and to measure the routine sample directly using a primary reference procedure. This of course is not feasible.

The complete traceability chain as presented here is valid only for those measurable quantities, which can have a value, expressed in SI units. When primary or secondary calibrators are not available the traceability chain for many measurands in laboratory medicine ends at a lower level, e.g. at the manufacturer's standing measurement procedure. In a situation where a manufacturer detects a new diagnostic marker and defines the measurable quantity by establishing a measurement procedure for this marker, the manufacturer's measurement procedure will form the top of the traceability chain. Nevertheless even in this simple situation the principles of the traceability concept are applicable.

An inevitable precondition for establishing traceable results to calibrators and control materials is the *analytical specificity* of the measurement procedures applied. Results of measurement cannot be traceable when the procedure applied partially is influenced by components, which are not consistent with the definition of the measurand.

Table 1 Reference procedures developed in the reference laboratories of the German Society of Clinical Chemistry

Electrolytes:	Calcium Chloride Lithium Potassium Sodium
Hormones:	Aldosterone Cortisol Estradiol-17β Estriol Progesterone 17-Hydroxyprogesterone Testosterone Thyroxine
Metabolites and substrates:	Cholesterol Creatinine Glucose Total Glycerol Uric Acid Urea Bilirubin
Drugs:	Theophylline Digoxin Digitoxin Lactate
Enzymes:	Alanine aminotransferase (ALT) Aspartate aminotransferase (AST) Creatine kinase (CK) γ-Glutamyltransferase (GGT) Amylase
Proteins:	Total protein Several plasma proteins

Traceability is not really a new fundamental concept in the field of laboratory medicine. Many years before the concept traceability had been mentioned in general chemical metrology, reference measurement procedures and reference materials had been established in clinical chemistry. Some basic experimental work for the development of reference measurement procedures and reference materials had already been undertaken in expert laboratories.

In the German proficiency testing system the use of reference measurement procedures for several measurands has been prescribed by legislation since 1988. As a result the "Reference Institute for Bioanalysis" has established reference procedures for electrolytes, metabolites and substrates, hormones and drugs as listed in Table 1. The reference procedures are now applied regularly for the setting up of target values in the control samples for internal and external quality assessment, for certifying matrix reference materials of the Institute for Reference Materials and Measurements of the European

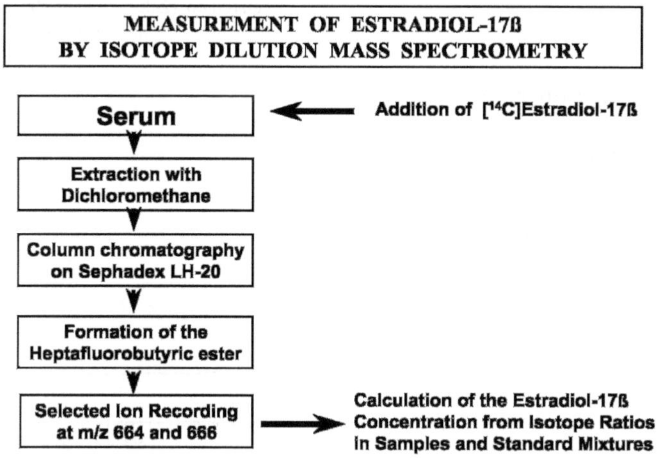

Fig. 2 Procedure for the measurement of estradiol-17β in human serum by IDMS

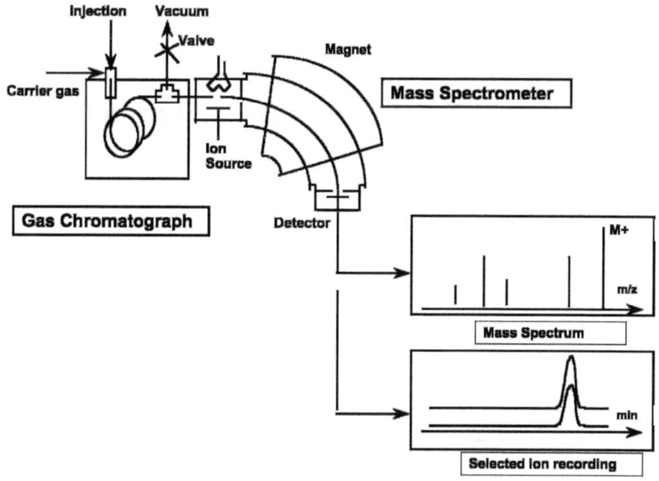

Fig. 3 Schematic drawing of a combined GC-MS instrument and its application for the recording of mass spectra and selected ion chromatograms

Union (IRMM) and for assigning values to manufacturer's calibrator and control materials.

Isotope dilution mass spectrometry—a principle of measurement suitable for establishing reference procedures

A long time before the concept of traceability became popular, the analytical principle of *isotope dilution mass spectrometry* (IDMS) had been developed and described for the first time in a clinical chemical reference laboratory in 1970 [5]. The technique was applied as a reference procedure for the measurement of estrogens in human body fluids. Ever since, isotope dilution mass spectrometry provides one of the most powerful tools for establishing reference procedure values in clinical chemistry. Meanwhile reference procedures for 16 different analytes have been developed using the analytical principle of the so-called 'primary method' isotope dilution mass spectrometry in the reference laboratories of the German Society of Clinical Chemistry (DGKC). These include creatinine [6], urea [7], cholesterol [8], total glycerol [8] and uric acid [9], as well as seven steroid hormones [10, 11, 12, 13], thyroxine [13] and the therapeutic drugs digoxin and digitoxin [in preparation].

The analytical principle of IDMS involved here is demonstrated using the measurement of estradiol-17β in human serum [13] as an example (Fig. 2):

To a serum sample containing about 250 pg estradiol-17β, 250 pg ^{14}C-labelled estradiol-17β is added. The two steroids are extracted and cleaned by column chromatography on Sephadex LH-20. Next, the isotope ratio of the non-labelled and the labelled estradiol-17β derivatives is measured by isotope dilution mass spectrometry. The analytical results are calculated from the isotope ratios determined in each sample and in a series of standards containing defined mixtures of the labelled and the non-labelled steroid.

As demonstrated in the schematic drawing of the combined gas chromatography-mass spectrometry instrument in Figure 3 the purified and derivatised samples are injected into a capillary column for gas chromatographic separation. They are transported through the column using helium as carrier gas and elute into the mass spectrometer at a retention time, which is characteristic for the substances under investigation. In the ion source of the mass spectrometer the substances are converted into positively charged molecular ions as well to smaller fragment ions. These are separated in a magnetic or a quadrupole field. With the conventional technique of gas chromatography-mass spectrometry complete mass spectra can be recorded, showing the molecular ions and fragment ions in a substance characteristic pattern. For the quantitative application applied here, a different technique applies: The ion separation system of the instrument is adjusted to record two masses, one characteristic for the non-labelled and one for the labelled substance under investigation. The two masses are monitored continuously during gas chromatography.

As a result, two chromatograms are recorded simultaneously after processing a serum sample as shown in Figure 4. Although the sample, extracted and chromatographically cleaned, contains hundreds of accompanying components from the biological matrix in addition to estradiol-17β and the steroid labelled with ^{14}C, it is almost exclusively these two that show up during gas chromatography when the mass spectrometric detector is adjusted to these specific masses. The two peaks are quantified by computer-assisted integration. The mass

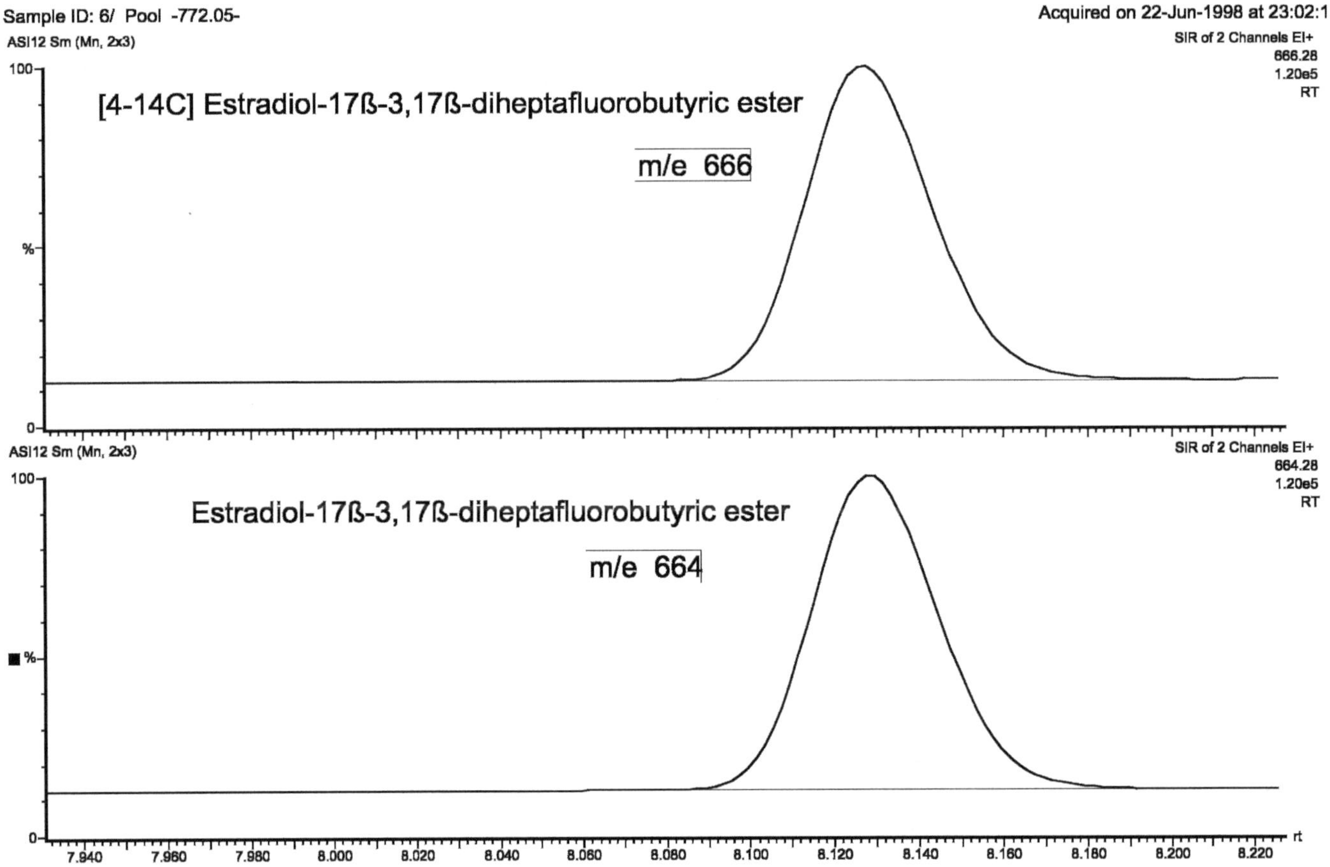

Fig. 4 Selected ion recording of the derivatives of 14C-labelled and the non-labelled estradiol-17β after processing of a human serum sample

spectrometer is used here as a substance characteristic detector with a specificity adjustable to the substances to be detected by selecting appropriate masses. The accuracy of this analytical process is achieved by means of the high specificity of mass spectrometry in combination with capillary gas liquid chromatography and the exact control of recovery that underlies isotope dilution.

Traceability of measurement results for metabolites and substrates

Target values obtained by reference procedures such as isotope dilution mass spectrometry are in use since 1988 serving as a basis for the evaluation of participants' results in the German proficiency testing system. It may be of some interest to know how the introduction of the concept of traceability improved the performance of diagnostic procedures since 1988.

A look at the list of routine method target values for creatinine, uric acid, total cholesterol and total glycerol in the control material of one manufacturer issued before introducing the reference procedure concept in 1988 (Table 2) clearly shows that a large scatter of up to 30% existed, depending on which method or test kit was used. Although no uncertainties were reported at that time it is most likely that some of the values showed significant disagreement. This situation was particularly untenable considering the fact that only one value (or a small interval of target value $\pm u_c$) for creatinine concentration in serum can be the "true" one. Obviously any progress towards improving the comparability of analytical results from different laboratories is hindered as long as procedures with a known or even unknown bias are accepted.

This unsatisfactory situation became also obvious in external quality assessment: Two different samples were distributed in a routine ring trial of the DGKC for cholesterol to about 1300 laboratories in 1987 and the results were then displayed in a YOUDEN [14] diagram as shown in Figure 5. Each dot in this diagram represents the two results from one laboratory, whereby the result for sample A can be read from the abscissa and that for the sample B from the ordinate. A laboratory with its dot just in the middle of the screen is in full agreement with the target value, which here is the reference procedure value certified by an isotope dilution mass spectrometry

Table 2 Method-dependent target concentrations in a control serum for external quality assessment (used before 1988)

Creatinine	µmol/l	Cholesterol	mmol/l	Triglycerides	mmol/l	Uric acid	µmol/l
Enzymatic/PAP	151	CHOD-Iodide	4.02	Fully enzymatic (Behr.)	1.15	Fully enzymatic (Boer./Merck)	457
Enzymatic UV system	161	CHOD-PAP	4.30	Fully enzymatic (Merck)	1.34	UV-system (Boehr.)	476
Jaffe without deproteinisation (Merck)	168	CHOD-Katalase	4.61	Fully enzymatic (Roche)	1.30	UV-system (Merck)	539
Jaffe after deproteinisation (Boehr.)	177	Peridochrom	4.69	Enzymatic (Boehr.)	1.36	Phosphotungstic acid (Goed.)	583
Jaffe without deproteinisation (Boehr.)	189	Liebermann-Burchard	5.49				

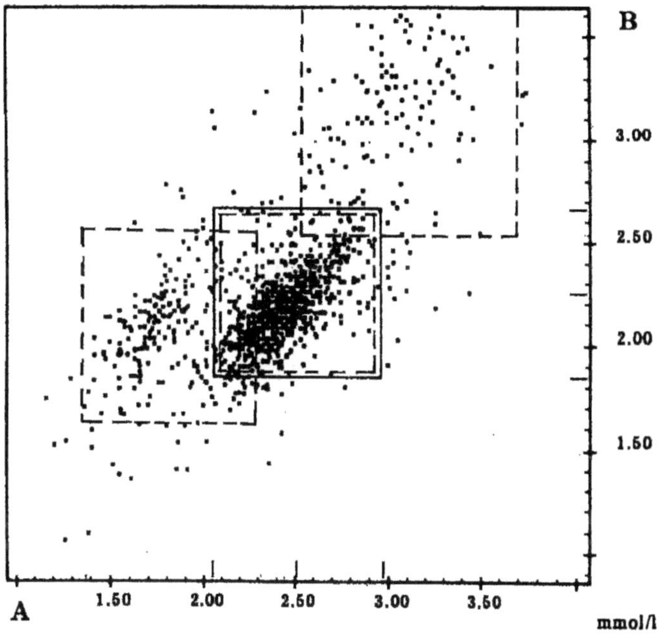

Fig. 5 YOUDEN diagram obtained after a ring trial for cholesterol analysis in serum conducted by the German Society of Clinical Chemistry (DGKC)

procedure. Participants' results from this survey for cholesterol in 1987 clearly show that three different groups of data have been reported according to three different procedures of cholesterol determination. The participants with relatively high cholesterol results had used the Liebermann-Burchard procedure, which was still in use in 1987. The group with low cholesterol values had applied the cholesterol oxidase/iodide method, and the data of laboratories using the cholesterol oxidase/para-aminophenazone (CHOD/PAP) method are situated in the middle of the screen. In 1987 participants' results were evaluated by comparison with the means of their peer group according to the different methodological principles used. Differences of up to 50% between the peer group target values could be observed for cholesterol measurements. Although no uncertainty data were reported for the different target values it may be assumed that the expanded uncertainties did not overlap. In view of the fact that there can be only one cholesterol target concentration interval (target value $\pm u_c$) for a serum, this situation was clearly untenable.

After introducing the reference procedure values for cholesterol, based on IDMS measurements, the different peer group target values have now been replaced by one reference procedure value, which in our case is represented as the exact middle of the screen. The corresponding limits of acceptance are shown as the solid square. As a consequence, methods with inherent systematic errors like the Liebermann-Burchard method and the cholesterol oxidase/iodide method disappeared off the market and today only procedures which are within the limits of acceptance with the reference procedure values established by isotope dilution mass spectrometry exist. In fact, until 1988 there was an unacceptably wide scatter of procedure-dependent target values for many clinical chemical parameters. In order to improve accuracy in clinical chemistry it was absolutely essential to replace these method-dependent target values with reference procedure values.

Traceability of measurement results for low-molecular hormones

The measurement of hormone concentrations in human body fluids has proved to be a valuable diagnostic tool in the field of clinical endocrinology. Thyroxine and the various steroids are the most commonly determined hormones and are usually measured by radio immunoassay (RIA) or by enzyme immunoassay (EIA) with a fairly high degree of sensitivity. However, a manufacturer's list of aldosterone-, cortisol-, progesterone- and estradiol-17β target concentrations in a commercial serum pool, as demonstrated in Table 3, indicates that given the same sample and using immunoassay, assigned values varied considerably from one test kit to another. For cortisol and aldosterone the results ranged between 100% and 200%. For progesterone and estradiol-17β determinations the results differed by a factor of 7. This is probably due to the different qualities of the antibodies and re-

Table 3 Test-kit specific target concentrations for steroid hormones in a commercial control serum

	Aldosterone pmol/l	Cortisol nmol/l	Progesterone nmol/l	Estradiol-17β Pmol/l
ABBOTT	121.9			
AMERSHAM		113.1		
BAXTER DADE DIR		104.8	2.16	396.4
BAXTER DADE AG ER				244.1
BAXTER DADE AD EXT				196.0
BECTON DICINSON		88.0		
BIOCLONE			1.91	
BIOMERIEUX			2.54	539.6
BIOTEX PREMIX	*99.4*	70.6	3.72	759.9
CAMBRID'GE MEDICAL		120.8	*0.86*	
CIBA CORNING		110.3		
CLINICAL ASSAYS		99.3		
CYBERFLUOR FIAGEN		88.2		
DIANOSTIC PRODUCTS	207.7	113.1	3.12	119.3
DUPONT RIANEN		135.1		
EURODIAGNOSTICS		115.8		
FARMOS DIAGN.		99.3	4.67	394.9
IMMUNCHEM COV. COAT		110.3	5.41	348.7
LEECO		113.1	2.99	144.2
MALLINCKRODT		88.3		
NML RIA		96.6		
NMS PHARMACEUTICALS			3.18	205.5
PANTEX IMMUNO DIRECT				143.1
PANTEX IMMUNO		118.6	4.13	190.8
PANTEX IMMUNOCOAT		132.4	*7.00*	154.9
PHARMACIA DELFIA		99.9		*790.0*
RSL	169.2		4.77	*117.4*
SCLAVO LISO PHASE	*277.4*	126.9	3.82	
SERONO		112.0		
SIBAR ELISA		121.3	1.27	
SORIN	165.9	*68.9*	2.86	139.5
SYVA EMIT		*137.9*		
TECHLAND RIA			4.77	
VITEK SYSTEMS		110.0		

Highest and lowest concentration values are given in italics

agents used in the various commercial test kits. What could a consensus value, which is still used as target value in many external quality control schemes, mean in such a context? A target value based on a consensus mean or median actually is of little use in judging test kits, which gave such variable results.

Using method-dependent assigned values for external quality control means having many different target values for the same analyte in the same control serum—a very impractical and, from a theoretical point of view, very unsatisfactory procedure which generates many different results for a substance of known molecular weight and with a defined number of molecules.

It therefore seemed imperative to establish a methodology which would provide the basis for the development of reference procedures. As a result, the target values for the collaborative surveys of the DGKC for steroid hormones have been determined by IDMS reference procedures since 1977.

The DGKC ring trial organisation had to reply to a complaint of a manufacturer who suspected that the bad performance of his customers in the proficiency system surveys for *progesterone* was due to commutability problems of the quality control materials used in the ring trials. The unsatisfactory performance of the test 44 became visible in the Youden diagram as the cloud above the limit of acceptance for sample B (lower progesterone concentration) as well as in the test kit specific bar graphs on the right hand site of the diagram (Fig. 6).

In order to validate the commutability of the DGKC control materials it was necessary to perform split sample measurements with patient samples using the test kit in parallel to the IDMS reference procedure for progesterone. The investigation revealed for both the patient sera and the ring trial results a considerable bias in relation to the reference procedure at low progesterone concentrations. (Fig. 7). The reason for the bad performance of the test was obviously a lack of specificity rather than a lack of commutability of the control materials. At even lower progesterone concentrations the bias increased up to 1000%. It should be noted that the kit manufacturer did unfortunately not issue any lower limit of determination for his measurement procedure.

For *cortisol* measurements an overview of the deviations of participants' medians in ring trials from the

Analyt	Methode
Progesteron	**Alle Methoden**

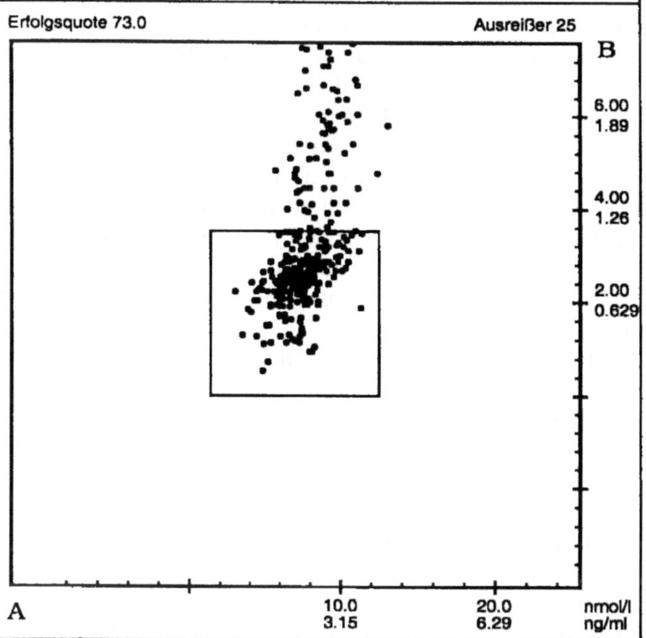

	A nmol/l	B	A ng/ml	B
Teilnehmerzahl	393			
Probe/Einheit	A nmol/l	B	A ng/ml	B
Mittelwert	7.67	3.13	2.41	0.985
Standardabweichung	1.57	1.42	0.494	0.447
Variationskoeffizient	20.5	45.4	20.5	45.4

Probe A (RMW= 6.94 nmol/l)

M	Kit	N	Min	16.P	50.P	84.P	Max
Alle		393	2.58	6.27	7.66	9.28	42.9
1	8	8	7.50		8.43		9.10
1	44	38	6.70	8.83	9.40	10.8	12.1
1	76	11	6.17		8.24		10.2
1	77	7	6.94		8.00		10.3
2	4	9	8.50		9.95		10.7
2	28	39	2.58	7.28	7.82	8.33	8.55
2	68	12	5.41	5.44	6.12	6.84	7.38
2	76	77	3.88	6.04	7.54	9.22	11.4
2	98	19	3.50	5.86	7.31	9.79	11.3
3	8	9	5.95		7.31		9.86
3	40	49	4.93	6.20	7.15	7.85	9.22
3	44	41	3.02	5.45	7.95	9.54	13.1
3	72	7	6.60		8.10		9.04

Probe B (RMW= 1.76 nmol/l)

M	Kit	N	Min	16.P	50.P	84.P	Max
Alle		393	0.541	2.16	2.80	5.41	26.8
1	8	8	2.80		3.04		3.61
1	44	38	2.01	5.65	6.64	8.15	9.86
1	76	11	2.86		8.84		13.7
1	77	7	2.73		7.70		19.0
2	4	9	2.78		3.09		3.43
2	28	39	0.541	1.79	2.40	2.69	2.99
2	68	12	1.21	1.66	2.16	2.44	2.51
2	76	77	1.14	2.35	2.70	3.37	4.77
2	98	19	1.27	1.52	2.38	3.17	4.77
3	8	9	2.42		4.45		5.41
3	40	49	0.731	1.91	2.35	2.80	3.37
3	44	41	1.94	2.51	3.50	4.45	6.04
3	72	7	1.30		2.54		4.10

Andere Kits:
1 20(1), 1 21(6), 1 24(5), 1 32(2), 1 41(1), 1 42(1), 1 50(3), 1 53(2), 1 68(3), 1 70(4), 1 98(1), 1 99(3), 2 12(4), 2 22(4), 2 35(6), 2 40(2), 2 41(1), 2 42(2), 2 44(1), 2 53(1), 2 74(1), 2 82(1), 2 99(3), 3 04(1), 7 76(1), 7 98(3), 7 99(4),

Fig. 6 YOUDEN diagram and test kit related evaluation of ring trial results for progesterone analysis in serum conducted by the German Society of Clinical Chemistry (DGKC)

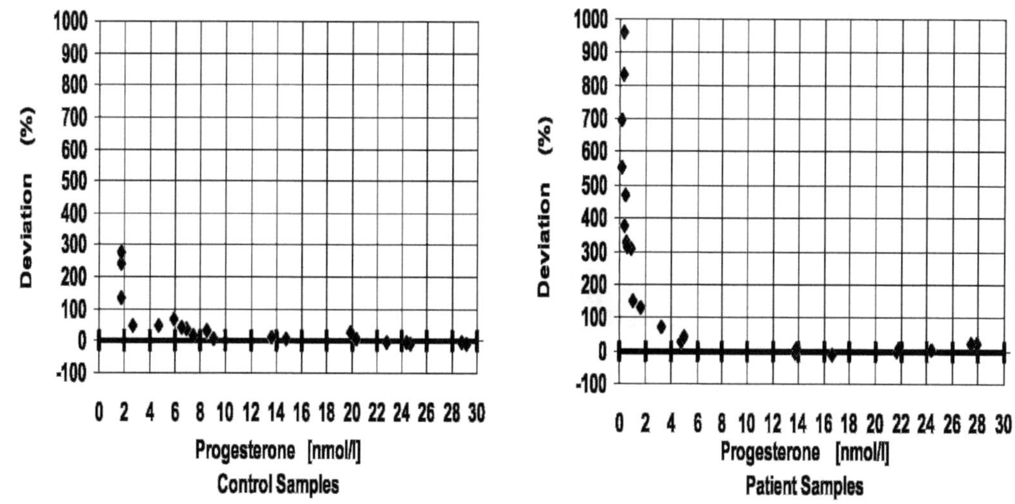

Fig. 7 Relative deviations from IDMS reference procedure values of results of test kit 44 for progesterone analysis depending on the progesterone serum concentration; *left*: for control samples; *right*: for patient samples

IDMS reference procedure values shows that the bias decreased from values of around 15% in 1994 to less than 10% in 2000 (Fig. 8).

The overall situation for the cortisol test kit performance looks quite satisfactory when the medians are compared to the reference procedure values. Today the average deviation is about 9%.

However, considerable differences can be observed when individual test kits are evaluated as shown in Figure 9. This becomes evident from a YOUDEN diagram obtained after a ring trial for cortisol. Test kit specific data are revealed in the YOUDEN diagrams as dark dots and compared to all results of some 400 participants (grey dots). It becomes evident that test kit 04, in particular, failed to fulfill the requirements given by the IDMS reference procedure values and the corresponding limits of acceptance (diagram on the right side). In contrast, the users of test kit 28 fall within the limits of acceptance with their results (left side diagram).

In ring trials for *aldosterone* usually a large number of participants cannot fulfill the requirements according to the directive of the Federal Medical Association in Germany as shown in Figure 10. The results for test-kit 50 are outside the limits of acceptance.

In general it can be stated that most of the available tests for aldosterone in serum measure more constituents not identical with aldosterone and the question arises whether it is justified to use the name 'aldosterone' for the analyte.

It may be suspected that the low clinical significance of aldosterone determinations, which is bemoaned by our clinical colleagues may be due to the poor perfomance of the available commercial tests rather than the validity of the quantity 'aldosterone' itself.

During the early years of external quality control, the accuracy of *unconjugated estriol* in serum proved to be astonishingly high as shown in Figure 11. This changed dramatically towards the end of 1981. Especially when the control samples contained conjugated estriol, the medians of the participants were significantly higher than the IDMS target values. As it turned out, just at that time a kit manufacturer who dominated the estriol-determination market in Germany started using a new antibody. This obviously gave rise to cross-reactions with the conjugated steroid. It was possible to convince the manufacturer that this problem needed correcting and, mainly

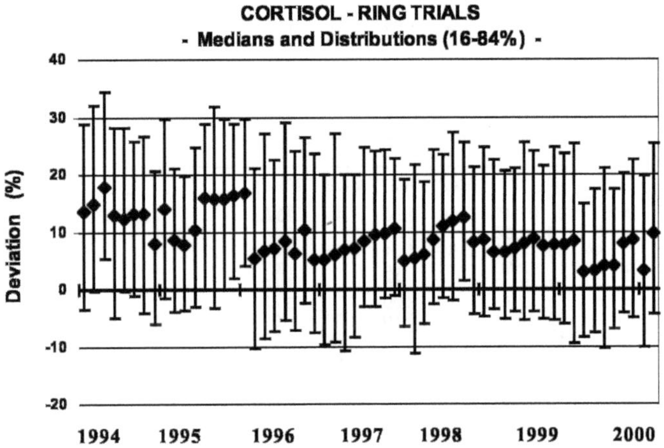

Fig. 8 Relative deviations of the medians of ring trial results for cortisol (*diamonds*) from IDMS reference procedure values and distributions of results (*columns*) from 1994 to 2000

Fig. 9 YOUDEN diagram of a ring trial for serum cortisol; the virtual *dot* in the middle of the diagram represents the IDMS target values for sample A and B and the *squares* demonstrate the limits of acceptance; *left*: results of test kit 28 emphasised by *dark dots*; *right*: results of test kit 04 emphasised by *dark dots*

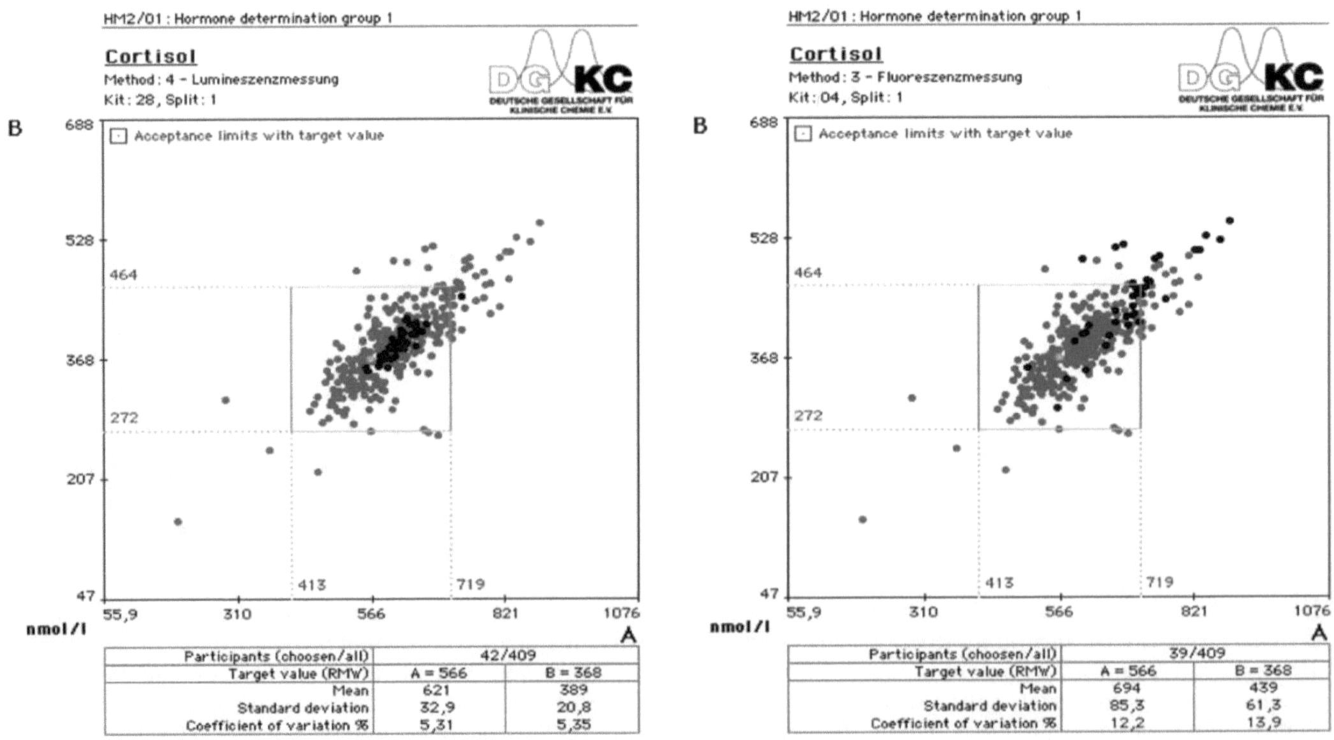

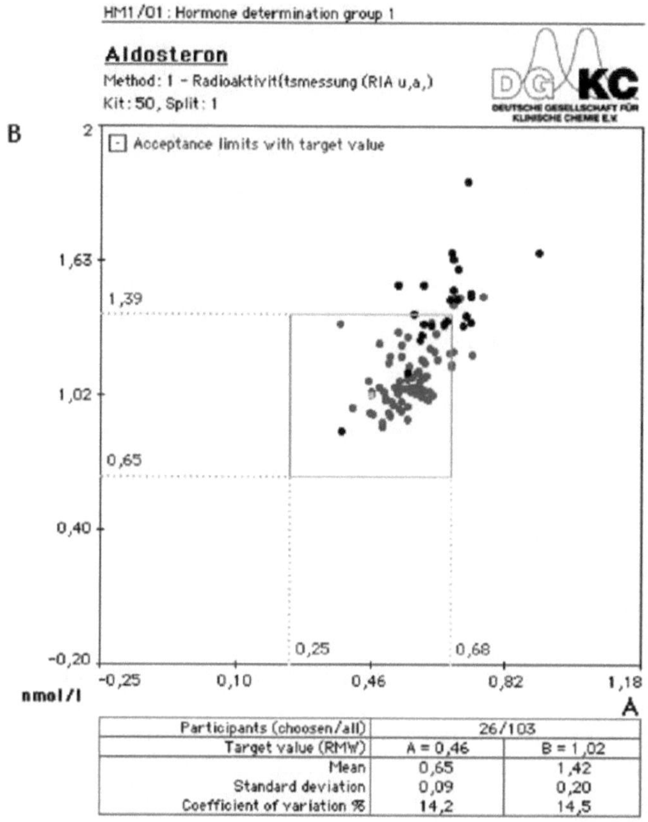

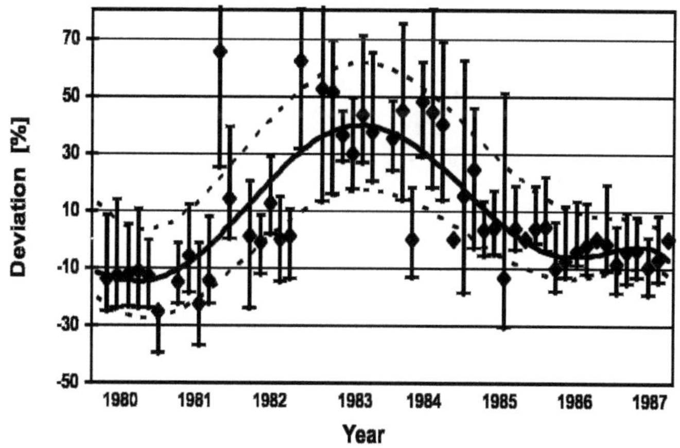

Fig. 11 Relative deviations of the medians of ring trial results for estriol (*diamonds*) from IDMS reference procedure values and distributions of results (*columns*) from 1980 to 1987

◀ **Fig. 10** YOUDEN diagram of a ring trial for serum aldosterone; the *virtual dot* in the middle represents the IDMS target values for sample A and B and the *square* demonstrates the limits of acceptance; results of test kit 50 emphasised by dark dots

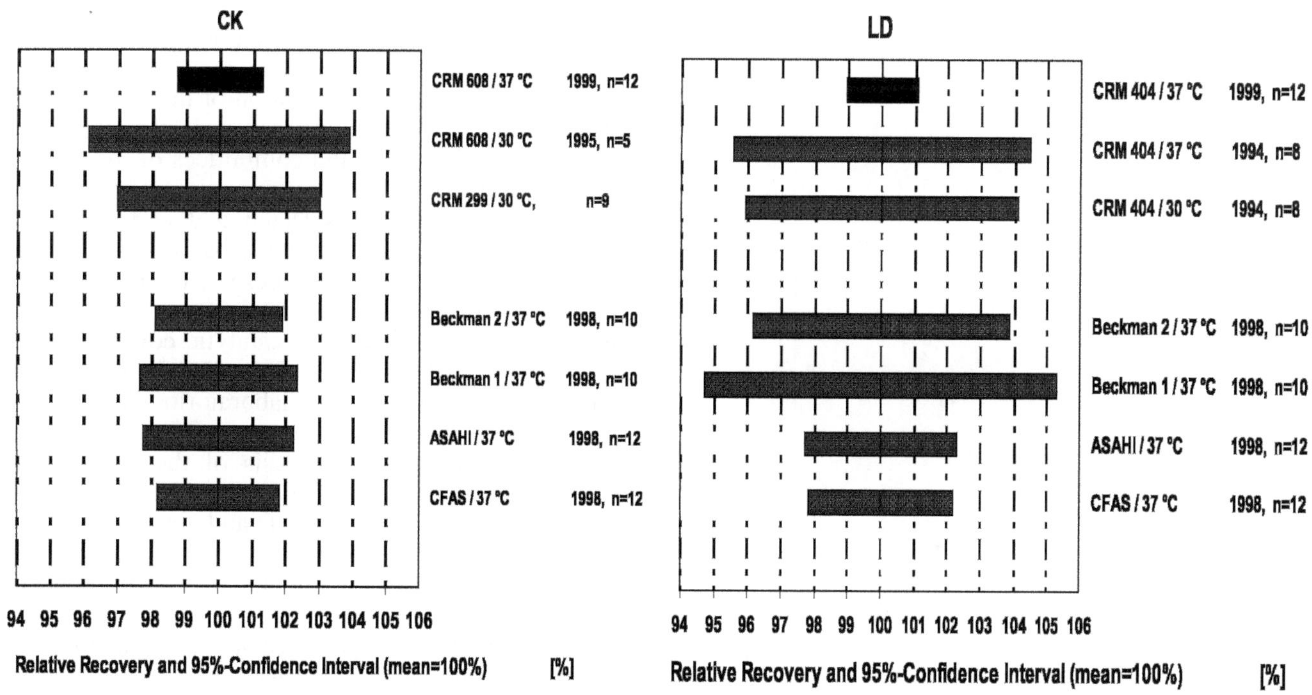

Fig. 12 95%-confidence intervals of the results from different reference laboratories for various certification experiments for creatine kinase (CK, left diagram) and lactate dehydrogenase (LD, right diagram). The top *bars* (black) show the interval for the most recent certification of the IRMM reference materials by using the 37°C IFCC reference procedures issued in 2002. The lower *bars* (grey) show 95%-confidence intervals of former certification campaigns for IFCC reference materials as well as commercial calibrators and control materials

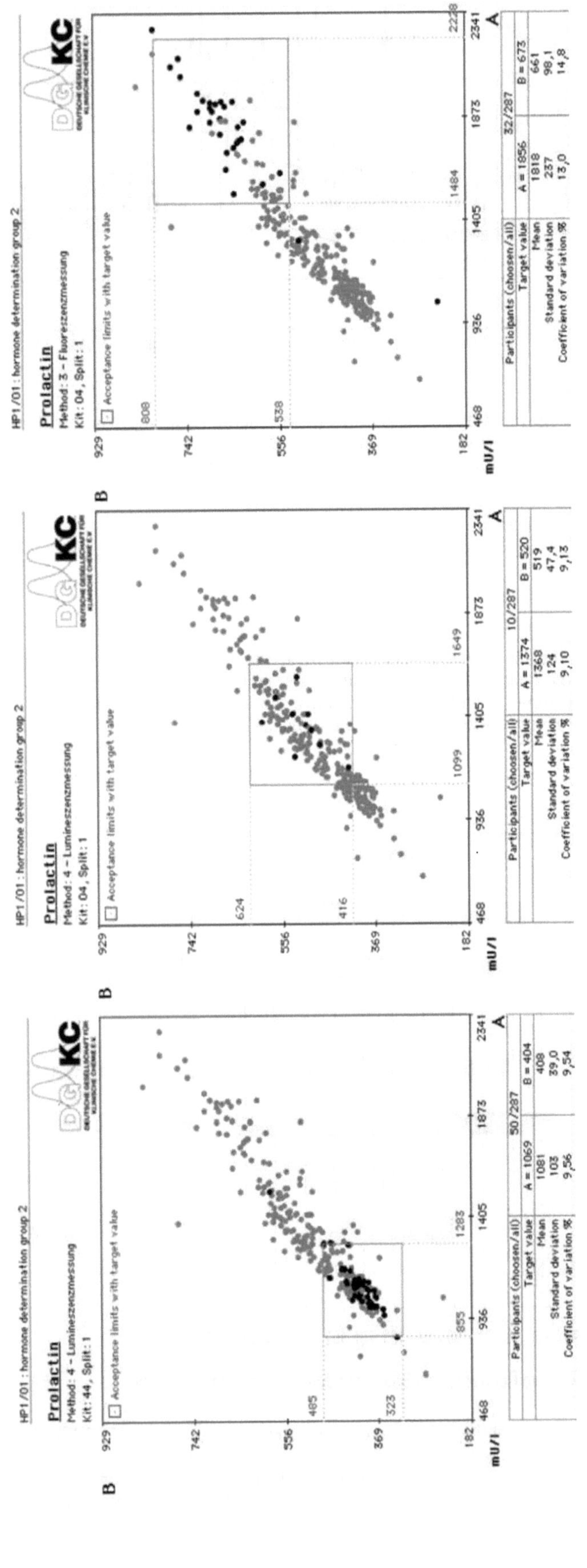

due to this, results have improved greatly since 1985. Estriol determinations are mainly used to monitor fetal well being in the last months of pregnancy. Since not only estriol but also estriol conjugates are higher in this period, we suspect that non-conjugated estriol was probably overestimated from 1981 to 1984 due to the test kit's lack of specificity not only in control samples but also in patient samples.

Manufacturers of diagnostic test kits sometimes argue that a lack of commutability of control materials and their inherent matrix effects are responsible for the deviations of routine results with respect to reference procedure values. Very often it is stated that the results obtained in patient samples are nevertheless true. Comparative measurements (not demonstrated here) reveal that the performance of many tests is even worse when they are evaluated by the use of patient samples.

Traceability of measurement results for non-SI traceable quantities

For *non-SI traceable quantities* the strategy for introducing traceability has to be different. This concerns a large number of analytes for which no defined molecular structure can be assigned, such as for many enzymes, proteo-homones, tumor markers and cardiac markers. The first and most important step must be the *definition of the quantity* before it is possible to establish reference systems (reference procedures, materials and reference network laboratories). Whenever possible, a global consensus on the definition of the measurand should be achieved. Consequently, definition of the measurand and establishment of reference systems is the objective of several working groups and committees of the Scientific Division of IFCC.

In many instances a selected and agreed reference measurement procedure forms the basis of the definition of the quantity and thereby represents the top of the calibration hierarchy. This is particularly true for establishing reference systems for the catalytic concentrations of *enzymes*. In 1999 members of the IFCC working group and some enzyme reference laboratories decided to establish new 37°C measurement procedures as 'IFCC reference procedures' on the basis of the existing 30°C IFCC procedures and to certify enzyme reference materials for ALT, AST, GGT, CK, LD and amylase in collaboration with the IRMM. The enzymes having IFCC reference measurement procedures have catalytic concentra-

Fig. 13 YOUDEN diagram obtained after a ring trial for prolactin; *left*: evaluation limits for the luminometric measurement using the test kit from manufacturer 44; *middle*: evaluation limits for the luminometric measurement using the test kit from manufacturer 04; *right*: evaluation limits for the fluorimetric measurement using the test kit from manufacturer 04; the respective participants' results are emphasised by *dark dots*

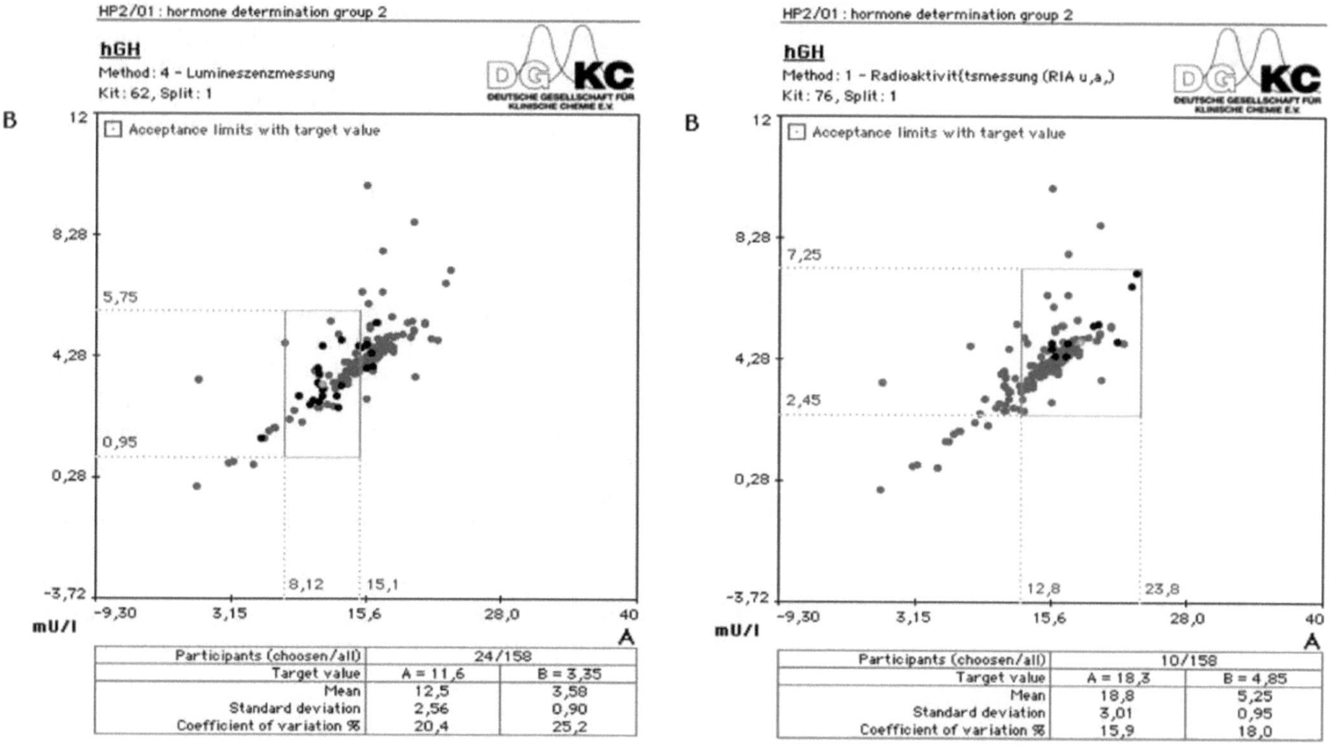

Fig. 14 YOUDEN diagram obtained after a ring trial for human growth hormone (hGH); *left*: evaluation limits for the luminometric measurement using the test kit from manufacturer 62; *right*: evaluation limits for the radiometric measurement using the test kit from manufacturer 76; the respective participants' results are emphasised by *dark dots*

tions with values which are traceable to the SI unit "katal" (=mol/s).

As shown in Figure 12 the certification campaign for creatine kinase (CK) and lactate dehydrogenase (LD) as examples demonstrates that:

(1) the 95% confidence interval of the laboratory results is less than +/- 1.5%; this investigation shows the excellent metrological performance of the participating laboratories from the Far East (Japan) to the Far West (California).
(2) the standard operating procedures, which were developed in the course of the study, can be used as a reference points for the definition of the measurands as the top of the traceability chain.

So far reference systems for the measurement of catalytic activity concentrations for different enzymes have been successfully established and can now be used for assigning traceable values to calibrators and control materials. The procedures have been published as IFCC reference procedures [15, 16, 17, 18, 19, 20, 21]. The IFCC enzyme project, which has been conducted together with the IRMM, could be regarded as a model for the development of reference systems in other fields of interest.

In Germany, the Federal Medical Association prescribes the use of IFCC enzyme reference procedures in clinical laboratory practice. Accordingly, the evaluation of participants' results in the proficiency testing system is based on IFCC reference procedure values.

In contrast to enzyme activities, the definition of the proteo-hormone measurands as well as of many tumor markers and cardiac markers is very critical and several aspects have to be regarded as it concerns the epitope to be detected, the sub-unit to be measured (β-chain or complete molecule) and finally the glycosidic structure of the molecule.

In view of this, it is not surprising that we find a large scatter of test kit dependent results in ring trials, e.g. for the measurement of Prolactin as shown in Figure 13. The data cover a factor of about three between the lowest and highest reported values. There is no doubt that here the individual tests detect different molecular entities of Prolactin. Actually, here different measurands are determined, which, to be correct, should have different names, e.g. Prolactin-A, Prolactin-B and -C. Accordingly, as long as a global agreement on one particular iso-form is missing, each of these different measurands—despite the fact that they share the same name—has to be judged separately. Consequently, different test kit specific target values and limits of acceptance have to be applied for the judgement of participants' results in external quality control.

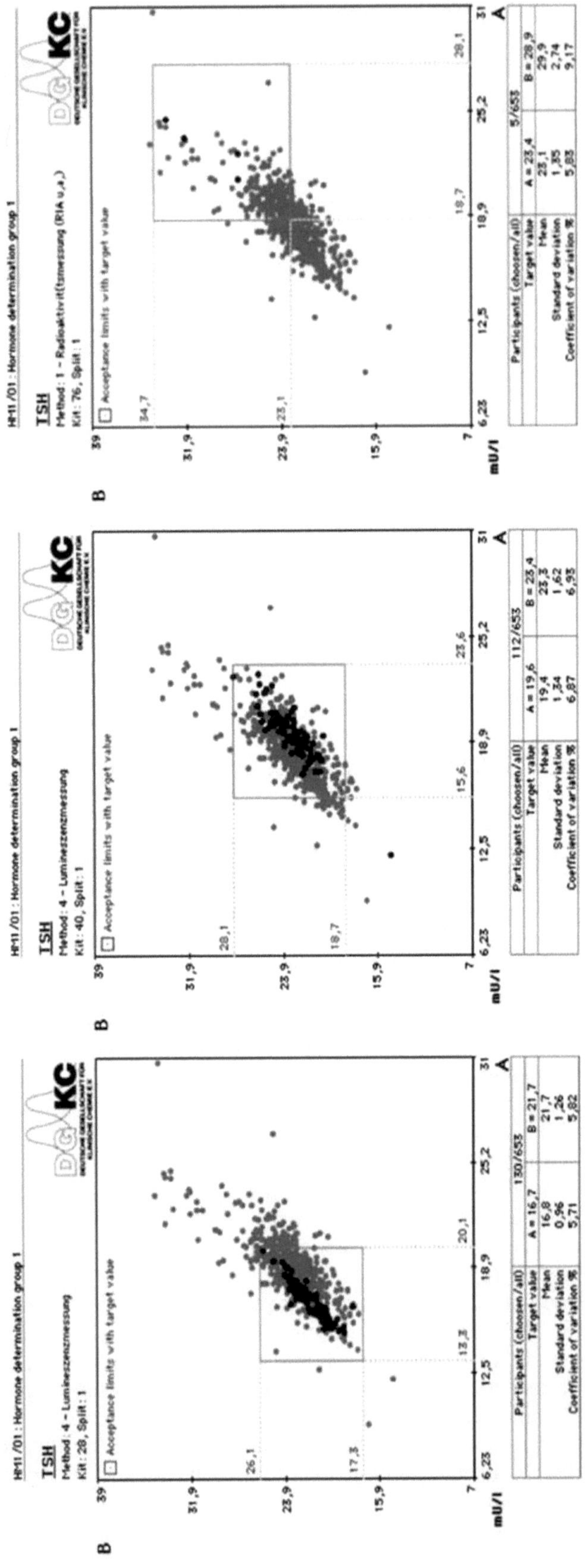

Figure 14 demonstrates ring trial results for human growth hormone where a particular wide range of results can be observed and where the highest values differ from the lowest by a factor of 10. Again, a test kit specific judgement is the only choice to evaluate participants' results.

This holds also for thyroid-stimulating hormone (TSH), where, as demonstrated in Figure 15, at least three different areas of acceptance have to be used depending on the test kits applied.

In summary it can be stated that for SI- traceable measurands the concept of traceability and the use of reference measurement procedures has been successfully implemented at least in the German external quality control scheme since 1988 although the full implementation of the traceability concept on a global basis still requires considerable effort.

For non-SI traceable quantities the predominant objective must be an agreement on the definition of these quantities on an international basis before reference measurement procedures can be developed and used for assigning target values in external quality assessment.

Fig. 15 YOUDEN diagram obtained after a ring trial for thyroid-stimulating hormone (TSH); *left*: evaluation limits for the luminometric measurement using the test kit from manufacturer 28; *middle*: evaluation limits for the luminometric measurement using the test kit from manufacturer 40; *right*: evaluation limits for the radiometric measurement using the test kit from manufacturer 76; the respective participants' results are emphasised by *dark dots*

References

1. Directive 98/79/EC of the European Parliament and of the Council of 27 October 1998 on in vitro diagnostic medical devices, OJ, 1998, No L 331
2. BIPM, IEC, IFCC, ISO, IUPAC, IUPAP und OIML: 1993, International vocabulary of basic and general terms in metrology (VIM)
3. BIPM, IEC, IFCC, ISO, IUPAC, IUPAP und OIML: 1993, Guide to the expression of uncertainty in measurement (GUM)
4. EN ISO 17511, In vitro diagnostic medical devices – Metrological traceability of values assigned to calibrators and control materials
5. Siekmann L, Hoppen HO, Breuer H (1970) Z Anal Chem 252:294–298
6. Siekmann L (1985) J Clin Chem Clin Biochem 23:137–144
7. Keßler A, Siekmann L (1999) Clin Chem 45:1523–1529
8. Siekmann L (1991) Eur J Clin Chem Clin Biochem 29 : 277–279
9. Siekmann L (1985) J Clin Chem Clin Biochem 23:129–135
10. Siekmann L (1974) J Steroid Biochem 5:727–732
11. Siekmann L (1979) J Steroid Biochem 11 : 117–123
12. Siekmann L, Breuer H (1982) J Clin Chem Clin Biochem 20 : 883–892
13. Siekmann L (1984) J Clin Chem Clin Biochem 22 : 551–557
14. Youden WJ (1959) Industrial Quality Control 25:133–137
15. Siekmann L, Bonora R, Burtis CA, Cerotti F, Clerc-Renaud P, Férard G, Ferrero CA, Forest JC, Franck PFH, Gella FJ, Hoelzel W, Jørgensen PJ, Kanno T, Kessner A, Klauke R, Kristiansen N, Lessinger JM, Linsinger TPJ, Misaki H, Mueller MM, Panteghini M, Pauwels J, Schiele F, Schimmel HG, Vialle A, Weidemann G, Schumann G (2002) Clin Chem Lab Med 40:631–634
16. Schumann G, Bonora R, Cerotti F, Clerc-Renaud P, Ferrero CA, Férard G, Franck PFH, Gella FJ, Hoelzel W, Jørgensen PJ, Kanno T, Kessner A, Klauke R, Kristiansen N, Lessinger JM, Linsinger TPJ, Misaki H, Panteghini M, Pauwels J, Schimmel HG, Vialle A, Weidemann G, Siekmann L (2002) Clin Chem Lab Med 40:635–642
17. Schumann G, Bonora R, Cerotti F, Clerc-Renaud P, Ferrero CA, Férard G, Franck PFH, Gella FJ, Hoelzel W, Jørgensen PJ, Kanno T, Kessner A, Klauke R, Kristiansen N, Lessinger JM, Linsinger TPJ, Misaki H, Panteghini M, Pauwels J, Schimmel HG, Vialle A, Weidemann G, Siekmann L (2002) Clin Chem Lab Med 40:643–648
18. Schumann G, Bonora R, Cerotti F, Férard G, Ferrero CA, Franck PFH, Gella FJ, Hoelzel W, Jørgensen PJ, Kanno T, Kessner A, Klauke R, Kristiansen N, Lessinger JM, Linsinger TPJ, Misaki H, Panteghini M, Pauwels J, Schiele F, Schimmel HG, Weidemann G, Siekmann L (2002) Clin Chem Lab Med 40:718–724
19. Schumann G, Bonora R, Cerotti F, Férard G, Ferrero CA, Franck PFH, Gella FJ, Hoelzel W, Jørgensen PJ, Kanno T, Kessner A, Klauke R, Kristiansen N, Lessinger JM, Linsinger TPJ, Misaki H, Panteghini M, Pauwels J, Schiele F, Schimmel HG, Weidemann G, Siekmann L (2002) Clin Chem Lab Med 40:725–733
20. Schumann G, Bonora R, Cerotti F, Férard G, Ferrero CA, Franck PFH, Gella FJ, Hoelzel W, Jørgensen PJ, Kanno T, Kessner A, Klauke R, Kristiansen N, Lessinger JM, Linsinger TPJ, Misaki H, Panteghini M, Pauwels J, Schiele F, Schimmel HG, Weidemann G, Siekmann L (2002) Clin Chem Lab Med 40:734–738
21. Siekmann L, Bonora R, Burtis CA, Cerotti F, Clerc-Renaud P, Férard G, Ferrero CA, Forest JC, Franck PFH, Gella FJ, Hoelzel W, Jørgensen PJ, Kanno T, Kessner A, Klauke R, Kristiansen N, Lessinger JM, Linsinger TPJ, Misaki H, Mueller MM, Panteghini M, Pauwels J, Schiele F, Schimmel HG, Vialle A, Weidemann G, Schumann G (2002) Clin Chem Lab Med 40:739–745

Accred Qual Assur (2004) 9:18–23
DOI 10.1007/s00769-003-0701-1

Mary M. Kimberly
Gary L. Myers
Randie R. Little

Clinical Laboratory Reference Networks

Received: 25 May 2003
Accepted: 24 August 2003
Published online: 21 October 2003

Presented at the CCQM Workshop on Traceability, 16–17 April 2002 at the BIPM, Sèvres, France

M. M. Kimberly (✉) · G. L. Myers
Division of Laboratory Sciences,
National Center for Environmental Health,
Centers for Disease Control and Prevention,
Atlanta, GA 30341–3724, USA
e-mail: mkimberly@cdc.gov
Tel.: +1-770-488-4683
Fax: +1-770-488-4192

R. R. Little
Departments of Child Health and Pathology,
University of Missouri School of Medicine,
Columbia, MO 65212, USA

Abstract The Centers for Disease Control and Prevention, or CDC, has a long history of providing traceability in clinical laboratory medicine. Early work was to develop reference methods for important clinical analytes. When the National Cholesterol Education Program issued recommendations for physicians and clinical laboratories for measurement of lipids and lipoproteins, CDC formed the Cholesterol Reference Method Laboratory Network (CRMLN) to provide manufacturers with access to the accuracy bases. The CRMLN assists manufacturers with calibration of diagnostic products to ensure traceability to higher-order technology. A certification program for manufacturers assures the clinical laboratory community that these products are accurate and precise. The CRMLN model for traceability has been applied to other networks, notably the National Glycohemoglobin Standardization Program.

Keywords Reference networks · Standardization · Lipids · Lipoproteins · Glycated hemoglobin

Introduction

The Centers for Disease Control and Prevention (CDC) has a history of establishing traceability in clinical laboratory medicine. From the 1960s through the 1980s, CDC focused on developing and evaluating reference methods for clinical analytes including cholesterol [1, 2, 3], triglyceride (TG) [4], high-density lipoprotein cholesterol (HDLC) [5, 6], low-density lipoprotein cholesterol (LDLC) [5, 6], glucose [7], uric acid [8], sodium [9], potassium [9], and digoxin [10]. The methods developed for cholesterol, TG, and HDLC have been used in the Lipid Standardization Program (LSP) [11]. The LSP serves two purposes. First, it ensures accurate, stable, and reproducible lipid and lipoprotein measurements within epidemiological studies and clinical trials. Second, it ensures comparability of lipid and lipoprotein results across different studies and trials in space and time.

In addition to developing reference methods and establishing the LSP, CDC has sponsored two conferences focusing on standardization and traceability. The first was the Conference on a National Understanding for the Development of Reference Methods and Materials, in 1977 [12]. This conference led to the formation of the Council of the National Reference System for Clinical Chemistry, which subsequently became the Reference System for the Clinical Laboratory, within the NCCLS (formerly the National Committee of Clinical Laboratory Standards). The second, in 1983, was the Second International Conference on Biomedical Laboratory Standardization. This conference led to the formation of the International Medical Laboratory Information System

(IMLIS), a database to provide comprehensive reference technology information for clinical laboratory science [13]. The sponsoring organizations were not able to continue funding the IMLIS, and the project ended shortly before the internet became widely available through desktop computing. However, this project can be considered a precursor to the work done now through the Joint Committee on Traceability in Laboratory Medicine to develop a comprehensive list of reference methods, reference materials, and reference laboratory networks.

The National Cholesterol Education Program

The National Cholesterol Education Program (NCEP) issued recommendations for physicians to assist them in evaluating risk for cardiovascular disease [14, 15, 16, 17]. These guidelines established medical decision points for risk assessment. In addition, the NCEP initiated a campaign to educate the public about the risk for high blood cholesterol. For these initiatives to be effective, measurement of lipids and lipoproteins needs to be comparable within and among clinical laboratories. Therefore, the NCEP also established performance criteria for clinical assays [18, 19, 20, 21]. They also recommended that total cholesterol (TC), HDLC, LDLC, and TG measurements be traceable to the CDC secondary reference methods. The rationale is that, through the LSP, CDC provides the accuracy base for the population studies used by the NCEP to determine the medical decision points. Therefore, for reliable risk assessment based on these medical decision points, clinical laboratory measurements need to be traceable to CDC. The CDC secondary reference method for cholesterol has been credentialed by NCCLS and has been compared to the National Institute of Standards and Technology's (NIST's) isotope dilution mass spectrometric (IDMS) primary reference method [6, 22]. There is a consistent bias of about 1.6% between the Abell-Kendall (AK) and IDMS methods. The CDC's secondary reference methods for HDLC and LDLC are the highest order methods available. The CDC secondary reference method for TG cannot be directly compared to the NIST primary reference method because they do not measure exactly the same analyte. However, CDC has also established an IDMS method for free glycerol. The sum of the CDC net glyceride value (by the secondary reference method) and the CDC free glycerol value (by IDMS) agrees well with the NIST total glyceride value, with an average difference of 0.4% for two levels of SRM 1951a [23].

CDC anticipated two problems with providing traceability to clinical laboratories. First, matrix effects limit the use of processed serum-based reference materials in standardization. One solution is to use fresh serum specimens for accuracy transfer and method comparisons. Second, CDC realized it could not standardize an estimated 100,000 clinical laboratories in the United States. A more practical approach to standardizing lipid and lipoprotein measurements is to ensure that diagnostic products are properly calibrated by the manufacturers and traceable to the CDC secondary reference methods. In 1989, CDC established the Cholesterol Reference Method Laboratory Network (CRMLN) to provide reference services to manufacturers [24, 25]. Because matrix effects complicate traditional approaches to assessing accuracy, the certification program offered by the CRMLN is based on analysis of fresh samples [26]. CDC believes that working with the manufacturers is the most effective means, with the greatest impact, of standardizing the measurement of lipids and lipoproteins.

CRMLN

The CRMLN comprises four United States and seven international laboratories that have established the CDC secondary reference methods or designated comparison methods for lipids and lipoproteins. The relevant reference methods are listed in Table 1 along with their metrologic properties. [11, 24, 27]. A list of the current members of the CRMLN can be found at the CRMLN web site [25]. A significant effort is required to standardize the CRMLN laboratories. First, performance criteria are established. As a rule, the CRMLN set bias and precision criteria half of that recommended by the NCEP for clinical laboratories as its first goal. As the CRMLN gained experience, the criteria were revised to those listed in Table 1. Second, method audits are performed for new laboratories and when members experience problems. Third, training for new laboratories is provided on-site at CDC.

The most extensive effort involves quality assurance. CDC provides common quality control (QC) materials and all members use standardized QC procedures. In addition, CDC regularly surveys the CRMLN laboratories to ensure they meet the required criteria. Initially, surveys were conducted monthly; however, as the laboratories gained experience and performance improved, the survey schedule was changed to bimonthly. These surveys use CDC frozen serum secondary reference materials that have been prepared using NCCLS C37-A protocol [28]. TC surveys include three levels analyzed in duplicate in two runs. HDLC surveys include four levels analyzed in duplicate in four runs. As an additional check on the quantitative step of the HDLC method, a low-total cholesterol material with cholesterol concentration <100 mg/dl is added to the TC survey. LDLC surveys include four levels analyzed in quadruplicate in four runs.

A designated comparison method for TG has been developed and is being evaluated in two CRMLN laboratories [29]. Once implemented, the monthly survey scheme

Table 1 Metrological properties of methods used by the Cholesterol Reference Method Laboratory Network (*CRMLN*). *2° RM*: Secondary reference method, *DCM*: designated comparison method, *IDMS*: isotope dilution mass spectroscopy, *AK*: Abell-Kendall, *CV*: coefficient of variation, *HDLC*: high-density lipoprotein cholesterol, *SD*: standard deviation, *LDLC*: low-density lipoprotein cholesterol, *NA*: not available

Analyte	Primary Reference Method (1° RM)	CRMLN Method, 2° RM or DCM	Bias of 2° RM vs 1° RM	Accuracy criterion (vs CDC)	Precision criterion
Total cholesterol	IDMS	AK (2° RM)	+1.6%	Bias≤1%	CV≤1%
HDLC	NA	Ultracentrifugation with AK (2 RM) or 50 K dextran-sulfate with AK (DCM)	NA	Bias≤1 mg/dL	SD≤1 mg/dl
LDLC	NA	Betaquantification with AK (2° RM)	NA	Bias≤2%	CV≤1.5%
Triglyceride	IDMS (for triglyceride and total glycerides)	Chemical extraction/ hydrolysis with enzymatic endpoint (DCM) (for net glycerides)	NA	Bias≤2.5% (tentative)	CV≤2.5% (tentative)

Table 2 Comparison of Cholesterol Reference Method Laboratory Network (CRMLN) traceability model to the International Organization for Standardization (*ISO*) model. *BIPM*: International Bureau of Weights and Measures, *NMI*: National Metrology Institute, *NIST*: National Institute of Standards and Technology (US NMI), *IDMS*: isotope dilution mass spectrometry, *ACL*: accredited calibration laboratory, *AK*: Abell-Kendall

ISO		CRMLN (Cholesterol)	
Responsible party	Service or activity	Responsible party	Service or activity
BIPM	SI unit	BIPM	SI unit
NMI	Primary reference method	NIST	IDMS
NMI	Primary calibrator	NIST	SRM 911b
ACL	Secondary reference method	CDC	AK
ACL	Secondary reference materials	CDC	SRM 1951a, two reference materials
ACL		CRMLN	AK
Manufacturer		Manufacturer	Fresh serum specimens
ACL or manufacturer	Secondary calibrator	Manufacturer	Secondary calibrator
Manufacturer	Routine method	Manufacturer	Routine method
End user	Routine calibrator	Clinical laboratory	Routine calibrator
End user	Routine method	Clinical laboratory	Routine method

for TG will include four levels that are analyzed in duplicate in four runs.

The CRMLN model for cholesterol fits into the International Organization for Standardization's traceability scheme [30]. Table 2 shows how the CDC and CRMLN define the various components of the hierarchy. The model for TG is similar to that for cholesterol, using SRM 1595 (tripalmitin) as the primary calibrator. The models for HDLC and LDLC are different because these heterogeneous analytes are not traceable to the SI unit, and primary reference methods do not exist. The reference methods at CDC are the highest order that can be obtained for HDLC and LDLC.

The CRMLN laboratories have demonstrated excellent performance over time. For TC, the average individual laboratory bias was 0.0% [with a standard deviation (SD) of the bias of 0.4%] for monthly surveys executed from September 1998 through February 2000 [24]. The average individual laboratory coefficient of variation (CV) was 0.3% (with a range of 0.0–1.3%) during the same period. For HDLC, the average individual laboratory bias was –0.1 mg/dl (with a SD of the bias of 0.2 mg/dl) for bimonthly surveys executed from July 2000 through January 2003. The average individual laboratory SD was 0.4 mg/dl (with a range of 0.1–0.9 mg/dl) during the same period.

Traceability for in vitro diagnostic (IVD) manufacturers

CDC and the CRMLN established a traceability scheme (Fig. 1). The CRMLN uses this approach in a certification program for manufacturers. In this program, NCCLS protocol EP9-A is used as a basis for comparison using fresh serum samples [31]. The manufacturer collects a minimum of 40 specimens and analyzes them in duplicate in five separate runs. The specimens are then shipped to a CRMLN laboratory for analytical and statistical analysis. When the NCEP performance criteria for bias and precision are met, the manufacturer is issued a Certificate of Traceability for the

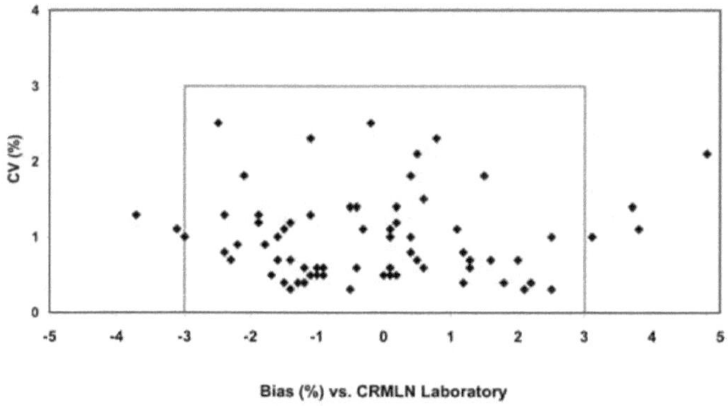

Fig. 1 The Centers for Disease Control and Prevention (*CDC*) approach to traceability link

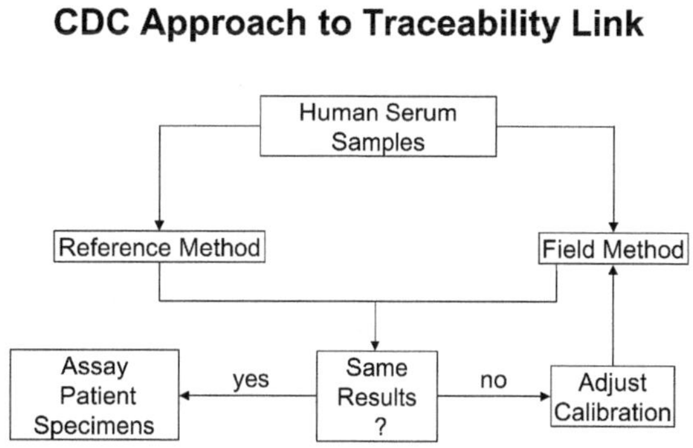

Fig. 2 Results of manufacturer comparisons for total cholesterol performed from January 2000 through March 2003. Sixteen manufacturers performed 79 comparisons. Bias versus the Cholesterol Reference Method Laboratory Network (*CRMLN*) laboratory is plotted on the *x*-axis and the coefficient of variation (*CV*) (%) is plotted on the *y*-axis. Vertical lines at –3% and +3% bias and the horizontal line at 3% CV indicate the National Cholesterol Education Program recommendations for accuracy and precision for clinical laboratories

specific instrument, calibrator, and reagent combination used in the comparison. The Certificate of Traceability is valid for 2 years. All certified systems are listed at http://www.cdc.gov/nceh/dls/crmln/crmln.htm. Manufacturers are encouraged to repeat the process every 2 years. Certification programs are available for TC, HDLC, and LDLC.

From January 2000 through April 2003, 16 manufacturers performed 79 comparisons in the TC certification program (Fig. 2). The majority of analytical systems met the NCEP performance guidelines and were issued a Certificate of Traceability. A proficiency survey conducted by the College of American Pathologists (CAP) in 1994 further demonstrated the impact of the CRMLN [32]. One fresh-frozen human serum sample was distributed as part of a CAP proficiency testing survey to a subset (578) of participants. CDC also analyzed the sample using the secondary reference method and set a confirmatory value of 4.251 mmol/l. The mean of the laboratories was 4.275 mmol/l, a difference of 0.57%. Although no "before" exists to this story (i.e. the CAP conducted this special survey after the CRMLN certification program had been in place for several years), the results demonstrate that the IVD products used by the participants obtained an accurate value on a sample simulating patient samples. We assume from this that most IVD manufacturers use the CRMLN to properly calibrate their products.

National Glycohemoglobin Standardization Program

The National Glycohemoglobin Standardization Program (NGSP) used the CRMLN model to establish a reference laboratory network to standardize glycated hemoglobin (i.e., HbA_{1c}) [33, 34]. The purpose of the NGSP is to standardize HbA_{1c} so that clinical laboratory results are comparable to the Diabetes Control and Complications Trial (DCCT) where relationships were established to mean blood glucose and risk for vascular complications. The NGSP is coordinated by the University of Missouri-Columbia, which also serves as the site for the Central Primary Reference Laboratory (CPRL) and the accuracy point. In addition, there are three Primary Reference Laboratories (two in the United States and one in Europe) and eight Secondary Reference Laboratories (four in the United States and four in Europe). These laboratories are monitored on a regular basis to maintain traceability to the CPRL. The NGSP provides services in three areas: (1) assistance with calibration for manufacturers, (2) certification for manufacturers and laboratories, and (3) assignment of target values for CAP proficiency testing for clinical laboratories. All services are based on the use of fresh (or fresh-frozen at –70 °C or below) blood samples. The criteria for certification are total imprecision of ≤4% (<3% for Level I Laboratory certification), and the 95% confidence interval of the differences between the test methods (manufacturers and clinical laboratories) and the SRL must fall within the clinical significant limits of ±1% HbA_{1c} (0.75% HbA_{1c} for Level I laboratory certification).

CAP surveys have demonstrated the impact of the NGSP on accuracy in HbA_{1c} measurements (Fig. 3). In 1993 laboratories reported results in different units (%HbA1c, %HbA1, %Total glycated hemoglobin and there was a large amount of variability between methods. Few of the method specific median values were comparable to DCCT results. In 2003, 98% of laboratories participating in the CAP survey (*n*>2000) reported using NGSP certified methods. For NGSP-certified methods, the method-specific medians were all within 0.5% of NGSP targets at all levels. Most method-specific, between-laboratory CV's for NGSP certified methods were <5%.

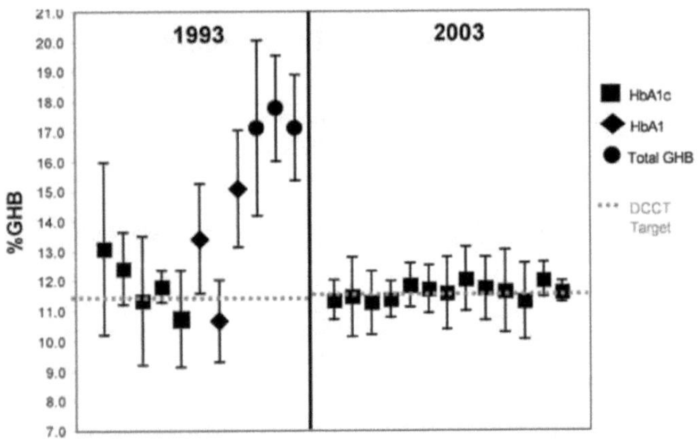

Fig. 3 College of American Pathologists proficiency testing survey results for HbA_{1c} before the initiation of the National Glycohemoglobin Standardization Program (NGSP) (1993) and 6 years after (2003). Each point represents a peer group using one of the following method types: HbA_{1c} (*square*), HbA_1 (*diamond*), total glycated hemoglobin (*GHB*) (*circle*). The *horizontal dotted line* is the Diabetes Control and Complications Trial (*DCCT*) target value, set by the Central Primary Reference Laboratory of the NGSP

At an international level, the International Federation of Clinical Chemistry (IFCC) established a working group on HbA1c standardization in 1995. The focus of this group has been the development of a higher order reference method and pure standards. An IFCC Laboratory Network that uses mixtures of purified HbA1c and HbA0 to calibrate two different reference methods has been established [35, 36]. The relationship between the IFCC and the NGSP networks has been established based on several sample comparisons (n=26 pooled specimens). The IFCC HbA1c results are significantly lower (approximately 1.3–1.9% across the relevant HbA1c range) than NGSP results. The NGSP will continue to monitor this relationship to ensure stability with a long-term goal of using this new reference method as the anchor for the NGSP. However, there has been much debate about which numbers should be reported worldwide—the accuracy-based IFCC numbers or the outcomes-based NGSP/DCCT/United Kingdom Prospective Diabetes Study numbers.

Conclusions

Clinical laboratory reference networks provide manufacturers access to traceability in the reference technology hierarchy. They are especially useful for methods that demonstrate matrix effects with processed calibrators, reference materials, or proficiency testing materials. In addition to serving the IVD manufacturing community, some of the CRMLN laboratories also provide proficiency testing programs for clinical laboratories using materials that have been value-assigned using the reference methods. Thus, through a wide range of services, clinical laboratory reference networks provide the clinical laboratory community with traceability to higher order methods and materials.

Acknowledgements The authors acknowledge the data management assistance of Mahnaz Dasti (Battelle Memorial Institute) and the expert technical assistance of the CDC Lipid Reference Laboratory under the direction of Parvin P. Waymack: Steven F. Ethridge, Charlene Griffin, and Sheldon Stribling (Battelle Memorial Institute).

References

1. Duncan IW, Mather A, Cooper GR (1988) The procedure for the proposed cholesterol reference method. CDC, Atlanta, Ga.
2. Cooper GR, Smith SJ, Duncan IW, Mather A, Fellows WD, Foley T, *et al* (1986) Clin Chem 32:921–929
3. Abell LL, Levy BB, Brodie RB, Kendall FE (1952) J Biol Chem 195:357–366
4. Lofland HB (1964) Anal Biochem 9:393–400
5. Myers GL, Cooper GR, Winn CL, Smith SJ (1989) Clin Lab Med 9:105–135
6. Cooper GR, Myers GL, Henderson LO (1991) Eur J Clin Chem Clin Biochem 29:269–275
7. Neese JW, Duncan P, Bayse D, Robinson M, Cooper T, Stewart C (1974) Clin Chem 20:878
8. Duncan P, Gochman N, Cooper T, Smith E, Bayse D (1980) Development and evaluation of a candidate reference method for uric acid in serum or plasma. CDC, Atlanta, Ga.
9. Carter RJ, Mather A, Dawson CD, Turner WE, Neese JW, Bayse DD (1980) The development of a CDC reference method for determining sodium and potassium in serum. CDC, Atlanta, Ga.
10. Myrick JE, Hannon H, Powel MK, Agnese ST, Shaw W, Fast DM, Spierto FW, Burtis CA, Bayse DD (1983) Assay manual for the quantitation of digoxin by radioimmunoassay with the candidate reference method. CDC, Atlanta, Ga.
11. Myers GL, Cooper GR, Greenberg N, Kimberly MM, Waymack PP, Hassemer DJ (2000) Standardization of lipid and lipoprotein measurements. In: Rifai N, Warnick GR, Dominiczak MH (eds) Handbook of lipoprotein testing, 2nd edn. AACC, Washington, D.C., pp 717–748
12. Boutwell JH (ed) (1978) A national understanding for the development of reference materials and methods for clinical chemistry, proceedings of a conference. AACC, Washington D.C.
13. Lines J, Heeren J (2002) IFCC celebrating 50 years. IFCC, Milano, Italy

14. National Institutes of Health (1989) Report of the expert panel on detection, evaluation, and treatment of high blood cholesterol in adults. Bethesda, Md. NIH publication no: 89–2925
15. National Institutes of Health (1993) Second report of the expert panel on detection, evaluation, and treatment f high blood cholesterol in adults. Bethesda, Md. NIH publication no: 93–3095
16. National Institutes of Health (2001) Third report of the expert panel on detection, evaluation, and treatment of high blood cholesterol in adults. Bethesda, Md. NIH publication no: 01–3670
17. National Institutes of Health (2003) http://www.nhlbi.nih.gov/about/ncep/index.htm
18. National Institutes of Health (1990) Recommendations for improving cholesterol measurement: a report from the Laboratory Standardization Panel of the National Cholesterol Education Program. Bethesda, Md. NIH publication no: 90–2964
19. Warnick GR, Wood PD (1995) Clin Chem 41:1427–1433
20. Bachorik PS, Ross JW (1995) Clin Chem 41:1414–1420
21. Stein EA, Myers GL (1995) Clin Chem 41:1421–1426
22. Ellerbe P, Myers GL, Cooper GR, Hertz HS, Sniegoski LT, Welch MJ, White E (1990) Clin Chem 36:370–375
23. National Institutes of Standards and Technology (1997) Certificate of analysis standard reference material® 1951a for lipids in frozen human serum. Gaithersburg, Md.
24. Myers GL, Kimberly MM, Waymack PW, Smith SJ, Cooper GR, Sampson EJ (2000) Clin Chem 46:1762–1772
25. Centers for Disease Control and Prevention (2003) http://www.cdc.gov/nceh/dls/crmln/crmln.htm
26. Miller, WG (2000) Matrix effects in the measurement and standardization of lipids and lipoproteins. In: Rifai N, Warnick GR, Dominiczak MH (eds) Handbook of lipoprotein testing, 2nd edn. AACC, Washington D.C., pp 695–716
27. Kimberly MM, Leary ET, Cole TG, Waymack PW (1999) Clin Chem 45:1803–1812
28. NCCLS (1999) Preparation and validation of commutable frozen human serum pools as secondary reference materials for cholesterol measurement procedures. NCCLS, Wayne, Pa. NCCLS document C37-A
29. Franzin M, Ferrero CA, Ceriotti F, Carobene A, Martini R, Guerra E (1999) Clin Chem Lab Med 37:S260
30. Dybkaer R (2003) Accred Qual Assur 6:46–52
31. NCCLS (2002) Method comparison and bias estimation using patient samples; 2nd edn. NCCLS, Wayne, Pa. NCCLS document EP9-A2
32. Ross JW, Miller WG, Myers GL, Praestgaard J (1998) Arch Pathol Lab Med 122:587–608
33. Little RR, Rohlfing CL, Wiedmeyer H-M, Myers GL, Sacks DB, Goldstein DE (2001) Clin Chem 47:1985–1992
34. University of Missouri School of Medicine (2003) http://web.missouri.edu/~diabetes/ngsp.html
35. Jeppsson J-O, Kobold U, Barr J, Finke A, Hoelzel W, Hoshino T, *et al* (2002) Clin Chem Lab Med 40:78–89
36. Finke A, Kobold U, Hoelzel W, Weycamp C, Jeppsson JO, Miedema K. (1998) Clin Chem Lab Med 36:299–308

Accred Qual Assur (2004) 9:24–32
DOI 10.1007/s00769-003-0715-8

Philip Taylor

One way of disseminating reference values with demonstrated traceability and demonstrated uncertainty to field laboratories: IMEP

Received: 17 October 2003
Accepted: 20 October 2003
Published online: 19 December 2003

Presented at the CCQM Workshop on Traceability, 16–17 April 2003, at BIPM, Sèvres, France.

Electronic Supplementary Material
Supplementary material is available for this article if you access the article at http://dx.doi.org/10.1007/ s00769-003-0715-8. A link in the frame on the left on that page takes you directly to the supplementary material.

P. Taylor (✉)
Institute for Reference Materials and Measurements,
European Commission-JRC, 2440 Geel
e-mail: philip.taylor@cec.eu.int
Tel.: +32-14-571 605
Fax: +32-14-571 863

Abstract An operational interlaboratory comparison programme is described which disseminates SI-traceable reference values to laboratories worldwide. These reference values have an uncertainty and traceability that is demonstrated at the highest metrological level. Participating laboratories can use these values to establish the degree of equivalence of their measurement results and can use this to support their measurement capability claims, e.g. towards third parties. The programme has been run by the Institute for Reference Materials and Measurements (IRMM) since 1988, in the first phase as an awareness programme. Currently, IRMM is focusing its efforts on educational aspects of metrology via a collaboration with the European Co-operation for Accreditation, national metrological institutes (NMIs) and interested academic networks. The viewgraphs used are presented in the "Electronic Supplementary Material" of this ACQUAL issue.

How does the International Measurement Evaluation Programme (IMEP) disseminate traceability?

The invitation to the speakers of the Comite Consultatif pur la Quantite de Matiere (CCQM) Workshop on Traceability was to address the question "How to disseminate traceability?".

In the IMEP programme[1], "SI-traceable values" with a full measurement uncertainty according to the *Guide to the expression of uncertainty in measurement* (GUM) are disseminated by IRMM to field (and other) laboratories by means of appropriately prepared test samples. The uncertainties are the end-product of an evaluation process of all uncertainty sources which is as complete as possible. Yet, the resulting combined and expanded uncertainty must be sufficiently small for the intended use of the result, i.e. smaller than the expected interlaboratory spread of the participants' measurement results. After having measured these samples, the participating field (and other) laboratories can compare their measurement results with the "SI-traceable value" which is released after they have submitted their own results. Both the certified "SI-traceable" value and its certified measurement uncertainty as well as the participants' values with their declared measurement uncertainty are displayed in simple, comprehensive pictures.

Further operational details

The choice of measurement method is the participants' responsibility, thus respecting their scientific freedom. If the measurement result only overlaps part of the uncer-

[1] International Measurement Evaluation Programme, run by the Institute for Reference Materials and Measurements (IRMM) of the European Commission in Geel (so far, from 1988 to date, 20 different evaluation rounds on different matrix materials)

tainty of the SI-traceable value, the participant knows they may have a small, but not serious problem. If the participant's measurement uncertainty does not even encompass part of uncertainty of the certified SI-traceable value, the participant can conclude that they have a real problem. But, if the participant's measurement value overlaps the certified SI-traceable value, the participant can immediately conclude that their measurement result is *equivalent to an SI-traceable value*. On the basis of the picture, the participant's performance can be objectively assessed by any designated body (e.g. an accreditation body) because the assessment is made against an 'external' reference value, neither determined by the assessing body, nor by the participant.

It is shown that an SI-traceable result does not necessarily coincide with the median or average of a number of participants' results, thus demonstrating that a set of systematic errors from a set of laboratories is not necessarily 'normally distributed'. It is also shown that the use of certified reference materials (CRMs) does not automatically lead to 'correct' results and the same is true when different systems for quality assurance are applied.

The IMEP philosophy is to create an awareness of these issues within both the measurement and accreditation communities so that they can take appropriate action themselves. Key to this endeavour is the task of disseminating traceability which can be best described as: "to deliver SI-traceable reference values carried by real-life samples to interested laboratories in order to enable them to determine *the degree of equivalence* of their own measurement result and a certified SI-traceable value".

It follows that participation in the IMEP programme (or any other for that matter) does not itself ensure traceability of the participant's result. That must be done by the participants' themselves each time they perform a measurement. If the traceability of a participant's result has been established to the same common reference (in this case the SI) as the certified SI-traceable value, comparability of the two values has become possible and a 'degree of equivalence' (dependent on the uncertainty ranges) will be a natural consequence of this process.

Thus the responsibility of IRMM is illustrated: it must disseminate independent SI-traceable reference values with certified combined/expanded uncertainties, carried by test samples of a similar nature as those being measured routinely in the field laboratory. The values must be obtained by reference measurement procedures so that they can serve as independent and objective references for the measurement and accreditation communities alike. The uncertainties of these values must be small enough for the intended use of the reference value. If not, better measurement procedures must be developed in order to arrive at a better reference value (i.e. with a smaller combined/expanded uncertainty).

Thus the IMEP programme is a continual learning process for reference laboratories, field laboratories and accreditation bodies. This was recognized by European Co-operation for Accreditation on signing a co-operative agreement with IRMM to train assessors in important metrological issues.

The electronic supplement to this paper makes available the viewgraphs used in this presentation.

Conclusion

An integrated approach to disseminating traceability is described as it has crystallized from more than 15 years of practice with IMEP. It shows how SI-traceable values are disseminated for the benefit of the end-users, consistent with the title and objective of this CCQM workshop. It contains the essential elements needed: standardization, quality assurance, accreditation, metrology and education (SQAME). The key tasks of metrological and other reference measurement institutes as well as accreditation bodies in the dissemination and use of metrological traceability naturally follow from these 15 years of experience.

It is shown how reference values with a demonstrated traceability and demonstrated uncertainty (according to ISO-GUM) are disseminated by IRMM to field (and other) laboratories by means of appropriately prepared test samples. The reference values are established using internationally recognized measurement capabilities and are demonstrated to be equivalent at the international level. The uncertainties are the end-product of an exhaustive evaluation process, yet, the resulting combined and expanded uncertainty are sufficiently small for the intended use of the result (i.e. to be smaller than the expected interlaboratory spread of the participants' measurement results). After having measured these samples, the participating field (and other) laboratories can compare their measurement results with these reference values, which are released after they have submitted their own results. Both the reference value and its measurement uncertainty as well as the participants' values with their declared measurement uncertainty are displayed in simple, comprehensive pictures.

In IMEP there is no requirement for the participants to perform measurements according to standardized measurement procedures, at least for those cases where what one is trying to measure does not depend on how it is measured.

The participant can decide on which improvement action to undertake in his laboratory. If the participant's measurement uncertainty does not even encompass part of uncertainty of the reference value, the participant can conclude they have a real problem. But, if the participant's measurement value overlaps the reference value, the participant can conclude that the traceability of their measurement results is demonstrated, and that their result is equivalent at the global level. On the basis of the picture, the participant's performance can also be objectively assessed by a third party, e.g. an accreditation body, because the assessment is made against an 'external' solid reference value, neither determined by the assessing body, nor by the participant, nor by a consensus value derived from the results of all participants. This reference value does not necessarily coincide with the median or average of a number of participants' results, thus demonstrating that a set of systematic errors from a set of laboratories is not necessarily normally distributed. The principle *'trust is nice, proof is better'* is also demonstrated by the fact that that the use of CRMs or the use of a quality management system or standardized measurement procedures does not automatically lead to reliable results.

The aim of IMEP is to create awareness both in the measurement community as well as in the accreditation community about these issues, so that both communities can take appropriate improvement action. Key in this endeavour is the task of disseminating measurement traceability, which can best be described as: to deliver reference values (preferably traceable to the SI, and values carried by real-life samples) to laboratories in order to enable them to assess if their own measurement results are equivalent at the global scale.

It must be stressed that the participation in an interlaboratory comparison programme such as IMEP (or any other for that matter) does not itself ensure traceability of the participant's result. That can only be achieved by the participant, for the measurement they perform. Traceability is about establishing a valid equation describing the measurement, and defining the link of the quantities defined in this measurement equation to a same common reference (where possible the SI).

During a period when trust in designated competence was self-evident, IMEP started as an awareness programme, stressing the need for *demonstrated* competence. The future challenge for IMEP is to evolve into a system that also incorporates educational aspects, to enable improvement. For this reason an EA-IRMM collaboration agreement was signed in February 2001. In the framework of this collaboration, different ways of assessing measurement performance are discussed, also stressing the need for the training of technical assessors in the basics of measurement science (metrology), on top of participation in IMEP.

In line with the objectives of the CCQM Workshop, we have tried to suggest how measurement traceability can be disseminated to field laboratories. An important anticipated evolution in this respect is the fact that the stakeholders dealing with accreditation/metrology/education/standardization will need to collaborate much more closely to achieve measurable progress.

The following figures show the PowerPoint presentation illustrating the points made in this article and the presentation is available as Electronic Supplementary Material on the Web page.

Fig. 1, Fig. 2, Fig. 3, Fig. 4, Fig. 5, Fig. 6, Fig. 7, Fig. 8, Fig. 9, Fig. 10, Fig. 11, Fig. 12, Fig. 13, Fig. 14, Fig. 15, Fig. 16, Fig. 17, Fig. 18, Fig. 19, Fig. 20, Fig. 21, Fig. 22, Fig. 23, Fig. 24, Fig. 25, Fig. 26, Fig. 27, Fig. 28, Fig. 29, Fig. 30, Fig. 31, Fig. 32, Fig. 33, Fig. 34, Fig. 35, Fig. 36, Fig. 37, Fig. 38, Fig. 39, Fig. 40, Fig. 41, Fig. 42, Fig. 43, Fig. 44, Fig. 45, Fig. 46, Fig. 47, Fig. 48, Fig. 49, Fig. 50, Fig. 51

Acknowledgements The author is indebted to P. De Bièvre for his valuable comments.

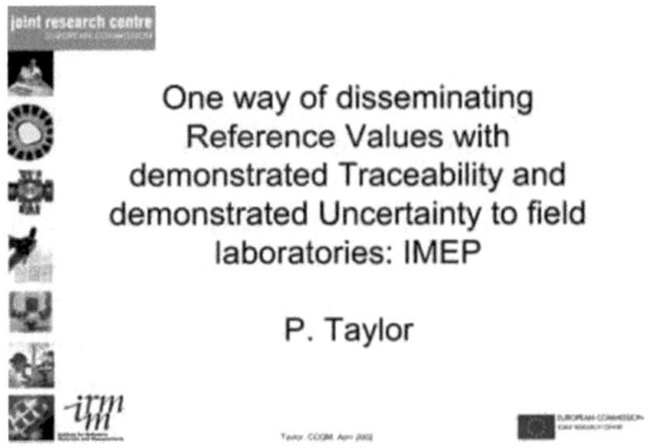

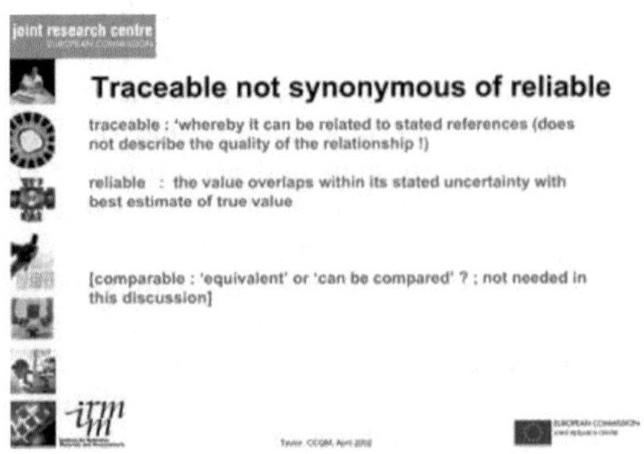

What I try to do in next minutes ...

- IMEP : why ?
- IMEP : our mental model
- IMEP : how did we go about ? Traceability ?
- IMEP : who is involved ?
- IMEP : does it help ?
- IMEP : future ?

Taylor, CCQM, April 2002

1988-to date :
IMEP as awareness programme

- there is a problem
- the need & usefulness of an external reference value

Taylor, CCQM, April 2002

in 1988 : in professional circles

Disbelief !

Taylor, CCQM, April 2002

Completed IMEP® Rounds

IMEP® Round	Title	Time Period	Publication
IMEP-1	Li in Human Serum	1988	Fres.Z.Anal.Chem (1988)332:718-721
IMEP-2	Cd in Polyethylene	1990-91	Fres.J.Anal.Chem (1993)345:310-313
IMEP-3	Trace elements in Water	1991-93	Accred.Qual.Assur (1996)1:71-82
IMEP-4	Trace elements in Bovine Serum	1991-95	Accred.Qual.Assur (1999) 3:447-458
IMEP-5	Fe in Human Serum	1991-94	Scand.J.Clin.Lab.Invest (1993)53 suppl 212,38
IMEP-6	Trace elements in Water	1994-95	Accred.Qual.Assur (1998)3:56-68
IMEP-7	Trace elements in Human Serum	1997-98	IRMM report GE/R/SIM/2/98 Accred. Qual. Assur(1999) 4:463-472
IMEP-8	$n(^{13}C)/n(^{12}C)$ and $n(^{18}O)/n(^{16}O)$ in CO_2	1997-99	IRMM report GE/R/IM/38/99 EUR 19060 EN
IMEP-9	Trace elements in Water	1998-99	IRMM report GE/R/IM/15/99 EUR 18724 EN
IMEP-10	Trace elements in Polyethylene	1997-98	IRMM report GE/R/SIM/11/98
IMEP-11	Metals in Car Exhaust Catalysts	1998-99	IRMM report GE/R/IM/20/99 EUR 18735 EN
IMEP-13	Trace elements in Polyethylene	1999-2000	IRMM report GE/R/IM/10/2000 EUR 19562 EN
IMEP-14	Trace elements in Sediment	1999-2000	IRMM report GE/R/IM/15/2000 EUR 19595 EN

Taylor, CCQM, April 2002

blind samples measured by participants ...

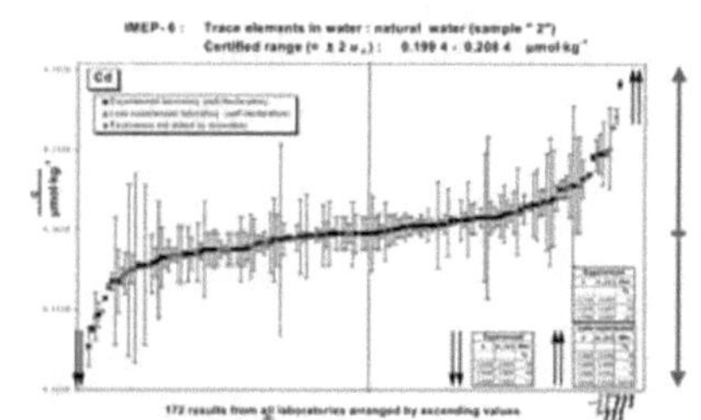

Taylor, CCQM, April 2002

Ongoing IMEP® Rounds

IMEP® Round	Title	Time Period	Status of the project
IMEP-12	Trace elements in Water	2000-2001	Samples available/establishment of reference values ongoing
IMEP-16	Pb in Wine	2000-2001	Samples distributed to participants
IMEP-17	Trace and minor constituents in Human Serum	2000-2001	Solicit for interest for participation Sample preparation ongoing
IMEP-19	Cd in Rice		Solicit for interest for participation Samples available

www.imep.ws

Taylor, CCQM, April 2002

we can do this !

Claim

true ?

Taylor, CCQM, April 2002

Our mental model ?

- What are our (explicit/implicit) assumptions
- check (experimentally) on occasions whether our model is correct !
- Change if needed !

Taylor, CCQM, April 2002

IMEP : our mental model

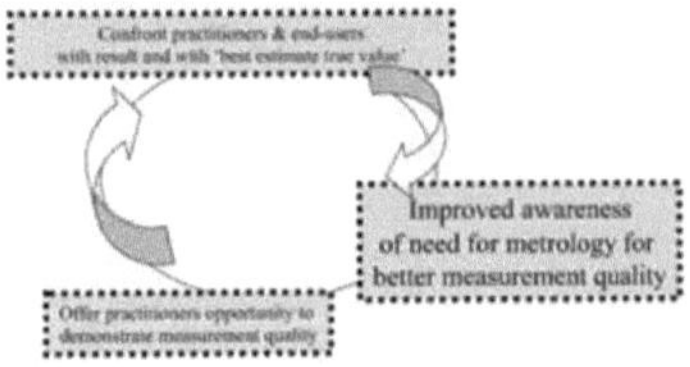

Taylor CCQM April 2002

IMEP basics

Taylor CCQM April 2002

IMEP reference value

- At least 2 laboratories contributing
- laboratories are NMI signatories or laboratories with demonstrated competence
- requested from contributors : measurement certificate and report, explaining traceability and uncertainty
- uncertainty of contributing certifying lab : set a priori
- uncertainty of the IMEP reference value
  - ISO-GUM
  - 'appropriately small'

Taylor CCQM April 2002

Contributing certification partners in IMEP

Organisation	country
University of Agricultural Sciences (BOKU-Vienna)	AUSTRIA
National Research Council Canada (NRC)	CANADA
IRMM	EU
Bundesanstalt für Materialforschung und -prüfung (BAM)	GERMANY
Deutsche Gesellschaft für Klinische Chemie-DGKC	GERMANY
DGKC	GERMANY
Mainz University	GERMANY
PTB	GERMANY
National Institute for Environmental Studies (NIES)	JAPAN
National Metrology Institute of Japan, NMIJ	JAPAN
National Research Centre for CRM (NRCCRM)	P.R. CHINA
Korea Research Institute for Standards and Science (KR	REP. OF KOREA
C.I.E.M.A.T.	SPAIN
Universidad de oviedo	SPAIN
SP	SWEDEN
Swiss Fed. Inst. of Technology	SWITZERLAND
NMi-Van Swinden Laboratorium	THE NETHERLANDS
NIST	USA
Laboratory of the Government Chemist (LGC)	UNITED KINGDOM

Taylor CCQM April 2002

IMEP value assignment : consists of *a set of operations* resulting in value assignment

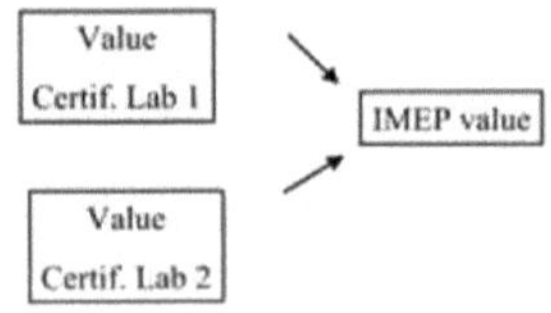

Taylor CCQM April 2002

IMEP value : SI traceable ? only if contributing values are SI traceable

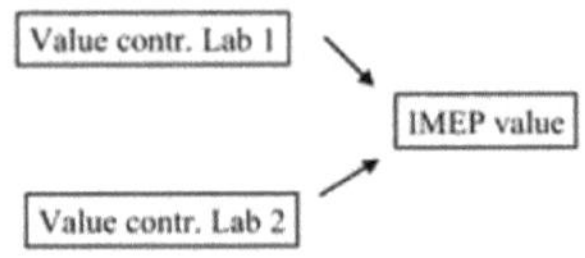

Taylor CCQM April 2002

IMEP value : reliable ?
Depends on
1: reliability input quantities
2: reliability assignment process

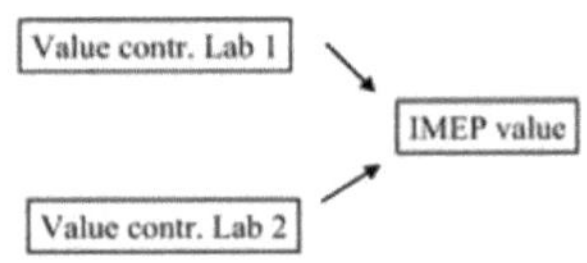

Taylor CCQM April 2002

IMEP certified test samples uncertainty contributions

- From measurement
- from measurement of homogeneity
- from measurement of stability

Taylor CCQM April 2002

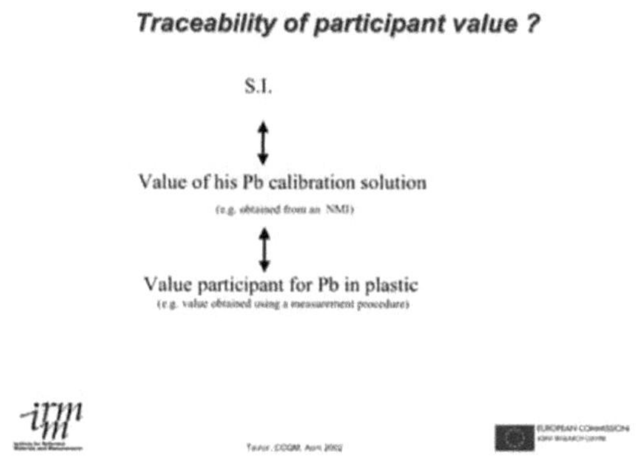
Traceability of participant value ?
S.I.
Value of his Pb calibration solution
(e.g. obtained from an NMI)
Value participant for Pb in plastic
(e.g. value obtained using a measurement procedure)

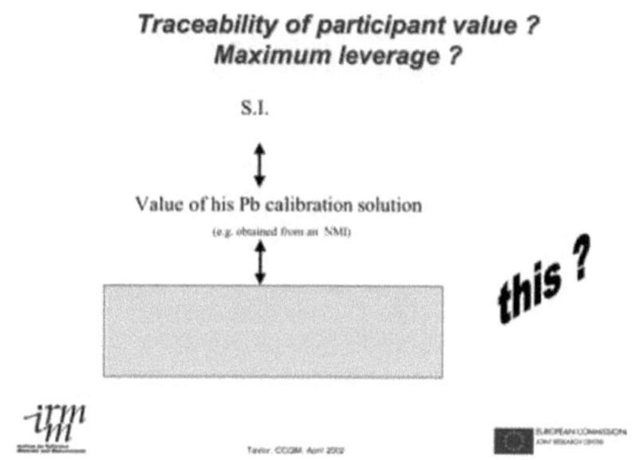
Traceability of participant value ?
Maximum leverage ?
S.I.
Value of his Pb calibration solution
(e.g. obtained from an NMI)
this ?

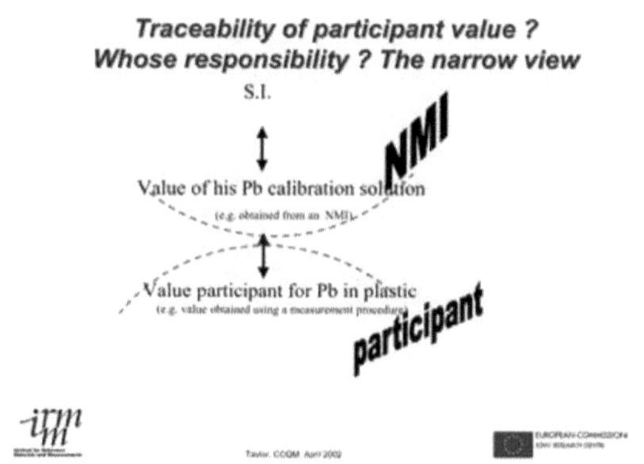
Traceability of participant value ?
Whose responsibility ? The narrow view
S.I.
NMI
Value of his Pb calibration solution
(e.g. obtained from an NMI)
Value participant for Pb in plastic
(e.g. value obtained using a measurement procedure)
participant

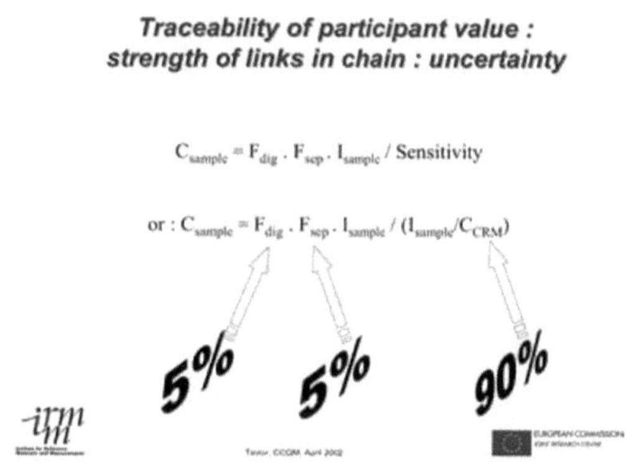
Traceability of participant value :
strength of links in chain : uncertainty
$C_{sample} = F_{dig} \cdot F_{sep} \cdot I_{sample} / \text{Sensitivity}$
or : $C_{sample} = F_{dig} \cdot F_{sep} \cdot I_{sample} / (I_{sample}/C_{CRM})$
5%
5%
90%

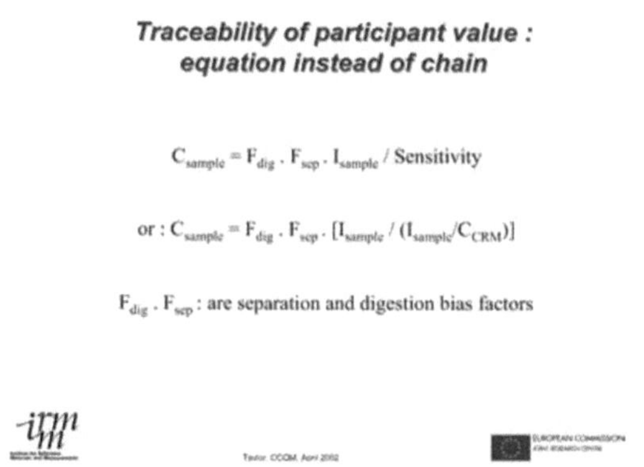
Traceability of participant value :
equation instead of chain
$C_{sample} = F_{dig} \cdot F_{sep} \cdot I_{sample} / \text{Sensitivity}$
or : $C_{sample} = F_{dig} \cdot F_{sep} \cdot [I_{sample} / (I_{sample}/C_{CRM})]$
$F_{dig} \cdot F_{sep}$: are separation and digestion bias factors

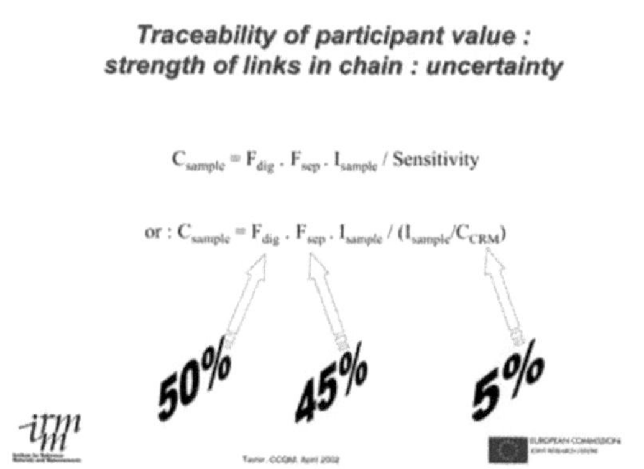
Traceability of participant value :
strength of links in chain : uncertainty
$C_{sample} = F_{dig} \cdot F_{sep} \cdot I_{sample} / \text{Sensitivity}$
or : $C_{sample} = F_{dig} \cdot F_{sep} \cdot I_{sample} / (I_{sample}/C_{CRM})$
50%
45%
5%

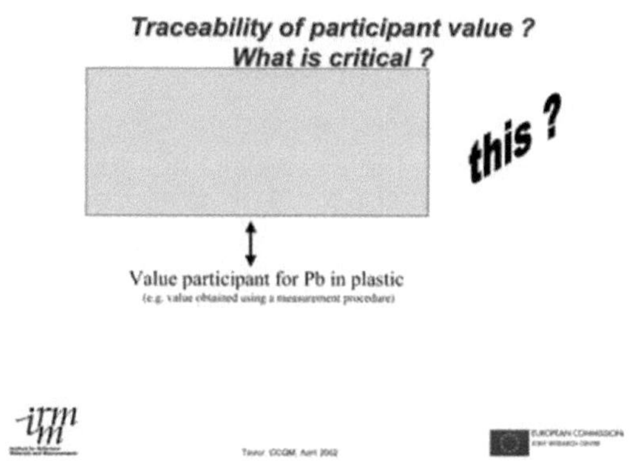
Traceability of participant value ?
What is critical ?
this ?
Value participant for Pb in plastic
(e.g. value obtained using a measurement procedure)

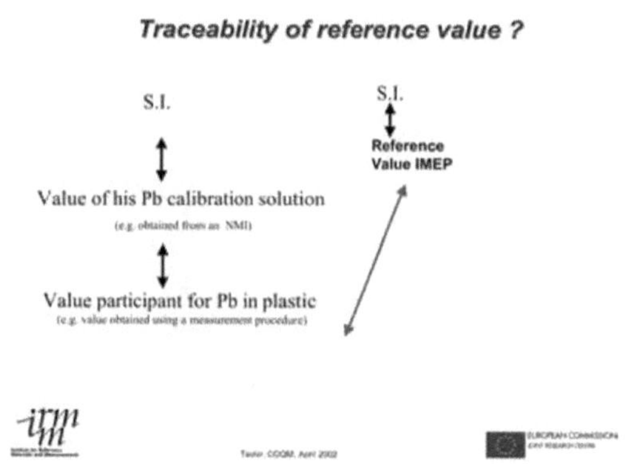
Traceability of reference value ?
S.I.
S.I.
Reference
Value IMEP
Value of his Pb calibration solution
(e.g. obtained from an NMI)
Value participant for Pb in plastic
(e.g. value obtained using a measurement procedure)

how to explain this to people ?

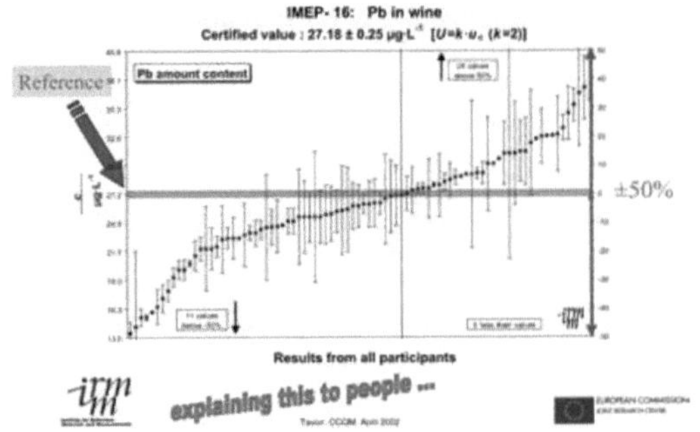

IMEP communication policy

- Use of 'external reference', metrologically sound
- availability of international metrology community

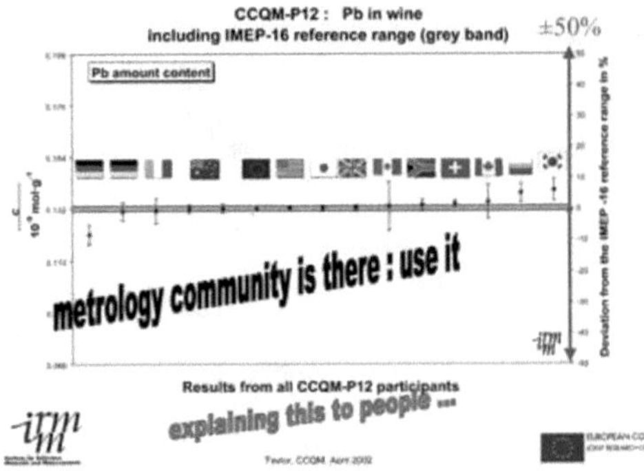

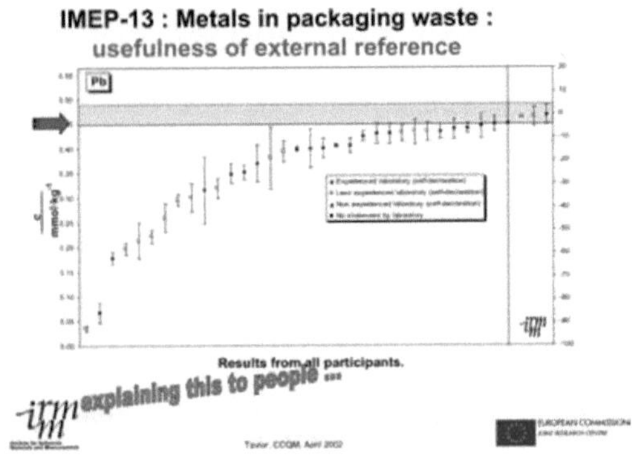

who's involved ?

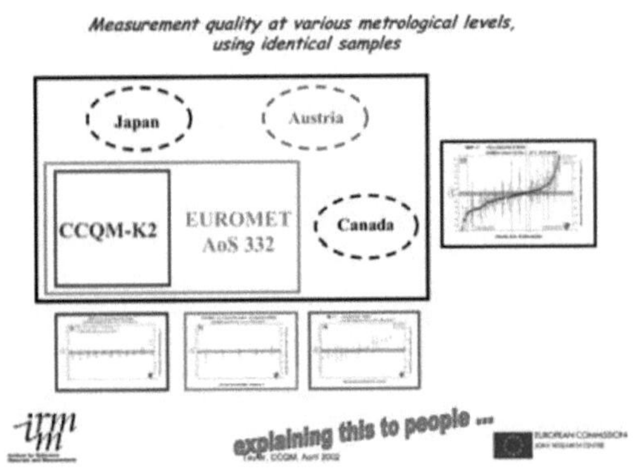

partners involved in the dissemination process

- Metrology organisations
- network of 'regional' & 'sector' coordinators (e.g. in clinical, food, ...)
- collaboration with European Accreditation
- educators

does it have an impact ?

honnest answer please ...

Taylor, CCQM, April 2002

No ...

IMEP- 12: Trace elements in Water **use of QM system**

Certified value :40.78·10⁻⁹ ± 0.82·10⁻⁹mol·L⁻¹ [U=k·u_c (k=2)]

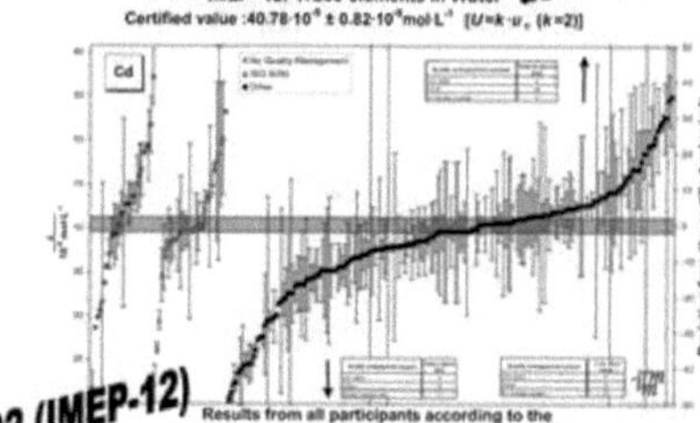

2002 (IMEP-12)

Results from all participants according to the Quality Management System used (self-declaration).

Taylor, CCQM, April 2002

Impact ? ... Yes

- Chemistry Metrology programmes receive more attention
- stronger voice of metrology organisations
- 'end user' attention (control bodies, enforcement agencies, ministeries ...)
- interaction with accreditation

Taylor, CCQM, April 2002

No ...

IMEP- 12: Trace elements in Water **accreditation**

Certified value :40.78·10⁻⁹ ± 0.82·10⁻⁹mol·L⁻¹ [U=k·u_c (k=2)]

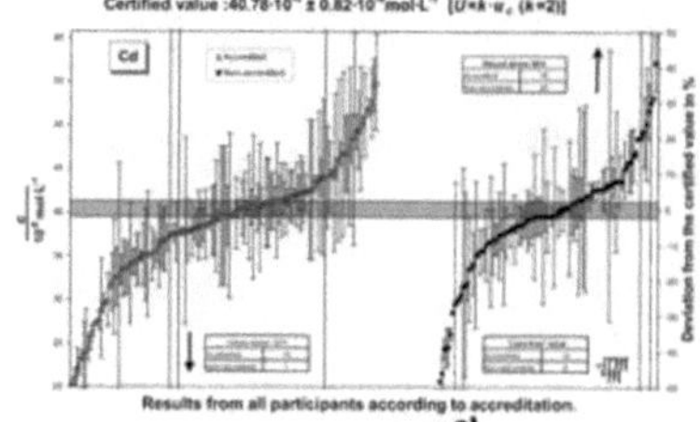

Results from all participants according to accreditation.

2002 (IMEP-12)

Taylor, CCQM, April 2002

beware of complacency !

Taylor, CCQM, April 2002

but metrology will solve this ...

Taylor, CCQM, April 2002

No ...

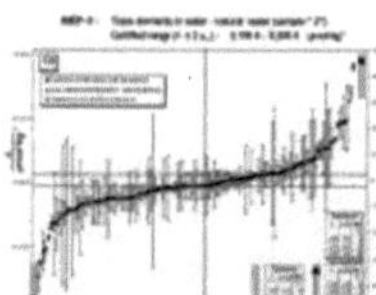

±50%

1994 (IMEP-6)

2002 (IMEP-12)

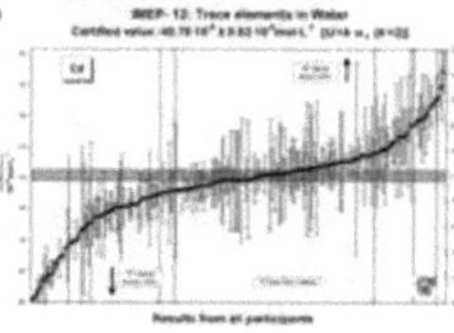

Results from all participants

Taylor, CCQM, April 2002

No ... : use of CRMs

IMEP- 12: Trace elements in Water

Certified value :40.78·10⁻⁹ ± 0.82·10⁻⁹mol·L⁻¹ [U=k·u_c (k=2)]

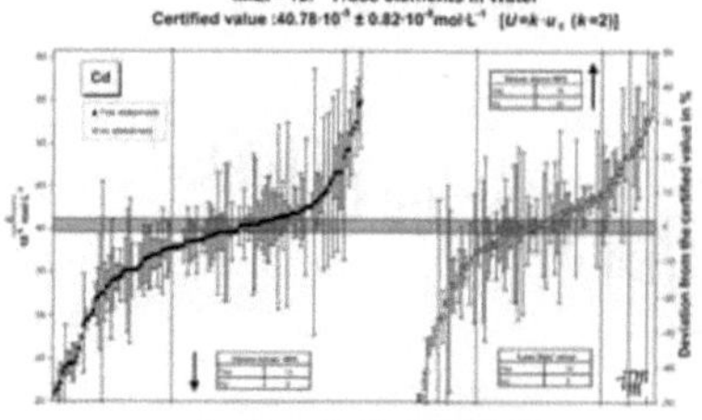

Results from all participants by use of CRMs

2002 (IMEP-12)

Taylor, CCQM, April 2002

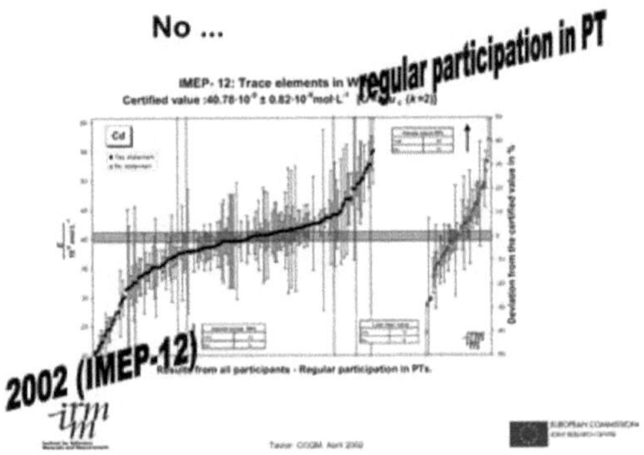

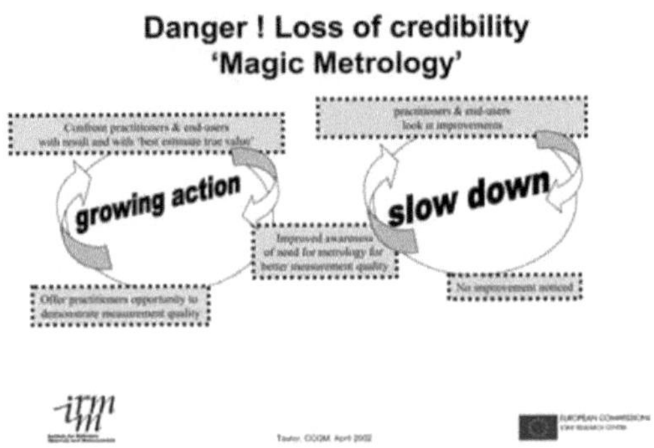

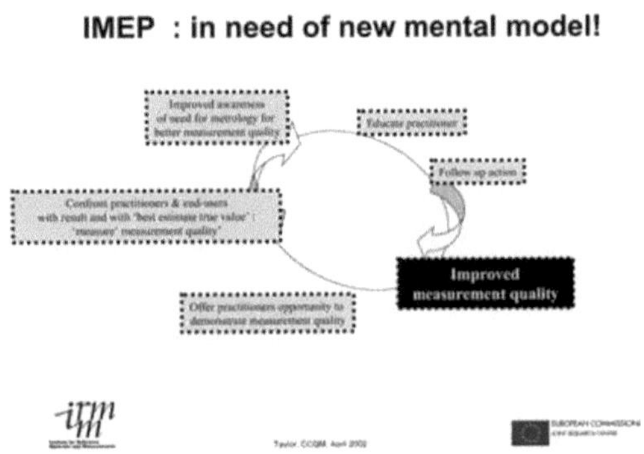

but that's not job of NMIs !

correct ...

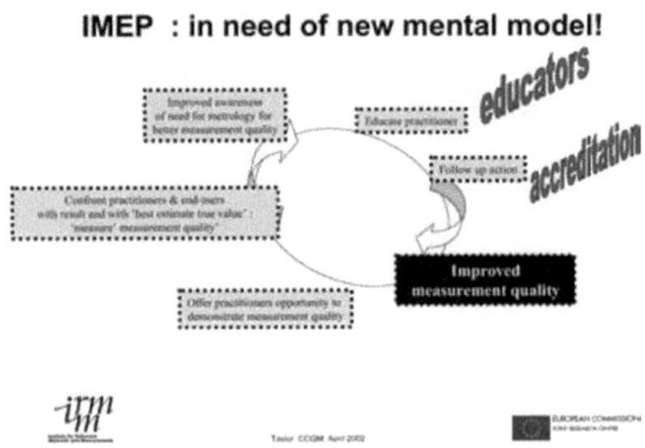

Outlook : IRMM initiative (EU 5FWP Thematic Network) in close collaboration with EUROMET & EA

- improve corrective action follow up by accreditation
- systematise process of learning after participation to an ILC
- design metrologically correct courses
- train trainers
- Train technical assessors accreditation

Epilogue ...

And by the way, lot's of the 'integration' comments made on IMEP, can be translated to 'international metrology activities', ...

S.Q.A.M.E.

- Standardisation
- quality assurance
- accreditation
- metrology
- education

Accred Qual Assur (2004) 9:237–238
DOI 10.1007/s00769-003-0748-z

Emil Völkert

Implementation of traceability – needs and perspective of the in-vitro-diagnosticum industry

Received: 7 June 2003
Accepted: 27 November 2003
Published online: 24 January 2004

Presented at BERM-9 – 9th International Symposium on Biological and Environmental Reference Materials, 15–19 June 2003, Berlin, Germany

E. Völkert (✉)
Chairman of EDMA WP Standardisation,
Roche Diagnostics GmbH,
Sandhofer Str. 116,
68305 Mannheim, Germany
e-mail: emil.voelkert@roche.com
Tel.: +49-621-7591656
Fax: +49-621-7594583

Abstract Manufacturers support the concept of traceability. However, only a small number of the medically relevant measurands can be traced to the highest metrological order. In many cases, the measured substances are heterogeneous mixtures where traceability can be established only to either an international conventional reference measurement procedure or to a manufacturer's own in-house reference system. The traceability concept needs to be seen in the context that the results of medical laboratories are not an aim per se, but are meant to provide useful medical information to clinicians, and that pre- and post-analytical steps may also contribute significantly to errors. There is a need for the further development of suitable reference measurement systems, but in view of the multitude of tasks and limited resources, priorities need to be set.

Keywords Traceability · Reference systems · International standards

Report

The In-Vitro-Diagnostic (IVD)-Directive [1] requires manufacturers to assure traceability of assigned values to calibrators and trueness control materials to reference measurement procedures and/or reference materials of a higher order, where available. This is a legal requirement for products marketed in Europe, and manufacturers are interested in applying the principle of traceability on a global scale, because it allows products to be marketed world-wide. The benefit to patients and users is perceived as the direct comparability of laboratory measurement results over regions and time.

For the concept to be applied in its ideal complete form, up to the highest possible metrological order, it must comprise the unequivocal description of the measurand – which is commonly a heterogeneous mixture – and of the detailed measurement procedures in human samples. It also requires the availability of a suitable reference material. Suitability should be based upon the definition of the analyte in the measurand. The assignment of the values must be performed by laboratories with appropriate qualifications, very often by laboratories of metrological institutions or other specially qualified laboratories. The framework for these requirements is outlined in several EN/ISO standards [2, 3, 4, 5, 6].

However, while in clinical laboratories some 1000–1500 medically relevant measurands are presently examined, the calibration can be made traceable for only about 60 of them to the highest metrological order (examples here include glucose and cholesterol).

The situation is much more complex in the majority of cases, where the measured substance is in fact a mixture of several components (isoforms; glycosylation). In several instances the species of interest depends on the medical application. Ferritin species for the determination of anaemia, for example, are different from those for monitoring tumors; therefore the intended *medical* application must be kept in mind when considering traceability. In such cases WHO International Standards may be applied, although these materials are not always suitable for these purposes. They do not necessarily guarantee that patient results obtained with kits from different manufacturers are comparable, even when the calibration is traceable to the same WHO material, mainly because of the heterogeneity of the material, and because of the inherent differences in specificity of (monoclonal) antibodies when used in the examinations. Other problems are possible lot-to-lot variations in some materials. For this reason international conventional reference measurement procedures with appropriate international conventional reference materials are needed, which serve as surrogate material for the "analytes" existing in human samples.

In cases where no suitable reference material or reference measurement procedure exists, the manufacturers need to establish their own reference system based on a suitable, reproducible and stable (manufacturer's) working calibrator.

Results from medical laboratories do not serve a purpose per se, but provide useful information to clinicians, helping to decide on diagnosis, remission or recurrence of disease in patients. But other factors such as patient condition, pre- and post-analytical influences play an important role in this process as well, and these need to be taken care of too.

Improvement in the metrological area may lead to a shift in reference intervals, which needs to be taken into account by the clinicians. This shift in the framework of the physician interpreting the results is sometimes met with resistance if no additional medical benefit arises. This may complicate the acceptance of this concept of traceability in its possible practical consequences.

Nevertheless, manufacturers are interested in the further development of reference materials and reference measurement procedures suited for application in human serum for medical purposes. Therefore, industry appreciates the efforts started by the Joint Committee of Traceability and Laboratory Medicine (JCTLM), an initiative supported by the Bureau International des Poids et Mesures (BIPM) and the International Federation of Clinical Chemistry and Laboratory Medicine (IFCC) in order to develop suitable reference materials and measurement procedures, as well as to collect information on the activities of reference laboratories. In view of the limited resources and large efforts connected with these activities, clear priorities need to be set. Projects need to take into account the clinical importance of the analyte, consider the technical difficulties that must be overcome, and, most importantly, decide whether improvement of the metrological side is reflected in a gain of medically relevant information.

References

1. European Parliament (1988) Directive 98/79/EC of the European parliament and of the Council of 27 Oct 1998 on in vitro diagnostic medical devices. Off J Europ Com L 331/1 (7 December 1998)
2. International Organization for Standardization (2002) Metrological traceability of values assigned to calibrators and control materials (ISO FDIS 17511:2002). International Organization for Standardization, Geneva, Switzerland
3. International Organization for Standardization (2002) Metrological traceability of assigned values for catalytic concentration of enzymes in calibrators and control materials (ISO FDIS 18153:2002). International Organization for Standardization, Geneva, Switzerland
4. CEN (1999) In vitro medical devices – measurement of quantities in samples of biological origin – presentation of reference measurement procedures (EN 12286:1999). CEN, Brussels, Belgium
5. CEN (1999) In vitro diagnostic systems – measurement of quantities in samples of biological origin – description of reference materials (EN 12287:1999). CEN, Brussels, Belgium
6. International Organization for Standardization (2000) Laboratory medicine – requirements for reference measurement laboratories in laboratory medicine (EN ISO 15195:2000). International Organization for Standardization, Geneva, Switzerland

Accred Qual Assur (2004) 9:198–204
DOI 10.1007/s00769-003-0747-0

Gill Holcombe
Richard Lawn
Mike Sargent

Improvements in efficiency of production and traceability for certification of reference materials

Received: 13 June 2003
Accepted: 27 July 2003
Published online: 20 January 2004

Presented at BERM-9—9th International Symposium on Biological and Environmental Reference Materials, 15–19 June 2003, Berlin, Germany

G. Holcombe · R. Lawn · M. Sargent (✉)
LGC Limited,
Queens Road, Teddington, Middlesex,
TW11 0LY, UK
e-mail: ms@lgc.co.uk
Tel.: +44-20-89437360
Fax: +44-20-89437360

Abstract Issues of current interest to certified reference material producers are addressed. Alternative strategies for certification of matrix reference materials are discussed and the benefits of adopting a flexible, cost-effective approach are described. The difficulty of undertaking homogeneity testing where certification is to be carried out with definitive techniques capable of providing very small measurement uncertainty is discussed. Methodology is described which combines conventional screening of the candidate material for homogeneity with an additional, precise assessment of homogeneity based on isotope dilution mass spectrometry measurements. A systematic procedure for evaluating the commutability (horizontal traceability or scope) of matrix reference materials has been evaluated and shows that in some circumstances matrix effects may be less pervasive than is generally believed. This offers the possibility, especially for trace analysis applications, of more efficient use of existing reference materials without compromising measurement reliability. Vertical traceability of matrix reference material data is of growing interest but is difficult to achieve with present interlaboratory certification exercises. A modification is described which attempts to address this issue. It also offers the possibility of improved identification of outliers and reduced variation of data between the participating laboratories.

Keywords Matrix reference material · Traceable · Commutability · Homogeneity · Certification

Introduction

LGC has been involved in the production of both pure substance and matrix reference materials (RMs) for about 15 years and during that time over 200 materials, the majority relating to food and environmental measurements, have been produced and certified. Whilst the pure materials have been certified largely on the basis of in-house measurements, the matrix materials have usually required a collaborative approach, involving interlaboratory measurement studies. The latter approach is extremely important and versatile for the production of matrix RMs, but it does raise a number of issues. Among these are the selection of appropriate procedures for assigning certified values and uncertainties, establishing meaningful traceability, identifying suitable participant laboratories and data processing.

In view of these issues, we have over several years undertaken a substantial research programme to develop definitive methods appropriate for in-house certification of matrix RMs, particularly for analytes at trace levels. These definitive measurement methods, most of which use isotope dilution mass spectrometry (IDMS), been the subject of extensive validation, including CCQM key comparisons and pilot studies involving other national measurement institutes. Hence, we are now able to augment interlaboratory data with data obtained at LGC using these very accurate measurements. We have also used this additional data to investigate some of the issues mentioned already concerning interlaboratory measurement studies. This work on development and validation of definitive methodology and its application to certification of matrix RMs has been widely reported [1, 2, 3, 4, 5, 6, 7]; in this paper the emphasis lies with the relative merits of the alternative certification strategies and use of our definitive measurements to facilitate improvements to the interlaboratory approach.

An aspect of matrix RMs which is of considerable importance is the question of "commutability" or "horizontal traceability". This refers to the scope of the materials, i.e. the extent to which a matrix RM of a particular composition may reliably be used to evaluate a measurement procedure that is applied to a routine test sample of a different composition. The differences in composition between a reference matrix and a routine test sample matrix must not cause the two materials to behave differently when a particular analytical method is applied. At present, the extent to which this is true is largely a matter of expert judgement based on knowledge of the measurement application. A better and more systematic understanding of the factors affecting horizontal traceability will enable users to select appropriate matrix RMs more reliably and producers to target their production activities more efficiently.

Certification strategies

As mentioned already, two main certification strategies are used by LGC and also the majority of other RM producers. The extent to which each is applied varies extensively, depending on both the application area and the preferences of individual producers. The main strengths and weaknesses of each approach are summarised in Table 1.

The authors have evaluated these approaches at considerable length, partly to judge whether one or the other has a clear advantage. Our main concern, however, has been to arrive at the most efficient and cost-effective means of producing RMs which are fit for the purposes of users. The outcome of this evaluation is that both approaches have specific and different merits and should be used as appropriate. In our view, there are some applications for which one or the other approach is clearly preferable, but in many cases using a combined approach on the same material is advantageous. By adopting this flexible certification strategy we can achieve not only cost-effective production but also provide users with a wide range of useful data in addition to the conventional "certified result". For example, a definitive IDMS measurement often provides a certified value with very low uncertainty and bias, so that users have a very narrow "target to hit" when validating their methodology. This is particularly helpful, for example, with applications where there is poor agreement between several widely used routine methods. Conversely, many important applications rely on "method-specific" data obtained using industry-standard methods, for example, fibre content or extractable heavy metals. In these cases a definitive method may not exist or such a result may not be relevant to the RM user.

As a further example, many RM users will be validating routine methodology which is quite different from the definitive techniques such as IDMS. If they experience problems with their method it is very helpful to have available additional data from other laboratories which used the same or alternative routine methods. This type of information is readily obtained from interlaboratory studies. In the past LGC and many other RM producers have not routinely supplied such information with their materials but we are now working with users to address this issue. Discussions with users also indicate that they value access to all available information about a material, not just fully certified results with stated uncertainty. Many of our materials could include such information, for example, "indicative" values obtained during feasibility studies on the material. We are presently considering how best to provide such information in a way which unambiguously indicates its status. A uniform approach to addressing this problem could usefully be developed by producers through ISO REMCO or another appropriate forum.

This flexible approach to certification has also led us to take a more holistic view of RM production. Interlaboratory certification studies frequently highlight significant measurement problems, sometimes in a large number of participating laboratories. The tendency has been to make participants aware of this situation through the certification report but to take little other action. In future we aim to provide better value for money to both

Table 1 A comparison of certification strategies

Interlaboratory comparison	Definitive methods
Large number of results by methods familiar to users	Fewer results, often by "specialised" methods
Method validation and uncertainty poorly defined	Methods well characterised with uncertainty budgets
Present approaches do not provide traceable data	Traceability of data well defined and accepted
Ideal when well-established methods are widely used	Ideal when problems exist with routine methods
Applicable to all common analytical applications	Not applicable to "method-specific" applications

funding agencies and users by using interlaboratory certification studies as a means of working with laboratories to identify and resolve measurement problems, the certified material being just one of several outcomes. With this type of study, the availability of definitive reference values for key analytes is of particular value. In order to facilitate these new approaches, and to identify efficient ways of disseminating all the useful data from our certification programme, we are establishing a series of sector-based RM user networks for UK laboratories.

Certification by interlaboratory measurement studies

Selection of participant laboratories

It has frequently been suggested that the organisers of interlaboratory certification studies should only accept "reliable" laboratories as participants and one obvious selection criterion would be to use only accredited laboratories. However, an evaluation of data obtained from several studies organised by LGC over the past 7 years indicates that accredited laboratories are not necessarily more reliable than nonaccredited laboratories. For example, Fig. 1 shows that in an interlaboratory study to determine magnesium in water, accredited laboratories (marked with an asterisk) were just as likely to produce results with a large deviations as nonaccredited laboratories.

Attempts to correlate analytical performance with other seemingly indicative laboratory characteristics, such as participation in proficiency testing schemes, regular use of certified RMs, number of years of experience and number of samples analysed per year were all equally unsuccessful. Therefore, in the absence of any simple and obvious means of identifying and preselecting only reliable laboratories as participants in certification studies, an investigation was undertaken of the validity of adopting the consensus mean (after outlier elimination) from an interlaboratory study as a certified value.

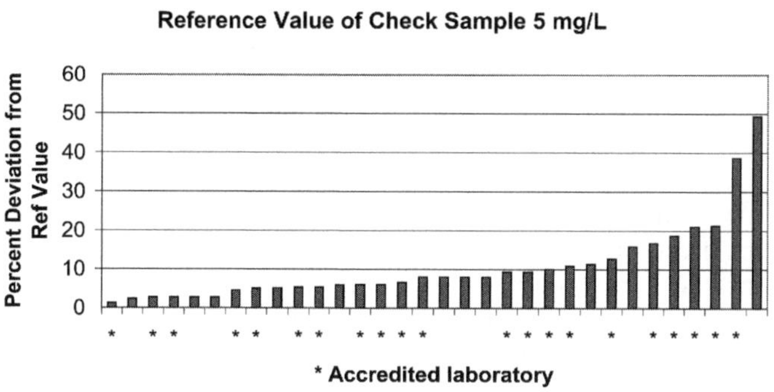

Fig. 1 Variation from reference value for determination of Mg in water

Validity of the interlaboratory consensus mean as a certified value

For a selection of materials and analytes, values were determined at LGC using high-accuracy IDMS methodology that had been the subject of international comparisons through the CCQM. Some typical results obtained are shown in Table 2.

The IDMS results generally confirm the validity of the interlaboratory approach to the certification of matrix RMs, since the agreement between the IDMS values and the corresponding interlaboratory means is largely encouraging. Some potential problem areas are highlighted where the difference between two values, although not large, is significant when the uncertainties are taken into account. The uncertainties of the interlaboratory means are calculated as the 95% confidence intervals:

$$U = t \times \frac{standard\ deviation}{\sqrt{n}}.$$

Such uncertainties are consistently larger, and often markedly so, than those provided by the IDMS measurements and this issue is discussed in subsequent sections of this paper.

However, depending on the intended application of the RM, larger uncertainties may sometimes be acceptable. The smaller uncertainty provided by IDMS-certified data may not always be necessary in routine environmental monitoring work, where a consensus-certified material may well be fit-for-purpose and provide a cost-effective approach to RM certification. The resource-intensive approach of IDMS may then be properly confined to the certification of RMs for critical applications.

The selective but regular use of high-accuracy IDMS methodology in conjunction with interlaboratory studies is another essential application. This will enable a picture to be built up of those analyte/matrix combinations that may be reliably certified using consensus data and those requiring more detailed and extensive characterisation.

Table 2 A comparison of isotope dilution mass spectrometry (*IDMS*) values and interlaboratory consensus values. *n* is the number of laboratories

Matrix	Analyte	Values	
		IDMS	Interlaboratory mean
River water	Pb	5.2±0.2 μg/l	4.8±0.4 (*n*=30)
Drinking water	Fe	236±4 μg/l	227±11 (*n*=19)
Estuary water	Cd	101±2 μg/l	104±11 (*n*=10)
Estuary water	Ni	186±3 μg/l	207±57 (*n*=11)
Sewage sludge	PCB 101	35.7±0.7 μg/kg	31.6±2.0 (*n*=10)
Strawberry leaves	Fe	820±30 mg/kg	710±40 (*n*=22)
Human serum	Creatinine	21.7±0.5 mg/kg	22.5±0.5 (*n*=20)

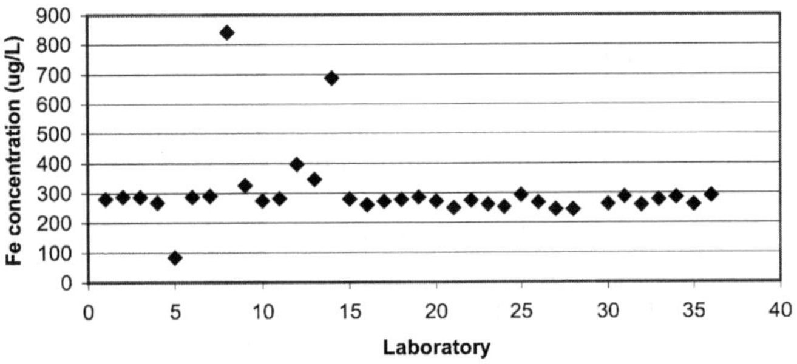

Fig. 2 Determination of iron in river water—laboratories using in-house calibration standards

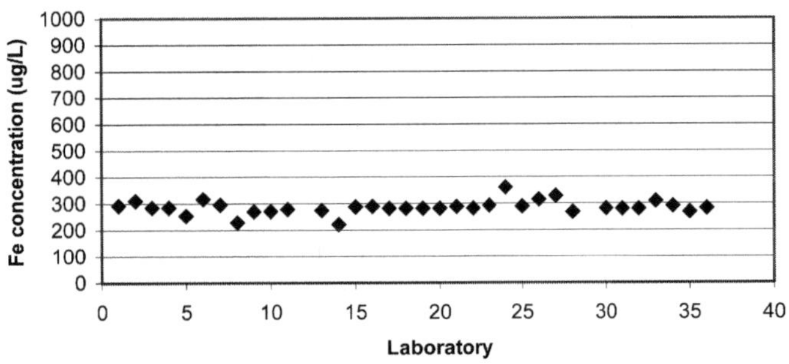

Fig. 3 Determination of iron in river water—laboratories using LGC-supplied calibration standard

Traceability of a consensus mean

An important requirement of quality standards such as ISO 17025 and ISO Guide 34 is that test results or values assigned to RMs should be traceable, preferably to national standards. This requirement extends to data obtained in interlaboratory certification studies and, as one approach to meeting the requirement, LGC has recently modified the way in which it organises these studies. In addition to sending each participant laboratory a sample of the candidate RM, an accurately prepared and verified instrument calibration solution has also been provided for analysis. The data reported for the solution may be used to normalise the results obtained for the matrix sample. This ensures that all of the laboratories' results are traceable to a common measurement standard that is of known and high quality. This has often had a significantly beneficial effect on the quality of the data returned, as Figs. 2 and 3 illustrate.

Figure 2 shows the results reported by all laboratories in a characterisation study of iron in river water, where each laboratory used its own in-house iron standard for instrument calibration. In contrast, Fig. 3 shows the results obtained when each laboratory's result was recalculated using data reported for the LGC-supplied calibration solution.

Table 3 Iron concentrations in river water

Type of value	Interlaboratory calibration standard	Number of laboratories	Iron concentration (μg/l)
Interlaboratory mean	In-house	30	275±7
Interlaboratory mean	LGC-supplied	34	286±9
IDMS (by LGC)	–	1	287±5

Comparison of Figs. 2 and 3 shows the marked improvement in between-laboratory variation that is obtained when all results are traceable to the common measurement standard supplied by LGC. A quantitative evaluation of the data shows that where laboratories used their own calibration standards the coefficient of variation is 41% and five out of the 35 laboratories are Grubbs outliers. The corresponding values when traceability to the LGC-supplied standard is established are 11% and one out of 35, respectively. This observation indicates the importance of establishing the quality of the calibration standards used in interlaboratory studies and strongly suggests that such matters as the source, preparation, storage and use of calibration standards are a "problem area" in a number of laboratories. Addressing such problems by provision of a calibration standard of known provenance will result in smaller uncertainties (95% confidence intervals) and will confer traceability on interlaboratory consensus mean values.

Additional evidence of the benefits obtained when the traceability of interlaboratory data to a common, reliable measurement standard is established is shown in Table 3. The interlaboratory consensus mean obtained when all laboratory data are traceable to the common standard is in very good agreement with the value obtained by the application of definitive IDMS methodology.

Uncertainty assignment

As described already, the expanded uncertainty of a consensus value is often calculated as the 95% confidence interval, which entails dividing the standard deviation of the laboratory means by the square root of n, the number of laboratories. Whilst this is an approach suggested in ISO Guide 35 when individual laboratory uncertainties are not available, if the number of participant laboratories is large, the uncertainty estimate could perhaps become unrealistically small. In such circumstances it may be necessary to limit n to some upper value, regardless of the actual number of data points, although currently there appear to be no recommended procedures for this. This is an issue that could usefully be considered in future interlaboratory certification activities.

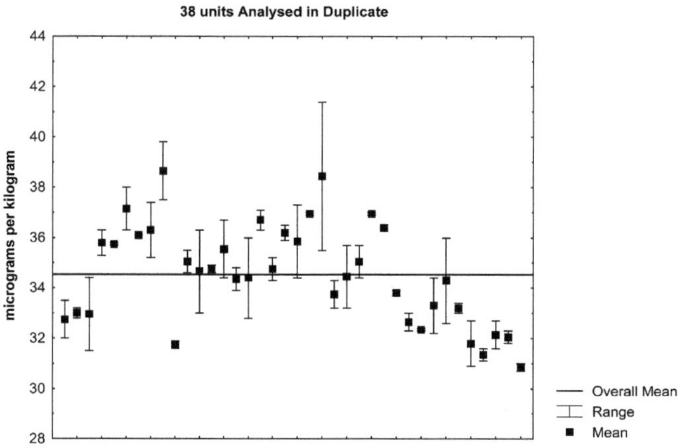

Fig. 4 Homogeneity data: sewage sludge/pentachlorobiphenyl (PCB congener 101)

Homogeneity testing

Homogeneity testing is a significant part of the RM production process and often involves the analysis of a large number of units in duplicate for the analytes of interest, to detect any small differences between units. An ANOVA is carried out and either the observed between-unit variation or the method variation, whichever is the larger, is used to obtain an estimate of the uncertainty due to any possible nonhomogeneity. It is important that the analytical method used is of high precision so that the homogeneity assessment is not confounded by measurement effects. However, identifying a method with a suitably high precision that can be maintained over the length of time required to analyse a large number of units under repeatability conditions can be difficult, with adverse effects on the homogeneity uncertainty estimate.

Homogeneity data obtained on a sewage sludge material when 38 units were analysed by one laboratory in duplicate for the PCB congener 101 (2,2′,4,5,5′-pentachlorobiphenyl) are shown in Fig. 4.

The analytical method used, which involved sample extraction followed by cleanup and determination by gas chromatography (GC)/electron capture detection (ECD), had a repeatability standard deviation of 1.3 μg/kg; therefore a contribution of at least ±2.6 μg/kg to the uncertainty (k=2) of the certified value must be expected when using this method to assess homogeneity. Such a value will make the major contribution to the total uncertainty (U) of the certified value, especially where the latter is based on high-accuracy IDMS measurements, which have an expanded uncertainty of only ±0.7 μg/kg (Table 2).

$$U=\sqrt{0.7^2+2.6^2}=\pm 2.7\,\mu g/kg$$

Therefore, combining a certified value uncertainty determined by IDMS with a homogeneity uncertainty determined using "routine" GC/ECD methodology will largely destroy the benefits of the IDMS technique. In these circumstances a different approach to homogeneity testing is required, which entails combining the IDMS measurements used to characterise the PCB content of the material with the homogeneity assessment. An evaluation of the uncertainty budget of the IDMS procedure indicates that the precision component (repeatability standard deviation) is 0.25 μg/kg, leading to a homogeneity contribution to the total uncertainty of ±0.5 μg/kg, so that

$$U=\sqrt{0.7^2+0.5^2}=\pm 0.9\,\mu g/kg.$$

The approach adopted requires the analysis of ten units in duplicate, randomly selected from the entire batch, using the IDMS method.

An alternative strategy for utilising the precision of isotope dilution measurements has been applied to certification of fuel oil RMs for sulfur content. In this case a series of candidate certified RM (CRM) replicates was evaluated for homogeneity using an "abridged" IDMS technique. The samples were isotopically spiked in the usual way and the ^{32}S-to-^{34}S ratio precisely measured by inductively coupled plasma MS. The isotope ratio obtained for each replicate was then normalised on the basis of an assumed ^{32}S concentration and the observed/expected ratio was plotted (Fig. 5). This approach is less time consuming than the full IDMS determination which was subsequently used for certification of the material (Fig. 6). The 2.1% residual standard deviation (RSD) obtained for replicate determinations is indicated in Fig. 5, demonstrating that the sample homogeneity is not a major source of uncertainty for the certified value.

Whilst it may not always be possible to use IDMS to assess the homogeneity of all materials for all analytes of interest, because of limited IDMS resource availability, the approach may be applied selectively to verify the suitability of bulk material preparation and subdivision procedures. Once IDMS has demonstrated that a particular bulk material preparation procedure is capable of producing material of sufficient homogeneity, the preparation procedure may be applied to similar material types without further IDMS measurements, but perhaps using more routine methodology to confirm that no "gross" inhomogeneity effects have unexpectedly occurred.

Commutability

A question of increasing importance to both suppliers and users of RM is the scope of a matrix RM, i.e. the extent to which a reference matrix of a particular type (e.g. a sediment) may properly be used to validate methods used for the routine analysis of test sample matrices of a different type (e.g. a soil). Relating different matrix types in this way is sometimes referred to as com-

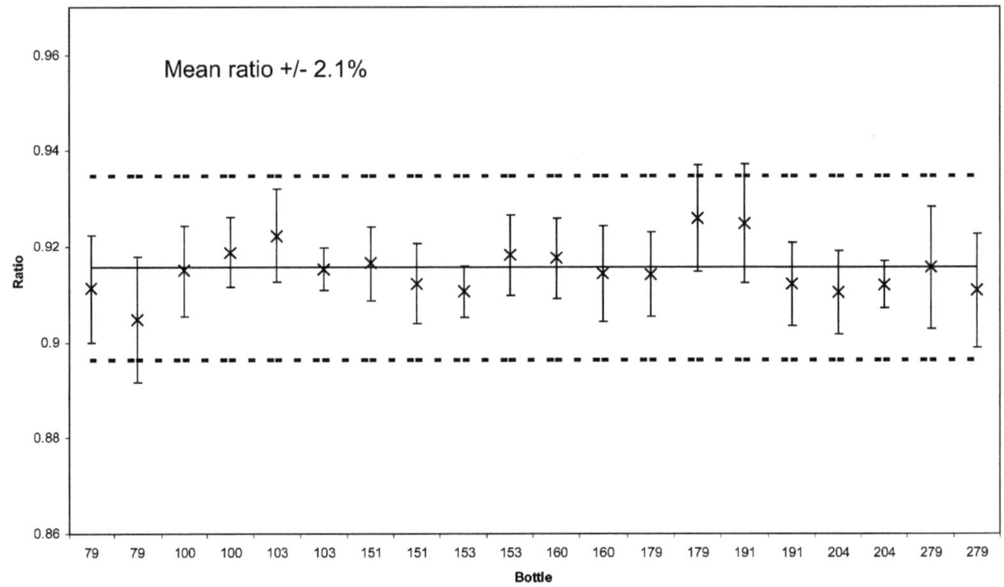

Fig. 5 LGC3023 (S in fuel) homogeneity testing by abridged isotope dilution mass spectrometry (*IDMS*)

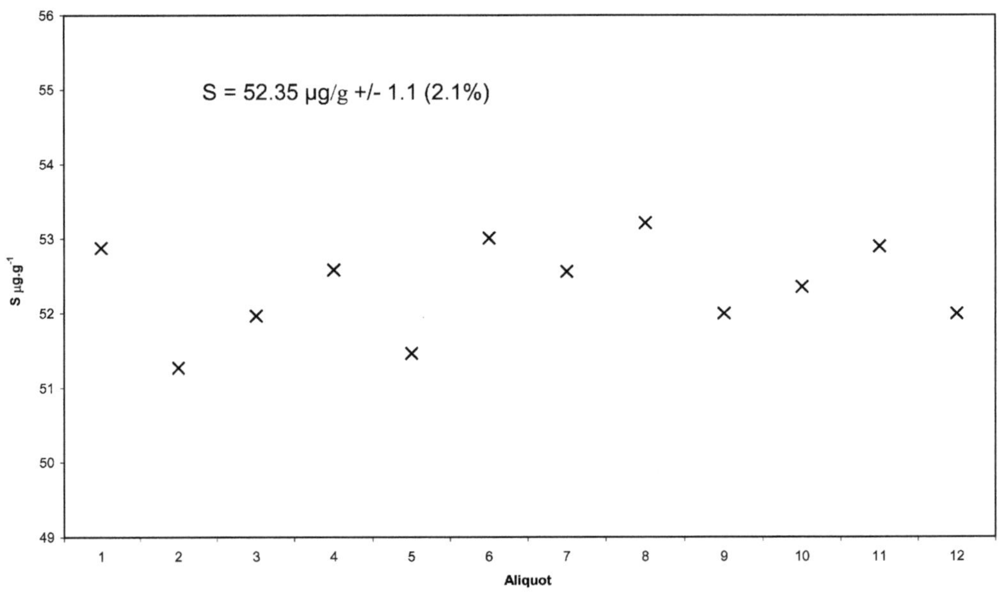

Fig. 6 LGC3023 (S in fuel) certification by IDMS

mutability or horizontal traceability. Commutability or horizontal traceability may be considered to exist between two matrices if they behave identically, within the uncertainties of the data, during a particular analytical procedure. If such behaviour could be demonstrated, one type of matrix could then reliably be used as a RM for the routine analysis of the other matrix type. In an attempt to investigate this topic, some interlaboratory studies organised for the certification of candidate matrix RMs were augmented by the distribution to participant laboratories of a second different matrix for which a reference value was already available.

For each matrix, the percentage deviation of each laboratory's result from the reference value was calculated and the values were plotted on a graph. An example of such a graph for two matrices, a soil and a river sediment, analysed for acenaphthene, is shown in Fig. 7.

Evaluation of the data by linear regression shows that the slope and intercept are not significantly different from 1 and 0, respectively. This is the situation that would be expected if the two materials behaved identically during analysis, so horizontal traceability between these two matrices is demonstrated. In practical terms this conclusion embodies the observations in Fig. 5, namely that individual laboratories produce consistently low or high results for both materials. Thus the laboratory producing a result of about +200% for the river sediment has also produced a similarly high result for the

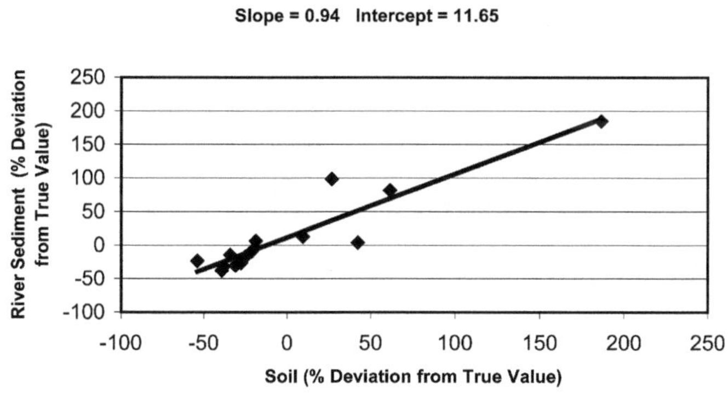

Fig. 7 Horizontal traceability: acenaphthene in soil and river sediment

soil. Such a laboratory would find a soil matrix an appropriate reference matrix for a sediment matrix and vice versa. The data in Fig. 7 support the general conclusion that the critical experimental factors affecting the determination of acenaphthene in soil and river sediment (e.g. extraction techniques, cleanup efficiency, instrument calibration, etc.) are of equal significance in both matrix types, i.e. the two materials do behave in a similar manner during analysis.

Whilst this conclusion is valid for data of the type described here (where the deviations of the results from the reference values are large), where laboratories are consistently producing results of very low bias (e.g. deviations of less than a few percent), a correlation between two matrix types is likely to be much harder to demonstrate. Thus for such laboratories and such analytical methods, matrix RMs that are a very close match in composition to the routine test samples are likely to be required.

Conclusions

The two widely used strategies for RM certification offer different strengths and in some circumstances one approach demonstrates a clear advantage over the other. For the majority of certification campaigns, however, using a combination of interlaboratory study and certification by definitive methods can bring significant benefits. Adopting a flexible approach to certification, including working with participating laboratories to improve their methodology and providing a wider range of useful data with the CRM, offers better value for money to funding agencies or users. The use of definitive, high-accuracy methods for certification greatly increases the difficulty and cost of undertaking homogeneity testing with commensurate precision. It has been shown that IDMS can be effectively employed for this purpose, either in an "abridged" form for separate homogeneity tests or by combining high-precision homogeneity testing with certification.

The commutability (horizontal traceability or scope) of matrix RMs is a critical issue in judging the suitability of a material for validation of a specific measurement procedure. In most applications, assessment of commutability relies on expert judgement but a more systematic approach would be of considerable valuable. A methodology by which the scope of a matrix RM can be evaluated systematically has been investigated and appears to show promise. It was also found that under some circumstances the scope may be wider than previously anticipated. With regard to vertical traceability, the requirements of standards such as ISO 17025 for traceable chemical measurement results has led to increasing demand for traceable matrix CRM data. This is relatively straightforward for certification by definitive methods but requires modification of existing interlaboratory certification schemes. It has been demonstrated that the RSD of the consensus mean value can be reduced by normalisation of data from the participating laboratories using a traceable, pure substance instrument calibration solution provided by the organising institute. The question then arises as to whether this approach, used in conjunction with a matrix quality control sample to check method validity, does indeed confer traceability of the consensus value to the organising institute. This concept is proposed as a topic for future discussion.

Acknowledgements The work described in this paper forms part of the Valid Analytical Measurement Programme, which is funded by the UK Department of Trade and Industry. The data for the figures demonstrating IDMS measurements of sulfur were provided by Ruth Hearn.

References

1. Catterick T, Fairman B, Harrington CF (1998) J Anal At Spectrom 13:1109
2. Wahlen R, O'Connor G, Fairman B (2001) In: Gelpi E (ed) Advances in mass spectrometry. Wiley, Chichester, pp 963–964
3. Wolff Briche CSJ, Harrington C, Catterick T, Fairman B (2001) Anal Chim Acta 437:1
4. Evans P, Wolff Briche C, Fairman B (2001) J Anal At Spectrom 16:964
5. Sargent M, Harrrington C, Harte R (eds) (2002) Guidelines for achieving high accuracy in isotope dilution mass spectrometry (IDMS). Royal Society of Chemistry, Cambridge, UK
6. Wolff Briche CSJ, Carter D, Webb KS (2002) Rapid Commun Mass Spectrom 16:848
7. Sargent M (2003) Accred Qual Assur 8:480

Accred Qual Assur (1998) 3:270–277

Reports and notes on experiences with quality assurance, validation and accreditation

M. Buzoianu
Hassan Y. Aboul-Enein

Traceable measurements in clinical laboratories

Received: 8 January 1998
Accepted: 21 April 1998

M. Buzoianu
National Institute of Metrology, Sos. Vitan-Bârzesti No. 11, 75669 Bucharest, Romania

H. Y. Aboul-Enein (✉)
Biological and Medical Research Department, MBC-03, P.O. Box 3354, Riyadh 11211, Saudi Arabia
Fax: +966-1-4427858,
e-mail: enein@kfshrc.edu.sa

Abstract Reliable, traceable and comparable measurements provide the rational basis for evaluation of the quality of a result and the starting point for recognized laboratory accreditation in any national area. Modern medical diagnostics and treatment involve rapidly rising numbers and types of clinical laboratory measurements, that are reliable. Therefore, the basic principles to be followed to assure the traceability of clinical measurements as required by the Romanian Laws of Metrology are reviewed. Main sources affecting the quality of the unbroken chain of calibrations that relate the measurements back to appropriate measurement standards are discussed. Examples of how to achieve traceable measurements in clinical laboratories are presented. Details of specific uses of reference materials, measuring instruments and standard measurement methods are also discussed.

Key words Traceability · Quality · Measurement uncertainty · Clinical reference materials · Clinical photometric system

Introduction

The increased attention paid to the concept of traceability and its implementation in the world of chemical measurements has been one of the major goals of metrological activity in recent years. In Romania traceability of chemical measurement has been closely connected with the accreditation of analytical chemistry laboratories. However, this concept has only recently been adopted in this country for clinical measurements. Within this framework an attempt is made to review what traceability means in terms of clinical measurements and what is now being done by the Romanian National Institute of Metrology (INM) to develop the principles of traceability in spectrophotometrical measurements performed in clinical chemistry laboratories.

The International Organisation of Standardization (ISO) guide "International Vocabulary of Basic and General Terms in Metrology" [1] defines traceability as "the property of a result of a measurement or the value of a standard whereby it can be related to stated references, usually national or international standards, through an unbroken chain of comparisons all having stated uncertainties".

Practically any user of a specific photometric device attempts to calibrate it against suitable reference standards existing in a recognized calibration laboratory. In turn the calibration laboratory should calibrate their standards according to those laid down by laboratories of the national measurement system. Thus, traceability is closely related to the dissemination of units. In Romania the assurance of uniformity and traceability of measurements, as well as the dissemination of units, is co-ordinated by the Romanian National Bureau of Metrology (BRML). Some aspects of traceability and its assurance within a calibration laboratory have been previously discussed by Buzoianu and Aboul-Enein [2].

The strength of the chain leading from the measurand of the sample being analysed in a clinical laboratory, up to a unit of the Systeme International (SI) or to the value of a recognized measurement scale, depends upon the way of evaluating the measurement uncertainty and scale of this uncertainty. In clinical laboratories evaluation of measurement uncertainty is a subject of great interest and it is now the focus of INM activity. We present some examples of evaluating the uncertainties of measured values in clinical measurements starting from the potential error sources. The meaning of measurement uncertainty for evaluation of the quality of the traceability chain for clinical spectrophotometric results is also discussed for different analytes. Starting from a general traceability scheme, practical aspects of traceable measurements are presented.

Legal measurement principles in clinical analysis

Uniformity of clinical measurements is the main goal of legal metrology norms and regulations issued accordingly to the Romanian Laws of Metrology [3]. To assure the necessary accuracy and uniformity within this framework, all instruments should be calibrated against the National System of Standards of Romania. A principle scheme for dissemination of units in Romania is described in Fig. 1. Accordingly, the organizational scheme of the national calibration activity for clinical measurements is shown in Fig. 2.

Traditionally, INM is involved in research, measurement and consultation for developing measuring technique. All of these aspects regard the realization, maintenance and the dissemination of units which are defined on the basis of the SI system. In recent years INM has also gained some experience in the field of human health related measurements. Thus, according to the Laws of Metrology, medical instruments (including spectrophotometers and photometers) are subject to metrological control. Since the suitability of such instruments with regard to medical application is specified by law, their metrological performance is checked by pattern approval tests and initial and periodic verification within legal metrology activities.

Calibration of medical instruments, as a set of operations establishing the relationship between the values indicated by the instrument and the values realized by standards, is accomplished by INM laboratories, area organized calibration laboratories of BRML, or by recognized calibration laboratories for medical instruments, as shown in Fig. 2. Validation of such instruments includes instrument testing and calibration: specify the intend use, a test to determine if the specifications are met and documentation. More details on out-

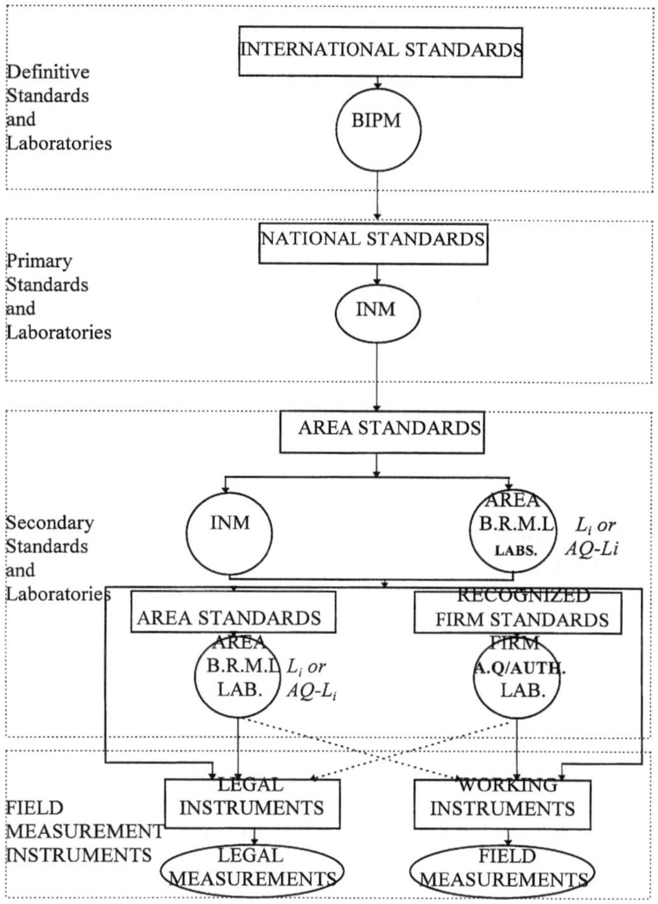

Fig. 1 Legal dissemination of units in Romania

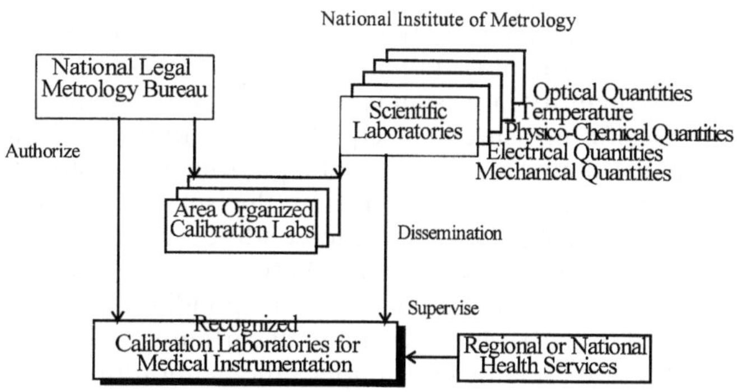

Fig. 2 Organizational scheme of the national calibration service

comes of validation of photometric systems used in clinical laboratories are presented by Buzoianu and Aboul-Enein [4].

In clinical chemistry few analysts pay attention to the question of reliability of the analytical result they produce, due to the idea that if the directions for conducting the analyses are followed the true value necessarily results. So, practical clinical measurements are commonly more precise than accurate (the deviation of a result relative to the true value is greater than the range of repeated measurements) in the same laboratory at different times or for different operators. The spread of values measured on the same sample also proved to be very large for most of the different analytes in blood, serum and other biological fluids [5]. Consequently, measurement principles for uniformity of spectrophotometric results in clinical laboratories concentrated on calibration, validation and traceability of such measurements. In this respect, different standard reference materials (RMs), methods and several metrological norms for the calibration of most diverse types of medical instruments have been issued [5].

On the calibration of photometric systems for clinical analyses

Photometric measurements performed in clinical laboratories use advanced chemical and biochemical methods and diverse instrumentation. Most analyses performed in clinical laboratories are based on spectrophotometric methods using photometric systems, such as absorption photometers, atomic absorption spectrophotometers and flame photometers. Typically, the result is expressed as mass concentration of analyte in solution (mg/dl), molar concentration (mmol/l), or catalytic concentration of enzyme activities in solution (U/l) [6].

Analyses are performed in accordance with standardized methods issued under the responsibility of a Technical Committee within the Health Ministry. Usually such measurements rely on a comparison of the measured quantity in the unknown sample with the same quantity in a "standard", i.e. an RM, according to a specific measurement equation [6], after calibrating the instrument. Calibration of a photometric system for clinical analyses usually means the set of operations that establish, under specific conditions, the relationship, within a specified range, between values indicated by the instrument and the corresponding values assigned to the RMs at the stated uncertainty. Calibration of the photometer itself implies the calibration of wavelength and absorbance scale by means of proper wavelength and absorbance RMs [5], traceable to national standards. A calibration of the instrument is still needed in concentration units to check the indicated provided value. The measurement result is then verified by application of that method of measurement to a certified reference material (CRM). Both the comparator – a photometric device with narrow or wide bandwidth, and the RMs should thus be validated.

In clinical laboratories both narrow and wide bandwidth instruments are used. A comparison between the absorbance uncertainty evaluated on ten types of absorption photometric devices of different bandwidth and the corresponding concentration uncertainty estimated on two analytes commonly measured (calcium and glucose) is illustrated in Fig. 3.

Obviously a small absorbance uncertainty is caused by the lowest concentration but there are many other sources of error. In this respect, it is the authors' opinion that calibrating and validating the metrological performances of photometric systems is a necessary condition but not on its own sufficient to achieve traceability in this field. In fact, a measurement uncertainty budget takes into consideration all uncertainties due to the way in which instrumentation is used, the CRMs and calibration of the system.

When a photometric system is calibrated in concentration units in many clinical chemistry applications, a linear curve is established usually using only one standard (concentration reference) solution. In these situations the legitimacy of the linear curve should be questioned. Available guidelines generally advice that a reasonable linear range be examined, i.e. a minimum of five points is recommended and these points should be sensibly spaced.

Some aspects on the calibration of flame photometers, blood analyte analysers and photometers for clinical analysis will be discussed.

Using five types of absorption photometric systems commonly employed in clinical laboratories, problems associated with the calibration of such instruments have been depicted. Monoelement CRMs [4] in the range of concentration indicated by the method of measurement applied were used in these situations, and as far as possible, calibration procedures agreed by clinical laboratories have been followed. Each CRM used was repeatedly measured (ten times at least). Whenever possible, a linear calibration curve was fitted. Then a correction factor for calibration and the uncertainty of this factor were determined. The degree of compatibility between the measurement result and the certified value of the CRM was tested in each situation. This algorithm is illustrated in Fig. 4. Accordingly, results on calibration are shown in Table 1.

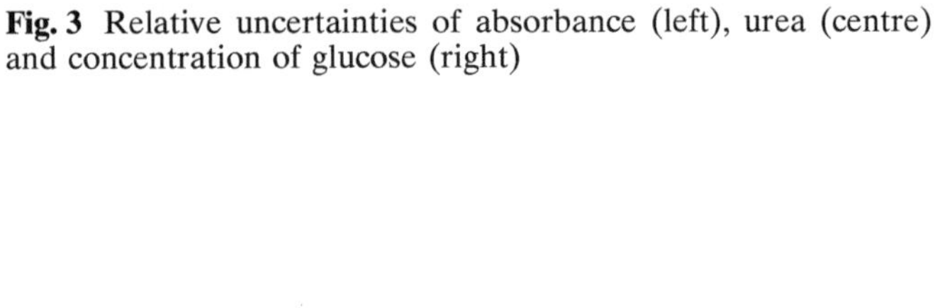

Fig. 3 Relative uncertainties of absorbance (left), urea (centre) and concentration of glucose (right)

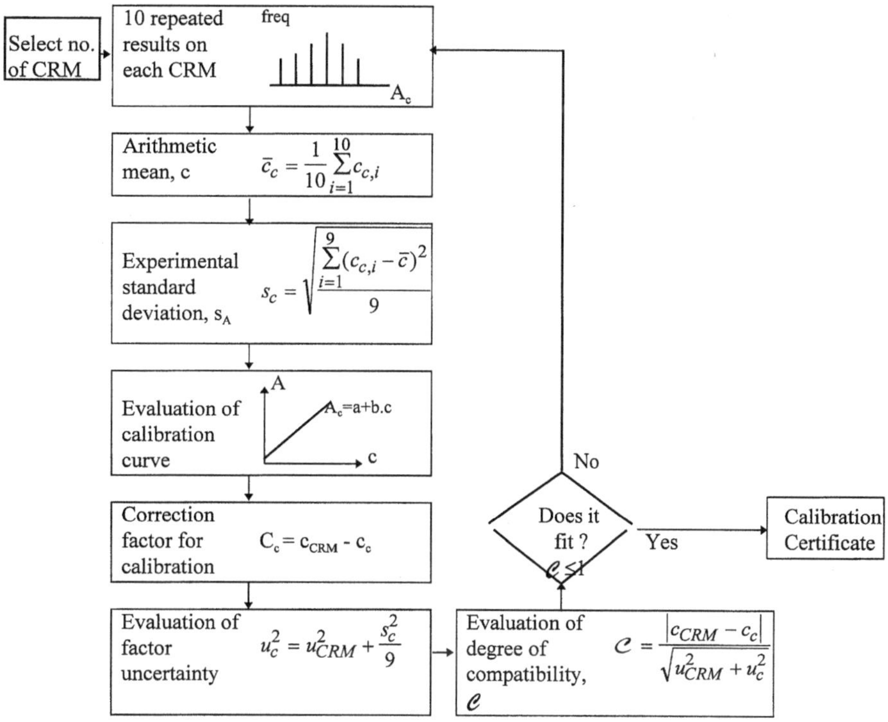

Fig. 4 The algorithm of evaluation and treatment of measurement results on calibration of photometric systems

Note that instruments 1–4 were photometric devices with less then 0.01 absorbance accuracy evaluated against reference neutral filters at 546 nm and A = 1.000, traceable to INM. The bandwidth provided by the interference filters equipping the absorption photometers was within the range of 4–10 nm. Instrument 5 was a 10 nm bandwidth photometric device with less than 1.0% absorbance linearity, evaluated at 405 nm and 500 nm, against liquid absorbance RMs type 16.02 and 16.03 [5]. Enzymatic colorimetric methods for determination of glucose and urea were used. The *o*–cresoftalein colorimetric method was used for calcium determination.

Uncertainty of the CRMs, assigned with 0.95 probability (2σ limit), U_{CRM}, usually exceeded three times the standard deviation of the repeated measurements, s_c. The uncertainty of the correction factor, u_C, had the same magnitude as u_{CRM}. Note that the uncertainty of calibration, evaluated as the square sum of standard deviation of the mean value estimated and the estimated corrections [7], did not exceed 5.5% of the nominal value of the glucose, 7.5% for urea and 7% for Ca.

The same algorithm was used to calibrate flame photometers and blood analyte analysers for Na, K and Ca determination. The results of calibrating such instruments are also presented in Table 1. Instrument 6 was an ion selective electrode analyser for Na/K/Cl with a 1.5% coefficient of variation at a 95% confidence interval. Finally, instruments 7–10 were flame photometers, validated against monoelement concentration CRMs [5] in accordance with legal metrological regulations. In these cases calibration uncertainty did not exceed 3.2% for Na or 3.6% for K.

Starting from the above determined calibration curves, the legitimacy of the hypothesis of linear adjustment was tested. Three reference solutions of known concentration were measured against the calibration curve. On each instrument three independent repeated observations were made. If c_{ij} denotes the ith observation (i = 3) on the jth instrument (j = 5), the best concentration estimate of the reference solution is the arithmetic mean $\bar{c}$ of the ij observation. The experimental standard deviation of the mean $s(\bar{c})$ is a measure of the uncertainty of $\bar{c}$, as an estimate of the concentration of the reference solution only if the instrument-to-instrument variability of observations is the same as the variability of observations made on a single instrument. Using analysis of variance individual random effects in the measurement have been evaluated according to the ISO guide [7]. Consequently, the consistency of the within instruments variability and between variability of observations were investigated by comparing the estimate of the within component of variation with the pooled estimate of variance obtained from the individual values by means of an *F*–test. Since the *F*–test was less than the tabulated value (3.71 for $\nu_1 = 3$, $\nu_2 = 10$, 95%) no difference between the two variances was concluded. Furthermore, we compared the individual mean observations with the known concentration. For instance, the dispersion of the individual values around the overall mean, of mean values around the overall mean and dispersion of mean values around the known

Table 1 Results on the calibration of photometric systems

	Analyte	c_{CRM} (mmol/l)	U_{CRM} (mmol/l)	c_c (mmol/l)	s_c (mmol/l)	C (mmol/l)	u_C (mmol/l)	u_{cal} (mmol/l)	C
	Glucose	5.55	0.30	5.51	0.07	+0.04	0.15	0.17	0.18
Instrument 1	Urea	9.60	0.50	9.20	0.14	+0.40	0.25	0.32	0.98
	Calcium	2.50	0.13	2.45	0.09	+0.05	0.07	0.11	0.39
	Glucose	5.55	0.30	5.42	0.20	+0.13	0.16	0.26	0.43
Instrument 2	Urea	6.66	0.50	6.41	0.25	+0.25	0.26	0.37	0.56
	Calcium	2.00	0.13	2.12	0.11	+0.12	0.07	0.14	0.78
	Glucose	5.55	0.32	5.59	0.06	−0.04	0.16	0.17	0.17
Instrument 3	Calcium	2.00	0.10	2.08	0.09	−0.08	0.06	0.11	0.66
	Glucose	5.55	0.32	5.46	0.06	+0.09	0.16	0.17	0.38
Instrument 4	Urea	5.80	0.80	4.90	0.59	+0.90	0.44	0.80	1.00
	Calcium	2.00	0.10	2.11	0.09	−0.11	0.06	0.11	0.91
	Glucose	5.55	0.30	5.43	0.04	+0.12	0.15	0.30	0.36
Instrument 5	Urea	13.32	0.60	12.70	0.59	+0.62	0.36	0.71	0.80
	Calcium	2.00	0.13	2.05	0.08	+0.05	0.07	0.11	0.39
	Sodium	144.0	7.2	145.0	2.1	−1.0	3.7	2.2	0.24
Instrument 6	Potassium	4.20	0.23	4.30	0.08	−0.10	0.12	0.15	0.53
	Sodium	144.0	7.2	149.5	2.2	−5.5	3.7	4.6	0.63
Instrument 7	Potassium	4.20	0.23	4.30	0.06	−0.10	0.12	0.14	0.66
	Sodium	144.0	7.2	145.0	2.5	−1	3.7	4.5	0.17
Instrument 8	Potassium	4.20	0.23	4.25	0.10	−0.05	0.12	0.16	0.25
	Sodium	144.0	7.2	147.0	2.2	−3	3.7	4.39	0.53
Instrument 9	Potassium	4.20	0.23	4.30	0.06	−0.10	0.12	0.13	0.58
	Sodium	144.0	7.2	143.0	2.3	+1.0	3.7	4.35	0.18
Instrument 10	Potassium	4.20	0.23	4.10	0.10	+0.10	0.12	0.16	0.51

values have been calculated in the case of glucose determination, as shown in Table 2. Note that experimental variance of means around the known value of the RM was in good agreement with the concentration uncertainty assigned to the material. In this situation, any measurement result obtained from the calibration curve is traceable to the RMs.

Evaluation of uncertainty components in photometric measurements specific to clinical laboratories

The concept of traceability depends on a chain of standards (artefacts or measurements) linked back to the appropriate international primary standard series of calibrations (intercomparisons between two standards in the chain). A measurement result obtained through calibration against one of these standards will itself be traceable. The uncertainty of calibration and the measurement result will depend on the uncertainties of the values assigned to the standards in the chain and the measurement procedure used. Unless the measurement uncertainty of each transferred value is reliably known there is no way to estimate the accuracy of the standard being calibrated and hence the accuracy of standards further down the chain or of the ultimate measurement result which depends upon them. This is why concepts of traceability and measurement uncertainty are intimately linked and measurement uncertainty is the key component of traceability providing the quantitative measure of the quality of measurement data. But evaluation of all uncertainty components occurring when measuring different analytes in clinical laboratories becomes quite a difficult problem. In addition, standard analytical methods used in national clinical laboratories, currently lack any indication about bias, repeatability or sensitivity of the method.

Evaluation of uncertainty components in photometric measurement specific to clinical analyses, performed in INM, follows the ISO guide "Expression and Quantification of Uncertainty Measurements" [7], using RMs and experimental quantification.

The steps considered when evaluating uncertainty measurement components in clinical analyses are illustrated in Fig. 5.

In Table 3 uncertainty components are summarized including the magnitude and method of evaluation for the end-point determination of glucose, urea and calcium, along with potassium determination by flame photometry. Unknown samples consisted of sera-type materials, gravimetrically prepared under well-con-

Table 2 Summary of glucose concentration calibration data obtained on different systems

	Instrument 1	Instrument 2	Instrument 3	Instrument 4	Instrument 5
Mean value (mmol/l)	7.68 5.28 16.48	7.61 5.30 16.37	7.48 5.35 16.85	7.60 5.46 16.64	7.65 5.50 16.54
Standard deviation (mmol/l)	0.05 0.10 0.15	0.07 0.10 0.18	0.07 0.10 0.15	0.07 0.07 0.12	0.07 0.10 0.15
Experiment variance:	Evaluation	Degree of freedom	7.61	Overall mean mmol/l 5.38	16.57
of the individual values (around the overall mean)	$\Sigma\,(c_{ij}-\bar{\bar{c}})^2$	ij−1=14	0.0173	0.0282	0.0768
within the parallel measurements	$\Sigma\,(c_{ij}-\bar{c})^2$	i−1=2	0.0049	0.0081	0.0225
of mean values (around the overall mean)	$\Sigma\,(\bar{c}_j-\bar{\bar{c}})^2$	j−1=4	0.0058	0.0094	0.0256
of the overall mean around the known value of RM	$\Sigma\,(\bar{\bar{c}}-c_{CRM})^2$	1	0.0441	0.0004	0.0049
of means around the known value of RM	$\Sigma\,(\bar{c}_i-c_{CRM})^2$	3	0.0771	0.0135	0.0537
F-test			3.53	3.48	3.41

Table 3 Uncertainty components for typical examples of end-point and flame photometric determination

Uncertainty components	Evaluation of the uncertainty component	Uncertainty (rel)			
		Glucose	Urea	Calcium	Potassium
Due to the photometric measurement	Starting from calibration uncertainty and run-to-run variation	0.018	0.060	0.010	0.025
Due to the CRM	From calibration certificate	0.026	0.022	0.007	0.012
Due to volume measurement	Starting from calibration uncertainty and run-to-run variation	0.002	0.002	0.002	0.002
Due to correlation coef.	[2]	0.002	0.002	0.005	0.002
Combined uncertainty	As square sum of above uncertainty components	0.029	0.064	0.012	0.028
Overall uncertainty	k=2	0.058	0.128	0.025	0.056

trolled conditions. The uncertainty of the preparation of these materials is indicated in parentheses.

Note that overall measurement uncertainty did not exceed 6% for glucose and potassium determinations. For Ca the overall measurement uncertainty did not exceed 2.5% and for urea 13%. These values closely agree with measurement uncertainties reported for different accuracy control sera, with the exception of the urea determination.

This way of evaluating measurement uncertainty, starting from the potential sources of error, was very useful in identifying those components that have a large contribution in the overall uncertainty and in minimizing them as much as possible. Also, good agreement

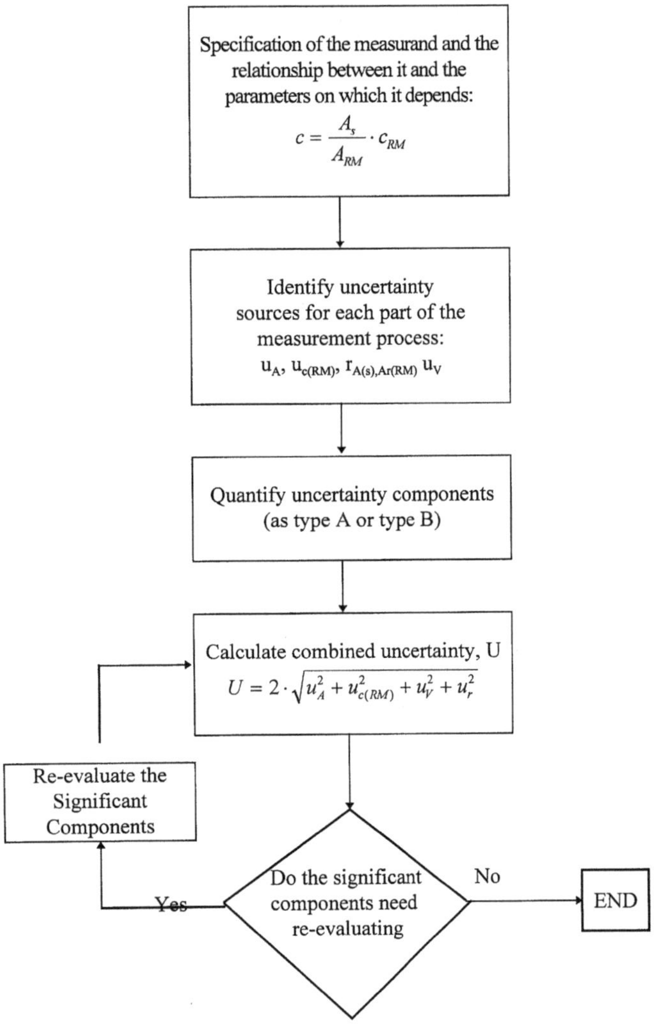

Fig. 5 Steps considered for estimation of uncertainty components

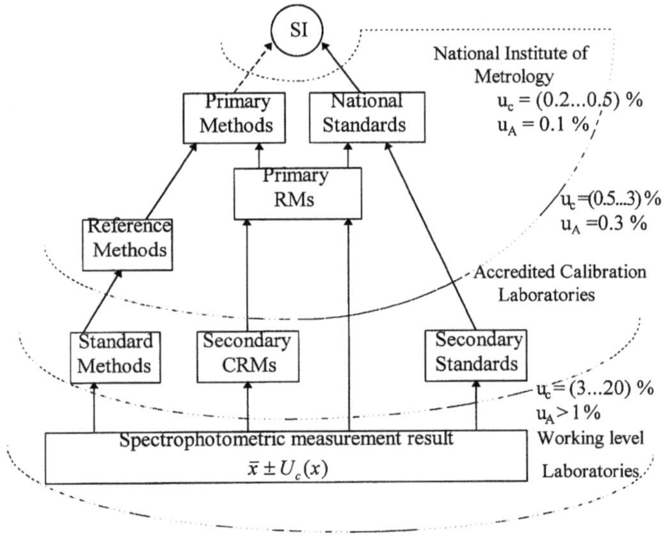

Fig. 6 Example of traceable measurements in clinical laboratories

was obtained between the uncertainty of calibration and the overall measurement uncertainty using two different approaches.

Practical aspects of traceable measurements

Assessment of the present situation regarding quality of spectrophotometric measurements [5] suggested the need for basic principles to be followed to assure the required traceability. Lately, some efforts have been focused on elaborating traceability schemes for clinical chemistry measurements. The purpose of any traceability scheme is to provide comparability, compatibility and consistency between the huge numbers of chemical measurements needed everyday, universally in clinical laboratories. On the other hand, the quality accreditation of clinical laboratories implies the introduction of the concept of traceability and care in attempting it.

To achieve the general traceability scheme of clinical measurements performed within the national area illustrated in [5] for photometric measurements, we need both the photometric system and the concentration standard solution to be traceable to the appropriate standards or RMs. In this respect is well known that traceability of chemical measurements involves concepts other than a direct comparison with physical standards [8]. This approach is shown in Fig. 6.

If traceability is to achieve its purpose in clinical laboratories, not only must an unbroken chain of calibration exist, but every calibration in the traceability chain must be carried out in a technically sound manner. The precise technical requirements that are appropriate for any calibration depend, among other features, on the uncertainties ratio between the standard and equipment involved.

For photometric measurement results, the ratio between the measurement accuracy of the photometer and the uncertainty of the upper standard used for its calibration is very important. Usually, this ratio should be of a minimum of 3. For physical standards used to calibrate photometric systems the ratio of 3 is most commonly followed. This rule generally applies also for weight and volume measurements performed in conjunction with the photometer.

In clinical chemistry, for concentration calibration of photometric systems, often this ratio does not exceed 1 or 1.2. It is well known that the higher the ratio, the stronger the calibration chain. Also note that the ratio between photometric uncertainty and concentration decreases from 3 to 1.5 higher up in the chain.

In cases where standardized methods are used in clinical laboratories to indicate bias, repeatability, or sensitivity, (for instance in enzymatic determinations) the measurement result is traceable to a reference method only if all the instruments involved in the method are appropriately calibrated against the proper physical standard.

If the traceability statement for the measurement result refers to a CRM or RM, the certificate of the material is needed to provide information on the method of measurement of the analyte(s), the uncertainty assigned and confidence level. In this respect national legal norms on absorption photometers for medical use require only the use of CRMs in the specific metrological activities.

Performing metrology in clinical chemistry relies on a constant study of the problems arising from the measurement process, but the idea of traceability is not sufficiently widespread in this field.

Conclusions

This paper has examined the role of calibration and evaluation of measurement uncertainty in clinical laboratories arising from the request for traceability assurance. To produce results which are accurate and reliable within the stated uncertainty, all uncertainties of the quality measurement process and the traceability chain should be demonstrated. Also, the quality of a spectrophotometric result depends critically on RMs and photometric systems whose traceability have been properly demonstrated.

References

1. International vocabulary of basic and general terms in metrology (VIM), 2nd edn (1993) International Organization for Standardization, Geneva, Switzerland
2. Buzoianu M, Aboul-Enein HY (1997) Accred Qual Assur 2:11–17
3. Romanian Law of Metrology (1993)
4. Buzoianu M, Aboul-Enein HY (1997) Accred Qual Assur 2:186–192
5. Buzoianu M, Fresenius J Anal Chem (in press)
6. Bronder TH, Rinneberg H (1992) Absorption photometers for medical use; metrological description of technical specifications; Proposal for international recommendation
7. Guide to the Expression of Uncertainty in Measurement (1993) International Organization for Standardization, ISBN 92-67-10188-9, Geneva, Switzerland
8. De Bievre P, Kaarls R, Peiser HS, Rasberry SD, Reed WP (1996) Accred Qual Assur 1:3–13

Accred Qual Assur (1998) 3:482–484

Reports and notes on experiences with quality assurance, validation and accreditation

Charles M. Beck II

A traceability protocol to the SI by gravimetric analysis

Received: 21 February 1998
Accepted: 17 August 1998

Charles M. Beck II (✉)
National Institute of Standards and Technology
Gaithersburg, MD 20899, USA
Tel.: +1 301 975 3847
Fax: +1 301 869 0413
e-mail: charles.beck@nist.gov

Abstract An example is presented of a traceability protocol for the measurement of a single-element strontium reference material solution, executed by a "primary" method of measurement for certification. The method of measurement is briefly described together with the measurement equation and the associated calculations for the estimation of uncertainties. This is followed by a discussion and estimate of each component of uncertainty associated with the measurement, together with a final estimate of uncertainty. The final estimate of uncertainty compares well with observed uncertainties for two previous laboratory measurements of the reference material.

Key words Certification · Gravimetric analysis · Measurement · Measurement uncertainty · Reference material · Traceability

Introduction

This example is of a traceability [1] protocol [2] for the chemical measurement of an element by a "primary" method of measurement [3]. It can be used for the certification of a single-element reference material by a national reference laboratory. This protocol relates to a very pure strontium nitrate solution, stabilized by 10% (by volume) nitric acid[1]. This solution is to be certified for the amount of strontium substance $n(\mathrm{Sr})$ per unit mass of aqueous solution $m(\mathrm{sol})$. The general measurement method described is based in part on the experience of certifying a currently available certified reference material (CRM) [4], Standard Reference Material (SRM) 3153a [5].

[1] Highly purified 10% (by volume) nitric acid is a standard analytical reagent equivalent to a solution of approximately 1.6 molality of HNO_3; that value is not critical compared with its freedom from trace contaminants.

Method of assay

The underlying purpose of this paper is to demonstrate the steps required in estimating the uncertainty of a gravimetric measurement, the value of which is traceable to the SI. Therefore, the intent of this brief description of the method of measurement is to that end and not just to be able to reproduce the Sr measurement.

The traditional method of measuring Sr by measuring the mass of precipitated $SrSO_4$, which is recommended in many textbooks, should not be used because $SrSO_4$ is volatile above 300 °C. An accurate measurement of Sr can be made using SrO as the chemical form for weighing. A measured mass of solution $m(\mathrm{sol})$, diluted to ~50 ml, is added slowly and with stirring to a stoichiometric excess (~5:1) of saturated ammonium oxalate solution (~0.35 mol/L), both solutions being at room temperature and previously adjusted to pH 8.5 with NH_4OH. The precipitated strontium oxalate, SrC_2O_4, is allowed to settle at room temperature for

~12 h. The resulting Ostwald-ripened [6] particles of SrC_2O_4 are quantitatively collected on fine-grain, ashless filter paper, the filtrate being reserved for subsequent Sr^{2+} determination. The precipitate is washed several times with saturated ammonium oxalate solution, diluted (1:1). The paper and precipitate are carefully dried and ignited to ~1100 °C, to constant mass (~3 h) in a tared, quartz (fused silica) crucible, to form stoichiometric strontium oxide of mass m(SrO), measured after cooling in air that is free of H_2O and CO_2. A platinum crucible must not be used since SrO reacts with platinum at elevated temperatures. All mass measurements must be buoyancy corrected. A small negative correction $\delta_1 m$(SrO) is applied for traces of other substances coprecipitated and determined by X-ray fluorescence spectrometry. Another small, but significant, positive correction $\delta_2 m$(SrO) is also applied for Sr^{2+} ions remaining in the filtrate and measured by flame atomic emission spectrometry (AES)[1]. On using AES or other types of spectrometry, no other ions should be detectable at a level greater than 10^{-6} mol/L in the filtrate or in any of the solutions used in the determination. In the chemical reaction of the determination, each Sr^{2+} entity is converted to one SrO, i.e., $n(Sr^{2+}) = n(Sr) = n(SrO)$.

Calculation of the measurement result

By division of the measured m(SrO) by the known molar mass M(SrO), the corresponding amount of substance n(SrO) is obtained. Thus the value of the concentration to be certified is:

$$\frac{n(\mathrm{Sr})}{m(\mathrm{sol})} = \frac{m(\mathrm{SrO})}{M(\mathrm{SrO})\, m(\mathrm{sol})}.$$

Note that m(SrO) is the mass of the SrO precipitate plus the mass of the Sr^{2+} ions in the filtrate, expressed as SrO [$\delta_2 m$(SrO)], minus the mass of the coprecipitated impurities in the SrO precipitate [$\delta_1 m$(SrO)].

Components of uncertainty of the measurement results

Uncertainty in M(SrO)

The molar mass of SrO has an established relative standard uncertainty of 1.1×10^{-4} which is almost entirely due to the variability in the isotopic composition of terrestrial strontium. This uncertainty is small and could be reduced further by one order of magnitude by a direct molar-mass measurement of the specific strontium in the solution, or by ascertaining that the source of the strontium had been free of major contamination by rubidium over a geologically significant period [7].

Uncertainty from SrO stoichiometry

Detectable changes in the mass of SrO variously heated in air are not observed and can be estimated confidently to be at a relative uncertainty level below 5×10^{-5}. Exact stoichiometry of SrO is generally assumed from long experience of consistent results. Nevertheless, the 1:1 ratio is confidently estimated to have a relative standard uncertainty of 0.7×10^{-4}. That statement of course includes any possible variability of the strontium valency manifested by strontium vacancies or interstitial ions.

Uncertainty of the gravimetric measurement

If the analyzed solution were perfectly pure, i.e., the compound SrO were pure, perfect, and free from adsorbed contaminas, and the chemical reaction proceeded perfectly, the relative standard uncertainty component derived solely from the measurement of the gravimetric ratio would be 1×10^{-4}. This assumes that a good analytical balance sensitive to ≅200 µg in a good environment, with a self-consistent set of external or built-in weights, is used for measuring the mass of the assayed portion of the solution. It is also assumed that a good analytical balance[2] sensitive to ≅3 µg in a good environment, with a self-consistent set of external or built-in weights, traceable to the kilogram, is used for measuring the mass of the SrO. The mass of the vessel holding the solution and that of the crucible containing the SrO should not exceed by more than twenty times m(sol) and m(SrO), respectively.

Uncertainties from possible departures from the ideal chemical compounds and reactions

Excellent laboratory conditions and expert handling are assumed for the estimation of these uncertainties.

Uncertainties associated with errors from contamination

Errors occur in transfer of the two solutions, to glassware, on the filter paper, on washing, on heating, on transfer to and on the balance; including adsorption effects of moisture or CO_2, perhaps forming $Sr(OH)_2$ or

[1] X-ray fluorence spectrometry was performed by P. A. Pella and flame atomic emission spectrometry by T. A. Butler, both of the NIST Analytical Chemistry Division.

[2] A microbalance can be used to advantage

$SrCO_3$, as well as occlusions or solid solutions in the SrO. Contamination of the crucible during the heat cycle should also be included. Such small contamination errors are estimated to add up to not more than a contributing relative standard uncertainty of 2×10^{-4}.

Uncertainties associated with errors from loss of chemicals

Errors resulting in the loss of chemicals occur during transfer in solution, by water evaporation when sampling the solution, reduction in Sr^{2+} by adsorption on the silica, or of SrC_2O_4 when filtering, and/or loss of SrO for instance by evaporation before weighing. The largest contributor to these possible but unobserved errors would be loss of SrC_2O_4 precipitate during the transfer to the filter paper. With good laboratory technique the total contribution to the relative standard uncertainty should not exceed 2×10^{-4}.

Uncertainties associated with the correction terms, $L_1 m(SrO)$ and $L_2 m(SrO)$

Both these corrections are themselves associated with uncertainties that are independent but probably partially off-setting. The larger of these corrections is evaluated to be about 1×10^{-3} with a relative standard uncertainty of $\pm 10\%$, so that these corrections should be included in the uncertainty budget as two relative standard uncertainties, one being $\sim 1 \times 10^{-4}$ and the other being $< 1 \times 10^{-4}$ (0.7×10^{-4} will be used as an estimate in subsequent calculations).

The budget of the relative uncertainty estimates (u_r)

Uncertainty in M(SrO): $u_r = 1.1 \times 10^{-4}$
Uncertainty from SrO stoichiometry:
$u_r = 0.7 \times 10^{-4}$
Uncertainty of gravimetric measurement:
$u_r = 1 \times 10^{-4}$
Uncertainty associated with errors from contamination:
$u_r = 2 \times 10^{-4}$
Uncertainty associated with errors from loss of chemicals:
$u_r = 2 \times 10^{-4}$
Uncertainties of correction terms:
$u_r = 1 \times 10^{-4} + u_r = 0.7 \times 10^{-4}$

The estimated combined relative standard uncertainty ($u_{c,r}$) is:

$$u_{c,r} = (12.2)^{1/2} \times 10^{-4} \cong 3.5 \times 10^{-4}$$

Discussion and Conclusion

It may be noted here that in the preparation of two individual lots of SRM 3153a, the experimental values of $u_{c,r}$ for the strontium assay on an aliquot of the bulk solution were 3.2×10^{-4} and 3.5×10^{-4} respectively [8]. These uncertainty data compare very well with the estimated $u_{c,r}$ of 3.5×10^{-4}.

It should be noted that for each of the lots of SRM 3153a mentioned above, the value of the expanded relative uncertainty, U_r, for the entire lot is considerably larger than the value for either the estimated $u_{c,r}$ of the Sr assay or the experimental $u_{c,r}$ of the Sr assay. This is true because U_r for the entire lot contains a "coverage factor" (k), and moreover, the value of $u_{c,r}$ used in calculating U_r is larger due to additional components of uncertainty resulting from bulk preparation, packaging, and transpiration of the solution through the container walls over time.

Careful consideration of the total analytical process was necessary prior to assignment of an estimated $u_{c,r}$ for the Sr measurement. This uncertainty is a critical part of the traceability of the measurement to the SI, because traceability has value only to the degree of the uncertainty, and one's confidence in the validity of that uncertainty.

Acknowledgements The author is indebted to a number of colleagues who have reviewed this paper and made valuable suggestions. Included in the list of colleagues are: K. R. Eberhardt (NIST), H. S. Peiser (NIST, retired), S. D. Rasberry (NIST, retired), T. W. Vetter (NIST), and R. L. Watters, Jr. (NIST).

References

1. ISO (1993) International vocabulary of basic and general terms in metrology, 2nd. edn. ISO, Geneva, Switzerland
2. De Bièvre P, Kaarls R, Peiser HS, Rasberry SD, Reed WP (1997) Accred Qual Assur 2:270–274
3. BIPM (1996) Comité Consultatif pour la Quantité de Matière, Report of First Meeting. Bureau International des Poids et Mesures, Sèvres, France
4. ISO Guide 30 (1992) Terms and definitions used in connection with reference materials, 2nd edn. ISO, Geneva, Switzerland
5. NIST Standard Reference Materials Catalog (1997–1998 edition) N. M. Trahey (ed) Special Publication 260, NIST Standard Reference Materials Program, Gaithersburg, MD, USA
6. Kolthoff IM, Sandell EB, Meehan EJ, Bruckenstein S (1969) Quantitative chemical analysis, 4th edn. Macmillan, New York, pp 220–225
7. IUPAC (1984) Element by element review of their atomic weights. Pure Appl Chem, IUPAC 56:741–742
8. ISO (1993) Guide to the expression of uncertainty in measurement. ISO, Geneva, Switzerland; see also, Taylor BN, Kuyatt CE (1994) Guidelines for evaluating and expressing the uncertainty of NIST measurement results. NIST Technical Note 1297, U.S. Government Printing Office, Washington, D.C., USA

Accred Qual Assur (1999) 4:203–206

Reports and notes on experiences with quality assurance, validation and accreditation

V.P. Antipin
A.A. Grigorieva

Reference samples for analysis of gas impurities in aluminium and titanium alloys: Features of production, certification and usage to ensure traceability of results

Received: 20 October 1998
Accepted: 9 November 1998

V.P. Antipin (✉) · A.A. Grigorieva
All Russia Institute for Light Alloys,
Bldg. 2 Gorbunova Street,
121596 Moscow, Russia
Tel.: +7-095-448 03 74
Fax: +7-095-446 18 01

Abstract The current state of production, certification and use of standard samples of aluminium- and titanium-based alloys with specified contents of gas impurities is described. A list of the certified standard samples with a specified gas impurity content which are available in Russia is presented.

Key words Traceability · Standard samples · Gas impurities · Aluminium alloys · Titanium alloys

Introduction

One of the general requirements for evaluation of the competence of test laboratories is the realization of traceability of measurement results [1]. Besides other measures, traceability is ensured by the availability of check and measuring standards. In the case of the determination of gas impurities in titanium- and aluminium-based alloys, reference samples (RSs) with a certified mass fraction of gas impurity to be determined serve as the above standards.

The need for RSs of metals with different gas constituents is governed by the effect of the impurities on the properties of the end material. In the case of titanium-based alloys, the oxygen and hydrogen contents need to be specified (nitrogen in titanium will be the theme of a special paper), while in the case of aluminium-based alloys only the hydrogen content needs to be specified. RSs are produced, as a rule, for a specific analysis technique. The main techniques used in the Russian aviation metallurgy industry for the determination of hydrogen in titanium- and aluminium-based alloys are vacuum heating and fusion in a gas carrier flow (GCF). Oxygen content in titanium-based alloys is determined, mainly, by neutron activation and reducing fusion in a GCF. All these techniques are generally accepted throughout the world as good analytical practice and correspond to the main criteria governing the choice of techniques for the analysis of critical products, i.e. reliability of correct results obtained in determination of gas impurities and stability of the analytical process. For convenient use of the above analysis techniques, the All-Russia Institute of Light Alloys (VILS) have produced VT16 titanium alloy RSs in the form of 2.5 mm diameter wire and aluminium alloy RSs as 10–12 mm diameter rods. Production of these RSs was carried out in accordance with the requirements imposed by the state standards of Russia for the production of the RSs [2–4], under the authority of the Russian State Committee for Standards (GOSSTANDART), beginning from the request for the proposal of production of RSs and terminating with the final approval of RSs and their introduction into the state register. Most of the samples produced at VILS are of the same standard as state reference samples (SRSs) and, hence, are fit for use in all branches of the national economy including spheres subject to state metrological control and supervision.

Manufacturing methods

All VILS's RSs are produced via plastic working of cast billets. The main problem in the production of RS material is the necessity to ensure metal homogeneity. Therefore, use was made of the results of numerous previous studies of metallurgical processes for the production of titanium and aliminium semiproducts. Based on these results, methods of melt processing, casting conditions and optimum ingot size, deformation techniques ensuring a minimum level of porosity and gas impurity segregation and attainment of homogeneous RS material were chosen.

Certification of standard samples

Studies concerning RS homogeneity were carried out at VILS according to GOST Standard 8.531-85 [3]. The inhomogeneity characteristics obtained (expressed as the standard deviation (s_n) of a random error component of inhomogeneity for samples with a preset mass) for all RSs were above one-eighth of the error value found via interlaboratory certification and, in accordance with standards [3], were taken into account during calculation of RS error characteristics.

Interlaboratory certification of the RSs was carried out by noless than 10 laboratories certified metrologically in the System in force in Russia. All laboratories had similar equipment for each analysis technique, worked according to unified certified control procedures (in most cases according to state standards for the applicable analysis techniques [5–7]) and were supervised by GOSSTANDART and analysis procedure developers.

Titanium-based alloys

For certification of the VT16 (Ti-Al-Mo-V) alloy standard sample with certified mass fractions of hydrogen and oxygen, use was made of three independent techniques for determination of each element: in the case of hydrogen determination – vacuum heating, fusion in a GCF, spectral-and-isotope balancing; in the case of oxygen – neutron activation analysis, reducing fusion in a GCF and vacuum heating. Instruments were calibrated using an existing set of titanium-based alloy SRSs with certified mass fractions of hydrogen and oxygen for spectral analysis (SRSs 1150-82P—1153-82P and SRSs 1437-78—1441-78, respectively).

Aluminium-based alloys

In the case of metrological certification of aluminium alloy RSs, the situation was complicated by the absence of aluminium alloy RSs with a certified mass fraction of hydrogen. On the whole, in the case of wrought aluminium alloys, determination of hydrogen, because of its very low concentration (0.1–0.4 cm^3/100 g, i.e. two orders of magnitude lower than those in titanium alloys), is complicated and has a number of features which influences the analysis procedure used and the processing of the results obtained. All existing analysis techniques used for the determination of hydrogen in solid aluminium alloy samples are based on the measurement of hydrogen liberated from a sample and they differ from each other depending on the hydrogen extraction techniques used (heating or fusion in vacuum; fusion in an inert GCF) and the hydrogen detection technique (measurement of gas pressure in a calibrated volume; mass spectrometric determination of the partial hydrogen pressure in an analytical system with continuous vacuum pumping; measurement of the partial hydrogen pressure in a gas carrier against its thermal conductivity). The main problem arising in all these techniques is the high value of the check experiment correction (CHEC) or, in other words, surface hydrogen. For example, in the case of the vacuum heating technique this value is 0.03–0.10 cm^3/100 g for most conventional wrought aluminium alloys and in some cases it can be 30–50% of the hydrogen content of alloys. More complicated problems arise when aluminium-lithium alloys are analysed, in this case the CHEC is about 0.20 cm^3/100 g. There are various analysis procedures which recommend different ways for reduction, stabilization and consideration of CHEC. Numerous studies carried out in various countries of the world have shown that of all the existing techniques used for the determination of hydrogen in aluminium-based alloys, vacuum heating ensures the most correct and reproducible results. This has become the reference and standard technique against which every technique is calibrated [8]. Two versions of the vacuum heating technique are used in Russia. The first is vacuum heating under conditions of gas accumulation with the use of an analyser against pressure (VH). This version is attractive because of the fact that it uses a McLeod manometer for direct measurement of the pressure of gas liberated from a sample in a calibrated volume, it is absolute and does not require calibration against the RS. In this version CHEC is determined via concurrent analysis of samples made from a rod degassed beforehand. The second version is vacuum heating under conditions of continuous pumping with the use of a mass-spectrometric analyser (VH-MS). It has a higher capacity and produces better reproducibility of the results, but requires regular calibration of the mass spectrometer. As it was shown in Ref. [9], the use of calibration against gaseous hydrogen, in the case of a simplified scheme of CHEC consideration [5], gives rise to results which are excessive by 0.04–0.06 cm^3/100 g for Al-Mg alloys and, on average, by 0.16 cm^3/100 g for Al-Li alloys. Reference [8] also illustrates the problem concerning CHEC consideration in the case of fusion in a GCF. This means that it is necessary to calibrate VH-MS and GCF instruments against the RS. Moreover, to ensure traceability and correctness of measurements, the RSs of various alloy systems with hydrogen contents similar to those in the real alloys are required. When the first types of RSs of the main aluminium alloys with a certified mass fraction of hydrogen were certified, the VH-MS and GCF techniques were excluded because of the absence of RSs. All results concerning studies on homogeneity of the RSs and on their interlaboratory certification were obtained via vacuum heating and vacuum fusion under conditions of gas accumulation employing an analyser against pressure. The vacuum fusion technique was used for corroboration of extraction completeness obtained via vacuum heating. Due to the development of RSs of main aluminium alloy systems, the VH-MS technique (without limitations regarding the chemical composition of the alloys to be analysed) and the GCF technique (for lithium-free alloys with a magnesium content of not more than 1–1.5 wt %) were used when the subsequent RSs were developed.

Long-term stability of standard sample composition

During the process of developing RSs studies concerning the stability of RS material were carried out by the laboratory developer within a year of production of the RSs. After the manufacture of SRSs, check measurements of the certified gas impurity content in samples chosen at random and kept under laboratory conditions were carried out periodically. For the very first type of RSs, the time limit for stability was set at 10–15 years. Within this period the samples showed stable certified values of hydrogen and oxygen corroborating long-term stability of RS material. Simultaneously, their homogeneity was tested, for example, for the very first types of RSs up to 80% of the RS material was analysed.

Nomenclature and use of standard samples of Titanium- and Aluminium-based alloys available in Russia

The list of existing SRSs prepared for production at VILS and their main characteristics are shown in Table 1. All RSs produced at VILS are unique in their own way. The set of aluminium alloy RSs has no analogue in the world and covers alloy systems of great importance for industry. Moreover, in the case of the main alloy systems (Al, Al-Mg, Al-Li), sets of RSs with various certified values of mass fraction of hydrogen are produced. They are in agreement with gas contents of various real groups of semiproducts. Of the well-known RSs, the VT16 alloy SRS is distinguished by the fact that mass fractions of hydrogen and oxygen are certified in the same material, while the certified values themselves ensure the possibility of analysis of, practically, all manufactured titanium alloy products. In accordance with the rules in force in Russia, the content of a certified component in the RS used for calibration of instruments should not be different from that of the sample being analysed by more than 2 times. In the

Table 1 Aluminium and titanium alloy state reference samples (SRSs) produced at the All-Russia Institute of Light Alloys (VILS)

No.	SRS number in the state register (number – year of production)	Alloy	Alloy system	Certified mass fraction of hydrogen, cm^3 / 100 g	Error of SRSs, cm^3 / 100 g (confidence coefficient is 0.95)	Rod (wire) dia., mm
1	3261-85	1010	Al	0.11	0.01	12
2	7220-96	1010	Al	0.13	0.01	12
3	7219-96	1013	Al	0.23	0.02	12
4	3262-85	1560	Al-Mg	0.22	0.02	10
5	prepared for production	1541	Al-Mg	0.31	0.02	12
6	3263-91P	1560	Al-Mg	0.42	0.03	12
7	6007-91	1201	Al-Cu-Mn	0.20	0.01	12
8	5060-89	1160	Al-Cu-Mg	0.17	0.03	12
9	prepared for production	1160	Al-Cu-Mg	0.19	0.01	12
10	7084-93	1450	Al-Li-Cu	0.28	0.03	12
11	7085-93	1420	Al-Li-Mg	0.86	0.05	12
12	3608-87	VT16	Ti-Al-Mo-V	0.0023 wt %	0.0003 wt %	2.5
				0.097 is a certified mass fraction of oxygen, wt %	0.006 error of SRSs, oxygen, wt %	

case of titanium alloys, the use of SRS 3608-87 allows one to correctly determine the oxygen content in a range of 0.05–0.20 wt % and the hydrogen content in a range of 0.001–0.005 wt %. An overwhelming majority of the results obtained during the determination of hydrogen and oxygen content of commercial titanium alloys fall within these ranges. In the cases when the hydrogen and oxygen contents fall outside the said ranges of concentration, company's reference samples (CRSs) with suitable certified values of the component to be determined are used for the calibration of instruments. Titanium alloy SRSs when prolongation of their terms of validity was not expedient because of a small amount of the material's remains are used, as a rule, as the CRSs.

As noted above, SRSs have no limitations in terms of fields of application and can be used for problems arising in the fields of state metrological control and supervision. All instruments for the determination of impurities in light alloys, both at VILS and at Russian aviation metallurgy works, should be certified when they are put into operation and subsequently checked periodically during their service-life. Certification should be done by using SRSs developed at VILS, which correspond to the range of alloys used for production of components and semiproducts. Uniformity of the instruments and the unity of the analysis procedures are the decisive factors ensuring traceability of the results obtained during determination of gas impurities in light alloys, as well as the unity and correctness of measurements within Russian aviation metallurgy.

The system of titanium alloy RSs (SRSs 3608-87 and CRSs), created at VILS, presently meets the requirements imposed on quantitative analysis of titanium alloys used for the production of components. As far as the system of the aluminium alloy SRSs is concerned, it should be noted that this is a system of RSs of the first generation. It ensures reliable quantitative analysis of hydrogen in conventional aluminium alloys where the level of hydrogen content is not below 0.10–0.15 cm^3/100 g. Development of techniques of deep degassing of melts for the production of high purity alloys with a hydrogen content of about 0.05 cm^3/100 g necessitates development of the aluminium alloy RSs with the level of hydrogen content below 0.10 cm^3/100 g. In this case two types of problems arise. The first problem is the necessity of severity of requirements for homogeneity of billets used for production of RS material, especially in the case of complex alloyed alloys. The second problem is the necessity of updating instruments and improvement of procedures for the determination of low hydrogen concentrations of aluminium alloys to improve accuracy and correctness of analysis and, in reference to the RSs, for reduction of error during their certification. At present, VILS is actively engaged in development of instruments and analysis procedures which comply with the requirements imposed. On their introduction into laboratories, these instruments and procedures will ensure certification of SRSs and VILS intends to begin the production of SRSs from aluminium alloys of high purity in terms of gas impurities.

References

1. ISO/IEC25 Guide. General requirements for evaluation of competence of calibration and test laboratories. ISO, Geneva
2. GOST Standard 8.315-91. Reference samples: Main provisions, procedure of development, certification, approval, recording and usage. GOSSTANDART, Moscow 1991 (Russia)
3. GOST Standard 8.531-91. Homogeneity of reference samples of dispersed materials. GOSSTANDART, Moscow 1991 (Russia)
4. GOST Standard 8.532-91. Reference samples of substances and materials. GOSSTANDART, Moscow 1991 (Russia)
5. GOST Standard 21132.1-81. Aluminium and aluminium alloys: Techniques for determination of hydrogen in solid metal. GOSSTANDART, Moscow 1991 (Russia)
6. GOST Standard 28052-89. Titanium and titanium alloys: A technique for determination of oxygen. GOSSTANDART, Moscow 1981 (Russia)
7. GOST Standard 24956-81. Titanium alloys: Determination of hydrogen in solid metal via vacuum heating. GOSSTANDART, Moscow 1989 (Russia)
8. Anyalebechi PN (1993) Foundry Trade J, June 4, pp 321–330
9. Antipin VP, Tulpakova RV, Danilkin VA. Improvement of correctness and capacity of hydrogen determination in aluminium alloys via vacuum heating with the usage of mass spectrometric analyser. (1994) In: Technology of processing of light and special alloys. Metallurgia, Moscow, pp 434–444

Accred Qual Assur (2000) 5:243–249

Steluta Duta

The use of certified reference materials in the Romanian traceability scheme

Received: 12 November 1999
Accepted: 10 December 1999

Invited paper presented at the 2nd Central European Conference on Reference Materials (CERM.2), 9–10 September 1999, Prague, Czech Republic

Steluta Duta
National Institute of Metrology,
Sos. Vitan-Barzesti 11,
75669 Bucharest, Romania
e-mail: office@inm.ro
Tel.: +40-1-334-5520
Fax: +40-1-330-1533

Abstract For ensuring the traceability and uniformity of measurement results, the main objectives of national metrology programmes in chemistry are to calibrate and verify measuring instruments, to evaluate the uncertainty of measurement results and to intercompare the analytical results, etc. The concept of traceability has developed recently in chemical measurements, thus, an attempt to implement the principles of metrological traceability especially by appropriateness calibration using composition certified reference materials (CRMs) is underlined. Interlaboratory comparisons are also a useful response to the need for comparable results. The paper presents some aspects and practices in the field of spectrometric measurement regarding the metrological quality of the traceability by calibrating the instruments using suitable and reliable CRMs. The uncertainty of results, as a measure of the reliability that can be placed on them, has been adequately described in different documents and, as a consequence, some examples of evaluating the measurement uncertainty are described. The relationship between uncertainty and traceability, as two fundamental concepts of metrology which are intimately linked, is underlined.

Key words CRMs · Spectrochemical measurements · Traceability

Introduction

International comparability and traceability of measurements to stable references are required in measurements for environmental monitoring and protection, international trade, clinical practice, health and safety, and industrial production. In this respect, this paper presents some practical aspects of traceability using certified reference materials (CRMs) and some examples regarding the uncertainty evaluation in spectrochemical measurements.

There are several possibilities to provide traceability of spectrochemical results. It is possible to establish and confirm the traceability of measurement results by traceable calibration of the measuring instruments against recognized standards.

Evaluation of the calibration uncertainty component is the most important component: the uncertainty of results depends on the uncertainty value of the CRMs used for the calibration and the quantitative relationships between uncertainty and traceability, which are two fundamental concepts of metrology which are intimately linked. In this way the traceable instrument calibration is an important step in assuring the traceability of spectrochemical results.

Some aspects of traceability in spectrochemical measurement

One of the most important tasks of National Metrology Institutes is to assure the traceability of measurement results. Traceability is defined as: "The property of the results of a measurement or the value of a standard whereby it can be related to stated references, usually national or international standards, through un unbroken chain of comparisons all having stated uncertainties" [1].

There are several possibilities to provide traceability of chemical measurements to SI units [2, 3]:

- Traceability, which is generally applied in metrology, can be illustrated by a hierarchy of standards: at the top there is the national standard, traceable to the SI units, which realizes a specific unit, followed by the reference standard and the working standard. The results of measurements carried out using one of the standards of this hierarchy are comparable. In chemical measurements it is possible to transfer the metrological hierarchy of standards to reference materials (RMs) which are the standards of chemical composition.
- Another way of providing a link between chemical laboratories and the SI units can use RMs, reference methods and standard measuring devices which are made available by National Metrology Institutes.
- A third possibility of making traceability available to chemical laboratories is that reference laboratories act as the link to the SI. These laboratories must have demonstrated in high-level international comparisons that they are capable of producing SI traceable measurement results in their specific field and that they are able to transmit the traceability to other laboratories in the field.
- Another possibility to establish the traceability of amount of substance measurements is to use the following primary methods indicated by the Comité Consultatif pour la Quantité de Matière (CCQM): isotope dilution with mass spectrometry, coulometry, gravimetry, titrimetry, determination of freezing-point depression, and methods which provide a direct traceability to SI units.

In Romania the dissemination of the units (Fig. 1) has been performed in accordance to the national regulations [4]. Any field measurement laboratory should try to link itself to the national standards by calibrating the instruments against national recognized standards from accredited laboratories. In turn, these laboratories link their own standards by calibrating them against the proper ones existing in Institute National of Metrology (INM), which are traceable to international standards. The calibration to establish and confirm the traceability measurement results to national or international standards is essential because the traceability involves a chain of standards linked back to the appropriate superior standards through a series of calibrations. The above described vertical traceability should be completed by horizontal traceability which is achieved by interlaboratory comparisons.

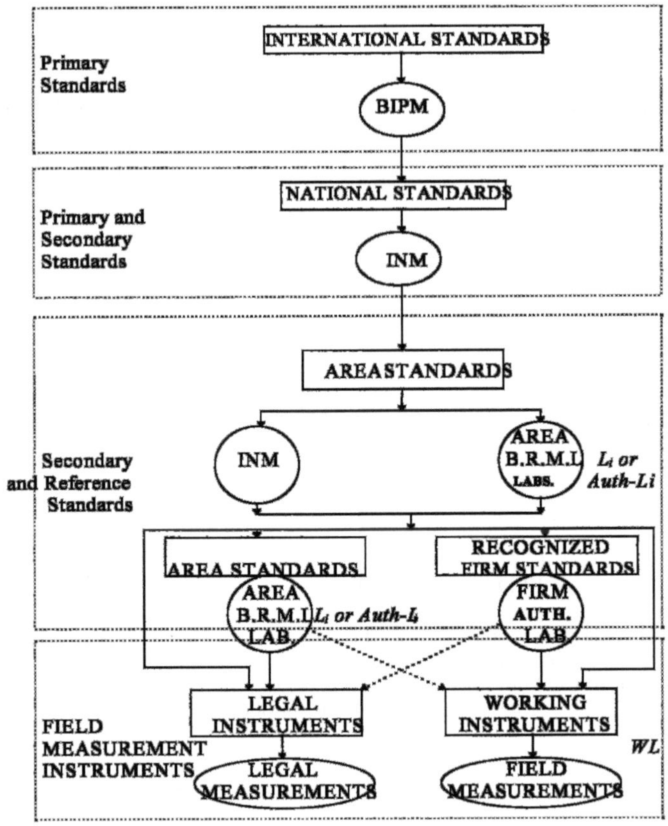

Fig. 1 The dissemination of the units in Romania

In the case of chemical measurements, CRMs, as standards of chemical composition, can be effectively introduced into the calibration process. The traceability of the certified value of RMs, as an essential part of the certification process, is as important as that of chemical measurements. In this respect, the metrological approach of titrimetry used for characterization of spectrometric RMs introduced into the calibration process of spectro(photo)meters is presented here.

Practical examples of traceability using CRMs

In accordance with [5] there are many types of calibration procedures: (1) a method which produces the anticipated result by performing a calculation defined on the basis of the laws governing the physical and chemical parameters; (2) a method which compares the content of the sample to be analysed to a set of calibration samples of known content, using a detection system for which the response (ideally linear) is recognized in the relevant working area; and (3) a method by which the sample to be analysed is compared to a set of calibration samples, using a detection system which has to be recognized to be sensitive not only to the content of elements or molecules to be analysed, but also to differences of matrix.

In this context, some aspects regarding the method which compares the content of the sample to be analysed to a set of calibration samples of known content is presented here. This method implies the use of the standards generally consisting of a determined quantity of analyte "diluted" in a large quantity of diluent (non-matrix standards).

Instrument calibration in spectrometry

Calibration means [1] the operation of establishing, under well-specified conditions, the relationship between the values of a quantity indicated by an instrument or a measurement system or and the value of a measure or RM, or the corresponding values of standards.

Usually, the calibration of a spectro(photo)metric instrument is a set of operations that establishes the relationship between the values indicated by the spectro-(photo)meter (absorbance) and the corresponding concentration values assigned to the spectrometric RMs.

In analytical spectrometry there are many types of calibration curves which are set up by measuring spectrometric reference solutions. The measurements yield a curve of absorbance versus concentration, and the points between the data of the reference solutions are interpolated by fitting a suitable curve, which normally follows the Beer-Lambert law and which gives rise to a straight line through the origin of the coordinate system. The measurement conditions and the results of the calibration curve evaluations in the case of chromium and lead measurements by electrothermal atomic absorption spectrometry are presented in Table 1.

The results regarding the evaluation of the calibration curves from Table 1 give rise to equations, which most accurately define the linear, quadratic and cubic calibration curves presented in Fig. 2 for chromium measurements and in Fig. 3 for lead measurements. The graphs and Eqs. 1 (linear), 2 (quadratic) and 6 (cubic) were obtained using a VARIAN AA 250 PLUS atomic absorption spectrometer and the graphs and Eqs. 3 (linear), 4 (quadratic) and 5 (cubic) were obtained using a PERKIN-ELMER 3300 atomic absorption spectrometer.

Even though in many analytical applications presented in this table, the correlation coefficient r has acceptable values (above 0.995) for the quadratic and cubic calibration curves, a spectrometric instrument, which is calibrated in concentration units, usually uses a linear curve which is established using several spectrometric RMs which are effectively introduced into the calibration process.

For instance, the manganese determination by molecular absorption spectro(photo)metry can be made using different types of instruments which have various technical performances. Some results are shown (Fig. 4) for manganese concentration measurements with a DR 2000-wide bandwidth 8 nm (series 1); a Hewlett Packard 8452 A, bandwidth 2 nm (series 2) and a Specord M 40-narrow bandwidth 1 nm (series 3).

Even though linearity tests are satisfactory (correlation coefficient r is above 0.995) for characterizing the spectro(photo)meter performance, in most of the cases, the curves show that the increase of the spectral bandwidth causes an apparent decrease in absorbance from the true absorbance. The accuracy of the spectro(photo)metric results is related both to the performance of the instrument and to the uncertainty due to the linear calibration curve (of the instrument) and, therefore, this uncertainty component must be evaluated.

Evaluation of the calibration uncertainty

A linear calibration curve of spectro(photo)meters is given by the relationship $c=(A-a)/b$ where: A is the

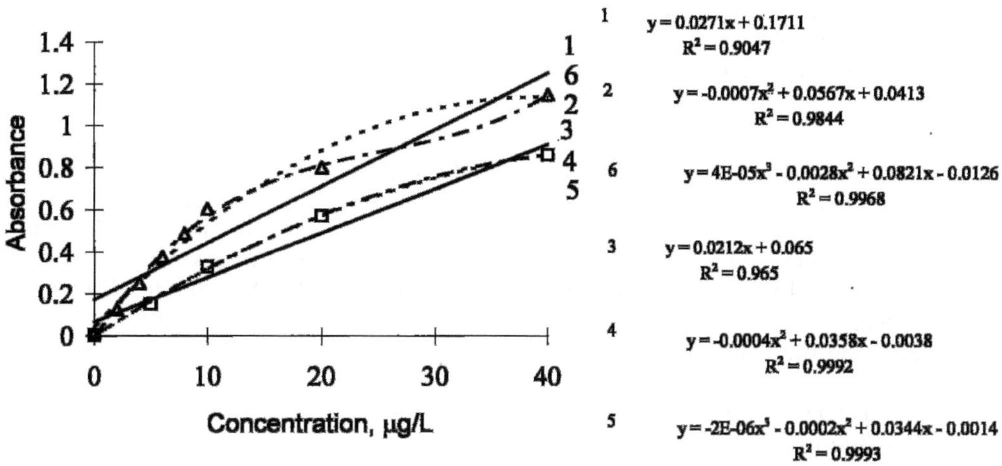

Fig. 2 The calibration curves for chromium measurement by atomic absorption spectrometry

Table 1 Measurement conditions and results of the calibration curve evaluation

Measurement conditions		Chromium: VARIAN AA 250 PLUS	Chromium: PERKIN-ELMER 3300	Lead: VARIAN AA 260 PLUS	Lead: PERKIN-ELMER 3300
Wavelength (nm)		357.9	357.9	283.3	283.3
Bandwidth (nm)		0.2	0.7	0.5	0.7
Lamp current (mA)		7.0	25.0	8.0	10.0
Type of graphite furnace		GTA-9x	HGA-600	GTA-9x	HGA-600
Working conditions of furnace: temperature (°C), time (s), temperature-time profiles (°C/s)		°C, s	°C, s, °C/s	°C, s	°C, s, °C/s
		85, 5	120, 50, 10	90, 5	110, 30, 22
		110, 37	165, 30, 1	110, 40	700, 30, 14
		150, 5	20, 15, 1	150, 10	20, 15, 1
		1000, 10	2500, 5, 0	400, 12	2300, 5, 0
		2600, 3.4	2600, 5, 1	2100, 3	2700, 3, 1
Volume of sample (µl)		20	20	20	20
Matrix modification					10 µl orto-phosphoric acid
Background correction				D_2	D_2
		Evaluation of calibration curves			
Calibration date		c (µg/l), A	c (µg/l), A	c (µg/l), A	c (µg/l), A_i
c – concentration		0, 0.0123	0, 0.003	0, 0.002	0, 0.256
A – corresponding absorbance		2, 0.1244	5, 0.151	5, 0.097	10, 0.437
		4, 0.2507	10, 0.328	10, 0.181	40, 0.765
		6, 0.3739	20, 0.571	15, 0.252	80, 1.009
		8, 0.4866	40, 0.761	30, 0.468	100, 1.166
		10, 0.6050		50, 0.682	
Types of calibration	Coefficients of curves				
Linear	a	0.1711	0.0650	0.0331	0.3279
	b	0.0271	0.0212	0.0135	0.0087
		$R^2=0.9047$ ($r=0.9512$)	$R^2=0.9650$ ($r=0.9823$)	$R^2=0.9906$ ($r=0.9952$)	$R^2=0.9744$ ($r=0.9871$)
Quadratic	P	0.0413	−0.0038	0.0047	0.0047
	Q	0.0567	0.0358	0.0182	0.0182
	R	-7×10^{-4}	-4×10^{-4}	-9×10^{-5}	-9.0×10^{-5}
		$R^2=0.9844$ ($r=0.9922$)	$R^2=0.9992$ ($r=0.9984$)	$R^2=0.9998$ ($r=0.9999$)	$R^2=0.9933$ ($r=0.9966$)
Cubic	P	−0.0126	−0.0014	0.0042	0.2576
	Q	0.0821	0.0344	0.0184	0.0198
	R	-2.8×10^{-3}	-2×10^{-4}	-1×10^{-4}	-2×10^{-4}
	S	4×10^{-5}	2×10^{-6}	2×10^{-7}	1×10^{-6}
		$R^2=0.9968$ ($r=0.9984$)	$R^2=0.9993$ ($r=0.9996$)	$R^2=0.9998$ ($r=0.9999$)	$R^2=1.000$ ($r=1.000$)
Relative standard deviation (%)		0.3	1.4	4.9	2.7
Detection limit (µg/l)		0.08	0.10	0.05	0.30
Chemical sensitivity (µg/l)		0.07	0.2	0.25	1.2

absorbance, c is the concentration, and *a and b* are the parameters of the linear curve. The uncertainty in a predicted value A_m – absorbance using linear regression – to a given value c, concentration, can be estimated in several ways: by calculating the variance and covariance, by estimating the correlation coefficient r, by evaluating the calibration data or by other methods [6].

Some aspects regarding the way to evaluate the uncertainty due to the linear calibration curve by evaluating the calibration data and by estimating the correlation coefficient are presented below.

The linear calibration uncertainty is estimated by the interval that can be expected to encompass a large fraction of the distribution of values that could be reasonably attributed to the linear curve. This interval, indicated in Fig. 5, is due to the linear adjustement of the concentration values used to determine the regression line and obtained values of the absorbance.

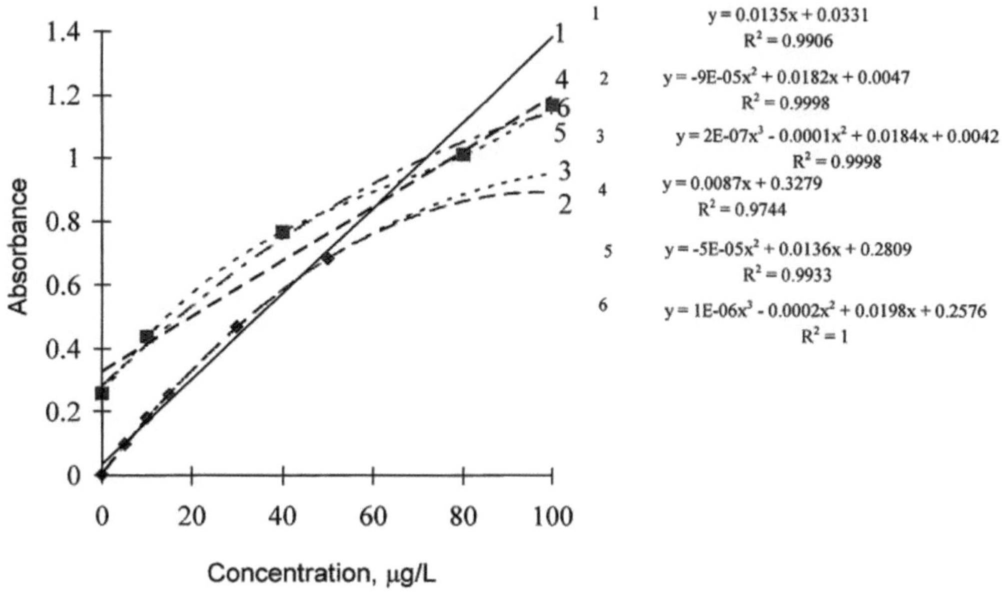

Fig. 3 The calibration curves for lead measurement by atomic absorption spectrometry

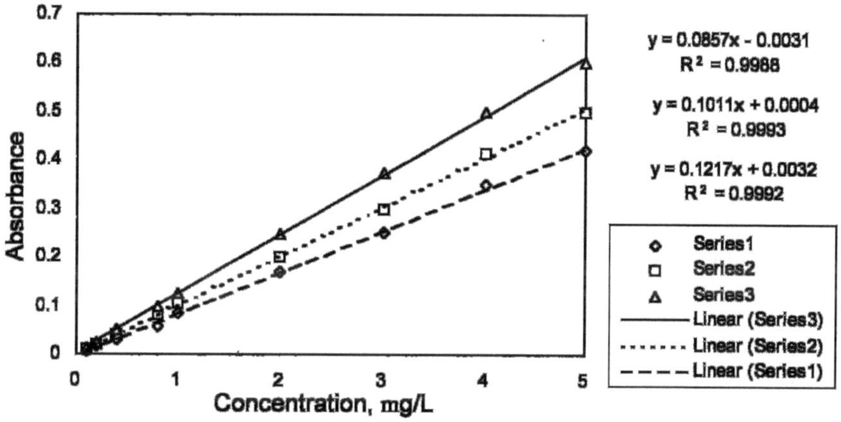

Fig. 4 The calibration curve for concentration solution of manganese

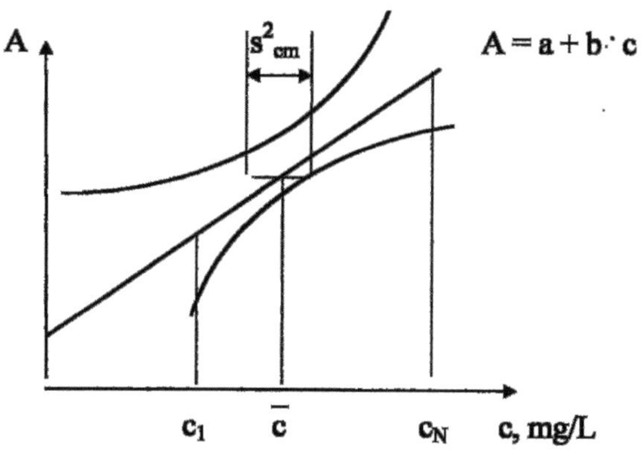

Fig. 5 The confidence interval of linear calibration curve

The evaluation of the linear calibration uncertainty s_{c_m} can be performed against: the standard deviation of the linear calibration curve, s_0; the slope of the curve, b; the number N of CRMs used for calibration curve; the number n of replicates; and the average absorbance signals $\bar{A}_m$ of the sample and of the CRMs ($\bar{A}$) using the calibration curve.

Some results of the calibration uncertainty evaluation, due to the linear calibration curve of copper determination c_m by molecular absorption spectro(photo)-metry using a Cecil 2020 instrument are illustrated in Table 2. Note that at the end of the linear range (0–10) mg/l the calibration uncertainty is bigger than in the middle of the linear range: for a concentration of 0.987 mg/l copper the uncertainty component due to the calibration is 3% and for a concentration of at 6.010 mg/l copper the uncertainty component due to the calibration is 0.56%.

In addition to the uncertainty due to the linear regression which was 0.034 for 6.010 mg/l copper, the overall uncertainty of the instrument calibration includes the uncertainty due to the photometic measurement and the uncertainty due to the CRMs. The overall calibration uncertainty was 0.036 for 6.010 mg/l copper.

Table 2 The evaluation of the linear calibration uncertainty

Sources of uncertainty	Method of evaluation	Estimations and experimental results
Linear regression	$s^2_{c_m} = \frac{s_0^2}{b^2}\left[\frac{1}{N}+\frac{1}{n}+\frac{(A_m - A)^2}{b^2 \sum_i (c_i - c)^2}\right]$	c_m / s_{cm} (mg/l): 0.987 / 0.030; 1.980 / 0.026; 3.995 / 0.023; 6.010 / 0.034 $A = 0.002 + 0.1217\ c$ $a = -0.002$, $s_a = 0.0042$, $t \cdot s_a = 0.0098$, $b = 0.1217$, $s_b = 0.0011$; $t \cdot s_b = 0.0025$, $r = 0.9999$, $s_0 = 0.0042$
Spectro(photo) metric method	Against physical stnadards (optical filters)	0.010 in accordance with $\Delta A/A = \Delta c/c$ (Lambert-Beer law)
CRMs used for calibration	From the CRM Certificate	$U_{certified}/3^{1/2} = 0.006$
Combined uncertainty	Square sums of components	$u_{cal} = 0.036$ for 6.000 mg/l Cu
Overall uncertainty	$k = 2$	$U_{cal} = 2 \cdot u_{cal} = 0.072$

Even thought a significant difference does not exist between the linear regression uncertainty and the overall uncertainty, this approach takes into account all sources of uncertainty and underlines the link between the field measurement results and the values of the standards used for the instrument's calibration. The ratio uncertainty between the CRMs and the photometer involved, gives the strength of the traceability link. Moreover, the evaluation of the overall uncertainty in spectrochemical measurements must take into account the steps of the spectrometric measurement process.

For each point of the process, the associated standard uncertainties below need to be estimated: u_s – for sampling, which includes uncertainty due to the chemical preparation u_p; u_M – for reproducibility of the analytic spectro(photo)metric system, which includes the dilution factor, the weight of the sample etc; u_{CRM} – for the value of the calibration standards; u_R – for reproducibility of the calibration, and u_{DA} – for the suitability of the method of calibration, which includes the data treatment. In this respect the evaluations of uncertainty of the spectrometric measurement are illustrated in Table 3 for copper determination in water by atomic absorbtion spectrometry and by molecular absorption spectro(photo)metry. The overall measurement uncertainty for copper in water was 3.6% using a Varian AA 250 PLUS atomic absorption spectrometer and 4.8% using a Specord M40 C.Z. Jena molecular absorption

Table 3 The evaluation of the overall measurement uncertainty by linear calibration of the instrument

Method measurement: AA Spectrometry and UV-VIZ spectrophotometry
Mathematical model: Linear curve $c = (A - a) \cdot f/b$
Low of uncertainty propagation: $u_c^2 = u_s^2 + u_M^2 + u_R^2 + u_{MRC}^2 + u_{DA}^2$

	Instrument			
Estimation	Varian AA 250 PLUS	Unicam Solaar 939	Specord M40 C.Z. Jena	Spectrophotometer DR 2000
a	0.0165	0.0322	−0.002	0.0015
b	0.0632	0.0821	0.1218	0.1217
r	0.9991	0.9989	0.9999	0.9998
s_0	0.0093	0.0162	0.0021	0.003
u_s	0.03	0.03	0.03	0.03
u_M	0.03	0.03	0.03	0.03
u_R	0.06	0.10	0.08	0.07
u_{MRC}	0.006	0.006	0.006	0.006
u_{DA}	0.00	0.00	0.00	0.00
A_m	0.142	0.193	0.230	0.242
c_m	1.985	1.961	1.910	2.006
u_c	0.073	0.109	0.091	0.082
I_c	0.25	0.39	0.98	0.02

spectro(photo)meter. The compatibility of results was evaluated as the compatibility index I_c which was acceptable (less than 1) in all of the cases presented in Table 3.

This approach of considering the potential sources of error, leads to the identification of the components having a significant contribution, and, therefore, to the decrease of their effects.

Conclusion

This paper presents some aspects regarding the uncertainty evaluation and traceability assurance of spectrochemical results using spectrometric RMs.

In this approach, calibration uncertainty is an important component of the traceability chain and uncertainty of results depends on the uncertainty of the certified values of RMs used for the calibration. Thus, the results are traceable to the standards used for the instrument calibration. The traceability of certified values of RMs is as important as that of spectrometric measurements. Therefore, it is necessary to use the spectrometric RMs that are characterized in a metrological manner. In this framework, the uncertainty and traceability, as two fundamental metrological concepts, are intimately linked.

References

1. VIM-ISO (1993) International vocabulary of basic and general terms in metrology, 2nd edn. ISO, Geneva
2. De Bièvre P, Kaarls R, Peiser HS, Reed WP (1997) ISO/REMCON 422:1–39
3. Dube G (1997) Metrology in chemistry. PTB-Bericht W-68:1–10
4. MO 147 (1998) Romanian Law of Metrology: Dissemination of the units in Romania.
5. ISO Guide 32 (1997) Calibration in analytical chemistry and use of certified reference materials. ISO, Geneva
6. EURACHEM (1995) Quantifying uncertainty in analytical measurement, 1st edn. EURACHEM

Accred Qual Assur (2001) 6:55–60

Petra Spitzer

Traceable measurements of pH

Received: 14 June 2000
Accepted: 22 August 2000

Presented at the International Conference on Metrology – Trends and Applications in Calibration and Testing Laboratories, 16–18 May 2000, Jerusalem, Israel

P. Spitzer
Physikalisch-Technische Bundesanstalt, Bundesallee 100, 38116 Braunschweig, Germany
e-mail: petra.spitzer@ptb.de
Tel.: +49-53-5923322
Fax: +49-53-5923015

Abstract The primary method for pH is based on the measurement of the potential difference of an electrochemical cell containing a platinum hydrogen electrode and a silver/silver chloride reference electrode, often called a Harned cell. Assumptions must be made to relate the operation of this cell to the thermodynamic definition of pH. National metrology institutes use the primary method to assign pH values to a limited number of primary standards (PS). The required comparability of pH can be ensured only if the buffers used for the calibration of pH meter-electrode assemblies are traceable to these primary pH standards. To assess the degree of equivalence, comparisons of primary measurement procedures for pH were organized in co-operation with EUROMET. Typical results will be presented. In 1998, the Consultative Committee for Amount of Substance (CCQM) decided to include the field of pH in its working programme. The first key comparison for this quantity was recently carried out on two phosphate buffer solutions.

Keywords Metrology in chemistry · Traceability · pH · Key comparison

Introduction

pH is the chemical parameter most frequently measured. Accurate pH measurements are needed in many areas, among which public health care, environmental protection and biotechnology are the most important ones. Thus, there is a huge demand for traceable measurement results for pH to ensure quality control and comply with the technical requirements.

The users of pH meters thus need calibration solutions of long-time stability which are traceable to primary pH standards pH(PS) related as closely as possible to the definition of pH. Although pH measurements are carried out on a large scale, the problems posed by the traceability of pH have not yet been adequately solved.

A hundred years ago, in 1909, Soerensen of the Carlsberg Laboratory in Copenhagen [1] defined pH in terms of the concentration with a scale of 0–14 (at 25 °C) which he derived from the ionic product of water ($K_W = 10^{-14}$ mol × dm^{-3}). Some years later, Lewis introduced the concept of activity, and in 1923 Debye and Hückel published their theory for strong electrolyte solutions. On the basis of this knowledge, Soerensen and Linderstroem-Lang [2] suggested a new pH definition in terms of the relative activity of hydrogen ions in solution:

$$\mathrm{pH} = -\lg a_\mathrm{H} = -\lg (m_\mathrm{H} \gamma_\mathrm{H} / m^0) \qquad (1)$$

where a_H is the relative (molality-based) activity, γ_H the molal activity coefficient of the hydrogen ion H^+ at the molality m_H in mol kg^{-1}, and m^0 a standard state chosen equal to 1 mol kg^{-1} of hydrogen ions.

Equation (1) involves the single ion activity of the hydrogen ion, and it might be said that thus the problems commenced. Activities of individual ions can never be measured without non-thermodynamic assumptions being made.

The traceability of pH measurements

The International Union of Pure and Applied Chemistry (IUPAC) recommendation [3] for the definition of pH scales has formed the basis for the standardisation of pH measurements since 1985. IUPAC recommended two different approaches to derive the pH values of pH standard buffer solutions. They yield two different pH values for one solution [4].

The prerequisite for the mutual acceptance of analytical data such as pH is comparability. Comparability requires the complete evaluation of the measurement uncertainties which in turn are based on traceability to recognised references. The need for traceable pH measurements and the confusion resulting from the ambiguous IUPAC recommendation led to various international initiatives being taken.

In 1997, IUPAC formed a Working Party on pH to develop a new pH concept. This work is now in its final stages and the final draft is just being reviewed. There is hope that the new recommendation will soon be accessible to the parties interested.

Numerous national and international standards on pH are still applicable. Following increasing demands for quality assurance in laboratories, a European standard is needed.

In 1999, a Working Group on Instrumentation in Electrochemical Analysis (WG 5) was created by the Technical Committee – Laboratory Equipment of the European Committee for Standardisation (CEN/TC 332). The standard relates to requirements for how to establish traceability between pH measurements performed by the user and the primary reference method using hydrogen electrodes. The revised IUPAC draft for pH is intended to serve as a basis for the new European standard on pH. It has been clearly stated that this standardisation work will not duplicate the work already completed by IUPAC or by the International Electrotechnical Commission (IEC).

The primary method for the measurement of pH

After extensive studies of buffer solutions and suitable electrochemical cells, Bates and his co-workers [5] suggested a conventional procedure, the Bates-Guggenheim convention [6], to assign pH values to standards. If this convention is used with an estimate of its uncertainty, traceability to the SI can also be established for pH.

The primary method for pH is based on the measurement of the potential difference of the electrochemical cell without a liquid junction involving a selected buffer solution, a platinum hydrogen gas electrode and a silver/silver chloride reference electrode, often also referred to as a Harned cell.

$$\mathrm{Pt\,|\,H_2\,|\,buffer\ S,\ Cl^-\,|\,AgCl\,|\,Ag} \quad \text{Cell I}$$

As a liquid junction potential is avoided, the cell potential consists merely of the electrode potentials of the hydrogen and the silver/silver chloride reference electrode. Chloride at known concentrations, m_{Cl}, must be added to the (chloride-free) buffer solution to use the silver-silver chloride electrode in cells without transference as a reference. This is different from silver/silver chloride reference systems with fixed potentials used for example as standard references in single-rod glass electrodes.

The application of the Nernst equation for the reaction of cell (I) yields the potential difference E_I (corrected to 101325 Pa – partial pressure of hydrogen gas) given by Eq. (2). E_I can be rearranged to give the so-called acidity function so that there are only measurable quantities on the right side of Eq. (3).

$$E_I = E^0 - k\lg(m_H\gamma_H m_{Cl}\gamma_{Cl}) \qquad (2)$$

$$-\lg(a_H\gamma_{Cl}) = (E_I - E^0)/k + \lg(m_{Cl}) \qquad (3)$$

E^0 is the standard potential in V of the silver/silver chloride electrode and γ_{Cl} the activity coefficient of the chloride ion. The Nernstian slope k in V is given by Eq. (4):

$$k = RT\ln 10/F \qquad (4)$$

where R is the molar gas constant in J mol^{-1} K^{-1}, F the Faraday constant in As mol^{-1} and T the thermodynamic temperature in K.

The standard potential difference of the Ag/AgCl reference electrode E^0 is determined in cell (I) filled with HCl at a fixed molality. For the molality of 0.01 mol kg^{-1}, the values for the mean activity coefficient of the HCl are given in [7] at various temperatures.

The measurements to get the cell potential E_{Ia} of cell I filled with HCl and E_I of cell I filled with buffer are performed simultaneously. The difference $\Delta E = E_I - E_{Ia}$ is therefore independent of the standard potential difference.

To obtain the pH, it is necessary to evaluate the activity coefficient of the chloride ion. So the acidity function is determined for at least three different molalities m_{Cl} of added alkali chloride. In a subsequent step, the value of the acidity function at zero chloride molality, $\lg(a_H\gamma_{Cl})^0$, is determined by linear extrapolation. The activity of chloride is immeasurable. The activity coefficient of the chloride ion at zero chloride molality, γ^0_{Cl}, is calculated using the Bates-Guggenheim convention (Eq. 5) which is based on the Debye-Hückel theory. The convention assumes that the product of constant B and ion size parameter a are equal to 1.5 (kg mol^{-1})$^{1/2}$ in a temperature range 5 to 50 °C and in all selected buffers at low ionic strength (I <0.1 mol kg^{-1}).

$$-\lg \gamma_{Cl}^{0} = -AI^{1/2}/(1+BaI^{1/2}); \quad \boldsymbol{Ba} = \mathbf{1.5}\,(\text{kg mol}^{-1})^{1/2} \tag{5}$$

A is the Debey-Hückel constant (limiting slope) in (kg $mol^{-1})^{1/2}$ and I the ionic strength of the buffer solution in mol kg^{-1}.

The various steps for the assignment of pH(PS) to the primary pH reference buffer are summarised in Fig. 1. In this figure also the main sources of uncertainty for the primary method for pH are mentioned.

The extrapolation to zero chloride molality is assumed to be linear provided the change in ionic strength on addition of chloride is less than 20%.

For a measurement of pH with cell (I) to be traceable to the SI, an uncertainty for the Bates-Guggenheim convention must be estimated. One possibility is to estimate a reasonable uncertainty contribution due to a variation of the ion size parameter. An uncertainty contribution of ±0.01 in pH should cover the entire variation. When this contribution is included in the uncertainty budget, the uncertainty at the top of the traceability chain is too high to derive secondary standards as used to calibrate pH meter-electrode assemblies.

For most measurements the contribution from the Bates-Guggenheim convention will therefore not be allowed for. Primary pH values stated without this contribution will be considered conventional.

The primary method is applied by national metrology institutes to assign conventional pH values to a limited number of primary standard (PS) buffer solutions in dilute aqueous solutions. The experimental details are given in [8, 9] where national standard measurement devices for pH in Denmark and Germany are described.

In order to improve the primary method for pH, investigations into solution theory and into the concept of single ion activity are necessary.

A model of electrolyte solutions which takes into account both electrostatic and specific interactions for individual solutions would be an improvement over the Bates-Guggenheim convention. It is hoped that the Pitzer model of electrolytes [10], which uses a virial

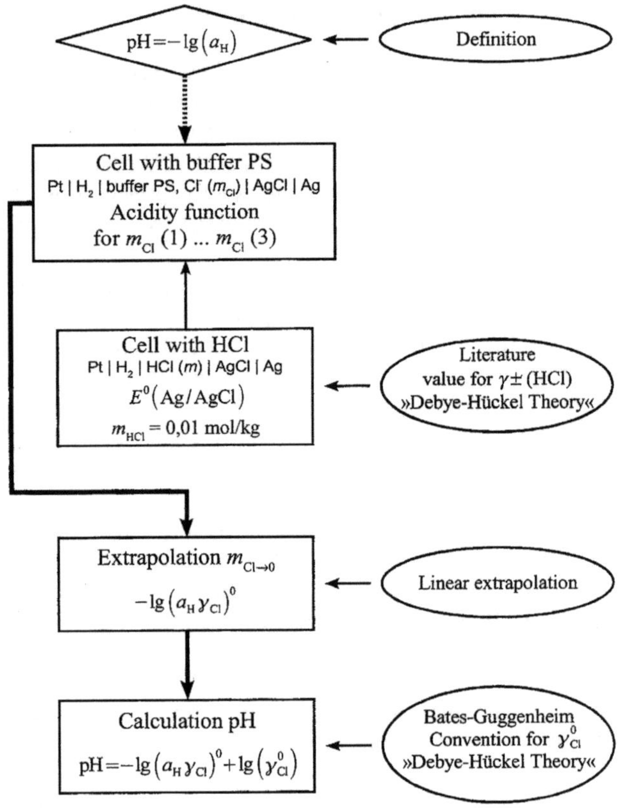

Fig. 1 Summary of the primary method for pH measurement

equation approach, will provide such an improvement. Until now, sufficient and reliable data are not available in the literature, so calculations for all buffer solutions of interest cannot be carried out. First limited work is being carried out on phosphate and carbonate buffers [11]. For the Pitzer approach an uncertainty must be estimated too.

Primary pH reference materials were chosen, see Table 1, which can be easily prepared and have a reproducible purity of preparation. Batch-to-batch differences in purity, however cannot be avoided. The

Table 1 Values of pH primary standards (PSs) for primary standards at 25 °C (PTB materials are chosen as an example)

Primary standard (PS)	PTB Primary reference material	pH-(PS)
Potassium hydrogen tartrate (sat. at 25 °C)	PTB-TA 00	3.557
Potassium dihydrogen citrate, 0.005 mol kg^{-1}	PTB-CIT 00	3.775
Potassium hydrogen phthalate, 0.005 mol kg^{-1}	PTB-PHT 00	4.008
Disodium hydrogen phosphate, 0.025 mol kg^{-1} + potassium dihydrogen phosphate, 0.025 mol kg^{-1}	PTB-PHOA 00	6.865
Disodium hydrogen phosphate, 0.03043 mol kg^{-1} + potassium dihydrogen phosphate, 0.008695 mol kg^{-1}	PTB-PHOB 00	7.416
Sodium tetraborate decahydrate, 0.01 mol kg^{-1}	PTB- BO 00b	9.182
Sodium hydrogen carbonate, 0.025 mol kg^{-1} + sodium carbonate, 0.025 mol kg^{-1}	PTB- CAR 00	10.014

pH(PS) values are valid, therefore, only with provision of a certificate for the specific batch.

The pH reference materials were selected also to cause small liquid junction potential <0.01 in pH if the pH of the buffer solution prepared from this material is measured in cells with transference [12]. The molality of the primary buffer solutions are kept at ≤0.1 mol kg^{-1} for the same reason [13]. Furthermore, the primary buffers have a long-time stability of stored solid material (>3 years), except solid borax buffer material. Borax buffer (0.1 mol kg^{-1}) has a restricted stability of about 2 years only [14].

The pH(PS) values listed in Table 1 are examples derived from Physikalisch-Technische Bundesanstalt (PTB) certificates for primary pH reference materials.

The typical measurement uncertainty for the determination of pH(PS) using cell(I) is $U=0.003$ ($k=2$) at 25 °C.

Consistency of primary pH buffer solutions

For the measurement results to be recognised at the international level, it is necessary to demonstrate the equivalence of the national traceability structures, including national measurement standards, with the aim of a mutual recognition of national measurement standards and certificates.

To evaluate the degree of equivalence of the national primary measurement procedures for pH, the first key comparison for this quantity was recently carried out by the CCQM on two phosphate buffer solutions. These experiments were piloted by the PTB Germany and involved another ten metrology institutes. A first evaluation of the results obtained shows that the majority of the results agree within the uncertainty stated by the participants. The draft B for this comparison will be available soon.

In the past, comparisons of primary measurement procedures for pH were carried out in co-operation with EUROMET [15–17], with the aim of improving the uniformity of pH measurements in Europe.

The results obtained in the measurement of five different buffers demonstrated a high degree of comparability for the measurements carried out at different laboratories. At 25 °C the pH value of the respective buffer agreed within $U=0.005$ (k=2). The evaluation of the results did not furnish evidence for significant effects of the cell design. Typical results are presented in Figs. 2–4.

The uncertainties stated by the participants were evaluated according to the *Guide to the Expression of Uncertainty in Measurement* (GUM) [18].

The participants were: GUM; Central Office of Measures, Poland; National Office of Measures (OMH), Hungary; Physikalisch-Technische Bundesanstalt (PTB), Germany; Danish Primary Laboratory for pH Measurement (DPL) c/o Radiometer Medical A/S, Denmark; National Institute of Standards and Technology (NIST), USA.

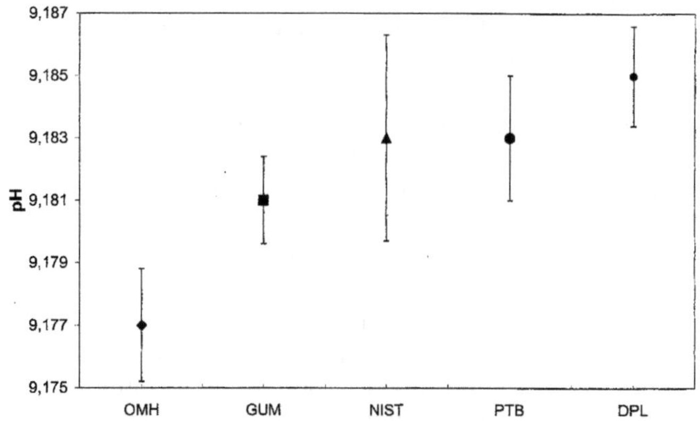

Fig. 2 EUROMET comparison 424 [19]. Buffer: sodium tetraborate decahydrate, 0.01 mol kg^{-1} (T = 25 °C)

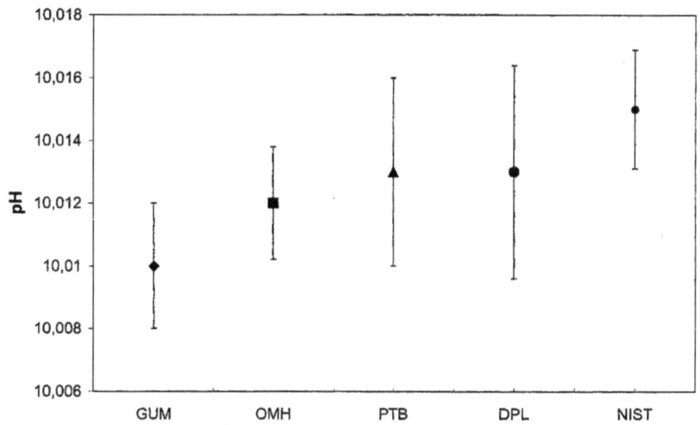

Fig. 3 EUROMET comparison 424 [19]. Buffer: sodium hydrogen carbonate, 0.025 mol kg^{-1} + sodium carbonate, 0.025 mol kg^{-1} (T=25 °C)

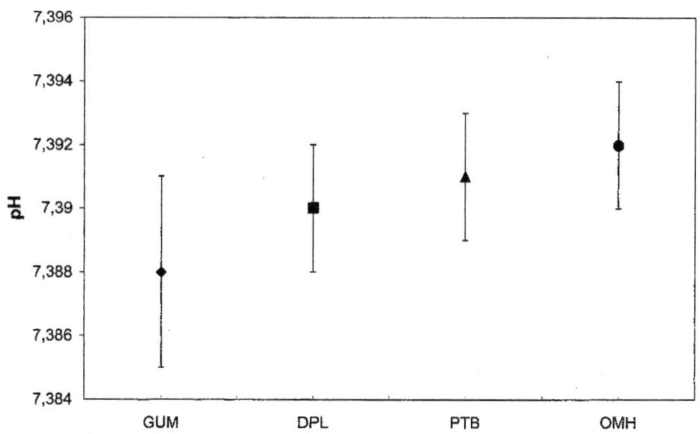

Fig. 4 EUROMET comparison 370 [18]. Buffer: disodium hydrogen phosphate, 0.03043 mol kg^{-1} + potassium dihydrogen phosphate, 0.008695 mol kg^{-1} (T=37 °C)

Secondary standards and secondary methods for pH measurement

In most applications, the use of a high-accuracy PS for pH measurement will not be justified if a traceable secondary standard of sufficient accuracy is available. It is therefore recommended to derive secondary pH standards, pH(SS), from the pH(PS) buffer solutions.

Deviating from the primary method for pH, measurements for deriving SSs are carried out in cells, separating the solutions by a diffusion-limiting or liquid junction device. Liquid junction potentials forming as a result cannot be determined directly and vary with the composition of the solution forming the junction and the geometry of the junction device. The uncertainty due to the liquid junction potential can be estimated from independent measurements or from theoretical assumptions.

Secondary pH reference materials can be derived from the PS buffer solutions by different measurement procedures, which provide results for:
pH(SS) of the same nominal composition as pH(PS)
pH(SS) of different composition
pH(SS) not compatible with platinum hydrogen electrodes.

To achieve highest metrological quality, it is strongly recommended to derive SSs from PSs of nominally the same chemical composition. Liquid junction potentials are largely minimised when buffer solutions of nominally the same chemical composition are separated from one another in a strictly isothermal cell (II) containing two platinum hydrogen cells at exactly the same hydrogen pressure [19].

$$\mathrm{Pt}\,|\,\mathrm{H_2}\,|\,\text{primary buffer (PS)}\;\|\;\text{secondary buffer (SS)}\,|\,\mathrm{Pt}\,|\,\mathrm{H_2} \qquad \text{cell II}$$

The primary and the secondary buffers are separated by a liquid junction device, preferably a glass disk of fine porosity. Under these conditions, the contribution of the liquid junction potential to the cell voltage is very small. The increase in uncertainty is also very small.

SSs derived from measurements in cell I

Buffer material that does not fulfil all the criteria for primary pH reference materials but to which pH values can be assigned using cell I are considered to be pH(SSs).

An example of such a secondary buffer is acetic acid for which a consistent chemical quality is hard to achieve. Calcium hydroxide and potassium tetraoxalate do not fulfil the criteria for a primary pH reference material because the contribution of hydroxyl or hydrogen ions to the ionic strength is significant. Also, the zwitterionic buffers [20] (e.g. HEPES and MOPSO) and the nitrogen bases of the type BH^+ (e.g., tri-hydroxymethyl aminomethane, TRIS) are excluded as primary pH reference materials because either the Bates-Guggenheim convention is not applicable, or the liquid junction potentials are high.

Calibration of pH meter-electrode assemblies

Routine pH measurements are carried out using pH meter-glass electrode assemblies. If the platinum/hydrogen electrode is replaced by a glass electrode cell, often designed as single-rod or combination electrode, the measurements of pH are affected by various random and systematic effects producing uncertainties of unknown magnitude. Hence, the glass electrode cell must be calibrated against standard buffer solutions traceable to primary pH standards. The choice among the methods should be made according to the uncertainty required for the application.

According to the number of standards used, the calibration procedures can be subdivided into:
Single-point calibration
Two-point calibration
Multi-point calibration.

In most routine applications, glass electrode cells are calibrated by the two-point or bracketing procedure, using two secondary (or primary) standards with values that "bracket" the range in which the unknown lies.

Multi-point calibration will be recommended if minimum uncertainty and maximum consistency are required over a wide range of pH(X) values [21, 22]. The calibration function of the electrode is then calculated by linear regression of the difference in cell voltage results from the standard pH values. This calibration procedure is also recommended for characterising the performance of electrode systems.

For single-point calibration using one standard, the calibration function is assumed to be a straight line defined by the intercept and the theoretical slope factor of the cell. In order to obtain the overall uncertainty of measurement, uncertainties of the respective pH(PS) or pH(SS) values must be taken into account.

Conclusion

The quantity pH is used to characterise the acidity of a system, but also in speciation [22] because of the importance of the species H^+ for controlling the chemical equilibrium. The required comparability of pH can be ensured only if the buffers used for the calibration are traceable to primary pH reference materials. pH(PS) and pH(SS) values of primary and secondary reference

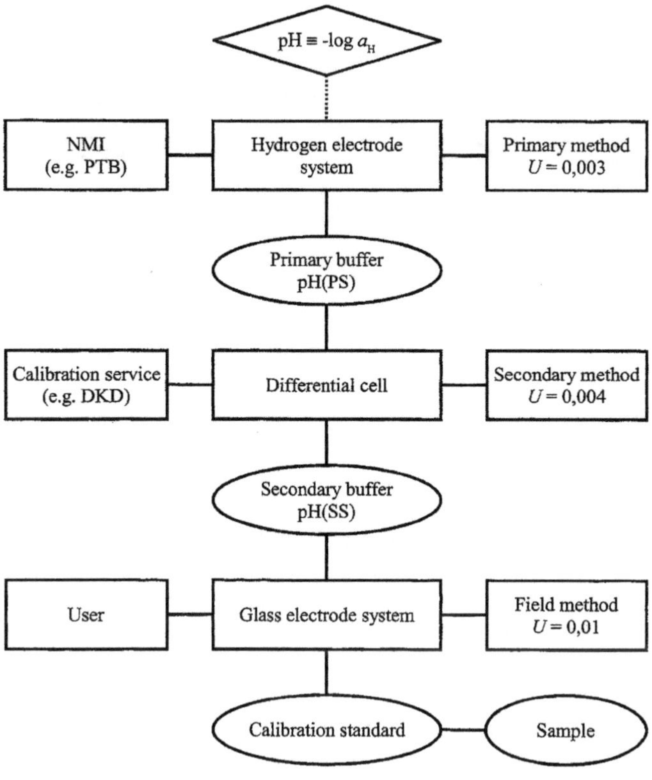

Fig 5 Traceability chain for pH in Germany. The uncertainty stated is the expanded uncertainty with a coverage factor $k = 2$. The uncertainty due to the Bates-Guggenheim convention is not taken into account

buffer solutions, respectively, were shown to be traced back as closely as possible to the thermodynamic definition of the pH.

Future improvements of the concept of single ion activity, e.g. the Pitzer treatment will open up the possibility for pH values to be traceable to the SI with acceptable uncertainties for calibration purposes.

The hierarchical approach to the traceability of pH measurements and pH reference materials is consistent with the agreed approach to traceability for metrology in chemistry [23]. Uncertainties stated for the primary method and for all subsequent measurements permit the uncertainties for all steps to be linked to the primary reference material.

For several years a traceability chain for pH measurements has been available within the German metrological infrastructure. By the choice of buffer solutions certified by the German Calibration Service (DKD) for the calibration of the pH meter-electrode assembly, traceability to the national standard is guaranteed for the user in the way shown in Fig. 5.

For the mutual recognition of measurement standards and certificates, it is necessary to demonstrate the equivalence of the national traceability structures, including national measurement standards. The first key comparison for pH took place in 1999 and the results will be available soon.

Acknowledgements The author would like to thank Ralf Eberhardt for technical assistance.

References

1. Soerensen SPL (1909) Biochem Z 21:131–199
2. Soerensen SP, Linderstroem-Lang K (1924) R Trav Lab Carlsberg 15:1–40
3. Covington AK, Bates RG, Durst RA (1985) Pure Appl Chem 57:531–542
4. Baucke FGK (1997) The definition of pH. Proposal of improved IUPAC recommendations. In: Spitzer P (ed) PTB Report W-68:Traceability of pH measurements.PTB, Braunschweig, pp. 10–21
5. Bates RG (1973) Determination of pH. Theory and practice. 2nd edn. Wiley, pp. 15–103
6. Bates RG, Guggenheim E (1960) Pure Appl Chem 1:163–168
7. Bates RG, Robinson RA (1980) J Solution Chem 9:455
8. Kristensen HB, Salomon A, Kokholm G (1991) Anal Chem 63:885
9. Spitzer P, Eberhardt R, Schmidt I, Sudmeier U (1996) Fresenius J Anal Chem 356:178–181
10. Pitzer KS (1991) Activity coefficients in electrolyte solutions, 2nd edn. CRC Press, Boca Raton, Fla., p. 91
11. Covington AK, Ferra MIA (1994) J Solution Chem 23:1–10
12. Naumann R, Alexander-Weber C, Baucke FGK (1994) Fresenius J Anal Chem 349:603–606
13. Bates RG (1973) Determination of pH. Theory and practice. 2nd edn. Wiley, p. 95
14. Naumann R, Alexander-Weber C, Baucke FGK (1994) Fresenius J Anal Chem 350:119–121
15. Spitzer P (1996) Metrologia 33:95–96
16. Spitzer P (1996) Metrologia 34:375–376
17. Spitzer P (2000), Metrologia 37 85–86
18. (1993) Guide to the expression of uncertainty in measurement. ISO, Geneva
19. Baucke FGK (1994) J Electroanal Chem 368:67–75
20. Good NE et al. (1966) Biochem J 5:467
21. Baucke FGK, Naumann R, Alexander-Weber C (1993) Anal Chem 65:3244–3251
22. Meinrath G, Spitzer P (2000) Microchim Acta 135:
23. King B et al, (1999) Accred Qual Assur 4:3–13

Accred Qual Assur (2001) 6:120–129

Paul E. Holland
Peter T. Woods
Chris Brookes
Ian Uprichard
Paul Quincey
Rod Robinson
Hansa D'Souza
Martin Milton
Brian Goody
William Bell
Des Alphonso

The development of gas standards and calibration techniques for measurements of vehicle, aircraft and industrial emissions, natural gas, occupational exposure and air quality

Received: 23 March 2000
Accepted: 18 August 2000

Paper based on a talk given at the Gas Analysis Symposium & Exhibition 1999, 7–9 November 1999, Eindhoven, The Netherlands

P.E. Holland (✉) · P.T. Woods
C. Brookes · I. Uprichard · P. Quincey
R. Robinson · H.D'Souza · M. Milton
B. Goody · W. Bell · D. Alphonso
Centre for Optical and Environmental Metrology, National Physical Laboratory, Teddington TW11 0LW, UK
e-mail: Paul.Holland@npl.co.uk
Tel.: +44-181-943 7174
Fax: +44-181-943 6755

Abstract The National Physical Laboratory (NPL) is involved in the dissemination of nationally traceable standards to which measurements of air quality, occupational exposure and air pollution source emissions, and natural gas analyses, can be referenced. This has required the development of national primary gas standards using absolute gravimetric and other techniques, and the development of dynamic calibration techniques for gaseous species which would be unstable in high-pressure cylinders. The methodology used for preparing gas standards gravimetrically is described, together with the rigorous quality assurance measurements and consistency checks which are used to demonstrate their accuracy and stability. The uncertainty budget assigned to these standards will also be summarised. NPL primary standards are used to certify traceable 'secondary' gas standards which are disseminated so as to ensure the accuracy of gas analysis measurements. Examples of the applications of these secondary standards are presented. The gas standards are employed in proficiency testing of industrial stack-testing organisations, and results of the initial rounds are presented. NPL gas standards are also now being used as the basis of the United Kingdom Environment Agency's new type-approval and certification scheme for continuous industrial stack-emission analysers. A recent important international initiative, in the field of gas analyses, is the agreement by national standards laboratories across the world to demonstrate the equivalence of their calibrations, by means of key comparisons between them. These worldwide key comparisons are complemented in Europe through the EUROMET initiative which seeks to establish the equivalence and comparability of calibration standards held at national standards laboratories across Europe. Examples of these intercomparisons are presented.

Keywords Traceability · Gas standards · Calibration · Quality assurance · Proficiency testing · International comparisons

Introduction

The National Physical Laboratory (NPL) has a responsibility through the United Kingdom's Department of Trade and Industry's (DTI's) National Measurement System programme for the improvement of gas concentration measurements, including gaseous air pollutants, both through the provision of nationally traceable standards and the development of suitable measurement methods. Gas standards provide accurate and traceable references for United Kingdom and foreign measurements of air quality, occupational exposure and air pollution source emissions.

Development of standards and measurement techniques at NPL

NPL is the focus of the United Kingdom's National Measurement System for physical measurements and its remit includes aspects of environmental measurements, which form an important component of the United Kingdom's Valid Analytical Measurement (VAM) programme. The VAM programme, currently undertaken jointly by NPL, the Laboratory of the Government Chemist (LGC) and AEA Technology, forms part of the Government's wider policy to ensure that measurements of all kinds are accurately made through the framework of the United Kingdom's 'National Measurement System'. A VAM programme is currently underway and its main features in the area of underpinning a wide range of measurements of gaseous concentration, including gaseous pollution measurements are described in this paper. NPL is also involved in the development and use of remote and in-situ spectroscopic techniques for monitoring industrial air pollution and stratospheric ozone, together with the trace gases associated with its depletion, but these will not be covered in this paper.

NPL gas concentration standards: overview

NPL has a well-established facility for the production of primary gas concentration standards by absolute gravimetric techniques. Standards of a range of different gases with widely differing concentrations are prepared in carefully selected passivated containers by the accurate consecutive weighings of the constituent gases. The concentrations of the gas standards prepared in this manner, which are expressed in absolute molar units, are traceable to the primary standard of mass.

Rigorous quality assurance procedures ensure the accuracy of these primary standards. Initially, all the parent gases are analysed comprehensively for impurities using sensitive gas chromatographic and spectroscopic instruments. These parent gases are then used to prepare the primary standards, employing a range of methodologies and gravimetric facilities. The various primary standards produced in this way are compared between themselves, to verify their internal consistency.

Secondary standards are also produced, which are certified with respect to the primary standards. These are made in-house, using the same analytical instrumentation, blending facilities and parent gases which are used to produce the primary standards. They are certified directly and individually against a range of NPL's primary standards over a period of time, to verify the stability of their concentrations. They are then disseminated to United Kingdom industry, government bodies and research organisations in order to provide the required measurement accuracy and national traceability. Secondary standards are also disseminated to other countries to provide them with the ability to carry out accurate measurements.

An important extra element of the production of gas standards is that comparisons are carried out, where possible, with the standards laboratories of other countries, in order to establish international uniformity in gas concentration measurements in Europe and elsewhere.

Quality assurance of NPL gas standards

High-sensitivity instrumentation is employed to assure the purity of the parent gases used to prepare the standards, to ensure that no chemical or physical reactions take place during any stages of the preparation and dilution process, and to demonstrate the longer-term stability of the standards after preparation. Two main types of instrument are used:

A high-resolution fourier-transform interferometer (FTIR) which measures the spectral absorption due to the presence of infrared-active gases across the entire spectral region from 1 to 15 μm. This is interfaced to a multi-pass optical-absorption gas cell, which provides a path length of up to 120 m to enable gas detection with high sensitivity.

A number of gas chromatographic instruments, with different detectors to enable a wide range of gases to be measured with high sensitivity (typically a few parts in 10^9 by volume, or significantly better than this where cryogenic pre-concentration is used).

Validation of primary gravimetric standards

The procedure used to validate the absolute concentrations of the range of NPL gravimetric standards has several stages, as indicated below:

1. The instruments and techniques outlined in the Section on Quality assurance of NPL gas standards are used to check the concentrations of all gaseous constituents in the standards other than the specified minor and major components. This verifies that no impurities or extraneous species have become entrained into the cylinders during the preparation

process and that no subsequent chemical reactions have taken place.

2. All the sources of uncertainty, both Type A and B which occur during the complete preparation process are itemised and quantified. These are then combined in a square-root-sum-of-squares manner to produce a combined uncertainty, which is then used to produce an expanded uncertainty with the use of an appropriate k factor. The value of this uncertainty will depend on a number of factors including the purity of the component gases, the weighing procedure and the uncertainties arising from the relative molar masses.
3. Each new primary standard is intercompared against two or more existing primary standards using an instrument with sufficient repeatability to demonstrate that their measured and gravimetric values are consistent within the expanded uncertainties indicated in (2.) above. The intercomparison procedure involves using the primary standards to bracket the new primary standard which is treated as an unknown. If the analytical value obtained for the new standard differs significantly from its gravimetric value, (typically $\pm$ 0.2%) it indicates an inconsistency in the new standard and it is discarded.
4. Standards with similar concentrations from different families are also intercompared using the procedure outlined in (3.). These measurements are carried out over the entire concentration range in order to demonstrate the consistency of the complete set of standards of a given component mixture.
5. Repeated intercomparisons are carried out in the manner outlined in (3.) above to demonstrate the long-term stability of the set of standards.
6. New batches of standards are produced regularly. These are intercompared with each other and also with older standards using the procedure outlined in (3.) to confirm the accuracy of the overall process.
7. NPL's gas standards are regularly intercompared with those of other national standards laboratories. A comprehensive measurement exercise carried out with the National Institute of Standards and Technology (NIST) USA [1] demonstrated the consistency of the CO/N_2 and CO_2/N_2 standards prepared by NPL and NIST over a wide range of concentrations. A similar intercomparison exercise was also completed on the concentration range of NO/N_2, C_3H_8/air and C_3H_8/N_2 standards. Similar intercomparisons have also been carried out with the Netherlands MeetInstituut (NMi), as part of a EUROMET agreement [2].
8. A series of international intercomparisons are being organised by the Consultative Committee for Amount of Substance (CCQM) under the Consultative Committee for Weights and Measures (CIPM). These are carried out to investigate the international uniformity of gas standards produced by selected laboratories (see Section on International comparisons and intercomparability).

Nationally traceable reference gas standards

NPL, in line with other national standards laboratories, retains its primary standard gas mixtures (PSMs) in-house. These primary standards are disseminated, however, through different types of calibration gas mixtures. These disseminated standards are known at NPL as 'Primary Reference Gas Mixtures', 'Secondary Gas Standards' and 'Certified Gas Mixtures'. An NPL leaflet has been prepared which explains the differences between these types of traceable standards and which also explains the relationship of these with the different types of standards produced by other national metrology institutes (NMIs). The main type of gas standards disseminated by NPL, are however, secondary gas standards and the procedures used for preparing and certifying these are outlined below.

Preparation procedure

Secondary gas standards are prepared in-house using the parent gases of the same specifications as those employed for the production of the primary gravimetric standards. This ensures that the gases used are of certified purity and contain no species that would affect the certified concentration values or be detrimental to the stability of the mixtures. The purity of the parent gases are checked against the manufacturers specification using gas chromatography and FTIR techniques.

The secondary gas standards are blended using the same apparatus as that employed to prepare the primary gravimetric standards. Precise measurements of the gas pressure and gas mass are carried out at each stage in the process and these enable the concentrations of the mixtures to be produced to within $\pm 1\%$ of the concentration of the appropriate NPL primary standards. Up to three mixtures with the same nominal concentrations can be produced together. The mixtures are subjected to stability checks before their concentrations are certified with respect to NPL primary standards.

Certification process

The gravimetrically prepared secondary gas standards are checked soon after preparation to confirm that their concentrations are as expected. They are then allowed to stand for a defined period before final certification is carried out. Following this period, all secondary standards are certified individually against NPL's gravimetrically prepared primary gas standards.

Apparatus

An automatic gas analysis system (AGAS) is used to certify the concentrations of secondary standards. The AGAS has dedicated gas analysers for each of the species used in the standards. Personal computers are used to control AGAS which allows zero gas, primary gas standards and the unknown secondary gas standard to be directed in rapid succession into the appropriate analyser. The AGAS then records the analysers response for each of the gases and subsequently uses these values to calculate the unknown secondary standards concentration, and other statistical information.

Experimental procedure

The certification procedure involves bracketing the concentration of the secondary standard between adjacent concentrations of several different pairs of primary gravimetric gas standards. This assumes that a linear algebraic interpolation of the analyser response can be made between the concentrations of these bracketing standards. However, the responses of all analysers exhibit some nonlinearity in their behaviour as a function of gas concentration. The magnitude of this nonlinearity will depend both on the analyser employed and on the concentrations of the gases being analysed. Therefore, to allow for this, the response of each analyser is measured over the required range of concentrations be generating a five-point calibration curve with appropriate known concentration standards.

The problem of analyser nonlinearity is overcome by choosing the difference in the concentrations of the two bracketing standards to be small enough that the uncertainty in assuming a linear response is small compared with the uncertainty of the overall certification process. The concentrations of the bracketing standards are generally chosen to be within $\pm 4\%$ of each other, with the concentration of the secondary standard between them, in order to produce an uncertainty of less than $\pm 0.1\%$ relative of value. This procedure is verified regularly by comparing the analytical results obtained with sets of three NPL gravimetric standards with their gravimetric values.

Gas standards routinely available

Table 1 summarises the range of gases and concentrations that are routinely supplied to customers. These include:

- Gas standards for measurements of gaseous pollutants emitted by vehicle and aircraft engines. These comprise binary mixtures of carbon monoxide in nitrogen, carbon dioxide in nitrogen, nitrogen monoxide in nitrogen, sulphur dioxide in air, hexane in nitrogen, and propane in nitrogen or air.
- Tertiary gas standards for the provision of traceable measurements to the United Kingdom's vehicle emissions testing programme (the "MoT test"). These comprise specific concentrations of mixtures of carbon monoxide, carbon dioxide and propane in a diluent gas of nitrogen.
- Ethanol in air mixtures as standards for the new generation of evidential breath alcohol analysers.
- Natural gas standards used for the determination of physical quantities of natural gas, including calorific values, Wobbe index, and density.
- Gas standards containing methane in a diluent gas of air or nitrogen, to provide traceable, accurate flammability measurements.
- Multi-component hydrocarbon standards to provide accurate calibration of instruments (generally gas chromatographs) used to monitor the concentrations of a wide range of volatile organic hydrocarbon compounds (VOCs) in ambient air. These standards currently contain 30 different hydrocarbon species that are important to photochemical ozone formation, with concentrations ranging down to a few parts per billion by molar value. They are disseminated widely in the United Kingdom and the rest of Europe as calibration standards, and as test mixtures for assessment of the quality of international ambient hydrocarbon measurements (often under the auspices of the European Commission – EC).
- Gas standards for the calibration of air quality monitors containing sulphur dioxide, nitrogen monoxide or nitrogen dioxide at ambient concentrations.

The above standard mixtures contained in cylinders are supplemented by several gas measurement facilities which can provide dynamic calibrations of gas mixtures and of gas monitoring instruments. These include an on-line facility which injects gas dynamically into a passivated multipass optical gas cell, where the gas concentration is certified spectroscopically. Some of the gas mixtures which can be certified by these dynamic blending facilities are given in Table 2.

Uncertainty analysis

Uncertainty of the gravimetric procedure

The uncertainty in the accuracy of any given gravimetric standard is obtained from the uncertainties associated with the weighing procedure used to produce the standard, those which arise from the purity of the gases

Table 1 Range of secondarystandards routinely supplied to customers

Application	Gaseous species	Concentration range	Uncertainty (95% confidence limits)
Industrial emissions	Sulphur dioxide in nitrogen	1000 & 100 ppm	±1%
	Sulphur dioxide in air	250 ppm	±1%
	Carbon monoxide in nitrogen	15%–10 ppm	±0.6%–1%
	Carbon dioxide in nitrogen	15%–0.5%	±0.6%
	Oxygen in nitrogen	22%–1%	±1%
	Propane in nitrogen	10%–500 ppm	±0.6%–1%
	Propane in air	1%–0.3 ppm	±0.6%–1.5%
	Methane in air	2%–1 ppm	±1–1.5%
	Nitric oxide in nitrogen	10%–1 ppm	±1–±2%
	Hexane in nitrogen	1000 & 100 ppm	±2%
	Toluene in air or nitrogen	100 ppm	±2%
	Multi-components for industrial emissions and waste incineration	1000–10 ppm	±1%
Vehicle and aircraft emissions	Carbon monoxide, Carbon dioxide, Nitric oxide, Propane multi-components in nitrogen.	15%–500 ppm	±0.4%
	Hexane in nitrogen	1000 ppm	±2%
Landfill gas and gas flammability monitoring	Methane in air	2%–1000 ppm	±1%
	Propane in air	1%–1000 ppm	±1%
Occupational exposure	Hydrogen sulphide in nitrogen	25–15 ppm	±2%
	Benzene in nitrogen	5 ppm	±1%
	Dichloromethane in nitrogen	100 ppm	±1%
	Benzene, toluene and xylene in nitrogen	~1 ppm	±5%
Air quality	30 component hydrocarbons in nitrogen (C2–C9)	10–1 ppb	±3%
	Sulphur dioxide in air or nitrogen	500–50 ppb	±1%–±3%
	Nitric oxide in nitrogen	1 ppm – 150 ppb	±2%
	Nitrogen dioxide in air	500–10 ppb	±2%–±5%
	Benzene toluene and xylene in nitrogen	10 ppb	±5%
	Carbon monoxide in air	20 ppm	±1%
Industrial process control	Oxygen in nitrogen	100 ppm	±1%
	Methane in nitrogen	22%–10%	±1%
	Carbon monoxide in nitrogen	10%–1%	±0.4%
Natural gas	C_1–C_6, N_2 and CO_2 (11 component mixture)	99%–20 ppm	±1%–±2%
Odour	Hydrogen sulphide in air	1–20 ppm	±3%
	Ethyl mercaptan in air	200 ppb	±3%
	n-Butanol in air	60 ppm	±3%
	1-Pentene in air	5 ppm	±3%

used and those from the relative molecular masses of its constituents. The uncertainty due to the sequential dilution process is also incorporated into the uncertainty budget.

Uncertainties arising from the weighing procedure

The sources of uncertainties arising in the weighing procedure have been grouped together into the following categories:
Balance repeatability
Thermal drift
Time drift
Draught instability
Location of cylinder on balance
Mass piece uncertainty
Resolution of balance
Sensitivity of balance
Buoyancy correction
Expansion of cylinder due to pressure
Mechanical handling of cylinder.

Each of the categories has values assigned from experimentation or suppliers specifications. These value comprise both Type A (those which can be assessed by statistical methods) and Type B (those which are as-

Table 2 Range of traceable calibration facilities available for gas concentration instruments

Application	Species	Concentration range	Accuracy confidence limits)
Industrial emissions/process control	Hydrogen chloride in nitrogen	1000–20 ppm	±2%
	Nitrogen dioxide in air	500–10 ppm	±3%
Open-path/Cross-duct monitors	Benzene, sulphur dioxide, nitrogen oxides, ozone, methane	100 ppm–20 ppb	±2%
Gas monitor calibration and product certification facilities	Range of gaseous species under different environmental conditions	1000–10 ppm	±2%
Air quality	Sulphur dioxide in air	1 ppm–50 ppb	±3%
	Nitrogen monoxide in nitrogen	100 ppb–5 ppm	±2%
	Nitrogen dioxide in air	1 ppm–50 ppb	±3%
	Butadiene in air		
	Benzene in air	100–10 ppb	±3%
	Ozone in air	10 ppb–1 ppm	±3%–±2%

sessed using other methods) uncertainties. The probability distributions are determined for each of the categories and an appropriate devisor is then used depending on the distribution assigned.

Uncertainties arising from gas purity

Purity analyses are performed for all parent gases used in the preparation of both primary and secondary gas standards. These purity measurements are required by the Gravcalc software. The sequential dilution of one standard to produce the corresponding lower value in the hierarchical structure incurs a greater relative uncertainty as the concentration is reduced, so that the relative uncertainty increases each time a dilution is made. It is a necessary condition of the NPL uncertainty estimation procedure that the sum of all component mole fractions are equal to 1, and under these conditions it is assumed that perfect correlation applies.

Table 3 Shows the range of gas concentrations which are currently tested for type-approval tests within MCERTS

Species	Range	Units
SO_2	0–10000	mg m^{-3}
CO	0–1000	mg m^{-3}
CO_2	0–20	% volume
NO	0–3000	mg m^{-3}
NO_2	0–3000	mg m^{-3}
HCl	0–2000	mg m^{-3}
TOC	0–70	mg m^{-3}
O_2	0–25	% volume
H_2O	0–45	% volume

Uncertainties arising from component relative molar masses

The relative molar masses of the gaseous components and the associated uncertainty in the relative molar masses are calculated from tables of atomic weights. The relative molar mass and uncertainties are combined with the gas purity and weighing uncertainty using the Gravcalc software.

Uncertainties in the certification of secondary standards

In order to obtain a value for the concentration of a secondary standard, data from each of the standards, zero gas and unknown is collected sequentially from the AGAS system. The concentration of the secondary standard is then obtained from these results using the following formula:

$$X_{SS} = R_1 + \left((R_2 - R_1) \left(\frac{Y_{SS} - y_1}{y_2 - y_1} \right) \right)$$

where R_1 = concentration of low value NPL primary standard, R_2 = concentration of high value NPL primary standard, y_1 = stabilised analogue output for low concentration primary standard, y_2 = stabilised analogue output for high concentration primary standard, Y_{SS} = stabilised analogue output for secondary standard.

This process is repeated between 6 and 10 times (depending on the concentration of the mixtures and the analytical uncertainty arising from the measurements) to form the mean value for the certified concentration of the secondary standard. Two such results are obtained on at least two different days with one pair of primary gas standards, and the complete procedure is repeated using a different pair of standards. The concentration of the secondary standard is then obtained by taking the mean of these four values.

The statistical uncertainty arising from the analytical measurement is derived from the automatic data collection procedure noted before. The AGAS computer performs a standard error analysis and produces both a mean and the standard error of the mean associated with that value. A computer program is used to combine the uncertainties from the primary gravimetric process with the uncertainties produced from the standard deviation of the instrument's response for each of the gas mixtures.

Thus, the uncertainty in the certification of the secondary standard σ_x^2, is:

$$\sigma_x^2 = \left(\frac{1}{R_2 - R_1}\right)^2 [(R_2 - X_{SS})^2 \sigma_{R1}^2 + (X_{SS} - R_1)^2 \sigma_{R2}^2] + \frac{1}{(y_2 - y_1)^4} \{(y_2 - y_1)^2 \sigma_Y^2 (Y_{SS} - y_2)^2 \sigma_{y1}^2 + (Y_{SS} - y_1)^2 \sigma_{y2}^2\}$$

where R_1 = low reference primary gas standard gravimetric value, R_2 = high reference primary gas standard gravimetric value, X_{ss} = Calculated concentration of secondary standard from linear interpolation, σ_{R1} = gravimetric uncertainty of low reference gas, σ_{R2} = gravimetric uncertainty of high reference gas, y_1 = instrument response to low reference standard, y_2 = instrument response to high reference standard, σ_{y1} = standard deviation of the response of instrument to low reference standard, σ_{y2} = standard deviation of the response of instrument to high reference standard, Y_{ss} = Instrument response to secondary standard, σ_Y = standard deviation of the response of instrument to secondary standard.

Measurement anomalies due to nonlinearities of the gas analysers have been determined and assigned a value of ±0.1% (at 68% confidence level). These values are added to the combined standard uncertainties after correcting for probability distribution as square root sum of squares.

The expanded uncertainty U is obtained by multiplying σ_x by an appropriate coverage factor (k) as specified in the *Guide to the expression of Uncertainty in Measurements* (GUM). Thus $U = k\sigma_x$ and $X_{ss} \pm U$.

Instrument type-approval testing

A facility for the testing and/or type-approval of continuous industrial emission-monitoring (CEM) instrumentation has been established. This is a comprehensive laboratory-based calibration and test facility which is primarily targeted on gaseous monitoring instruments and can be used to evaluate, for example, a CEM's accuracy, linearity, detection limit, drift, cross-sensitivity, repeatability, effects of sample gas pressure and temperature, and the influence of environmental conditions (e.g. ambient temperature from −40 °C to +60 °C).

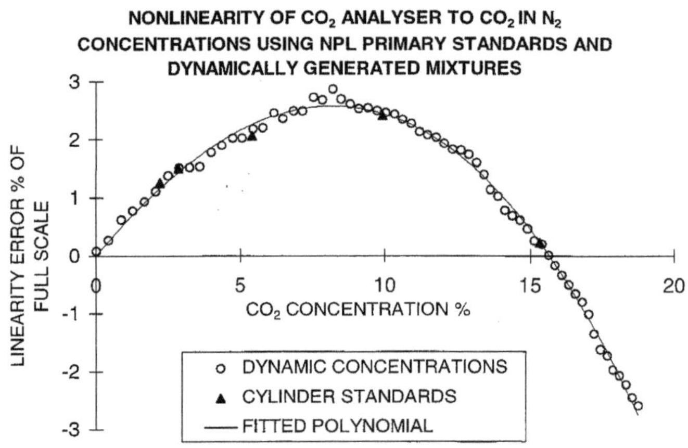

Fig. 1 Nonlinearity graph for CO_2 analyser

As part of this a dynamic blending system has been developed which enables gas mixtures with large throughput rates to be delivered for testing with known concentrations with a measurement uncertainty of about ±0.2% relative of concentration over a dynamic range of 50 to 1. An example of a linearity test on a CEM carried out with this dynamic blending system is shown in Fig, 1. This is a graph showing the CEM's departure from linearity. The same graph shows the results obtained using NPL primary gravimetric gas standards which confirm the accuracy of the blending system and the nonlinearity of the CEM.

The NPL testing facility is used to perform all the laboratory tests specified within the Environment Agency's MCERTS Scheme for CEM systems. This MCERTS Scheme is to be extended to cover **ambient** air quality monitoring instrumentation particularly that required within the EC Air Quality Framework and Daughter Directives, in the near future.

Results of a trial United Kingdom proficiency testing (PT) scheme using stack-emission gases

The results of the first round of a trial gaseous measurement PT scheme has been carried out by the NPL with the United Kingdom's Source Testing Association (STA). Eighteen STA member companies took part in the scheme, which involved the round-robin measurements of a number of nationally traceable, standard gas mixtures which had accurately known concentrations.

PT schemes provide a way of assessing the performance of laboratories by a series of regular interlaboratory comparisons. In a typical PT scheme a test sample or material is sent for analysis to all participating laboratories. The results of the analyses are compared to assigned values of the samples. The assigned value may be a 'true' known value or in some cases, where a 'true' value is unknowable, it is based on the consensus mean of all the results from the laboratories. The set of results are reported anonymously and, in addition, each participant is made aware of their own results. In this way participants are able to assess their performance in relation to other laboratories. The key feature of a PT scheme is that it should be carried out regularly, and that a degree of improvement is then looked for in both poorly performing laboratories and in the overall performance of all participants.

In setting up a trial PT scheme it was decided to initially focus on measurements of gaseous components using CEM equipment. This has the benefit that the test samples, in this case NPL standard gas mixtures, each have a known, 'true' value, and that the analysis of these is nondestructive, in the sense that the same gas mixture can be analysed by more than one laboratory.

Sixteen organisations returned results after participating in the scheme. In practice more than one sample of each gas was circulated to different participants, each with a known concentration, traceable to NPL primary gravimetric gas standards. The purity of each sample was also checked to ensure no potentially interfering substances were present in the cylinders.

The participants were told the nominal concentration of the cylinders they received, but not the absolute value. The results of these analyses were returned to NPL. Participants were given the option to report the measurement uncertainty that they assigned to the results obtained. Not all participants took part in all tests and very few reported uncertainties.

The results of each analysis were expressed as percentage differences from the true value (Fig. 2). This allowed a comparison to be made between the different gas samples used at each nominal concentration. Overall the results have been encouraging, with most participants reporting results within 10% of the true value.

There are a number of ways in which the results of PT schemes can be interpreted. The most straightforward technique is to examine the percentage differences of the reported results from the true value, as has been applied above to this PT scheme. It is then left to the participants to gauge how well they have performed. One disadvantage of this approach is that the percentage deviation will depend on the species being measured; for example analysis of 1000 ppm CO might be expected to give better uncertainty than analysis of 500 ppm NO. Using this technique it is difficult to compare results from the analysis of different gases within a PT scheme.

An alternative interpretation of PT scheme results involves the calculation of a performance score for each result. This is usually based on comparing the results achieved against an assigned target standard deviation, σ. The simplest form of this is the 'z score'. This is calculated by dividing the deviation of each result from the true value by σ, thus:

$$z = \frac{x - T}{\sigma}$$

where: z = z score, x = value obtained by participant, T = true value for test sample, σ = assigned value for standard deviation.

This provides a z score for each result which can be compared with other z scores from analyses of the same sample and with analyses of different species. If a suitable value of σ is chosen for each species then the z score also provides a method of deciding decision limits for the PT scheme. In general, if all results are normally distributed about the true value of the test sample and

Fig. 2 Summary of results of the National Physical Laboratory Source Testing Association (NPL/STA) proficiency testing (PT) scheme for gaseous analysis

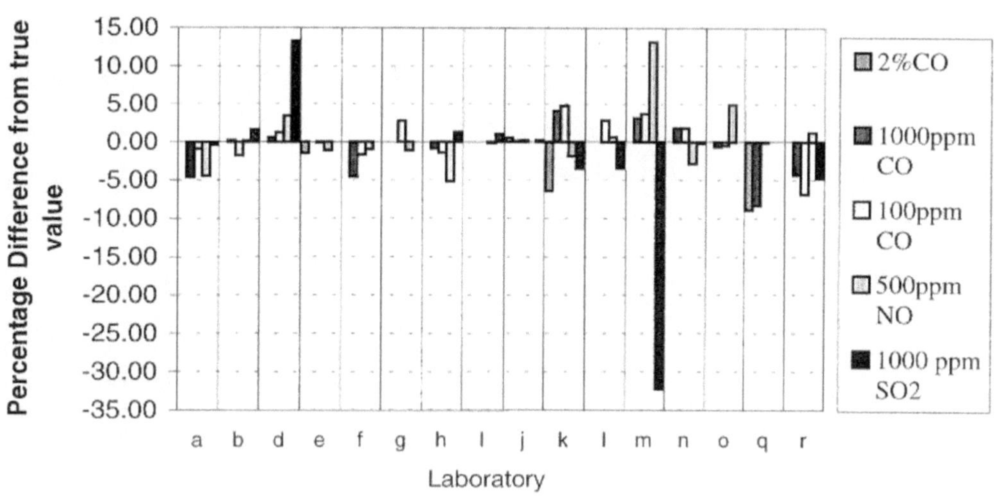

a reasonable value of σ has been chosen, then few ($<5\%$) of the z scores should lie outside ± 2. Z scores lying outside ± 3 would be strongly indicative of a real bias in the reported value, rather than random uncertainty. From this it is possible to apply a classification as follows:

$\|z\| \leq 2$	satisfactory
$2 > \|z\| < 3$	questionable
$\|z\| \geq 3$	unsatisfactory

These limits allow each participant to judge their own performance and can be used to indicate potential problems. Figure 2 gives the results obtained in this pilot study expressed in terms of percentage deviations of each laboratory's results from the true value. The target standard deviation is usually taken to be a value which is fit for purpose for the measurements being made. As an example, z scores have been calculated for the results obtained during this trial PT scheme. The values of uncertainty given in Annex 3 of the Hazardous Waste Incineration Directive have been used, in an ad-hoc manner, to derive σ. The z scores calculated from these are for example only, and a satisfactory score in this test should not be taken as compliance with the requirements of the Directive; the figures in the Directive were used purely to illustrate the use of z scores. Figure 3 summarises the z scores calculated in this way. It can be seen that most results fall into the acceptable category, with only one analysis falling into the unacceptable category. If a z score approach is to be used in a subsequent ongoing PT scheme then target standard deviations should be agreed for all species tested.

The results of this trial PT scheme show the usefulness of such an exercise. However, for it to truly count as a PT scheme, with all of the associated benefits that that would bring, the scheme should run on a regular basis. This would enable STA members and others to monitor their own performance against their peers, and hopefully provide a regular incentive to strive for quality. Having a formally managed and approved PT scheme would also provide a demonstration of the commitment of participants to increased quality in their results.

International comparisons and intercomparability

National Standards laboratories worldwide carry out comparisons with each other to demonstrate the accuracy and international uniformity of their primary standards. Such comparisons are becoming increasingly important to facilitate growing international trade.

As a result, the recently formed CCQM within the CIPM has established a formal intercomparison programme whereby a series of 'Key Comparisons' of 'amount of substance' measurements most are carried out between selected NMIs across the world. There have so far been 14 such 'Key Comparisons' involving these NMIs worldwide. Figure 4 gives an example of results obtained at an early stage for binary carbon monoxide in nitrogen primary gas standards. This example shows typically the types of results on intercomparability obtained:

- NMI results agree with the 'known' concentration values within their stated measurement uncertainties, but such uncertainties are very different between NMIs;
- Other NMIs show some disagreements with the 'known' concentration values within their stated

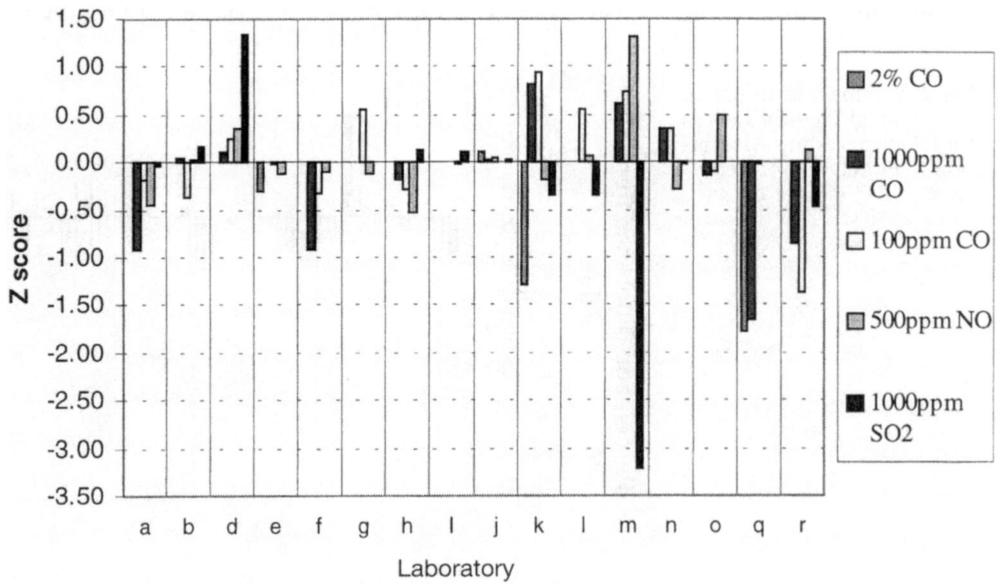

Fig. 3 Z scores for NPL/STA PT scheme for gaseous analysis

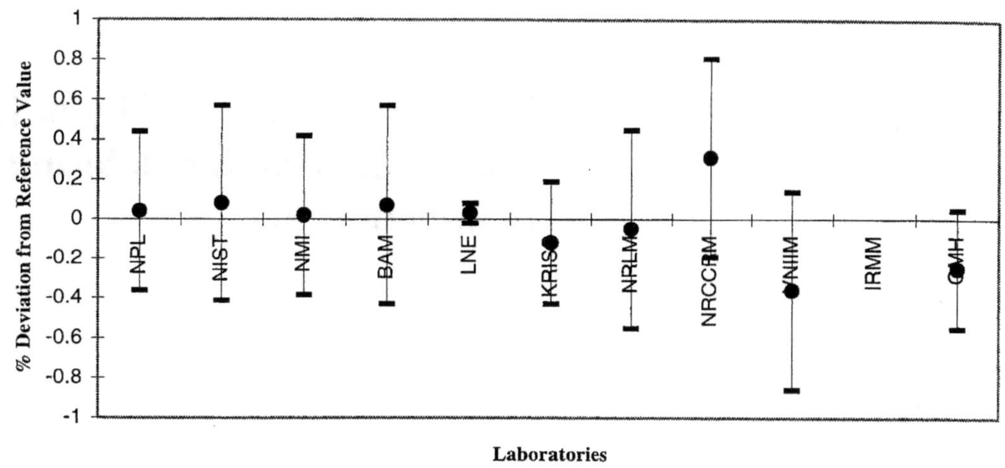

Fig. 4 International comparison of 1% carbon monoxide in nitrogen standards

measurement uncertainties, thereby indicating that these uncertainties may have been underestimated.

A further range of international comparisons are being carried out regularly to complement these worldwide 'Key Comparisons'. These are carried out in Europe under EUROMET. EUROMET aims to mirror and propagate the worldwide comparability of Key Comparisons to a wider range of European NMIs. A number of comparison projects on gas standards are underway organised by EUROMET. An example of a bilateral comparison between NPL UK, and NMi The Netherlands, has been published [2], a project to harmonise air quality measurements across Europe (HAMAQ) has recently been completed, and further international comparisons are underway [3].

References

1. Hughes EE, Davenport AJ, Woods PT, Zielinski WL (1991) Environ Sci Technol 25:671–676
2. International Comparison: Gas standards (1999) Metrologia 36:155–158
3. EUROMET Directory (1999): www.euromet.org

Accred Qual Assur (2001) 6:302–305

Ilona Šperlingová
Ludmila Dabrowská
Miloò Tichý
Jan Kučera

Problems of traceability of total protein and catecholamine determinations in human urine

Received: 12 November 1999
Accepted: 16 August 2000

Paper based on a talk given at CERM-2: 2nd Regional European Conference on Reference Materials, 9–10 September 1999, Prague, Czech Republic

I. Šperlingová (✉) · L. Dabrowská
M. Tichý
National Institute of Public Health,
Šrobárova 48, 100 42 Praha,
Czech Republic
Tel.: +420-2-6708 2675
Fax: +420-2-6731 12 36

J. Kučera
Nuclear Physics Institute, 250 68,
2 Řež u Prahy, Czech Republic
Tel.: +420-2-6617 2268
Fax: +420-2-68 57 003

Abstract Some problems arising in the establishment of the traceability of the certified reference material (CRM) CZ 6007a (total protein, creatinine and stress indicators) are discussed. Bovine serum albumin is recommended as a calibration standard for total protein determination

Keywords Reference material · Traceability · Total protein · Stress indicators

Introduction

Certified reference materials (CRMs) should be used to ensure comparability of results, traceability of measurements at different levels in the traceability chain and support implementation of legislation, standardization programmes, research programmes, accreditation of laboratories and industrial production processes [1].

The Czech Reference Material CZ 6007a for total protein and creatinine in human urine was prepared [2], and served as a preliminary batch for the preparation of a certified reference material (CRM) for the stress indicators adrenaline (A), noradrenaline (NA) and dopamine(DA)in human urine. Some major problems in the traceability of these different analytes are presented.

Methods

The material was prepared by freeze drying pooled, urine samples obtained from healthy volunteers. Sodium merthiolate was used as preservative agent.

The commonly used preservative sodium azide at a concentration of 1 mg/ml interferes with the protein determination lowering values of total protein concentration to 60–50% (Fig. 1). The interference is noticeable for concentrations of 0.2 mg/ml (Fig. 2). However, such low sodium azide concentrations have insufficient bactericidal effects.

Total protein was determined using one turbidimetric and five spectrophotometric methods [2]. A. NA and DA were determined by fluorimetric [3] and high performance liquid chromatography (HPLC) [4] methods.

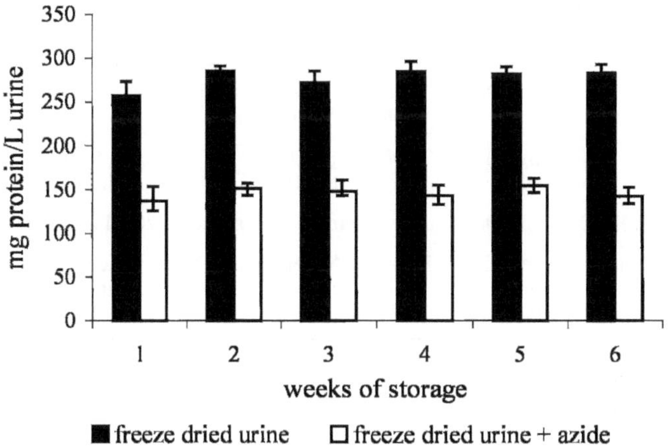

Fig. 1 Impact of sodium azide (NaN_3 1 mg/ml) on total protein concentration in urine determined by the Lowry method

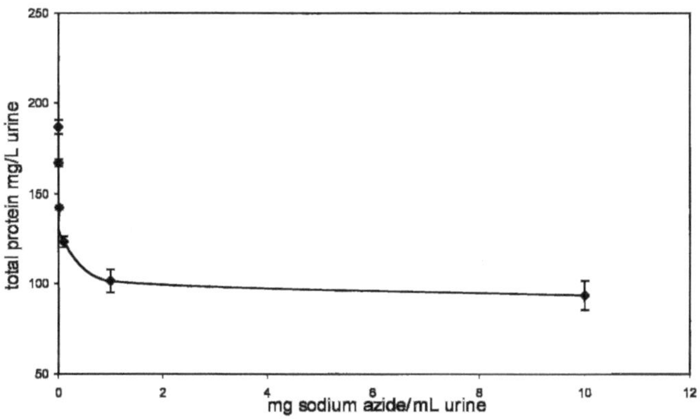

Fig. 2 Dependence of interference of the amount of sodium azide on the determination of total protein concentration by the Lowry method

The results were evaluated using analysis of variance (ANOVA).

Traceability problems in total protein determination

Several kinds of traceability problems occurred during the preparation of CRM CZ 6007a "Total Protein and Creatinine in Human Urine."

1. A reference material (RM) for traceability of total protein in human urine does not exist. The control materials used in clinical laboratories for calibration purposes are not unified. Some laboratories use bovine serum albumin as a calibration material (URINE-CHIMIE BIOTROL) [5], whereas others use a mixture of human serum albumin (70%) and globulin (30%) (LYPHOCHECK Quantitative Urine Control, BIO-RAD) [6]. The use of various protein calibration standards yields various results of

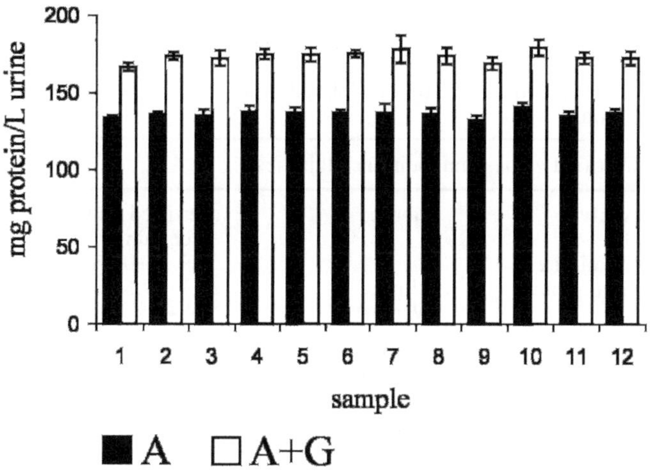

Fig. 3 Determination of total protein concentration (Lowry method): A, standard albumin; A+G, standard BIORAD –70% albumin +30% globulin

protein concentrations in urine (Fig. 3). The reactions in spectrophotometric determinations take place only with some proteins or their fragments (2).

In our experiments crystalline bovine serum albumin from Fluka, cat.no. 05470, (FLUKA albumin) was used for calibration and traceability purposes. The Fluka product was tested for traceability against the National Institute of Standards and Technology (NIST), USA Standard Reference Material (SRM) 927c (Total Protein Standard) bovine serum albumin (NIST albumin). Two different statistical techniques were used to evaluate traceability of the FLUKA albumin to the NIST albumin.

Calibration curves were constructed with the NIST albumin (5 concentrations in triplicate) and with the FLUKA albumin (5 concentrations in duplicate) in the concentration range of 50–250 mg/l. The measured values of individual concentrations fluctuated around the fitted lines, with a standard error of 0.007 of the measured absorbance. The difference between FLUKA and NIST albumin calibration lines was statistically insignificant, as evaluated by the *t*-test: $P=0.14 \gg 0.05$. The calibration lines differed only in the range of a random error. The FLUKA albumin was, thus, equivalent to that of NIST. Statistical evaluation was carried out using the regression analysis module of the statistical package SPSS, version 4.0.

The concentration of the FLUKA albumin, as determined in weighted samples of the product using the calibration line constructed with the NIST albumin, fluctuated around the weighted amount in the range of experimental error of the determination considered as the 95% tolerance limit. The data were evaluated using EXCEL 97.

Table 1 Six methods for the determination of total protein concentration in human urine were compared: the Lowry method, biuret method, methods using the dyes Commassie Brilliant Blue (G250), Ponceau-S, pyrogallol red and the turbidity method by Exton

Certified values and their uncertainities		
Analyte	Method	Certified value and uncertainity
Creatinine		9.34+0.18 mmol/l
Total protein	Biuret	217+17.5 mg/l
	Watanabe	168+8.9 mg/l
Information (noncertified) values		
Analyte	method	Noncertified value and uncertainity
Total protein	Lowry	141+2.6 mg/l
	Bradford	173+19.6 mg/l
	Pesce/Strande	108+1.7 mg/l
	Exton	79+3.4 mg/l

2. The value of total protein concentration in human urine depends upon the method used for its determination. Six methods were compared: the Lowry method, the biuret method, methods using the dyes Commassie Brilliant Blue (G250), Ponceau-S, pyrogallol red and the turbidity method by Exton (Table 1). The uncertainty of both certified and non-certified values is given as the 95% confidence interval. Valuation of the uncertainty occurring during the RM preparation (pipeting of urine and reconstitution of freeze-dried urine) was included. Certified and non-certified values were derived from interlaboratory comparison.
3. The effect of additives (e.g. a preservative) must be considered and evaluated (see Methods).

Traceability in stress indicator determination

LYPHOCHECK Quantitative Urine Control (BIO-RAD) [6] and ClinRep-Control (Merck/Recipe) [7] were used for traceability purposes for the determination of A, NA and DA concentrations in human urine.

Three modifications of the HPLC technique, and the fluorimetric method were used for A and NA determinations. Fluorimetric determination of DA was found to be unsuccessful.

The results of the analysis were compared with the mean values, and "acceptable ranges" of the commer--cial quality-control products provided by the manufactures. However, a definition of the "acceptable ranges" is not given. The mean values of the concentrations of NA and DA determined using the three modified HPLC methods agree well with the means and fall within the "acceptable ranges" of both control samples.

Table 2 Results of control A, NA and DA determinations in LYPHOCHEK Quantitative Urine Control (Lot. 620 51)

National Institute of Public Health Mean±SD (N=7)					
NA (mg/L)		A (mg/L)		DA (mg/L)	
40.4±2.1		10.4±0.8		62.9±0.9	
LYPHOCHEK Quantitative Urine Control (Lot. 620 51)					
Mean	Acceptable ranges	Mean	Acceptable ranges	Mean	Acceptable ranges
38	27.0–49	11.8	8.2–15.1		

Table 3 Results of control A, NA and DA determinations in ClinRep Urine Control (Lot. 719)

National Institute of Public Health Mean±SD (N=6)					
NA (mg/L)		A (mg/L)		DA (mg/L)	
HPLC method – 1. Modification					
60.9±1.5		17.9±1.1		165.9±3.5	
HPLC method – 2. Modification					
58.9±2.5		14.3±1.4		157.9±4.1	
HPLC method – 3. Modification					
61.8±3.1		8.8±1.7		160±5.5	
Fluorimetric method					
61.6±3.5		43.2±5.5		not done	
ClinRep Urine Control (Lot. 719)					
Mean	Acceptable ranges	Mean	Acceptable ranges	Mean	Acceptable ranges
57.2	45.7–68.7	20	16–24	151	121–181

The values of the NA concentrations determined using the fluorimetric method agree well with the means and fall within the "acceptable ranges" of both control samples as well. Only values of A concentrations acquired by the HPLC method recommended by ClinRep-Control (Merck/RECIPE) agree well with the means, and fall within the "acceptable ranges" of ClinRep-Control (Tables 2, 3).

Conclusions

Our results indicate the importance of using a unified protein standard for calibration and traceability. Having no RM to make a comparison against, we used bovine serum albumin (Fluka, cat. No. 05470) which was **traceable** to the NIST SRM 927c (Total Protein Standard) bovine serum albumin. Both preparations can be used for traceability.

Preservatives can influence the total protein determinations, so their effects must be eliminated.

The values of total protein and A concentrations in human urine are method-dependent. It is necessary to certify the mean values and their uncertainties individually for each method. These operationally defined certified values are thus valid only when the prescribed standard operation procedures are strictly followed.

Acknowledgements This work was supported by the Internal Grant Agency of the Ministry of Health of the Czech Republic (Grant No. 2760-3, NJ/4512-3).

References

1. Kristiansen N, Schimmel H, Müller M, Pauwels J (1999) Clin Chem Lab Med 37:112
2. Šperlingová I, Dabrowská L, Tichý M, Kučera J (1998) Fresenius J Anal Chem 360:756–760
3. Werner U (1975) Z Klin Chem Biochem 13:341–349
4. Merck/Recipe ClinRep Instruction Manual for determination of Catecholamines in Urine by HPLC
5. URINE-CHIMIE BIOTROL (1998) Lot 5338, Diagnostics MERCK-BIOTROL
6. BIO-RAD (1998) LYPHOCHEK Quantitative Urine Control, Lot No. 62051-Normal, Hercules, California. USA
7. ClinRep Urine Control (1998) Level I, Lot No. 719, Recipe Chemicals Wilhelm-Riehl

Accred Qual Assur (2001) 6:306–309

Michela Sega
Margherita Plassa
Elena Amico di Meane

Traceability in routine chemical measurements: an example of application in the determination of CO_2 at atmospheric concentration

Received: 16 March 2000
Accepted: 27 November 2000

Paper based on a talk given at the 3rd EURACHEM Workshop "Status of Traceability in Chemical Measurement", 6–8 September, 1999, Bratislava, Slovak Republic

M. Sega (✉) · M. Plassa
E. Amico di Meane
Istituto di Metrologia "Gustavo Colonnetti" del C.N.R., Strada delle Cacce 73, I-10135 Torino, Italy
e-mail: M.Sega@imgc.to.cnr.it

Abstract In routine chemical measurements traceability can be achieved by using analytical instruments calibrated against primary reference materials. In the present work the calibration of a CO_2 non-dispersive infrared (NDIR) analyzer with measuring range 0–2000 μmol/mol of CO_2 and a resolution of 5 μmol/mol is reported. A procedure with working reference gas mixtures (WRMs) has been adopted, which requires seven calibration points. Primary reference gas mixtures (PRMs) are used to validate WRMs in a narrower range around the average atmospheric CO_2 concentration value. In this range the relative uncertainty reached is of the order of some parts in 10^3 and the corrections are between 1 μmol/mol and 5 μmol/mol.

Keywords Traceability · Carbon dioxide determination · Calibration · Uncertainty

Introduction

Non-dispersive infrared analyzers are usually employed to determine carbon dioxide concentration at atmospheric levels, as they are stable, user friendly, and suited to continuous monitoring. At the Istituto di Metrologia "G. Colonnetti" (IMGC), as in other metrology laboratories, the determination of the CO_2 concentration in air is carried out for different purposes in mass, length, and environmental measurements. As NDIR spectroscopy is not a primary method of analytical measurement it does not provide direct traceability to the SI; it is hence necessary to refer the obtained results to traceable reference materials, namely PRMs of CO_2 in N_2 at appropriate concentrations.

The repeatability and short-term stability of the NDIR analyzers used at IMGC, i.e., Hartmann and Braun's URAS 10E, were tested in a previous work [1]. A procedure for calibrating NDIR analyzers with WRMs has been developed, which requires seven calibration points to establish the relationship between the analyzer output and the analyte concentration in the whole concentration range of interest. Since the traceability of measurements around the average CO_2 concentration is of particular concern, the measurements traceability is achieved by comparing PRMs with WRMs in a narrower range around the average value. In the present work the calibration of a CO_2 NDIR analyzer URAS 10E with measuring range from 0 μmol/mol to 2000 μmol/mol and a resolution of 5 μmol/mol is reported; the uncertainty budget is also evaluated [1–3].

Calibration

The calibration procedure suggested by the instrument manufacturer consists in periodically checking and adjusting the zero point and a span point by means of a zero reference gas (usually N_2) and of a span reference gas of suitable concentration. An uncertainty of 20 μmol/mol may be achieved in this way [1], i.e., a relative uncertainty of 5% at the average CO_2 concentration in IMGC laboratories (400–450 μmol/mol). According to our procedure linearity is not assumed a priori and the multipoint calibration covers the concentration range in IMGC laboratories. Five of the seven calibration points are equally spaced inside the measurement range, be-

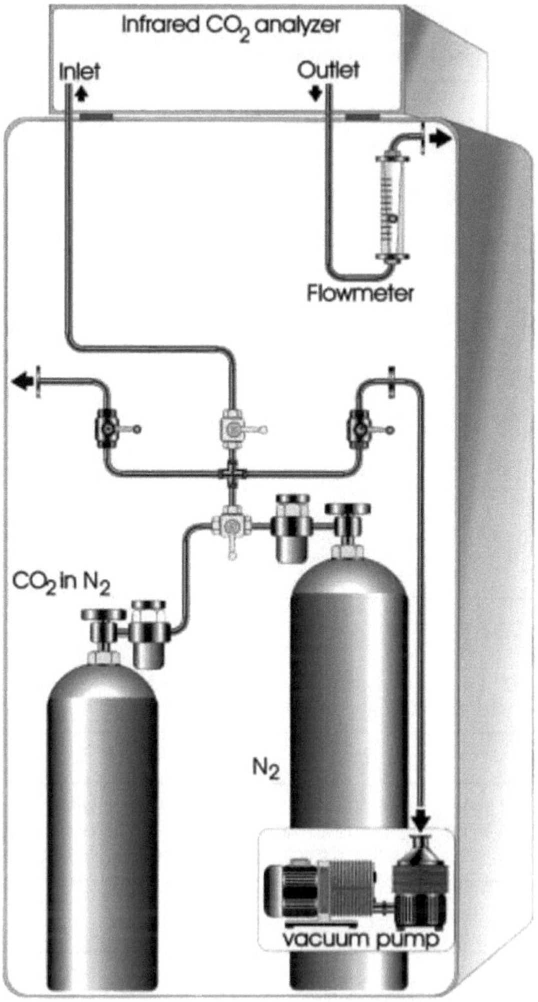

Fig. 1 Calibration facility for NDIR analyzers

tween 300 µmol/mol and 700 µmol/mol; one more is just below and another just above the range.

The WRMs are binary mixtures of CO_2 in nitrogen; for each one the concentration with its uncertainty are certified by the supplier. The comparison of WRMs with PRMs was carried out in the range 300–500 µmol/mol. PRMs used were supplied by a COFRAC accredited laboratory and were gravimetrically prepared mixtures of CO_2 in nitrogen. Each cylinder is accompanied by a certificate of analysis which reports the concentration and its uncertainty as provided by the analytical verification of the mixture.

According to [4] a preliminary evaluation was made to identify the sources of uncertainty responsible for the uncertainty budget. The effects of sampling at different heights, of water vapor interference, of instrument hysteresis had already been checked [1] and they give no significant contributions. On the other hand, the values of pressure/flow of incoming gas have been shown to be

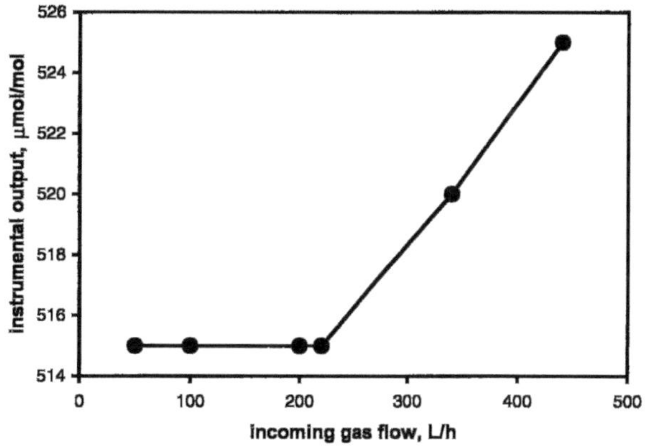

Fig. 2 Variation of the instrument output with the incoming gas flow

important influence quantities. This type of analyzer is equipped with a small diaphragm pump that drives the external air into the measurement cell. During the calibration the inlet pump is not used and the operator sets the overpressure/flow of the incoming gas: therefore the instrument to be calibrated must be characterized before, in order to carry out the calibration at the appropriate flowrate. For this reason the calibration facility, represented in Fig. 1, is equipped with a flowmeter, and a flow of 70 l/h was set as the operating condition. Figure 2 shows the behavior of the analyzer response as a function of the incoming gas flow.

The purpose of the calibration is to determine a polynomial correction, to be applied to instrument readings, and its uncertainty. The measurands are the polynomial coefficients α_i, arranged in a column vector $\boldsymbol{\alpha}$; for the uncertainty estimation, its variance – covariance matrix $\boldsymbol{\Psi}_{\alpha}$ is needed. For any CO_2 concentration, x_i, n instrumental readings y_{ij} are recorded (j=1–n, n=15).

The uncertainty estimation algorithm adopted [2, 3], based on the weighted least squares method, takes into consideration:

- The instrument repeatability, evaluated by replicating five times each measurement in three runs carried out in different days
- The instrument resolution
- The WRMs concentrations uncertainty, based on the certificate of the gas supplier and on the comparison with the PRMs
- The effect of covariances between WRMs concentrations (the hypothesis of full correlation, i.e., the worst condition, is assumed)
- The contribution of the model inadequacy, i.e., the capability of the assumed mathematical model to fit the instrument response.

Table 1 Results for WRMs validation. All values are expressed in μmol/mol. The uncertainties are expanded uncertainties for k=2

PRM concentration	PRM expanded uncertainty	Instrument reading	WRM concentration	WRM expanded uncertainty	Instrument reading
302.6	2.7	300	302	3	305
		300			305
		300			305
		300			305
		300			305
		300			305
		300			305
		300			305
		300			305
		300			305
503.4	4.5	500	504	5	505
		500			505
		500			505
		500			505
		500			505
		500			505
		500			505
		500			505
		500			505
		500			505

Results

The WRMs validation by comparison with PRMs was made using mixtures with the same CO_2 concentrations. The results are presented in Table 1. They show that the values certified by the WRMs supplier are in accordance with the PRMs values, and the set of seven WRMs could be used without any correction.

The relationship between the instrument output, y, and the input analyte concentration, x, being considered as non-linear, correction polynomials of the second order are fitted to the experimental curves:

$$x=y+d(y)=y+\alpha_0+\alpha_1 y+\alpha_2 y^2 \quad (1)$$

The model equation may be written in matrix form as follows:

$$\boldsymbol{d}=\boldsymbol{A}\,\boldsymbol{\alpha} \quad (2)$$

where

$$\boldsymbol{A}=\begin{pmatrix} 1 & y_1 & y_1^2 \\ 1 & y_2 & y_2^2 \\ \vdots & \vdots & \vdots \\ 1 & y_m & y_m^2 \end{pmatrix} \quad \boldsymbol{\alpha}=\begin{pmatrix} \alpha_0 \\ \alpha_1 \\ \alpha_2 \end{pmatrix}.$$

From the definition of $\boldsymbol{d}$ the variance-covariance matrix $\boldsymbol{\Psi}_d$ is evaluated, taking into account the variance-covariance matrices of the input data x_i and of the instrument readings y_i [1].

It is hence possible to obtain an estimate of $\boldsymbol{\alpha}$ by applying the weighted least squares method:

$$\boldsymbol{\alpha}=(\boldsymbol{A}^{T}\boldsymbol{\Psi}_d^{-1}\boldsymbol{A})^{-1}\boldsymbol{A}^{T}\boldsymbol{\Psi}_d^{-1}\boldsymbol{d} \quad (3)$$

Table 2 The values of coefficients α_i and of their uncertainties $u_c(\alpha_i)$, with the variance–covariance matrix $\boldsymbol{\Psi}_\alpha$

		$u_c(\alpha_i)$	$\boldsymbol{\Psi}_\alpha$		
α_0	-7.15×10^{-6}	4.8×10^{-6}	2.34×10^{-11}	-9.88×10^{-8}	9.16×10^{-5}
α_1	-1.33×10^{-2}	2.2×10^{-2}	-9.88×10^{-8}	4.70×10^{-4}	-4.29×10^{-1}
α_2	70.92	20.7	9.16×10^{-5}	-4.29×10^{-1}	428.43

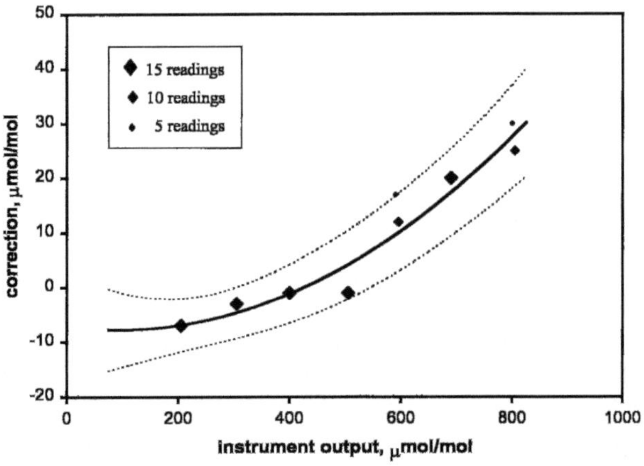

Fig. 3 Correction curve for the CO_2 analyzer with its expanded uncertainty band (k=2)

The corresponding estimate of the variance-covariance matrix is

$$\boldsymbol{\Psi}_\alpha=(\boldsymbol{A}^{T}\boldsymbol{\Psi}_d^{-1}\boldsymbol{A})^{-1} \quad (4)$$

from which the combined standard uncertainty $u_c(\alpha_i)$ of each coefficient α_i is computed. In Table 2 the values of

α_i, $u_c(\alpha_i)$, $\boldsymbol{\psi_\alpha}$ are reported. In Fig. 3 the correction curve and its uncertainty limits for the CO_2 analyzer are shown.

Conclusions

The relative uncertainty reached with a calibrated NDIR CO_2 analyzer URAS 10E (Hartmann and Braun), having a resolution of 5 μmol/mol, is of the order of some parts in 10^3 at the average CO_2 concentration in the atmosphere. The corrections in the range 300–500 μmol/mol are between 1 μmol/mol and 5 μmol/mol and they rise to >30 μmol/mol at higher concentrations within the instrument range.

In the present example the instrument resolution and the uncertainty of the gas mixtures compositions give the highest contributions to uncertainty, which is anyhow one order of magnitude lower than with a two-point calibration.

The long-term reproducibility of this type of analyzer is under evaluation; this part of the experiment will make it possible to complete the IMGC procedure by determining a suitable period for re-calibration or by suggesting an expansion of the uncertainty linked to the time elapsed since instrument calibration.

References

1. Plassa M, Mosca M, Sega M (1998) In: Chung MS (ed) Proceedings of the 16th International Conference IMEKO TC3/APMF '98, pp 183–191
2. Mosca M (1997) In: Halttunen J (ed) Proceedings IMEKO XIV World Congress, pp 157–163
3. Nielsen L (1998) Metrologia 35:115–118
4. EURACHEM (1995) Quantifying uncertainty in analytical measurement. London

Accred Qual Assur (2002) 7:529–537
DOI 10.1007/s00769-002-0522-7

D. Glavič-Cindro
M. Korun
B. Vodenik

Traceability of measurement results of the effective acquisition time in gamma-ray spectrometry implemented by the pulser method

Received: 2 March 2002
Accepted: 26 July 2002

D. Glavič-Cindro · M. Korun (✉)
B. Vodenik
"Jožef Stefan" Institute, Jamova 39, 1000 Ljubljana, Slovenia
e-mail: Matjaz.Korun@ijs.si
Tel.: +386-1-477 3900
Fax: +386-1-423 2120

Abstract Measurement of the effective acquisition time of a spectrum by the pulser method is described. The measurement results were verified up to count rates of 12000 s^{-1} at various settings of the pulse processing electronics and spectral shapes. Systematic effects of up to 2% were observed. The clocks in the spectrometers were calibrated by counting pulses generated by the DCF 77 signal with the frequency of 1 Hz. In 1-day measurements at low count rates a relative uncertainty of about 0.5 10^{-3} of the effective acquisition time was obtained. At counting losses up to 30% the relative uncertainty remained below 1.5 10^{-2}.

Keywords Gamma-ray spectrometry · Pulser method · Uncertainty · Traceability

Introduction

Test laboratories should maintain traceability of their test results in order to claim competence. Traceability of the results can only be demonstrated if all quantities entered into the calculation of the end result are traceable or verified. For measurement results traceability is achieved, as required by the ISO 17025 Standard [1], by calibration of the measuring equipment. If results of model calculations are entered into the calculation of the end result they must be verified as well. Here, verification replaces traceability because results of model calculation are, according to the definition of traceability [2], not traceable. Analogously to the definition of traceability, verification is the process of relating the results of a calculation with a stated reference. Correction factors and their uncertainties, which are extracted by comparing the results of model calculations with the stated reference, take into account the difference between the results and the reference, and establish the evidence for the accuracy of the calculations.

As test laboratories, gamma-ray spectrometry laboratories are usually engaged in determination of the activities of gamma-ray emitters in samples. In order to determine the activity present in a sample, beside the counting efficiency, the emission probability and the peak area, the effective duration of the counting time must be known. It is a well-established fact that the use of the time measured by the live-time clock in the ADC in activity calculations results in systematic effects [3] arising from the neglect of pile-up effects.

To cope with the pressures exerted on test laboratories in a competitive market, laboratories are forced, in order to cut the cost of labour, to introduce fully automatic measurement and spectral analysis procedures. In such a procedure the pulser method was implemented by the Gamma-ray Spectrometry Group at the J. Stefan Institute to fulfill the requirement for traceability of the effective acquisition time measurement results. The automatic procedure computes the effective acquisition time from traceable quantities by a verified algorithm. It calculates also the uncertainty of the effective counting time, arising from whatever source. The algorithms used for the calculation are reliable enough to yield reasonable results without the intervention of the operator over a broad range of counting conditions. In this contribution the calculation of the pulser peak area and its uncertainty are explained, the results of the verification measurements are presented and the calibration of the spectrometer clocks is described.

The pulser method

In gamma-ray spectrometry the activities of the gamma-ray emitters present in a sample are calculated from the areas of peaks in the spectrum. Therefore, the measurement of the effective acquisition time must take into account the possible influence of processing of pulses in the spectrometer on the peak areas. The effective acquisition time is given by the length of the time interval between the acquire start and acquire stop signals received by the spectrometer, minus the time periods when the spectrometer was not able to record pulses properly from the detector as counts in the spectrum due to any interference from other pulses. The resulting losses of counts from peaks are called counting losses and originate from the dead time of the ADC and in the pile-up effect. They can be measured by the pulser method, described by Debertin and Helmer [3].

By applying the pulser method the probability that a pulse generated by the pulser is registered as a count in the pulser peak is determined. Therefore, the number of pulses generated during the spectrum acquisition and the number of pulses registered in the spectrum as counts in the pulser peak must be known. The frequency of pulsers used in gamma-ray spectrometers is adjustable and to avoid any inadvertent change of frequency, the number of generated pulses is measured by counting. Usually, to diminish the influence of pulser pulses on the shapes of other peaks in the spectrum, the pulsers are operated at a frequency of 10 Hz. When spectrum acquisition is started, a computer-controlled counter for counting the generated pulses [4] is started as well. However, it should be mentioned that in our case both the ADC in the spectrometer and the counter are started by the controlling computer through a computer network. Since both starts are executed by consecutive computer commands and because of possible delays on the network there is no guarantee that acquisition and counting start at the same moment. For a similar reason the two stops may not be executed at the same time. Therefore, in order to take into account the possible difference between the acquisition and counting times, the counting time is recorded together with the number of generated counts. From the counting time, the number of generated counts is recalculated to the number of counts generated during the spectrum acquisition time. This number, although not used for calculation of the effective duration of the acquisition time, is needed for estimating its uncertainty.

Evaluation of the pulser peak area

The pulser peak is situated at the high-energy end of the spectrum in order to reduce the uncertainty of its area since in this spectrum region the background is low and varies slowly with energy. The shape of the pulser peak exhibits a central part where the majority of counts are registered and the tails characterized by the shape and strength which vary with the imperfection of the baseline restoration and with the count rate due to the pile-up effect. Usually, the shapes of the tails do not resemble the shape of the tails of a Gaussian function. Other peaks in the spectrum are wider and reside on a higher background and therefore their tails cannot be observed in the spectrum as readily as the tails of the pulser peak. As a consequence, their shape is less susceptible to effects influencing the tails. The evaluation of the pulser peak area with the same algorithm as the areas of other peaks results in systematic effects [5] and in poor reproducibility, since the width of the pulser peak region is sensitive to the details of it's tail shapes. This may result in cutting off the tails of the pulser peak or in erroneous determination of the background counts. Also, the uncertainty of the pulser peak area is evaluated assuming random registration times of pulses as opposed to the registration times of pulser pulses which are equidistant in time.

The traceability of time measurement results can be established only if the pulser peak area can be compared with a stated uncertainty to the duration of a traceable time interval. Since the uncertainty of the pulser peak area calculated with general-purpose, peak analysing programs is inadequate, in principle, and does not account for the variability of the pulser peak area induced by the variability of its tails, the pulse peak area is calculated by a separate program.

The program calculates the number of pulser pulses registered in the central region of the pulser peak by summing the counts registered there

$$n_{PC} = n_L + \sum_{i=i_L}^{i_H} N_i + n_H - n_B.$$

Here the channel numbers i_L and i_H represent the first and the last channel number of the central region, N_i denotes the contents of the i-th channel, n_L and n_H the correction terms, arising from the boundaries of the central region possibly not coinciding with integral channel numbers and n_B the number of counts of the continuous background within the region. The number of background counts subtracted from the peak area is determined from the continuous background at the low-energy side of the pulser peak at the distance where the contribution of the pulser peak to the spectrum is negligible.

The boundaries defining the central part of the pulser peak are set to the positions where the ratio between the contents of two successive channels, N_{i+1}/N_i, reaches its maximum on the low-energy side and its minimum on the high-energy side of the peak. The number of counts in the pulser peak is calculated by summing up the counts in its central region and the counts registered in its tails n_{TL} and n_{TH}:

$$n_P = n_{PC} + n_{TL} + n_{TH}.$$

The numbers of counts in the pulser peak low- and high-energy tails are calculated by assuming that at the boundaries of the central region the height of the tails is equal to the spectrum height, that the tails have exponential shape and that the decay constant describing the tails shape is given by the extreme values of the ratio. These assumptions imply that the number of counts in the low-energy tail is given by

$$n_{TL} = N_L / R_L + N_L / R_L^2 + \ldots = \frac{N_L}{R_L - 1}$$

and the high-energy tail by

$$n_{TH} = N_H R_H + N_H R_H^2 + \ldots = \frac{N_H}{R_H^{-1} - 1}.$$

Here N_L, N_H, R_L and R_H denote the height of the pulser peak and the extreme values of the ratios of the number of counts in successive channels at the boundaries of the central region, respectively.

The pulser pulses are generated at a constant frequency, so that the variance of the pulser peak area that results from the statistical nature of spectrum acquisition is approximated at counting losses well below 50% by

$$\Delta_S^2 n_P = n'_C - n_P + n_B,$$

where n'_C denotes the number of pulser pulses generated during the acquisition of the spectrum

$$n'_C = n_C + \nu (T_A - T_C).$$

Here n_C, ν, T_A and T_C denote the number of the pulses counted, the pulser frequency and the acquisition and counting time, respectively. The variances generated by the assumptions made in the analysis on the pulser peak shape include contributions from the uncertainties of the boundary positions $(\Delta_L N_L)^2$ and $(\Delta_H N_H)^2$, where Δ_L and Δ_H denote the uncertainties of the lower- and upper-boundary position of the central region, respectively, and variances due to the uncertainties of the number of counts in the tails $\Delta^2 n_{TL}$ and $\Delta^2 n_{TH}$. Assuming that these contributions are not correlated, the pulser peak area uncertainty is obtained as

$$\Delta^2 n_P = \Delta_S^2 n_P + (\Delta_L N_L)^2 + (\Delta_H N_H)^2 + \Delta^2 n_{TL} + \Delta^2 n_{TH}. \quad (1)$$

It should be mentioned that in general the contributions to the cumulative uncertainty originating in the number of counts in the tails are smaller than the contributions of the statistical uncertainty and the uncertainties of the boundary positions and that the latter become more important at high count rates.

The counting conditions, such as the count rate, spectrum shape and settings of the pulse-processing electronics influence the shape of the pulser peak. In Fig. 1 various shapes of the pulser peak measured at low count rates are presented. The spectra A and B were measured with the channel width of 0.1 keV on a low-volume germanium detector. They exhibit pulser peak shapes which resemble the shape of the Gaussian function. The spectra C, D and E were measured with a channel width of 0.33 keV on germanium detectors with efficiencies between 25 and 50% relative to that of a 3 × 3 inch NaI. In spectrum C the pulser shape exhibits nearly symmetric tails and represents a shape most frequently encountered in low-level measurements on semiconductor detectors. The spectrum D shows a pulse peak with an increased high-energy tailing. The presence of the tail indicates that the noise of the detector has increased and signals a degradation of the detector performance. In spectrum E the performance of the detector has degraded to the state where the resolution of the pulser peak has increased.

In Fig. 2 the shape of the pulser peaks measured at elevated rates are presented. The spectra are arranged in the sequence of increased fraction of pulser counts registered in the tails. It can be observed that, in general, at high count rates and long shaping times the tails of the peaks are more pronounced. As a consequence the uncertainty of the peak area becomes larger and the oppor-

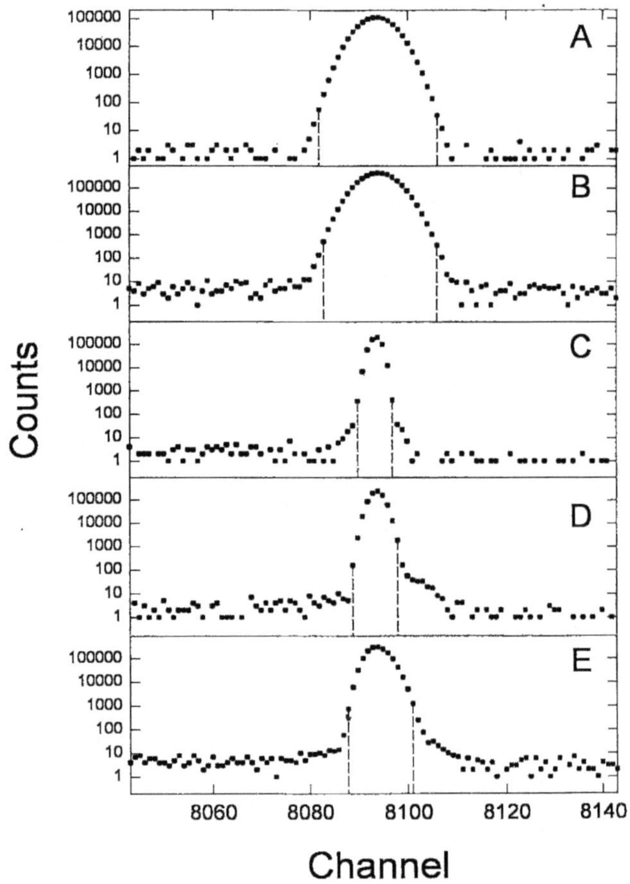

Fig. 1 Shape of the pulser peaks acquired at low count rates. The boundaries defining the central part of the pulser peak, rounded to the nearest channel number, are marked

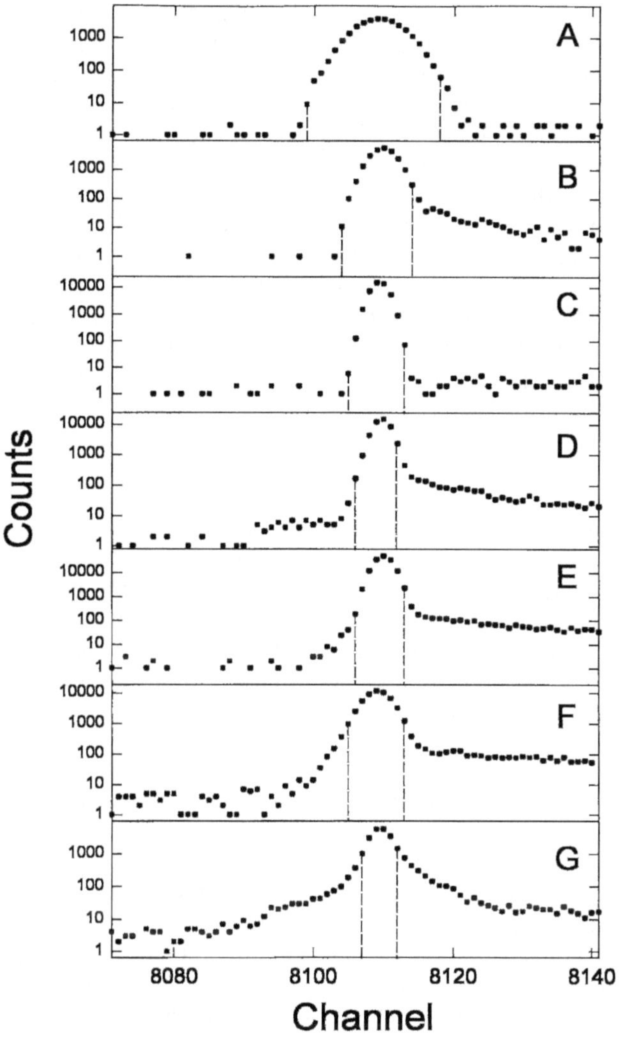

Fig. 2 Shape of the pulser peaks acquired at elevated count rates. The boundaries defining the central part of the pulser peak, rounded to the nearest channel number, are marked

tunity for systematic effects, originating from the assuming exponential shape of the tails, becomes greater, too.

The relative importance of the statistical uncertainty and the uncertainties of the boundary positions can be assessed from Table 1, where the quantities contributing to the uncertainty of the area of the central part of the peak are given for the shapes presented in Fig. 2. It can be observed that strong tails introduce small uncertainties in the peak boundaries because of the small relative statistical uncertainties of the channel contents near the boundaries. Nevertheless, they result in larger peak area uncertainties since they are multiplied by the large spectrum height at the boundaries. It can also be observed that the contributions of the uncertainties of the boundary positions are comparable to the contribution of the statistical uncertainty of the number of counts in the pulser peak.

The largest systematic effects in the pulser peak method occur because the peaks belonging to the registration of photons have a different shape and reside on a different background than the pulser peak. Therefore, the areas calculated may be affected by changes in the peak shape in different ways. The model of the pulser peak stabilizes the influence of the pulser peak tails on its area, since the tail properties are deduced from channel contents near the central part of the peak. The probability of systematic effects is smaller because the details of the tail shape cannot influence the pulser peak area, similarly to the tails of other peaks, where the details are smeared out by the counting statistics and hidden by the larger resolution. These effects are taken into account empirically as a result of the verification, resulting in correction factors and their uncertainties.

Experimental

To assure the traceability of the results of effective acquisition time measurements, two independent steps have to be performed: the clocks defining the time base of the measurements have to be calibrated and it must be demonstrated that the influence of counting conditions affecting the measurement results is compensated for properly by applying correction factors or by covering the discrepancies by uncertainties. The calibration of the clocks establishes the traceability of the results of measurements performed in ideal circumstances where the results of the effective time measurement are not affected by the conditions of counting, i.e. at low count rates.

Table 1 Contributions to the uncertainty of the area of the central part of the pulser peak at elevated count rates

Sp. N.	Shaping time [μs]	Count rate [s^{-1}]	Δ_L	N_L	$(\Delta_L N_L)^2$	Δ_R	N_R	$(\Delta_R N_R)^2$	n'_C-n_P	n_B	n_P	$\Delta n_P/n_P$ [%]
A	3	360	1.8	17	936	0.40	51	416	858	21	29584	0.16
B	1.5	2270	0.34	36	150	0.35	514	32364	1268	9	23315	0.79
C	8	1030	0.38	51	375	0.40	125	2500	2482	8	44529	0.16
D	3	4260	0.09	293	696	0.14	1881	69348	8669	6	42587	0.66
E	3	2240	0.18	726	17077	0.20	2476	245223	9153	7	147131	0.35
F	8	2920	0.10	697	4858	0.12	1351	26283	13294	20	50532	0.42
G	10	2830	0.06	834	2504	0.09	1299	12235	7828	13	19839	0.78

The verification of the effective acquisition time measurement results, i.e. the measurements of correction factors for the measurements of the effective acquisition time, which depend on the parameters characterizing the counting conditions, extends the traceability to the range of counting conditions where the correction factors are measured.

Calibration of clocks

The effective acquisition time for the counts registered in peaks as measured by the pulser method is calculated by

$$T_L = \frac{n_P}{n_C} T_C - n_P \cdot \tau_P,$$

where T_L, n_P, n_C and τ_P denote the effective acquisition time of the spectrum, the number of pulses registered in the pulser peak, the number of pulses, generated during T_C and the dead time of the spectrometer due to the registration of one pulser pulse, respectively. As mentioned in [6], measuring the effective duration of the acquisition time by a constant-rate pulser results in counting losses in the absence of the pulser pulses since the pulser pulses cannot interact with one another. In order to take into account at least partially the counting losses due to the presence of the pulser pulses, the effective duration of the counting time is corrected for the dead time due to pulser pulses. With a Wilkinson type ADC of 100 MHz frequency this time amounts approximately to100 µs and reduces the effective duration by 0.1% at a pulser frequency of 10 Hz. Although the effective acquisition time does not depend on the time provided by the spectrometer clock, the latter must nevertheless be calibrated since the acquisition time is used in the calculation of the uncertainty of the number of counts registered in the pulser peak.

To take into account the delays on the computer network the spectrum acquisition and the counting of generated pulses are controlled by two independent clocks. The calibrations of these two clocks establish the traceability chain to a national standard for the measurements of time intervals. The DCF 77 signal emitted from Mainflingen near Frankfurt/Main, which defines the legal time in Germany, is controlled by the PTB time standard based on atomic clocks. The receiver, a commercially available device [7], responds to the DCF 77 signal with a series of pulses 1 V high and 0.1 s wide with the frequency of 1 Hz. If the receiver loses the signal it continues to emit pulses and synchronizes with the atomic clock again when the connection is re-established. The pulses are counted by the ADC and the counter. Since all the pulses have nearly equal shapes they are stored in a narrow region in the spectrum. Registration of the pulses resulting from noise leads to registration outside this region. The elimination of noise at the input of the counter is achieved by using a single-channel analyser with the window set around 1 V. It should be noted that some ADCs cannot properly process pulses with a duration of 0.1 s. For the calibration of spectrometers with such ADCs the pulses are shortened by feeding them to a single-channel analyser or a triggerable pulse generator. From the number of pulses acquired in the spectrum and counted by the counter, which determines the number of seconds elapsed between starting and stopping the acquisition and counting, the calibration factors for the clocks in the spectrometer and counter are obtained. Assuming no correlation in time between the start and stop pulses and the pulses from the receiver, the uncertainties of the calibration factors are given by

$$\Delta c = \frac{2}{n\sqrt{6}}, \tag{2}$$

where n denotes the number of counts acquired in the spectrum or counted by the counter. It follows from Eqs. (1) and (2) that for a 1-day calibration of the clocks the relative uncertainty of the calibration factor is almost 3 orders of magnitude smaller than the relative uncertainty of the pulser peak area acquired at counting losses of 5.0 10^{-4} and a pulser frequency of 10 Hz.

Verification of effective acquisition time measurement

When using the pulser method the assumption is made that the counting losses affect the pulser peak area to the same degree as the peaks resulting from registration of photons. Since the pulser peak area is calculated using a different algorithm from the one used for the areas of other peaks in the spectrum, there is no a priori guarantee that the influences on the peak shape of the variations in energy, count rate, spectral shape and settings of the electronics affect the pulser peak area in the same manner as the peak areas of other peaks.

In order to see what effect the measurement of the effective acquisition time using the described method of calculation of the pulser peak area has on a constant count rate under varying counting conditions, measurements of the peak count rates from a ^{137}Cs source in a fixed counting geometry were performed. Simultaneously with this source another source with a different gamma-ray emitter was counted. The position of the second source was varied in order to vary the total count rate in the spectrum. To observe the influence of the spectral shape on the count rates from the ^{137}Cs source the measurements were performed twice, with the second source containing ^{241}Am or ^{60}Co. From these measurements correction factors describing the influence of the counting rate and spectral shape on the result of the effective acquisition time measurement were obtained. The correction factor is the count rate in the gamma-ray peak belonging to ^{137}Cs measured in specified counting conditions normalized to the count rate in the same peak measured in the absence of the second source. It should be observed that the uncertainty of the correction factor includes not only the uncertainty of the acquisition time but the uncertainty of the gamma-ray peak area as well.

The measurements were performed on three detectors. Two detectors, a Ge(Li) and a p-type, were connected to PGT 386 amplifiers. One of the amplifiers operated with a shaping time of 4 μs and the base-line restorer threshold set to VAR. With the second detector connected to the amplifier operating at 3 μs two sets of measurements were made: one with the base-line restorer threshold set to AUTO and the other with the base-line restorer threshold set to VAR. The thresholds of the base-line restorers set to VAR were adjusted manually to minimize the tails of the pulser peak at low count rates. The third detector was a low-energy detector connected to an ORTEC 573 amplifier operating with a shaping time of 6 μs and in the Gaussian mode. This amplifier was equipped with an automatic base-line restorer. The third detector was used in order to see how different amplifiers influence the performance of the pulser peak method.

The count rates were calculated as the number of counts registered in the full-energy peak reported by the peak analyzing procedure, divided by the effective duration of the acquisition time, determined from the area of the pulser peak. The number of counts in the x- and gamma-ray peaks and their uncertainties were determined by the Standard Peak Search and Hypermet programs, purchased from Canberra, as described in [8]. Since these programs only report statistical uncertainties of the peak areas, the systematic ones were only taken into account empirically by expanding the uncertainty according to the difference in the peak areas reported by the two programs. The uncertainties of the count rates were calculated by combining the uncertainties of the effective acquisition time and the peak areas. The total count rate in the spectrum was calculated by dividing the number of counts registered in the spectrum by the effective acquisition time calculated by the pulser method. To maintain clarity, the uncertainties of the total count rate generated by the uncertainties of the acquisition time are not indicated in the figures.

The dependence of the correction factor on the total count rate in the spectrum in the measurements with the p-type detector is presented in Figs. 3 and 4. Figure 3 shows a comparison of the dependences measured with the ^{241}Am source for measurements with the base-line restorer threshold set to AUTO and VAR. Figure 4 presents the dependencies of the correction factors when ^{60}Co was used as the second source. In Fig. 5 the dependence of the correction factor on the total count rate measured with the Ge(Li) detector is presented. The dependencies of the correction factors on the total count rate measured on the spectrometer with the Ortec 573 amplifier are presented in Figs. 6 and 7, with Fig. 6 pre-

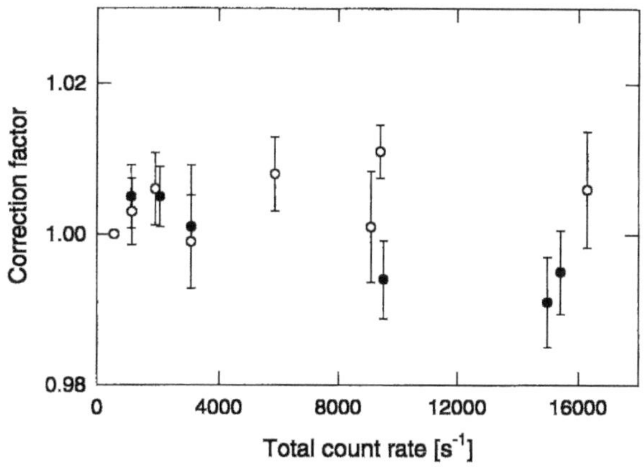

Fig. 3 Correction factor for the 662 keV peak as a function of the total count rate measured on a p-type detector with an ^{241}Am source with the base-line restorer threshold set to AUTO (full circles) and with the base-line restorer threshold set to VAR (open circles)

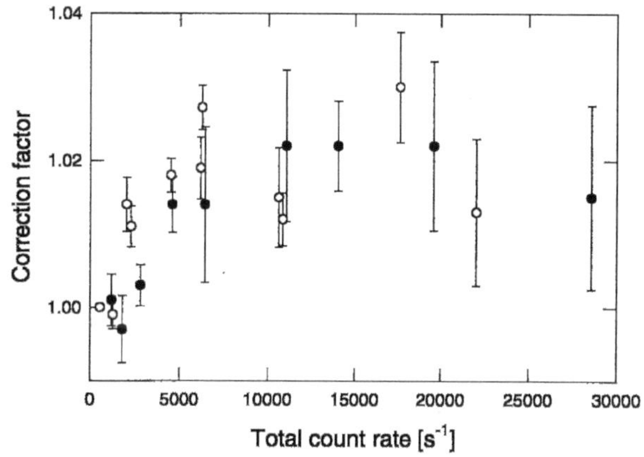

Fig. 4 Correction factor for the 662 keV peak as a function of the total count rate measured on a p-type detector with a ^{60}Co source with the base-line restorer threshold set to AUTO (full circles) and with the base-line restorer threshold set to VAR (open circles)

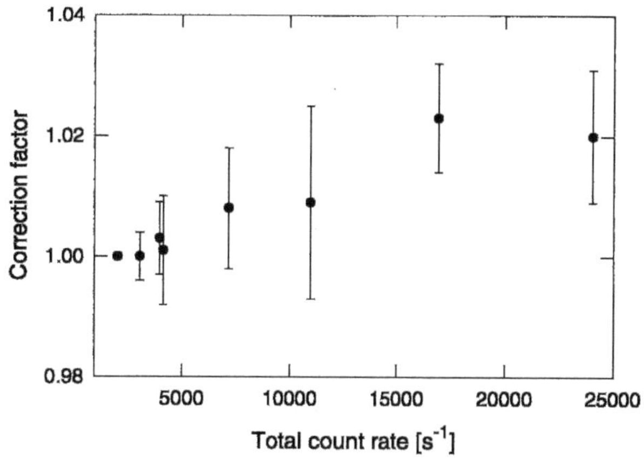

Fig. 5 Correction factor for the 662 keV peak as a function of the total count rate measured on a Ge(Li) detector with an ^{241}Am source with the base-line restorer threshold set to VAR

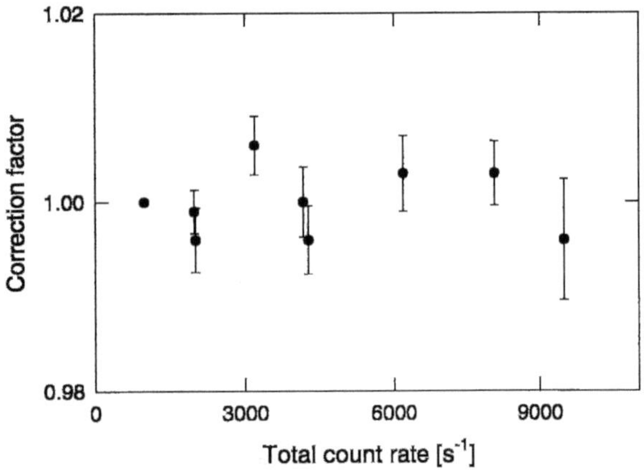

Fig. 6 Correction factor for the 662 keV peak as a function of the total count rate measured on a low-energy p-type detector with an ^{241}Am source

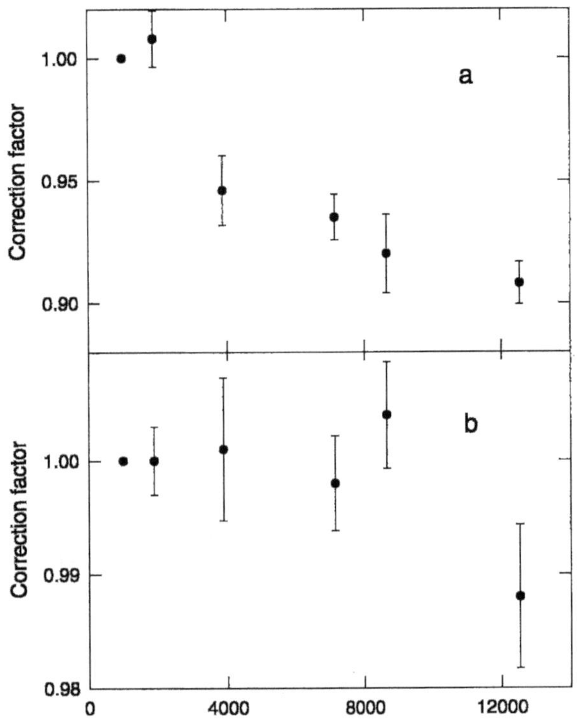

Fig. 7 Correction factors for the x-ray peak at 32 keV (**a**) and the correction factors for the 662 keV peak (**b**) as functions of the total count rate measured on a low-energy p-type detector with a ^{60}Co source

senting the dependence on the total count rate measured with the ^{241}Am source and Fig. 7 the dependence on the total count rate measured with the ^{60}Co source. Figures 6 and 7b present the dependencies of the correction factors for the gamma-ray line and Fig. 7a the correction factors for the x-ray line at 32 keV.

Discussion

The pulser peak method gives better estimates of the effective counting time than the methods implemented in the ADC converter which estimate the time when the gate of the ADC is closed electronically, since in the former the influence of the pile-up effect on the peak areas is taken into account. However, an automatic analysing procedure evaluating the pulser peak area introduces systematic effects which are caused by the distortion of the shape of the pulser peak. These systematic effects are reflected in the dependence of the count rate from a source located at a fixed position on the total count rate in the spectrum. The effects arise only partially from the difference between the calculation of the pulser peak area and the areas of other peaks in the spectrum. The other sources of systematic effects originate in the difference between the pulser peak shape and the shapes of other peaks in the spectrum and in the relatively low background near the pulser peak.

The presented measurements showed that, at the accuracy achieved, the influence of the setting of the restorer threshold on the PGT 386 amplifier on the correction factor could hardly be observed (Figs. 3 and 4). The spectral shape had a larger influence. In measurements with an ^{241}Am source the dependence on the total count rate was much weaker than in the measurement with a ^{60}Co source. In the latter case, the deviation of the correction factor from unity reached 2% at count rates above 5000 s^{-1}. A similar conclusion can be drawn for the second spectrometer with a PGT 386 amplifier. Here, a 1% deviation was reached at 10000 s^{-1} and a 2% deviation at 15000 s^{-1} (Fig. 5). A different dependence was measured with an Ortec 573 amplifier. Here, no deviation in the count rate in the cesium gamma-ray peak was observed either in measurements with the ^{241}Am source or with the ^{60}Co source up to a count rate of 10000 s^{-1} (Figs. 6 and 7b).

The deviation of the correction factor from unity measured with spectrometers with PGT 386 amplifiers is comparable with the deviations measured in [5]. There the authors reported a deviation of the peak count rate of approximately 2% at 1400 keV and 0.5% at 122 keV at a total count rate of 9000 s^{-1}. The measurement with the Ortec 573 amplifier yielded superior results. At the accuracy achieved, the deviation was not measurable up to 10000 s^{-1} with this amplifier. The difference in the performance of the measurements with different amplifiers originates, most probably, in the fact that the Ortec amplifier is of newer design.

The deviations of the correction factor from unity as a function of the total count rate requires the introduction of a rate- and spectrum-shape- dependent correction factor to be applied to the effective acquisition time measured on the spectrometers with PGT 386 amplifiers. Since the dependence of the factor on the spectrum

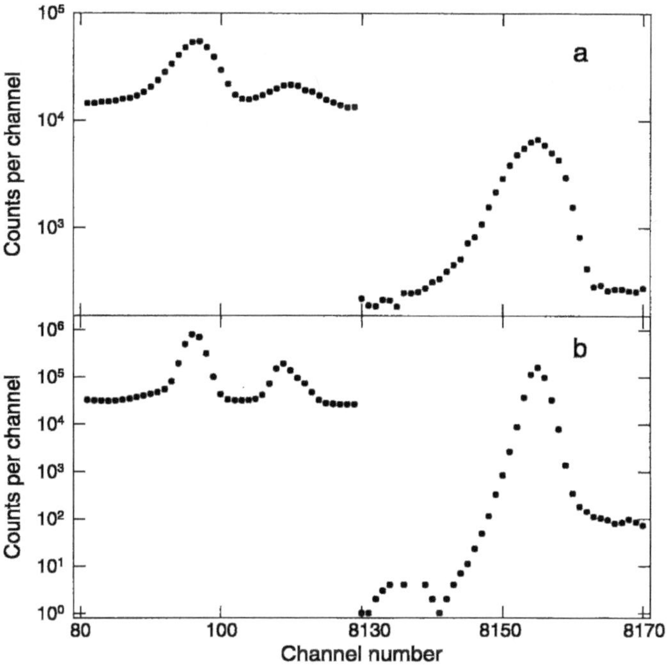

Fig. 8 Shapes of the x-ray peaks and the shape of the pulser peak at count rates of 840 s^{-1} (**b**) and 12500 s^{-1} (**a**) measured on a low-energy p-type detector with a ^{60}Co source

shape is too poorly known, the uncertainty should be increased to cover the variability. To maintain traceability in a broad variety of counting conditions a 1% uncertainty is added quadratically to the uncertainty of the effective counting time at total count rates larger than 5000 s^{-1} by the automatic spectrum analysis procedure. This uncertainty accommodates the systematic effects introduced by the pulser method in the measurement of the effective acquisition time.

The interval of count rates presented in the figures gives the approximate range where the described calculation of the pulser peak area gives reasonable results. It can be observed in the figures that at high total count rates the correction factor decreases. This is due to a systematic overestimation of the pulser peak area which originates in the overestimation of the number of counts in its high-energy tail. Namely, piled pulser pulses, which are registered above the pulser peak, due to the smaller resolution of the pulser peak, distort the high-energy slope of the pulser peak more strongly than the high-energy slope of the gamma-ray peak, resulting in overestimation of the counting time. It should be noted that by reducing the shaping time of the amplifier the useful range of the pulser peak method implemented by the described calculation of the pulser peak area can be expanded.

It should also be pointed out that the measured deviations of the correction factors from unity describe systematic effects of the count rates only in well-separated singlet peaks. If the peaks are not well separated larger systematic effects may originate in their area calculation. An example is given in Fig. 7a where the deviation for the correction factor for the x-ray peak at 32 keV is presented. The stronger deviations measured are due to deficiencies of the peak analysing programs in resolving overlapping peaks. In Fig. 8 the shapes of the ^{137}Cs x-ray peaks and the shape of the pulser peak at total count rates of 800 s^{-1} and 12000 s^{-1} in the measurements with the ^{60}Co source are shown. It can be observed that with an increasing total count rate the resolution in the spectrum worsens and the low-energy tail increases. As a consequence, the x-ray peaks overlap and the peak analysing program erroneously increases the continuum background at the expense of peak areas. The equivalent measurements of the count rate with the ^{241}Am source could not be made because of the interference of the 32 keV x-ray line with the gamma-ray line at 32.2 keV from ^{241}Am.

Conclusion

A robust implementation of the pulser method used for measurement of the effective acquisition time in gamma-ray spectrometry is described. The area of the pulser peak in the spectrum is calculated assuming exponential tails. With this assumption the search for the pulser peak start and end channels is substituted by the calculation of the boundaries of the pulser peak central region. In addition to a complete uncertainty budget for the pulser peak area, greater robustness of the calculation is also achieved as compared to calculation with programs designed for general peak analysis. Three sources of uncertainty are considered in the effective acquisition time calculation: the uncertainty of the pulser peak area, the uncertainty due to the systematic errors arising from different counting conditions and the uncertainty of the calibration of clocks.

At low count rates the relative uncertainty of the pulser peak area in 1-day measurements remains below 10^{-3}. At counting losses of 30% and at acquisition times of 1 h the relative uncertainty of the counting time does not exceed 1%. To cover the possible influence of systematic effects on the effective duration of the effective acquisition time originating in different counting conditions, a 1% uncertainty is included in the uncertainty budget at count rates exceeding 5000 s^{-1}.

Traceability of measurement results of the effective acquisition time was achieved by implementing the calibration of clocks with the DCF 77 radio signal controlled by the German time standard. By verifying the measurement results the traceability of the calibration is extended to the range of counting conditions where the deviations between the expected and measured count rates are covered by the introduction of a 1% uncertainty.

References

1. ISO/IEC 17025 (1999) General requirements for the competence of testing and calibration laboratories. International Standards Organization (ISO), Geneva
2. ISO (1993) International vocabulary of basic and general terms in metrology. International Standards Organization (ISO), Geneva
3. Debertin K, Helmer RG (1988), Gamma- and x-ray spectrometry with semiconductor detectors. North Holland, Amsterdam
4. Korun M, Lakovič G (1993) Computer controlled counter CCC-01CL, manual and documentation, JSI Report DP-6875, Ljubljana
5. Debertin K, Schötzig U (1977) Nucl Instr Meth 140: 337–340
6. Tenney FH (1984) Nucl. Instr Meth. 219: 165–172
7. Conrad Electronic (1994) DCF-77-Modul-Set, Conrad Electronic GmbH, Hirschenau, Germany
8. Glavič-Cindro D, Korun M, Vodenik, B (1999) Appl Radiat Isot 53: 237–241

Accred Qual Assur (2003) 8:461–466
DOI 10.1007/s00769-003-0684-y

Yoshito Mitani
Alejandro Perez-Castorena

Practical ways in establishing traceability in chemical and other measurements in Mexico

Received: 29 January 2003
Accepted: 2 August 2003
Published online: 6 September 2003

Y. Mitani (✉) · A. Perez-Castorena
Centro Nacional de Metrologia, Mexico,
Col. Centro, Apdo. Postal 1-100, QRO.
76000 Queretaro, Mexico
e-mail: ymitani@cenam.mx
Tel.: +52 442-2110560
Fax: +52 442-2110569

Abstract Practical ways to establish traceability in chemical measurements are examined to understand such diversified field measurements, which cover all principles of chemical and other measurements, and are considered to be applicable to measurements of materials properties in general. A description is given of several initiatives in Mexico to establish a comparable measurement and calibration capability and dissemination scheme. Additional efforts for establishing traceability in field measurements are described in order to achieve traceable measurements harmonized with other countries, with particular emphasis on the accreditation of analytical laboratories based on their technical competence.

Introduction

Most of the primary measurement standards are the realization of the SI units, and are under custody of each country's National Metrology Institute (NMI). The link between the realization of the SI units and primary standards is established through primary methods of measurement. These are methods which do not require any reference of the same quantity. Additionally, through a series of comparisons between NMIs, comparability of measurements among traceable measurement systems at international level are recognized by each country.

Following the worldwide effort to harmonize measurement capabilities among countries, as a consequence of the strong tendency of globalization of economies, the importance of implementing traceable chemical and other measurements has been recognized as one of the principal tasks of any NMI.

In Mexico this task was initiated by CENAM in 1992. Particularly in the field of metrology in chemistry, the strategy has been developed in a parallel way to the work of the Consultative Committee for the Quantity of Matter (CCQM), by adopting the definitions of primary method of measurement and primary reference material given in the first CCQM meeting in 1995 [1]. These definitions give to NMIs clear guidelines for establishing a traceability scheme to the SI base units.

This suggested to CENAM the development of activities for the establishment of traceable chemical measurements in Mexico in two stages: the first step from 1992 to 1997 [2] corresponded to the period of development of infrastructure and human resources of CENAM, which was possible thanks to the collaboration of other NMIs involved in developing reference materials and their later certification; and the second period, from 1998 to 2002, in which a limited number of certified reference materials (CRM) were developed and certified for industrial application as well as to meet normative requirements, and also some of the primary methods of measurement were declared as national standards in a Federal Register, known as DOF.

After the signing of the CIPM MRA in 1999, CENAM has also devoted a lot of effort to demonstrate its calibration and measurement capability (CMC), which is required to establish a comparable and internationally recognized national measurement system. This task has implied for CENAM a significant challenge and at the same time strong pressure. In Appendix B of the CIPM MRA the results of Key Comparisons organized by CCQM are compiled and made available at the BIPM Website http://www.bipm.org, which are considered as supporting evidence of metrological services listed in its Appendix C declared by each NMI and examined by CCQM.

Traceability in chemical and other measurements

The definition of traceability according to the International Vocabulary of Basic and General Terms in Metrology, is given as follow:

> Property of the result of a measurement or the value of standard whereby it can be related to stated references, usually national or international standards, through an unbroken chain of comparisons all having stated uncertainties

The traceability definition also can be interpreted according to [3] as follows: "a traceability chain is a chain of values linked by measurements which consist of comparisons of one value, ending in the comparison with the value of the unit we have chosen to express the result of our measurements", with of course all comparisons having stated uncertainties. This interpretation gives clarity in the meaning of traceability concept.

The main parts that support the traceability in chemical measurements are: primary analytical methods, reference materials and valid analytical methods suitable for some available instruments for a group of materials, according to their nature, range of measurements in a specific matrix. These elements should serve to establish an uninterrupted chain of comparisons in chemical measurements and its uncertainty estimation.

As a natural process, it has been attempted to apply the traceability concept for chemical measurements, for which two illustrative proposals have been recognized; to establish a traceability structure which can be set up locally, regionally or internationally, by describing the organizational scheme in a clear and general way as well as its application [3], and the other, to illustrate practical ways of establishing traceability of chemical measurement to SI units by indicating intermediate reference points and primary methods [4].

In new fields of metrological interest, such as IVD medical devices [5], where measurement of quantities in samples of biological origin is involved, the metrological traceability of values assigned to calibrators and control materials is identified by the traceability chain and calibration hierarchy. This approach agrees in general with that given in metrology in chemistry.

It is now well understood and widely accepted that a general scheme of traceability must enable one to represent the connection between the results obtained by the procedure of measurement, called field measurement in this article, of a routine laboratory, in terms of SI units, and by a series of measurements at intermediate reference points, which may be reference materials, reference instruments, reference measurement methods maintained by reference laboratories, which are finally compared to the values obtained at the highest metrological level by primary methods. This scheme is shown in the Fig. 1.

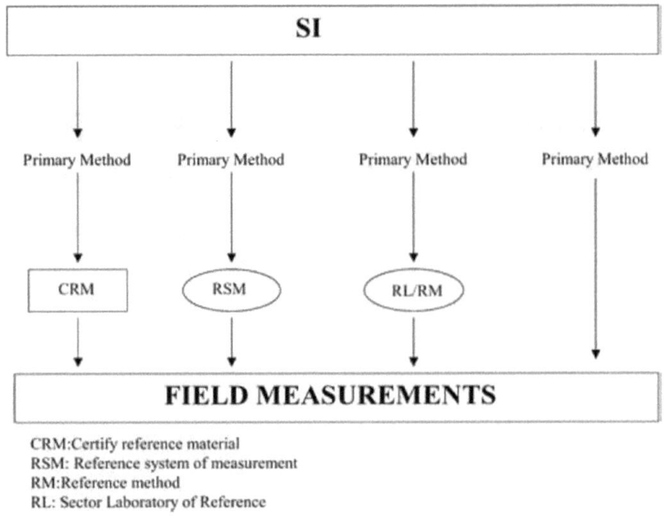

Fig. 1 Practical ways to establish traceability in chemical and other measurements

Then, the dissemination of the accuracy of the standards can be established in all the chemical and other measurements by the application of one of the following mechanisms [4]:

1. Use of reference materials traceable to SI. In the majority of measurements, the certified reference materials (CRM) traceable to SI are by far the best definable reference points and they are most frequently used as measurement standards in chemical and other measurements. These materials are the means of achieving reliable measurements and they are available from the internationally recognized organizations for a wide range of users.
2. Reference Systems of Measurement. This route of traceability is based on the development of a measurement system or instruments to create reliable intermediate reference points for measurements, which are susceptible to calibration.
3. Reference measurement methods, maintained appropriately by reference laboratories. Most measurements belong to this group, because there are still limited the availability of CRM and Reference System applicable to chemical and materials property measurements. These measurements should be carried out by laboratories that have competence in maintaining measurement methods supported by a series of measurements with demonstrable traceability to SI.

4. Primary methods applied directly to field measurements. This route corresponds to cases in which a field laboratory is able to establish primary methods to establish a direct link between their measurements and the SI.

The National Center of Metrology (CENAM) has been working to adopt the above mentioned mechanisms as practical ways to establish the traceability of all measurements carried out in Mexico.

Practices under development in Mexico

These four mechanisms are shown briefly according to [3, 4, 6], in the Fig. 1.

An example of the use of a reference material as an intermediate point in this traceability mechanism is the reference material DMR-160a (CENAM identification), sodium chloride, with its purity value assigned. This reference material is intended to be applied by field laboratories in chloride measurement, silver titration, and in all those analytical methods which require NaCl with a specific purity value. In Fig. 2 a complete traceability chain is shown by the use of reference materials to the SI units.

The purity value of this salt is assigned by a coulometric method, which is one of the potentially primary methods for amount of substance determination, according to the CCQM. The principle of this method is as follows: given an analyte dissolved in a solution, both the solution and the analyte contained in an electrochemical cell, the amount of electrical charge required to perform a chemical reaction in which the analyte of interest can be converted in another compound, is directly proportional to the amount of substance of this analyte.

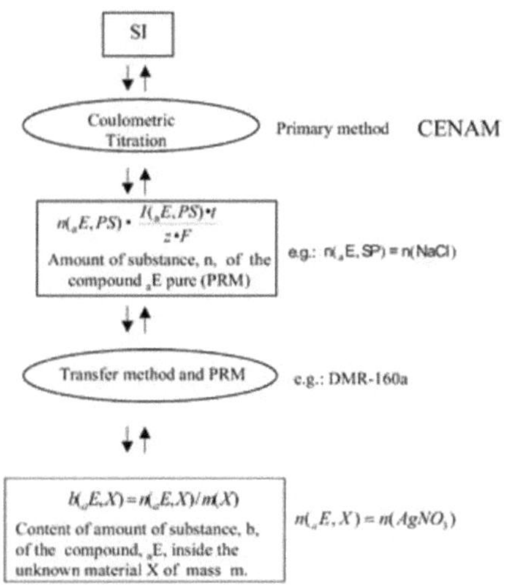

Fig. 2 Traceability chart with CRM in amount of substance

How to make CRMs available to field laboratories?

Actually, among internationally recognized providers, CENAM also provides a limited number of CRMs to the users, as one of CENAM's responsibilities established under the Mexican Federal Law of Metrology and Standardization. It has developed so far more than 200 CRMs which are expected to meet the domestic and regional needs in the categories of high purity chemical substances, organic and inorganic reference solutions, water, pH, electrolytic conductivity, food, fuels and minerals, among others. The list of available CRMs is updated monthly at CENAM Website http://www.cenam.mx by the office of MRTC Program.

In most field analysis in which separation techniques are the main difficulties, the traceability chain could not be accomplished easily by the use of calibration standards of a simple matrix. Consequently, either the validation of analytical methods or calibration by complex matrix reference materials is required. However, unless the process is clearly described with corresponding uncertainty, the validation process becomes a bottleneck for establishing a traceable measurement. Then, in most applications, the role of CRMs of a similar matrix becomes crucial in the quality of measurements.

However, due to the lack of availability of CRMs, many field laboratories make use of commercially available chemical substances, which are not normally accompanied by a certificate having enough information in accordance with ISO Guide 31.

Based on our recent assessment, we are recommending to field laboratories and commercial suppliers of chemicals to distinguish clearly CRM quality products with suitable certificates from other chemicals and reagent. This assessment has been requested by standardization authorities and is now under practice as a part of the formal recognition process of accredited testing laboratories that have to demonstrate their capability to conduct traceable measurements through usage of standards traceable to national standards of foreign countries, instead of national standards, in case they are not available in the country. This is very common for chemical measurements in Mexico, because only around 70 types of CRMs are available at CENAM, based on its actual capability, and analytical laboratories should look for other CRMs, which have demonstrated traceability to national standards of foreign countries.For historical reasons, there are many CRM providers who are not necessarily NMIs, but private companies or industrial associations that have been developing CRMs as the tools for their quality management; however, their traceability to national standards are sometime questionable. During the course of assessment, it was found that they normally do no declare uncertainty, and if they declare it they normally declare very small values for chemical components of substances without any supporting evidence.

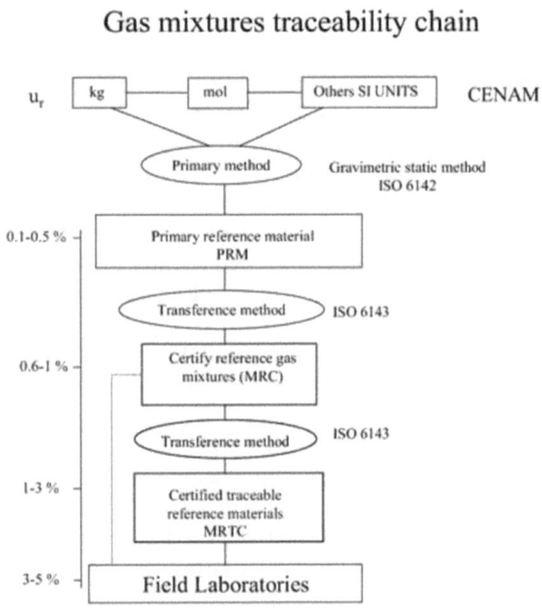

Fig. 3 Traceability chart with MRTC type CRM in gas mixtures

In order to combine the capability of these commercial producers and the capability of the certification of NMI, CENAM has launched a program called Certified Traceable Reference Materials, MRTC in Spanish. This is a similar initiative to NTRM of NIST. It is intended primarily to promote the capability of domestic industries to produce and certify CRMs in those fields where there exist enormous demand and absolute lack. This program is under development in the field of gas standards for vehicular emissions, Fig. 3, and pH measurements with domestic industries.

How to promote reference laboratories?

As was mentioned previously, most measurements are method dependent and it may be necessary to identify laboratories as reference laboratories that have competence and capability to carry out measurement with demonstrable traceability to SI. Consequently reference laboratories maintain reference procedures, which provide a reference point to field measurements. The competence of reference laboratories with respect to environmental conditions, staff, and management performance can be subjected to an accreditation process, while the reference method should be validated and also verified on the basis of documented reference procedures and on the results of parallel comparative measurement [7], in which the participation of NMI is essential to emphasize metrological robustness of the method.

In this context, complex reference materials, where they are technically feasible, are normally developed by NMIs for the validation of reference procedures, and help establish traceability of these measurements in the sense that these measurements are supported metrologically by traceable measurements to the SI units, and are capable of reproducing the value within the acceptable uncertainty. The so-called reference laboratories are expected to be capable of conducting the validation process along with NMI.

Method-dependent measurements can be grouped by sector. For example, in the clinical fields there are cases where some higher order reference materials are required for IVD methods, such as for determination of glucose in human serum. It is also required of reference laboratories in specific measurement methods. These issues are now under the responsibility of JCTLM (Joint Committee on the Traceability of Laboratory Medicine of CCQM). CENAM has developed a reference material for glucose and cholesterol determination in human serum, and certified by IDMS, which is under review by JCTLM for the use by reference laboratories in any country applying a reference method.

The important task is to develop reference laboratories in the country in each sector of importance. The metrological scheme for the definition of reference laboratories can be represented in Fig. 4.

Fig. 4 Proposed sectorial reference laboratories scheme to establish traceability in field measurements

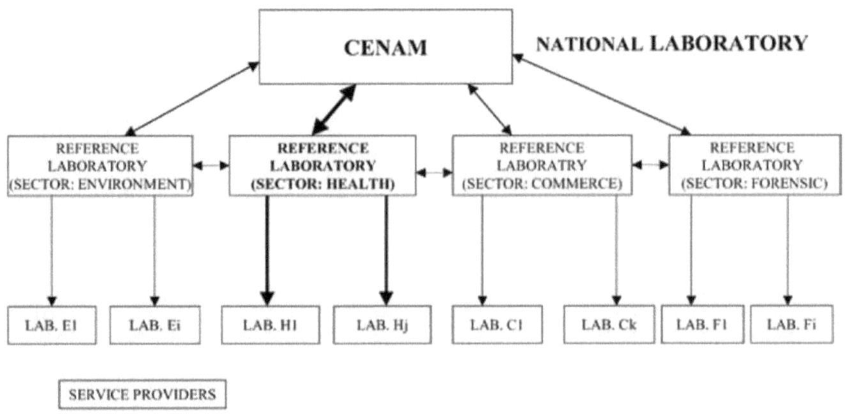

Sectorial reference laboratories and field laboratories

It is considered necessary to involve all governmental entities that have the responsibility in conformity assessment to regulations in the establishment of sectorial reference laboratories. From this standpoint, more collaboration is expected between CENAM and public sectorial laboratories, which are the technical authorities in the surveillance of mandatory standards and regulations. The idea is to give them a metrological responsibility in that sector called sectorial reference laboratories. Their functions are expected to be as follows:

- Establish traceability of their measurements to CENAM in all the quantities required in their field of responsibility
- Disseminate the accuracy of the national standards to the field laboratories by participating in the development of MRTC type CRMs
- Provide Proficiency Testing (PT) to field laboratories to establish comparability and reliability
- Develop and validate analytical methods in the field of responsibility
- Conduct type approval of measurement instruments used by the field laboratories for the conformity evaluation to the specific regulations under their responsibility

It is intended to share metrological responsibility in their respective level and fields between CENAM and reference laboratories, by maintaining coherent and comparable measurement capability among reference laboratories and consequently providing traceability to field laboratories, Fig. 5.

These tasks may deserve the highest priority of the government in the next few years, to extend collaboration programs to the fields of pharmaceuticals, clinical, health, environmental, agricultural and forensics, and also to look for modifying some part of the actual law to incorporate explicitly into the national metrological infrastructure, by assigning a specific metrological responsibility to these reference laboratories.

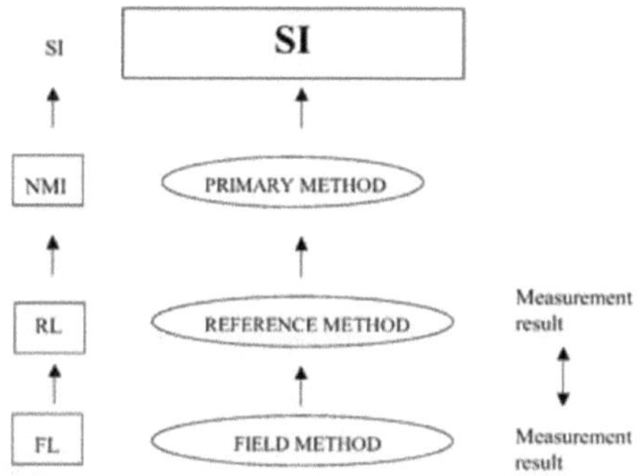

Fig. 5 Sharing metrological responsibilities with reference laboratories to provide traceability to field laboratories

Field laboratories

According to the metrology law in Mexico, accreditation of testing and calibration laboratories is under the responsibility of private accreditation bodies. However, due to the requirement of international comparability and needs for international recognition, the accreditation process based on the more transparent evaluation based on the competence of laboratories is under discussion between CENAM and EMA, an authorized accreditation body in Mexico. One of the collaborating items is to prepare technical guide for the review team based on the concept of measurement traceability and the uncertainty evaluation of field measurements. For the improvement of mutual understanding between the review team and laboratory members, a practical guide sheet to evaluate traceability elements and uncertainty sources is designed which enables both members to identify easily one of the practical ways of traceability described above for each measurement method in which the laboratory is applying for accreditation.

To improve measurement capability of field laboratories, CENAM has also been offering a PT scheme, not only because there are few PT providers in Mexico, but also due to the need to promote traceable measurement by the use of reference value provided by CENAM. Following the successful implementation of a PT program for environmental measurement laboratory assessment made by authorities of three local governments [8], similar efforts have been made to promote among laboratories who could be considered in the future as reference laboratories in food, petrochemical, clinical [9] and industrial sectors.

It is worth mentioning that the development of complex matrix reference materials such as PAH in soil has been successfully carried out under the collaboration program with PTB/BAM in Germany, by exchanging experts and use of complementary measurement capabilities [10].

Final remarks

From the practical point of view, we have identified several activities which may promote the dissemination structure of the accuracy realized in national standards. Since the tasks required are extremely diversified and demanding, it is suggested to identify a series of sectorial reference laboratories which could take part in the metrological responsibility with CENAM. This scheme is applicable to many countries, where a traceable metrological infrastructure is needed.

The availability of CRMs and MRTC type CRMs of CENAM depends on the number of RM producers and CENAM's own capacity not only in calibration and measurement, but also in management in production, certification and distribution of its CRMs in timely manner. These capabilities will be subjected to international peer review process this year, which is a part of CIPM MRA requirements.

References

1. Quinn TJ (1997) Primary methods of measurement and primary standards. Metrologia 34:61–65
2. White Paper (2000). Report of Governmental Transition, 1994–2000, Laboratories of Metrology in Chemistry Bildg. Q "Establishment of traceability in analytical measurements in Mexico". CENAM, October (in Spanish)
3. De Bièvre P (2000) The key elements of traceability in chemical measurement: agreed or still under debate? Accred Qual Assur 5:423–428
4. Richter W (1997) Primary methods of measurement in chemical analysis. Accred Qual Assur 2:354–359
5. ISO (1999) In vitro diagnostic medical devices-measurement of quantities in samples of biological origin—metrological traceability of values assigned to calibrators and control materials, pr EN ISO 17511. International Organization for Standardization, Geneva
6. De Bièvre P (2000) Traceability of (values carried by) reference materials. Accred Qual Assur 5:224–230
7. Siekmann L (2002) Measurement traceability in clinical chemical analyses. IVD Technol Mag
8. Ma Rocio Arvizu et al. (2000) Biological and environmental reference materials in CENAM. Fresenius J Anal Chem 370:156–159
9. Perez-Urquiza M et al. (2003) Use of reference value for the Proficiency Test scheme in the measurement of glucose and cholesterol in human serum. Accred Qual Assur (in press)
10. Perez-Urquiza M et al. (2003) International co-operation in the development and application of matrix reference materials case study: polycyclic aromatic hydrocarbons (PAH) and total petrol hydrocarbons (TPH) in soil. BERM8, Berlin

Accred Qual Assur (2003) 8:467–474
DOI 10.1007/s00769-003-0658-0

Ph. Charlet
A. Marschal

Benefits of the implementation of a metrological structure for water analyses

Received: 15 January 2003
Accepted: 3 April 2003
Published online: 25 July 2003

P. Charlet (✉) · A. Marschal
Bureau National de Métrologie—
Laboratoire National d'Essais,
1 rue Gaston Boissier,
75724 Paris Cedex 15, France
e-mail: philippe.charlet@lne.fr
Tel.: +33-1-40 43 39 80
Fax: +33-1-40 43 37 37

Abstract Traceability of measurements still often remains a non-operational notion for end-user chemists. A practical project, sponsored by the French Ministry of Research, has been conducted to evaluate and to demonstrate possible benefits of the implementation of a metrological structure for improvement of the quality of water analyses. LNE (Laboratoire National d'Essais) was in charge of the build up of the traceability chain in a concrete case of determination of some heavy metals (cadmium and lead) in a groundwater. Pure solutions for calibration and a matrix RM have been certified by LNE and then used by 46 labs (mainly French) in a inter-laboratory study. Results have shown a measurable bias in lead analysis in the groundwater for all methods in routine use by laboratories. This project has demonstrated the interest of a metrological approach for method calibration, method validation and estimation of measurement uncertainty.

Keywords Metrology · Traceability · Analytical chemistry · Certified reference material · Water analyses

Introduction

Much has been written in the past 10 years on traceability in chemical analyses but most of these contributions can be classified as "scientifically logical" or "politically correct". Less can be considered as "operationally" relevant and useful for end-users [1]. The purpose of this paper is to contribute to this topic by addressing practical aspects of the traceability of chemical measurements, considering routine analytical methodologies of field laboratories, in a specific case of environmental analysis.

One objective of this document is to help end-user chemists, who have no philosophical interest in knowing "if a bias must always be corrected"; but who need to establish which are the main sources of uncertainty of a measurement and what are their orders of magnitude in order to produce reliable and useful results.

For a large number of field chemists, the general meaning of traceability remains unclear, and it is often linked to an idea of mystery or magic, or even worse, to a notion of drudgery.

It may be useful to recall that for a chemist, there are three different types of traceability [2]:

- Material traceability, which is related to the processing history of a batch.
- Documentary traceability, which consist of finding raw data and all documents used before the issue of the analysis report.
- Metrological traceability, which has to ensure that the unit stated to the measured value is universal. This can be ensured by a logical succession of operations which can be the use of pure substances and reference material.

From the point of view of the user, all three types of traceability are important.

Due to more stringent regulations and particularly the new European Directive on Water (European Directive 2000/60/CE, 23 of October 2000), it is challenging for decision-makers to take the right decisions about management and restoration of water resources to protect public health and environment. In this context, there is a need to document and possibly to increase the reliability of measurements and therefore, as a first step, to improve the quality of chemical analyses. To achieve this goal, the enhancement of metrological traceability of measurement is required, this can be put into effect through the implementation of metrology principles in field laboratories.

At the end of 2001, a project, coordinated by LNE (Laboratoire National d'Essais), one of the four French Institutes of Metrology, was initiated to evaluate the possible benefits of the implementation of a metrological structure for environmental analyses. This project was granted by the French Ministry of Research (Direction of Technology), under the acronym of METREAU (for: Metrology of Water). It was conducted in the framework of RITEAU, a network of innovative technological developments in the field of water.

The METREAU project focused on the determination of some heavy metals, cadmium and lead, in a groundwater. Concentrations of metals have been chosen to correspond to current and possible limits of future legislation.

Apart from the metrological aspect, another objective of this project was the assessment of the natural variability of water characteristics and the effect of such a variability on the uncertainty of measurements. This paper does not cover this part of the project.

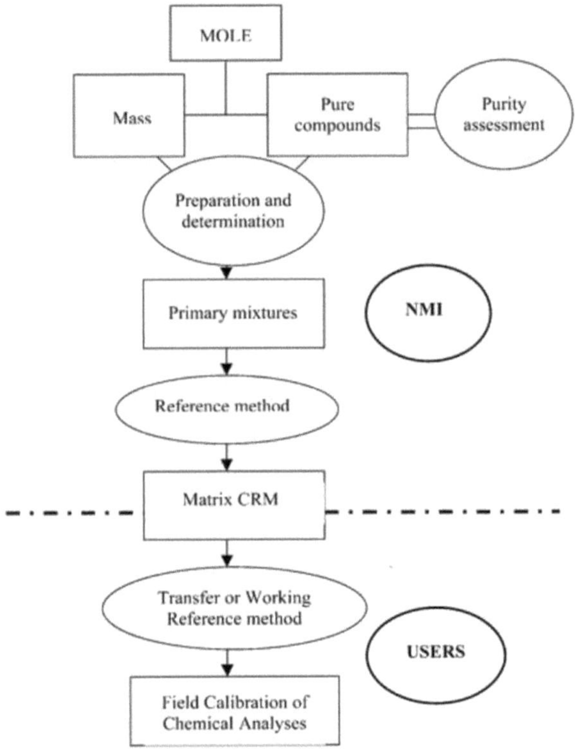

Fig. 1 Diagram of the traceability chain in analytical chemistry

Metrological traceability of chemical analyses

Figure 1 presents a classic flexible calibration scheme for chemical measurements. It also underlines some of the duties of a National Institute of Metrology [3, 4, 5].

Considering this diagram, it is important to emphasise a few points:

- In the field of environmental analyses, to achieve the required comparability and traceability of measurements, there is a clear need for matrix reference materials. These reference materials must present a sufficient matrix matching with real environmental samples, otherwise they are useless, due to a lack of commutability, that is to say their ability to demonstrate inter-assay properties comparable to real samples. This matrix match can be considered as a more important property of the RM than the level of uncertainty of the measurand concentration [6, 7]. It has been estimated that only 10% of the required matrix RM in the field of environment are currently available [8].
- Pure substances are obviously needed for calibration of the measurement stage of a method. Very often, commercial substances are used as pure standards and the level of uncertainty associated with the purity contributes to the uncertainty of the measurement. Of course, a bias can be introduced in the measurement if a commercial reagent or a in-house standard presents a deviation from the nominal value.

In their regular day to day practice, field laboratories use commercial reagents or prepare in-house solutions for the calibration of instruments, and they rely on purity assessment of producers. For method validation and even measurement uncertainty, field labs regularly participate in proficiency testing schemes. In such inter-laboratory comparisons, the reference value is usually obtained as the arithmetic mean of results of participants.

The metrological structure involves, as a focal point, the competences of a metrological institute, and can be schematised on Fig. 2.

Using primary methods, the metrological institute certifies standard solutions and matrix CRM and, therefore, ensures traceability of these "etalons". Methods which were performed in this project for pure substances characterisation (and presence of trace impurities) and certification of matrix CRM are high-accuracy titration and isotopic dilution mass spectrometry. The metrological lab can also provide the reference value of the sample used in a proficiency testing. Another important task is to assist field laboratories to establish the overall uncertainty budget of the measurement. For this purpose, LNE has organised several training courses for more than 10 years.

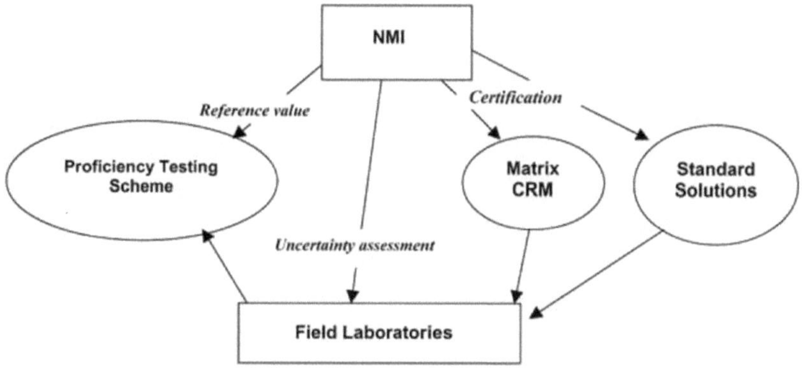

Fig. 2 Metrological structure set up for the Metreau project

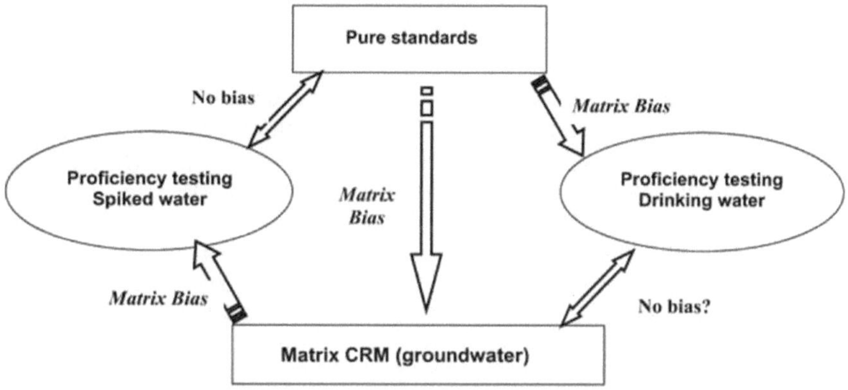

Fig. 3 Different ways to organise a proficiency testing in order to underline bias

In this metrological structure which has been set up for the METREAU project, different partners have been involved:

- LNE is the Metrological Institute
- Merck-Eurolab has prepared standard solutions of lead and cadmium at about 1 g/L
- BRGM (Bureau de Recherches Géologiques et Minières) is the French geological institute, it was responsible for choosing the groundwater site and collecting water samples
- BIPEA (Bureau Interprofessionnel d'Etudes Analytiques) is a proficiency testing provider
- LDAR (Laboratoire Départemental d'Analyses et de Recherches de Périgueux) is a field laboratory representative of end-users

Strategic approach

Considering the metrological structure, a proficiency testing scheme was organised with laboratories working in water analyses. It was decided to choose a deionised water spiked with heavy metals (at 5 and 20 µg/L for cadmium and lead, respectively) for the round robin test. A matrix CRM (groundwater containing cadmium and lead) was also sent to laboratories. Laboratories have been asked to analyse, in duplicate, the spiked water sample and the matrix CRM. They also had to analyse these samples using their usual calibration standard solutions first, and then pure certified solutions of cadmium and lead. The results of the two working ways, regular and metrological were compared.

An important objective of this project was to provide elements of method validation by estimating the bias, that is to say the difference between the measured value and the true value of measurands in samples. This can be underlined through a proficiency testing analysing either a spiked pure water or a matrix sample (drinking water), according to the diagram represented in Fig. 3.

A common way to assess the matrix bias is to analyse a drinking water (therefore containing a matrix) in the proficiency testing. For this analysis, laboratories calibrate their instruments using standard, commercial or in-house solutions. If a CRM is available, a possible matrix effect can be corrected by adjusting operational instrument parameters to "match" the certified value.

For the METREAU project it was chosen not to organise a proficiency testing with a drinking water to avoid a possible matrix gap with the matrix CRM. It is well known that, for the production of groundwater matrix CRM, it is difficult to choose a matrix sample representative of an average natural water. For water analyses, this is a major point to determine if a matrix CRM is appropriate. For the METREAU project, it would have been risky to get different matrixes since uncontrolled

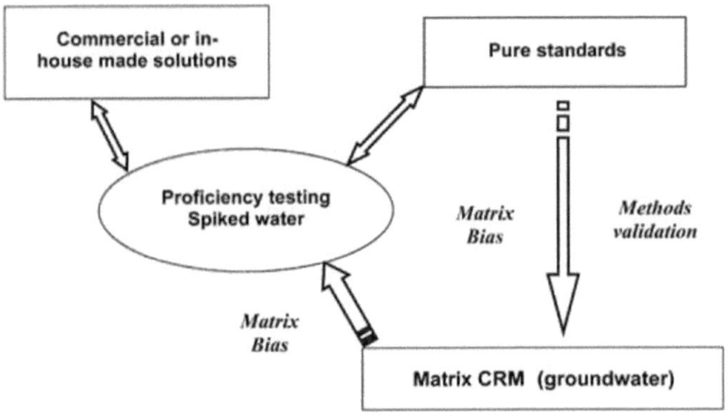

Fig. 4 Elements in the proficiency testing organised to evaluate matrix bias and method validation

factors (how similar is the matrix CRM to the drinking water) would have been introduced. Therefore, there was a risk of invalidating the whole demonstration of the project. An inter-laboratory comparison was organised for analysing a spiked pure water and a matrix CRM, since the diagram shows that there is a pseudo equivalency of matrix bias effects where they can be observed in a symmetric way, according to the type of inter-laboratory comparison (spiked or drinking water) carried out.

This choice of a spiked water sample allowed some targets to be set:

- Certified pure solutions were used to estimate the reliability of regular calibrations
- Matrix CRM was used to control the quality of routine determinations performed by laboratories, and so to evaluate the accuracy of analyses
- The magnitude of bias was estimated through the analysis of the matrix CRM using pure certified standard for calibration

The strategic approach can be summarised in a simple diagram presented in Fig. 4.

Certification of pure substances of cadmium and lead

Calibration solutions of cadmium and lead have been prepared from high purity (higher than 98%) cadmium and lead nitrate dissolved in nitric acid. A high accuracy titration method (EDTA complexometry reaction with photometric detection) was used to determine the final concentrations. Control of the level of impurities was carried out by ICP-MS:

- Lead concentration: 1.0020±0.0045 g/L at 20 °C (k=2)
- Cadmium concentration: 1.0118±0.0023 g/L at 20 °C (k=2)

Certification of the matrix reference material

The concentrations of the different ions contained in the collected groundwater are presented in Table 1. A high level of potassium ions can be noticed.

Concentrations of cadmium and lead were determined by isotopic dilution ICP-MS using two instruments: a quadrupole equipped with a collision cell and a high resolution (magnetic sector) ICP-MS. Results are presented in Table 2.

Results of lead analyses were confirmed by another NMI, the Swedish National Testing and Research Institute (SP) in Boras, Sweden where a bottle of the same batch was sent. SP has performed two parallel experiments, one at SP and one at LGC in UK, performing isotopic dilution ICP-MS. These results were not taken into account for the reference value.

A very good agreement was obtained between the two laboratories.

Stability testing (for 6 months) and homogeneity testing were carried out and confirmed by BRGM. The overall uncertainty was determined taking in account these tests and therefore including the homogeneity uncertainty and the stability uncertainty:

$$u_{CRM} = \sqrt{u_{car}^2 + u_{hom}^2 + u_{stab}^2}$$

The characterisation uncertainty u_{car} was evaluated, according to GUM, taking into account the mean of 16 results obtained with the two ICP/MS instruments (quadrupole and magnetic sector).

For the homogeneity testing, three independent analyses of cadmium and lead were performed on 20 bottles.

Table 2 Concentrations of lead and cadmium in the matrix CRM, determined by LNE (and confirmed by SP for lead)

NMI	Lead (μg/L)	Cadmium (μg/L)
LNE	19.78±0.32	4.62±0.06
SP	19.73±0.26	
SP on LGC apparatus	19.89±0.25	

Table 1 Major chemical species present in the matrix CRM

	Mg^{2+}	Ca^{2+}	K^+	Na^+	Cl^-	Br^-	NO_3^-	SO_4^{2-}
mgL	28.1	72.4	288.1	49.3	381.8	1.9	15.7	85.2

Fig. 5 Results of cadmium analyses in the spiked pure water of the proficiency testing using regular calibration solutions and certified solutions

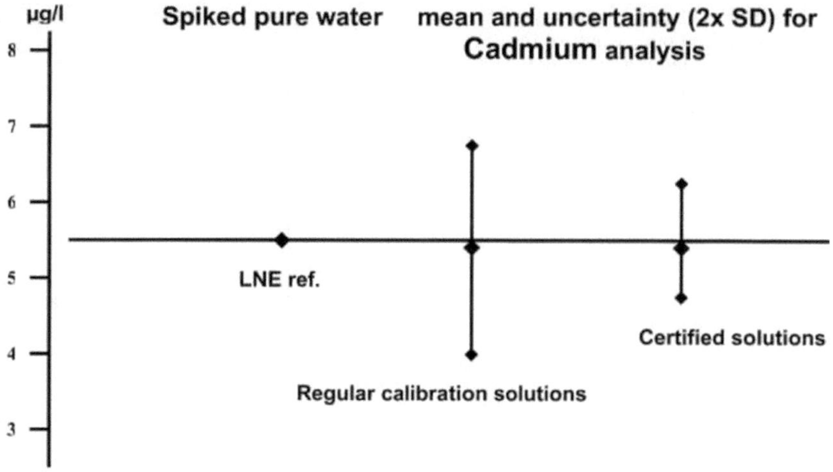

For homogeneity and stability testing, no significant effects (at 5% statistical risk) were observed.

The concentration values of the certified reference material were the following:

- Lead concentration: 19.8±1.0 µgLl at 20 °C (k=2)
- Cadmium concentration: 4.6±0.4 µg/L at 20 °C (k=2)

Results of the inter-laboratory comparison

Forty six field laboratories (mainly from France) have participated in the inter-laboratory comparison. They have analysed the water sample using their regular calibration solution and then have repeated this analysis using the certified standards. They also have analysed the matrix CRM. Laboratories have produced two results (duplicate) per sample. Techniques used were mainly atomic absorption spectroscopy with furnace but also ICP-OES and ICP-MS.

Analysis of water sample

The ISO 5725 standard was used to interpret the data. Even if the main purpose of this standard is related to the validation of a method, it can be used to evaluate some components of the measurement uncertainty. The homogeneity of the population of results, in terms of mean and standard deviation was determined using statistical tests (Cochran and Grubbs). A few laboratories were rejected after the tests. Tables 3 and 4 present the comparison of overall performance of laboratories when working with usual and metrological calibrations solutions.

Very few laboratories (3) were rejected by statistical tests and this number remained about constant when laboratories used certified calibration solutions. This means that no laboratories were using commercial or in-house prepared solutions which presented a bias. Results of laboratories were very close to the reference value and, on the whole, results were considered as very good by the PT provider, but it is important to recall that the water sample did not contain any matrix. By using certified calibration solutions, there is no improvement of the mean but for cadmium a significant improvement of repeatability and reproducibility SD was observed as shown on Figs. 5 and 6.

Table 3 Performances of the participants in the proficiency testing for cadmium analysis in the spiked pure water

	Cadmium (µg/L) with regular calibration solutions	Cadmium (µg/L) with certified calibration solutions
Reference value LNE	5.514±0.075	
Mean	5.375	5.406
Repeatability SD	0.208	0.121
Reproducibility SD	0.700	0.434
Number of labs after tests	42	41

Table 4 Performances of the participants in the proficiency testing for lead analysis in the spiked pure water

	Lead (µg/L) with regular calibration solutions	Lead (µg/L) with certified calibration solutions
Reference value LNE	21.54±0.48	
Mean	21.051	21.504
Repeatability SD	0.761	0.707
Reproducibility SD	2.442	2.509
Number of labs after tests	42	44

Fig. 6 Results of lead analyses in the spiked pure water of the proficiency testing using regular calibration solutions and certified solutions

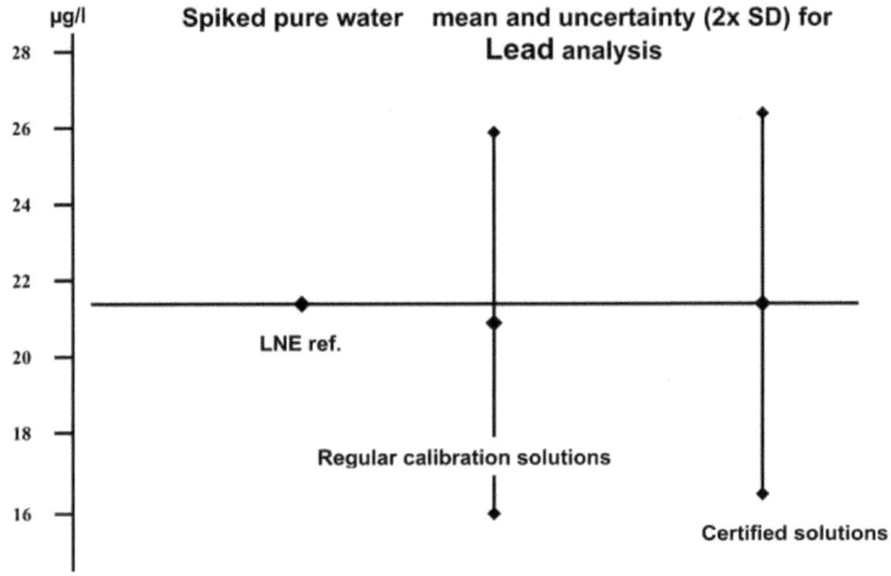

Fig. 7 Repartition of results for lead analysis in the matrix CRM by the 45 laboratories

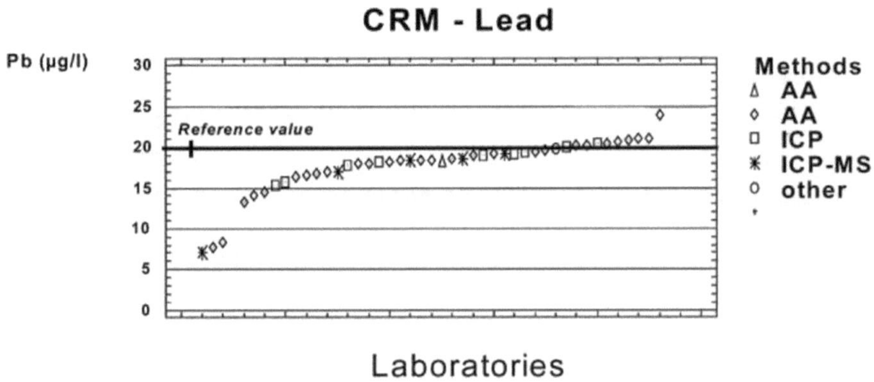

Table 5 Performances of the participants in the proficiency testing for cadmium and lead analysis in matrix CRM

	Cadmium (µg/L)	Lead (µg/L)
Reference value LNE	4.62±0.36	19.8±1.0
Mean	5.09 (4.60)	17.62
Repeatability SD	0.17	0.83
Reproducibility SD	3.35	3.56
Number of labs	45	45

Analysis of the matrix CRM

The groundwater matrix CRM has been analysed in duplicate by laboratories, using certified calibration solutions. Results are presented in Table 5.

In the case of lead, the mean of results is significantly lower than the reference value, by about 11%. A large number of laboratories (37 out of 45) have underestimated the concentration of lead as shown in the repartition curve of laboratories around the reference value (Fig. 7).

In the case of cadmium, one laboratory result has strongly modified the mean. By eliminating this outlier, the mean of laboratories was 4.60 µg/L, therefore very close to the reference value. The distribution curve of laboratories was symmetric around the reference value as shown on Fig. 8.

This bias in lead analysis was observed for all the methods used by laboratories: atomic absorption (flame and furnace), ICP (optical and MS), and other methods, as shown in Fig. 9. This bias was not observed for cadmium, and means of the different methods were very close to the reference value.

Complementary experiments were carried out at LNE to explain this bias observed for lead. Experiments were performed using Zeeman furnace AAS and ICP-MS on "reconstituted" matrix samples spiked with lead and cadmium. Strong matrix effects were obtained for both methods, and lead was underestimated by 15–30%. In

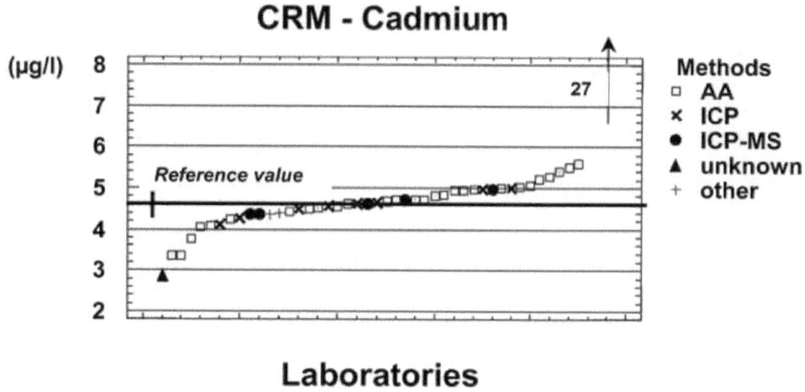

Fig. 8 Repartition of results for lead analysis in the matrix CRM by the 45 laboratories

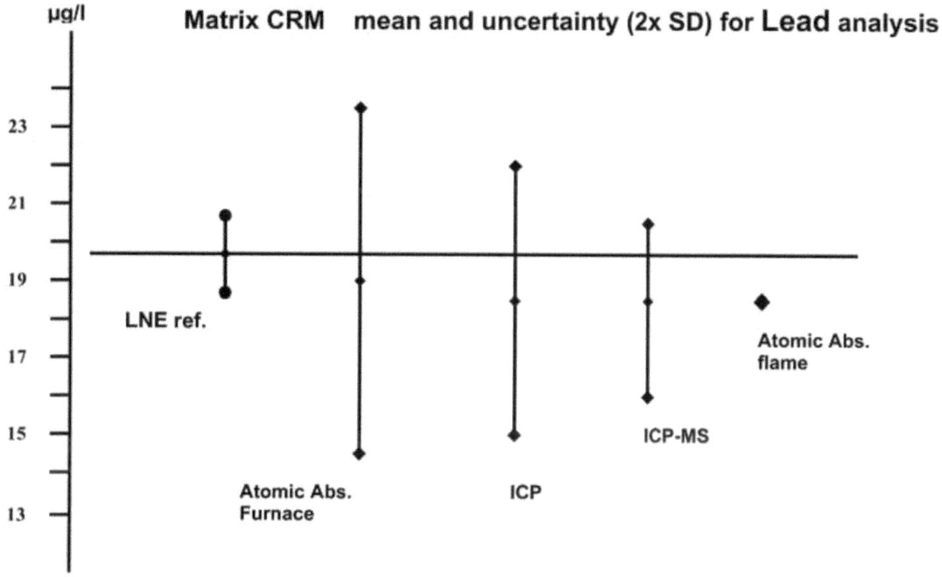

Fig. 9 Results of lead determinations according to analytical methods used by participant laboratories

ICP-MS, even when using a internal standard (Indium), a bias of 8% was observed.

No similar effect was observed for cadmium.

Conclusions

A metrological structure was implemented in the METR-EAU project to evaluate the benefits for laboratories of using certified pure solutions for calibration and matrix CRM for method validation.

In the field of environmental analyses, the traceability of results of laboratories is not established on a classic common base where "one unique" standard pure solution or certified reference material (produced by a NMI for instance) is used by all laboratories as a first basic step for calibration and method validation. Many laboratories ensure the traceability of their results by using commercial solutions or by preparing in-house solutions for the calibration of their instruments. The reliability of these calibration solutions is estimated by laboratories by calculating the contribution of the preparation stage of these standards (weight precision, dilution precision of commercial solutions, purity of standards, ...) on the total uncertainty of the measurement. Calculations demonstrate that this approach is reasonable and that other contributors are of major importance. In the case where traceability is materialised by laboratories using house-made or commercial solution, the common base between laboratories is then the formal definition of the concentration of metals in water.

The following results were obtained:

– The certified pure solutions of cadmium and lead have demonstrated the reliability of calibrations performed by laboratories. No bias was observed with the commercial or in-house solutions used by the participants of the proficiency testing. In this case, it demonstrates that there is no particular benefit for laboratories to carry out analyses using "metrologically" certified high purity solutions. This paper has demonstrated that the pragmatic approach cannot be caught out for these specific metals, in this specific matrix. It would be unreasonable to extend this conclusion to other measurands or other matrices without demonstration.

- On average, very good results have been obtained by laboratories during the Proficiency Testing Scheme for the pure water, for both the regular and metrological way of working. Using certified calibration solutions, some improvement of repeatability and reproducibility SD was observed, particularly in the case of cadmium analysis.
- Analysis of a matrix CRM has shown a noteworthy bias in lead analysis of about 10%. This bias was observed for all routine methods used by laboratories. Complementary experiments have indicated that a strong matrix effect was responsible for this bias. This underlined bias has demonstrated that for this matrix water, only a primary method can provide the true value of the sample of an inter-laboratory comparison. The mean value of laboratory results cannot be used as a reference value.

In conclusion, the METREAU project has shown that a metrological approach can bring valuable information on reliability of calibration of the measurement stage of a method and give prominence of estimation of possible bias for method validation. The development of metrology principles and concrete application of such an approach can be considered as a major mission of a National Metrological Institute.

Results of the METREAU project will be used to disseminate information to end-users regarding the state of the art on the quality and comparability of chemical determinations in their field. Possible benefits of the implementation of a metrological structure have to be appreciated by field laboratories in order to know if they are appropriate to their use and constraints.

Considering the present results, it is strongly advised to renew equivalent projects in other sectors and for other measurands.

Acknowledgements This work was part of the METREAU project financially supported by the French Ministry of Research and it was performed under the auspices of the Réseau RITEAU. The authors acknowledge all the partners of this project: Anne-Marie Fouillac and Stéphane Roy from BRGM, Marie Lafargue from BIPEA, Didier Lagoutte from Merck Eurolab, Laurent Ley from LDAR and the two experts from the scientific committee: Jean-Michel Mermet (University of Lyon) and Philippe Quevauviller (EU Commission). The authors are very grateful for the continuing support provided by our colleagues at LNE: Michèle Désenfant, Béatrice Lalère, Cédric Rivier and Guillaume Labarraque. The participation of Dr Bertil Magnusson from the Swedish National Testing and Research Institute (SP) was fully appreciated. The authors also acknowledge Dr Igor Maksimov from VNIIFRI for his advice in reviewing this manuscript.

References

1. Marschal A (1980) La mole : de la définition à l'utilisateur. Bulletin BNM no 39
2. Marschal A, Andrieux T, Compagnon PA, Fabre H (2002) Accred Qual Assur 7:42–49
3. Marschal A (1994) Traceability and calibration in analytical chemistry. Proc Symposium EUROLAB, Florence, Italia
4. Marschal A. (1997) Chemical metrology laboratories: their task, their tools, their contributions to biological and environmental analyses. Proc BERM-7, Antwerpen, Belgium
5. ISO Guide 32 (1997) Etalonnage en chimie analytique et utilisation des matériaux de référence certifiés. ISO, Geneva, Switzerland
6. EEE-RM Working Group (2001) The selection and use of reference materials, a basic guide for laboratories and accreditation bodies
7. Pauwels J, Lamberty A (2001) Fresenius J Anal Chem 370:111–114
8. Quevauviller Ph (2002) Matériaux de référence pour l'environnement. Editions TEC & DOC, Lavoisier, Paris

Accred Qual Assur (2004) 9:493–498
DOI 10.1007/s00769-004-0824-z

L. Brüggemann
R. Wennrich

Validation steps for traceability of linear calibrated chemical measurements

Received: 18 July 2003
Accepted: 2 March 2004
Published online: 7 May 2004

L. Brüggemann (✉) · R. Wennrich
Department of Analytical Chemistry,
UFZ Centre for Environmental Research
Leipzig-Halle,
Permoserstrasse 15, 04318 Leipzig,
Germany
e-mail: lutz.brueggemann@ufz.de
Tel.: +49-341-2352512
Fax: +49-341-2352625

Abstract Based on the new draft of the EURACHEM/CITAC Guide "Traceability in Chemical Measurement", this publication describes how traceability can be achieved for chemical measurements using a linear calibration function. Traceability can be accomplished without larger expenditure, if the measurement is calibrated on the basis of appropriate reference standards and the linear regression employed is selected and validated statistically in a suitable form. The determination of nickel in *aqua regia* eluates of sediment samples, employed for an ICP-OES measurement, is used as a practical illustration of this approach.

Keywords Traceability · Linear calibration · Method validation

Introduction

Comparability is a key property of chemical measurements. While results can be compared directly under repeatability conditions, a more general approach is needed to provide meaningful comparison to results of other measurements made at different times and places. This "comparability over space-and-time" is routinely achieved by linking the individual measurement results to some common, stable reference or measurement standard. Results are therefore correlated to that reference. This strategy of linking results to a reference is termed "traceability" [1, 2]. Traceability is a key property in metrology, and for this reason the traceability of results is even explicitly demanded in the international norm ISO 17025 [3].

As described in [4], a measurement is a set of operations with the object of determining the value of a quantity. The measurement includes a set of conditions and an equation from which the result is calculated using the values of the measured parameters. The implication is that if the values of all these parameters are traceable to stable references, the results will be consistent. However, this expectation is based on some assumptions; specifically, a functional relation between the amount of measurand and its response, freedom from overall bias, and absence of other significant effects. Method validation answers the question "are these assumptions valid?" by making experimental tests of the assumptions [5]. Where no other significant effects are found, the method now explicitly includes all of the factors known to require traceability. If all the identified factors are indeed made traceable to suitable references, the method can be expected to produce consistent results. Then the method is considered to be validated, and can be used without changes.

If a statement of comparability at any confidence level needs to be made then other information is essential. The uncertainty of the results is needed [6], because only results accompanied by measurement uncertainty are comparable. To obtain consistent and useful measurement results, it is important that both a chain of comparisons to reference standards, and the uncertainties associated with these comparisons, are established. These principles lead directly to the definition of traceability in the International Vocabulary of Basic and General Terms in Metrology (VIM) as "Property of the result of a measurement or the value of a standard whereby it can be related to stated references, usually national or international standards, through an unbroken chain of comparisons all having stated uncertainties" [4].

This definition also implies the duality between the hierarchical establishing of traceability and the corresponding cumulative effect of the uncertainties referring to the single levels of the calibration chain [7].

Activities needed to establish traceability

To establish traceability for a particular developed or selected method, the following four activities must be accomplished [2]:

- Demonstrating, by validation, that the calculation and measurement conditions include all of the influence quantities that significantly affect the result, or the value assigned to a standard
- Identifying the relative importance of each influence quantity – dictated by their quantitative effect on measurement results – in order to decide on the degree of control or calibration
- Choosing and applying appropriate reference standards
- Estimating the uncertainty

Only the last two aspects are considered here.

Choosing and applying appropriate reference standards

To make sure that all the values used in the measurement equation and all other fixed values used in the measurement are traceable, it is necessary to establish procedures for calibration of the measuring equipment or for controlling fixed values, and for ensuring the calibration, certification or control of all the references used in the measurement. Calibration, together with validated methods, is therefore the key to traceability. In practice, it is recognised that calibrated and certified reference standards are not always available, but it is always necessary to establish sufficient control through the appropriate choice of measurement standards.

Factors to be considered when assessing the appropriateness of a reference material include the following [2]:

- Matrix effects and other factors (measurand, measurement range, matrix match and potential interferences, homogeneity and stability, measurement uncertainty, certification procedures)
- Track record of both the producer and the material. For example, whether the reference material has been subjected to an interlaboratory comparison, cross-checked by use of different methods, or if there is experience of use in a number of laboratories over a period of years
- The validity of the certification and uncertainty data, including conformance of key procedures with recognised quality standards

Some or all of the requirements may be specified in the customer and analytical specification, but often it will be necessary for the analyst to use professional judgment and fitness for purpose criteria [8].

Uncertainty estimation

The minimum required for useful measurements is:

- Assessing the contribution of each reference value to the total uncertainty, or, if appropriate, complying with the equipment, calibration, and control requirements of the standard method (norm) in use
- Assessing the overall uncertainty in the result

Traceability for linear least squares calibration

In this case, two aspects, the formal establishment of traceability to reference standards and the formal estimation of uncertainty, can be treated in a general way.

Suppose we have a system where there is a linear relation between analyte concentration and response of the measurement system. We perform a linear calibration with k calibration levels of the analyte concentration, each represented by j, each of them with n_j repetitions. All together there are $n = \sum_{j=1}^{k} n_i$ single calibration measurements, each represented by i.

A simple linear regression model

$$y = b_0 + b_1 x, \tag{1}$$

where y represents the response of the measurement system for analyte concentration x, and b_0 (intercept) and b_1 (slope) designate the regression coefficients used. Estimates for the regression coefficients, B_0 and B_1, respectively, and for the other regression parameters (see Table 3) are calculated from the calibration data set $\{x_i, y_i\}$.

Since it is a characteristic feature of the precision of the prediction of the regression function (1) the estimated standard deviation of a predicted single value y_{pred} at position x_0

$$S_{y,pred} = S_e \sqrt{1 + \frac{1}{n} + \frac{(x_0 - \bar{x})^2}{\sum_i (x_i - \bar{x})^2}} \tag{2}$$

can be used, where S_e defines the residual standard deviation of the regression model.

The (predicted) concentration for the observed response y_{obs} is calculated by means of the inverse function to (1):

$$x_{\text{pred}} = \frac{y_{\text{obs}} - B_0}{B_1}, \tag{3}$$

which represents the measurement equation mentioned above.

An estimation of the measurement uncertainty of x_{pred}

$$u(x_{\text{pred}}) = \frac{S_e}{B_1}\sqrt{\frac{1}{p} + \frac{1}{n} + \frac{(x_{\text{pred}} - \bar{x})^2}{\sum_i (x_i - \bar{x})^2}}, \tag{4}$$

based on the calibration data, is then derived [6, 9], where the estimate concerns the mean prediction value for p repetitions.

Our task, to establish traceability, is then simplified. The results of the measurement Eq. (3) are directly traceable to the calibration solutions, because the regression coefficients B_0 and B_1 trace back to the analyte concentration of samples via the observed responses y_{obs} of the analyte concentrations on calibration. Therefore, the proper execution of regression is a crucial condition for establishing the traceability. The regression model used has to be carefully selected and validated.

The condition of variance homogeneity can be proven with the help of statistical tests (F-test for quotient of variances at the lower and upper end of the calibration range, or better, in the case of $n_j \geq 5$ by using the Bartlett-test, which includes all of the variances in the calibration range). If variance homogeneity is violated, a weighted least squares regression (WLS) should be used.

Normally, the uncertainties in the concentrations of the calibration solutions (variable x) are small in relation to the uncertainties of the response of the measurement system (variable y), so that the regression parameter can be estimated using ordinary least squares (OLS). In exceptional cases, the test quantity S_{ex}/S_x (S_{ex}^2 means the variance in the concentration of the calibration solutions for a particular calibration level, and S_x^2 indicates the total variance in concentration) can be calculated, and if $S_{\text{ex}}/S_x > 0.2$ the regression parameters should be estimated by orthogonal distance regression (ODR) [10].

Using an OLS-estimation, the sum of squares of the residuals

$$e_i = y_i - y_{i,\text{fit}} = y_i - (B_0 + B_1 x_i) \tag{5}$$

is minimised. By visual inspection of the residual plot, deviations in linearity, normality and variance homogeneity, can be recognised, as well as outliers.

The goodness of fit supplies the coefficient of determination for the regression

$$R^2 = 1 - \left(\frac{\sum_{i=1}^{n} (y_i - y_{i,\text{fit}})^2}{\sum_{i=1}^{n} (y_i - \bar{y})^2} \right), \tag{6}$$

which indicates the part of the variance explained by regression (this quantity is equal to the square of the correlation coefficient).

The suitability of the regression model should be proven by a special statistical lack-of-fit-test, which is based on an analysis of variance (ANOVA). Here the residual sum of squares of regression is separated into two components: the sum of squares from lack-of-fit (LOF) and the pure "error" sum of squares (PE, pure errors)

$$\sum_{i=1}^{n} (y_i - y_{i,\text{fit}})^2 = \sum_{j=1}^{k} n_j (\bar{y}_j - y_{j,\text{fit}})^2 + \sum_{i=1}^{n} \sum_{j=1}^{n_j} (y_{i,j} - \bar{y}_j)^2. \tag{7}$$

The following F-test is based on the means of the deviation squares concerned:

$$\text{MS}_{\text{LOF}} = \frac{1}{k-2} \sum_{j=1}^{k} n_j (\bar{y}_j - y_{j,\text{fit}})^2, \tag{8}$$

and

Table 1 Calibration data set for the calibration experiment of Example 1 (concentrations and responses are denoted by x and y, respectively)[a]

Probe	i	j	n_j	x	$x_{j,\text{M}}$	y
BW1	1	1	9	0	0.000	340.00
BW2	2			0	0.000	437.67
BW3	3			0	0.000	388.67
BW4	4			0	0.000	391.67
BW5	5			0	0.000	341.00
BW6	6			0	0.000	319.67
BW7	7			0	0.000	391.33
BW8	8			0	0.000	333.67
BW9	9			0	0.000	313.67
LKSD1/1	10	2	3	3.39	3.437	2098.33
LKSD1/2	11			3.39	3.437	2082.00
LKSD1/3	12			3.53	3.437	2172.33
LKSD2/1	13	3	3	7.18	7.330	3556.67
LKSD2/2	14			7.25	7.330	3685.33
LKSD2/3	15			7.56	7.330	3981.67
LKSD3/1	16	4	3	13.75	13.870	6713.67
LKSD3/2	17			13.76	13.870	6795.33
LKSD3/3	18			14.10	13.870	6751.33
LKSD4/1	19	5	3	10.01	9.910	4917.33
LKSD4/2	20			9.98	9.910	5022.33
LKSD4/3	21			9.74	9.910	4881.00
LKSD1/1+M4	22	6	3	13.38	13.433	6405.33
LKSD1/2+M4	23			13.53	13.433	6554.33
LKSD1/3+M4	24			13.39	13.433	6429.00
LKSD2/1+M4	25	7	3	17.16	17.127	8345.33
LKSD2/2+M4	26			17.12	17.127	8073.67
LKSD2/3+M4	27			17.10	17.127	7919.00
LKSD3/1+M4	28	8	3	23.49	23.530	11084.00
LKSD3/2+M4	29			23.29	23.530	10981.00
LKSD3/3+M4	30			23.81	23.530	11126.00
LKSD4/1+M4	31	9	3	19.66	19.803	9153.33
LKSD4/2+M4	32			19.80	19.803	9273.67
LKSD4/3+M4	33			19.95	19.803	9325.00

[a] k=9 calibration levels are indexed by j and there are n_j calibration measurements at level j; n=33 measurements altogether, indexed by i.

Table 2 Test of variance homogeneity within the calibration range of Example 1 using the F-test and the Bartlett test[a]

j	n_j	S_j^2	$(n_j-1)S_j^2$	$(n_j-1)\ln(S_I^2/S_j^2)$
1	9	1754.12	14032.96	14.398
2	3	2317.03	4634.06	3.043
3	3	47499.30	94998.60	−2.998
4	3	1670.44	3340.88	3.697
5	3	5386.51	10773.01	1.356
6	3	6411.48	12822.96	1.007
7	3	46579.87	93159.74	−2.959
8	3	5566.33	11132.67	1.290
9	3	7764.51	15529.02	0.624
$\sum$	33		260423.90	19.459

$\alpha=0.05$

F-test: H_0: $\sigma_{j_A}^2=\sigma_{j_B}^2$

Condition $S_{j_A} > S_j B : j_A = 8; j_B = 1$

$F_{\text{test}} = S_{j_A}^2/S_{j_B}^2 = 3.17$

$F_{nj_A-1,nj_B-1;\alpha} = 4.46$

$F_{\text{test}} \leqslant F_{j_A-1,j_B-1;\alpha}$, H_0 is not rejected

Bartlett-test: H_0: $\sigma_1^2=\sigma_2^2=\cdots=\sigma_k^2=\sigma^2$

Condition $n_j \geq 5$

$$S_I^2 = \frac{\sum(n_j-1)S_j^2}{\sum(n_j-1)} = 10851.00$$

$$c = \left(1+\frac{1}{3(k-1)}\right)\left(\frac{1}{n-k}\sum\left(\frac{1}{n_j-1}-1\right)\right) = 1.170$$

$$X_{\text{test}}^2 = (1/c)\left(\sum(n_j-1)ln\left(S_I^2/S_j^2\right)\right) = 16.6$$

$$X_{k-1;\alpha}^2 = 17.5$$

$X_{\text{test}}^2 \leqslant X_{k-1;\alpha}^2$, H_0 is not rejected

[a] H_0 denotes the underlying statistical hypothesis referring to the variances σ_j^2 of the calibration level j; S_j^2 and S_I^2 are the empirical variances and, respectively, the pooled variance; the appropriate test sizes and the constant c are calculated in the usual way.

$$\text{MS}_{\text{PE}} = \frac{1}{n-k}\sum_{i=1}^{n}\sum_{j=1}^{nj}\left(y_{i,j}-\bar{y}_j\right)^2. \quad (9)$$

H_0 (suitability of the regression model) is examined by means of the test quantity $F_{\text{test}}=\text{MS}_{\text{LOF}}/\text{MS}_{\text{PE}}$, which is compared with the corresponding value of the F-distribution with k–2 and n–k degrees of freedom at the significance level α. For $F_{\text{test}} > F_{\alpha;k-2;n-k}$ H_0 is rejected. A lack of linearity in the relation (1) can also be indicated by this test result. The LOF-test should be used in combination with the residual plot.

Example 1: Establishing traceability for the determination of nickel in aqua regia eluates of sediment samples

An analysis method based on a microwave-supported leaching using aqua regia was elaborated in order to determine the maximum acid-soluble proportion of nickel in sediment samples. During the development of the method, the first two activities mentioned above (compilation of significant influence quantities, and of their relative importance) were accomplished.

For calibration, a set of reference materials (LKSD-1 to LKSD-4, lake sediments, CANMET, Canada) were used that had associated data on their acid-soluble proportions. Three hundred milligrams of the sample materials were placed in the digestion vessels of the microwave digestion instrument (MULTIWAVE, Perkin Elmer). 1.2 ml HNO_3 (65%, suprapure) and 3.6 ml HCl (30%, suprapure) were added to these sediment samples. Additionally, the three samples of reference material were doped with multielemental standard solution Merck IV before the microwave digesting procedure (standard addition), which allowed the number of calibration levels to be enhanced to k=9. All of these samples were microwave-supported acid leached at maximally 220°C according to the following program: 8 min/500 W, 3 min/800 W, 20 min/1,000 W, 10 min/0 W. After centrifugation at 3,500 rpm for 20 min, and subsequent decanting, the resulting solutions were topped up with deionised water (Milli-Q, Millipore) to 50 ml, and afterwards used as the basis for the calibration of the ICP-OES (CIROS, Spectro AI). The determination of nickel was accomplished at λ=231.6 nm (background-corrected).

Table 1 shows the calibration data set. The variance in the x-values (concentrations) within each of the calibration levels is small compared to the total variance in the x-values (S_{ex}/S_x=0.0167<0.2), so OLS-regression using the data set $\{x_{j\text{M}}, y\}$ can be accomplished.

Variance homogeneity of the y-values (responses) over the calibration range was proven using both the F-test and the Bartlett-test (although its application is not completely

Table 3 Parameter estimates for the OLS-regression (Example 1)

Regression model	$y = b_0 + b_1 x$	
Number of data pairs	n	33
Parameter estimations		
Mean value x	$\bar{x} = \frac{1}{n}\sum_i x_i$	9.8582
Mean value y	$\bar{y} = \frac{1}{n}\sum_i y_i$	4866.192
Sum of squared deviations x	$Q_x = \sum_i (x_i - \bar{x})^2$	2120.115
Sum of squared deviations y	$Q_y = \sum_i (y_i - \bar{y})^2$	432183315.7
Sum of products of deviations	$Q_{xy} = \sum_i (x_i - \bar{x})(y_i - \bar{y})$	956759.47
Coefficient of determination	$R^2 = Q_{xy}^2 \big/ (Q_x Q_y)$	0.99903
Slope	OLS: $B_1 = Q_{xy}/Q_x$	451.277
Intercept	$B_0 = \bar{y} - B_1 \bar{x}$	417.419
Residual standard deviation	$S_e = \sqrt{\frac{1}{n-2}\left(B_1^2 Q_x - 2B_1 Q_{xy} + Q_y\right)}$	116.342

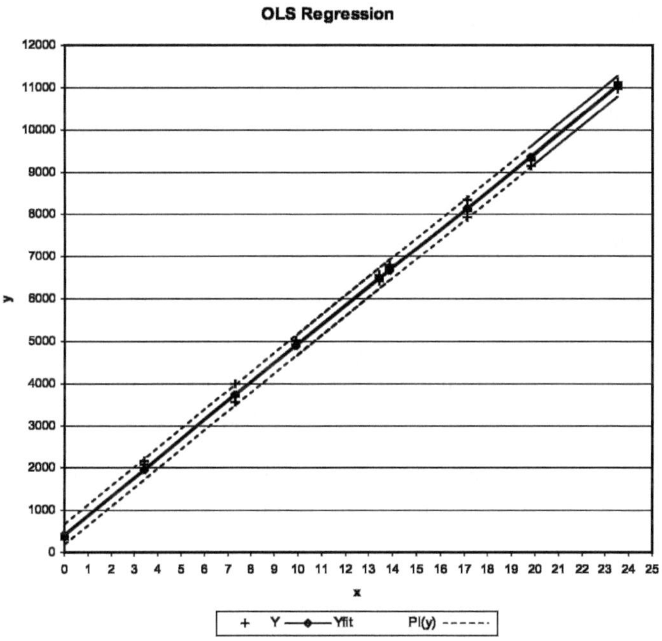

Fig. 1 Regression line and prediction interval PI(y) for the prediction of a single value y_{pred} at position x (based on Eq. 2) of the OLS-regression for the calibration of Example 1

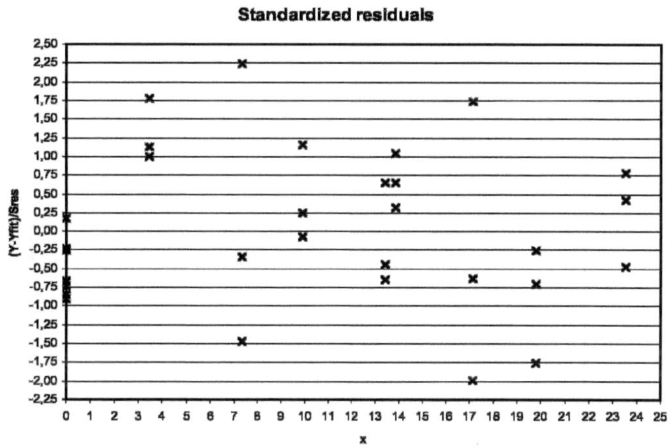

Fig. 2 Standardised residuals (given by Eq. 5) for the OLS-regression (Example 1)

correct because of the unfulfilled condition $n_j \geq 5$, see Table 2). In Table 3 the estimates of the regression parameters are given. In Fig. 1 the regression line and the prediction interval for the prediction of a single value are represented. The plot of the standardised residuals (Fig. 2) suggests no model deviation.

The ANOVA results for the regression are compiled in Table 4 . The F-test value is smaller than the associated critical F-value (there is no significant lack of fit for the OLS-regression). Figure 3 shows the absolute and relative measurement uncertainties for the calibration range. The uncertainties are attributed to the predicted concentrations (x), calculated by means of Eq. 4, and averaged per calibration level. It is apparent that the absolute uncertainties are smaller in the centre of the calibration range, while the relative uncertainties remain almost constant.

The resulting calibration function is traceable to two reference materials for which the acid-soluble parts of nickel are certified: the river clay sediment (LGC 6139, Laboratory of the Government Chemist, Teddington, UK) with a certified concentration of 38±1 mg/kg, and the soil BRM #04 (Bundesanstalt für Materialforschung und -prüfung, Berlin, Germany) with a certified concentration of 22.7±3.7 mg/kg (mean of three labs). Both materials comply with the requirements of reference materials, as mentioned above.

Using the certified materials as samples, nickel concentrations of 38.52±0.31 (n_j=3) mg/kg (LGC 6139) and 24.22±0.75 (n_j=3) mg/kg (BRM #04) were obtained. The uncertainties are smaller than the certified

Table 4 ANOVA results[a] for the OLS-regression of Example 1

Variation	SS (Sum of squares)		df		MS		F-test
Regression	$SS_{Reg}=\sum_{j=1}^{k} n_j\left(y_{j,\text{fit}}-\bar{y}\right)^2$	431763718.5	1	1	MS_{Reg}	431763718.5	
Residual	$SS_{Res}=\sum_{i=1}^{n}\left(y_i-y_{i,\text{fit}}\right)^2$	419596.7	$n-2$	31	MS_{Res}	13535.38	
Lack of fit	$SS_{LOF}=\sum_{j=1}^{k} n_j\left(\bar{y}_j-y_{j,\text{fit}}\right)^2$	159172.8	$k-2$	7	MS_{Lof}	22738.98	$F_t=\frac{MS_{LOF}}{MS_{PE}}=2.10$
Pure error	$SS_{PE}=\sum_{j=1}^{k}\sum_{l=1}^{nj}\left(y_{jl}-\bar{y}_j\right)^2$	260424.1	$n-k$	24	MS_{PE}	10851.01	$F_t\leqslant 2.42=F_{0.05;k-2,n-k}$, no significance for LOF
Total	$SS_{Tot}=\sum_{i=1}^{n}\left(y_i-\bar{y}\right)^2$	432183315.7	$n-1$	32			

[a] j are the k calibration levels, and SS and MS are the sum of squares and the mean sum of squares, respectively; df denotes the degrees of freedom.

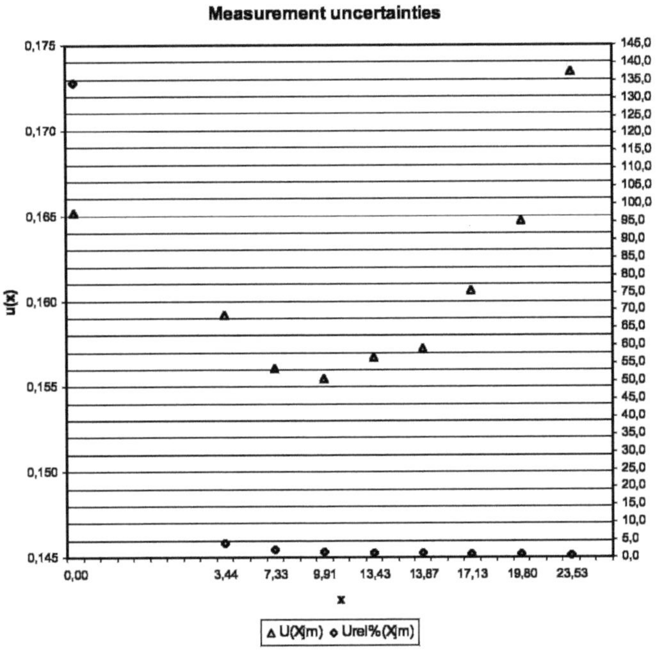

Fig. 3 Absolute and relative measurement uncertainties (left and right ordinate axes, respectively) for the calibration range in Example 1 (uncertainties are calculated via Eq. 4)

values, which is attributable to the fact that only one laboratory measured the samples, resulting in a good repeatability precision. Therefore, this application supplies a good comparability between the microwave-supported leaching and the conventional method of aqua regia elution.

Conclusion

For chemical measurements with a linear calibration function, traceability of results can be formally established without great expenditure if the calibration is based on suitable reference standards and the linear regression is performed as shown above and (statistically) validated. The use of reference materials as samples make it possible to establish the traceability of a new analysis protocol by using an existing analysis method.

References

1. De Bievre P, Kaarls R, Peiser HS, Rasberry SD, Reed WP (1996) Accredit Qual Assur 1:3–13
2. EURACHEM (2003) EURACHEM-Guide: Traceability in chemical measurement, Voting Draft. EURACHEM, Caparica, Portugal
3. ISO (2000) ISO 17025: General requirements for the competence of testing and calibration laboratories. ISO, Geneva
4. VIM:1993, International vocabulary of basic and general terms in metrology
5. Thompson M et al (2002) IUPAC technical report. Pure Appl Chem 74:835–855
6. EURACHEM (2000) EURACHEM-Guide: Quantifying uncertainty in analytical measurement, 2nd edn. EURACHEM, Caparica, Portugal
7. Hässelbarth W (1998) Rückführbarkeit analytischer Messungen. In: Günzler H (Hauptherausgeber) Analytiker Taschenbuch 19. Springer, Berlin Heidelberg New York, pp 45–74
8. EURACHEM (1998) EURACHEM-Guide: The fitness for purpose of analytical methods. EURACHEM, Caparica, Portugal
9. Brüggemann L, Wennrich R (2002) Accredit Qual Assur 7:269–273
10. Massart DL, Vandeginste BGM, Buydens LMC, De Jong S, Lewi PJ, Smeyers-Verbeke J (1997) Handbook of chemometrics and qualimetrics, part A. Elsevier, Amsterdam, pp 171–230

Accred Qual Assur (1998) 3:350

William P. Reed

Traceability, is it what we really want in our chemical measurements?

It is difficult to describe the qualities we want in chemical measurements without using the word "traceability". Just as the thought of traceability brings up the idea of relating measurements to national standards or perhaps to the SI units, the implementation of such traceability brings up a wide range of ideas on how this should be achieved. Some ideas encompass vast systems and interrelationships of chemical measurements and others are no more than simple declarations of traceability.

Perhaps in our desire to achieve traceability we have forgotten why it is most important. That is, traceability is supposed to provide measurement comparability (the ability to compare measurements) on a global level. Unfortunately chemical measurements are notoriously prone to errors caused by the interference of substances found in the sample being analyzed (matrix errors). Hence simply relying on a vertical chain of measurements linking chemical measurements to national standards does not guarantee comparability among measurements. The extra component that is missing is the horizonal intercomparision of measurements among laboratories using appropriate matrix materials. This component provides for comparability by developing and maintaining laboratory skills and in doing so examining and understanding the interferences in the chemical measurement process.

This issue describes some of the many activities taking place today in the United States to develop measurement systems that will hopefully provide for traceability and or comparability of chemical measurements. Much of the activity is taking place in the private sector where we are seeing multiple approaches to measurement traceability and comparability both in the measurement of chemical quantities and in the production of reference materials. The extent to which they are successful is yet to be seen but the incorporation of horizonal linkages has in the past proved to be an effective tool in providing for measurement comparability.

It is interesting to note that even at the highest levels, national standards laboratories designated by their governments to be responsible for disseminating the SI units for length, time (frequency) and voltage intercompare their measurements and generate formal statements of comparability even though the units can be made traceable to the SI from a single laboratory's measurements via first principles. Should not chemical measurements also undergo the same scrutiny, not only with the quantity, the mole, but also with natural matrix materials if we are to have universal comparability?

W. P. Reed
Gaithersburg, MD, USA

Accred Qual Assur 4:154–155

H. Steffen Peiser
Paul De Bièvre

Traceability in measurement – time for an update?

Abstract The present international definition of "traceability" is discussed and suggestions are made for a possible refinement of the definition.

Key words Traceability · Metrology in chemistry · Vocabulary in metrology

The current definition

Chemists in many parts of the world are increasingly using the language of measurement science (metrology). They express their measurements in a way that language is commonly used by physicists and engineers. Chemical analysts, in particular, often describe their results in terms of "traceability". But what is the precise meaning of traceability?

In a chapter on measurement standards (etalons) in the International Vocabulary of Basic and General Terms in Metrology (VIM), first published in 1993, *traceability* is as defined follows:

"Traceability is the property of the result of a measurement or the value of a standard whereby it can be related to stated references, usually national or international standards, through an unbroken chain of comparisons all having stated uncertainties."

"Notes:

1. The concept is often expressed by the adjective *traceable*.
2. The unbroken chain of comparisons is called a *traceable chain*.
3. (Applicable only to the French text) La manière dont s'effectue la liaison aux étalons est appelée *raccordement aux étalons*".

An unauthorized translation of the French text in the third note is as follows: "The manner in which the relationship with the measurement standards is accomplished is called connecting (or, perhaps, binding, or even bonding) to the standards ...".

Reference to standard measures is a practice that goes back to biblical times. A linear measure, the ell, was the length of a standard forearm (about 45 inches) and was used well into modern times. The standard for on a particular kind of stainless steel is one containing 18% chromium and 8% nickel. Thus, not only the units of the International System of Units (SI units), such as mass, temperature and density, but also compositional standards can be defined by attributes or represented by reference materials. Any of these measurement standards can be referenced in technology or commerce often as a contractual or mandatory requirement. "Traceability" and "traceable" are commonly used in relation to (the value of) a material standard.

What does the definition say?

The definition authorizes the use of traceability in the specialized language of measurement science. Since the result of every measurement is a value, all traceability relationships are between values. Undoubtedly, these values are expected to be multiples or sub-multiples of a specific SI unit – or other internationally recognized unit quantity or ratios of unit quantities. A specific reference to a standard is always required and the result of every measurement in a traceability chain must be stated, referenced or clearly understood. It must be assumed (although in the definition it is not explicitly stated) that for each pair of traceable values as well as for the uncertainties of all comparisons in the relevant chain, these values are stated in terms of the identical unit quantity or of ratio of quantities.

What is excluded from the definition?

The definition by no means requires equality of any of the values in a traceability chain, or even that they be numerically similar. The definition only demands that a measurement of the difference of the values (or the ratio between them) has been made with stated uncertainties. The virtue of similarity between values is generally reflected in lower uncertainties of measurement. The definition also does not demand that uncertainties should always be stated relatively to the magnitude of the value in the measurand for which traceability is sought or previously established in a chain. A departure from common practice, as for instance in the use of "absolute" uncertainties, presumably should be made clear within the traceability statement.

Traceability to an authority, institution, or laboratory remains undefined. The existing definition also does not authorize traceability relationships between objects. It has been so used, such as for a local standard or a measurement to that of an institution.

The limits of uncertainty and range of values of the quantity are not required in this definition. Thus, traceability at a level that is judged to be good for its purpose is recognized even when associated with an uncertainty much greater than the current state of the art. Traceability need not connect values from a lower to a higher recognized authority[1]. If this is intended but not self-evident, then the connection should be described.

The definition makes no mention of attributes of the measurand's material matrix or of that of the stated reference, although the ability to compare measurands does not only depend on the relevant quantity and its values, but also on other attributes of the sample and reference standard. The associated components of uncertainties must be kept in mind.

How might the definition be upgraded?

Currently a revision of VIM is under review. We suggest that the following possible amplifications and refinements be considered:

1. Since "traceable to the SI" is currently undefined, the phrase might be specifically defined as a traceability relationship to the appropriate SI unit by primary methods of measurement [as defined by the Comité Consultatif pour la Quantité de Matiére (CCQM) of the International des Poids et Mesures (BIPM)] and with nearly the smallest uncertainties achievable at the time of measurement.
2. Although the current definition clearly states that traceability is a relationship between values, in practice the term is not limited to this use. A revised definition could include a more emphatic statement that traceability refers exclusively to measurement values.
3. The failure of the definition to stipulate a limitation of range of the relevant quantity for a traceability statement might become a serious problem. Measurement methods, uncertainties and relative capabilities vary greatly over the range of values of a quantity. A laboratory that demonstrates traceability, say, to a national kilogram does not necessarily justify its traceability for mass measurements at the megagram level. The range of commonly measured values of several quantities exceeds 20 orders of magnitude. A given traceability relationship might require a limitation of range to which it is applicable, or include a dependence of uncertainty with varying ratio of value measured to the value in the reference standard.
4. The following or a similar statement in the revised definition might prove useful: "each of the related values and the uncertainty of the relationship must all refer to the same quantity on an internationally agreed measuring scale. Such a requirement is self-evident when the measurement consists of a value difference between a measurand and the quantity in a standard. It becomes a non-trivial requirement when the traceability measurement consists of a ratio of ratios of corresponding measurements".
5. If the measurands in a traceability chain are intrinsic attributes of materials, the traceability statement might be expected to limit permissible dissimilarities of matrices. (By an intrinsic attribute of a homogeneous material we mean one that is invariant with respect to sampling).
6. The third note in the current definition might also be stated in English after choosing tan English term, such as "connecting", or "binding".
7. Although methods of proof, evaluation or appraisal of traceability claims may be very important, they should not, in the opinion of the authors, be part of the definition in VIM, but worked out in internationally agreed guides or standards such as published by the International Organization of Standardization (ISO). In this regard a change of definition is not favoured by the authors. These issues should be addressed through international agreement in guides such as those issued by ISO.

[1] Recognized institutional capability to make measurements of the concerned quantity in the involved range.

Paul De Bièvre (✉)
Retired from the Institute for Reference Materials and Measurements (IRMM)
European Commission – JRC,
Duineneind 9, B-2460 Kasterlee, Belgium
Tel.: +32-14-851338
Fax: +32-14-853908
e-mail: paul.de.bievre@skynet.be

H.S. Peiser
Retired from the National Institute for Standards and Technology (NIST) 403 Russel Avenue Apt 313, Gaithersburg, MD 20877, USA

Papers published in this section do not necessarily reflect the opinion of the Editors and the Editorial Board

Accred Qual Assur (1999) 4:478–479

Wolfhard Wegscheider

On the existence of primary methods of measurement

Abstract There is much discussion in chemical metrology about the definition of primary methods of measurement, just as a couple of years ago there was debate about its predecessors, absolute methods and definitive methods. It is argued in this paper that the designation of certain methods as being primary only makes sense if there is an outstanding property identified that is common to all primary methods, and not present for all non-primary methods. The aim to identify primary methods should not blur our notion that it is the good practice of analytical chemistry that produces good results, not a particular method of analysis.

Key words Primary methods · Traceability · Uncertainty · National Measurement Institute

Professional satisfaction for the analytical chemist is to provide good results to her/his customers. It took outside economic pressures to persuade analytical chemists that results must be good, suitable for purpose, but no better.

Traceability [1] is required for results to be suitable for purpose [2] and has two indispensable ingredients:

- a recognized national or international reference
- a complete uncertainty budget tracing back to this reference.

It then follows that there can be a condition that precludes traceability, i.e. the non-existence of a standard and the lack of an uncertainty statement. No matter how rigorously the uncertainty is estimated and irrespective of the size of the estimate obtained, traceability cannot exist without an accepted reference.

It does not follow that a maximum size of uncertainty must not be exceeded before traceability sets in. Consequently, traceability is a discontinuous property of a result with respect to the existence of a standard at the end-point of the traceability chain, or can be thought of as a continuous property with respect to the size of uncertainty: it is uncommon, but not impossible to have results that are more traceable because they have smaller uncertainty than others.

As for primary methods of measurement, it is sensible to interpret "highest metrological quality" (at least) as "providing traceable results". Additionally, one could demand that results of the highest metrological quality should not only be traceable, but also carry as small an uncertainty as possible at a given point in time. It needs to be examined whether primary methods of measurement have an exceptional property that is not found in other methods of measurement, and one must – for reasons of scientific honesty – insist on a property inherent to the method, not its operation or, ever more restrictive, its operation within the walls of a National Measurement Institute. Otherwise, we should try to define just the primary operation or execution of a method, as it is futile to invent methods (in a Platonian sense) that cannot be practiced or are not practicable.

I also deplore the idea that a primary method of measurement in chemistry should be no more than a means of providing a link to another base quantity, such as mass. Chemical metrology is already obsessed with the characterization of ultra-pure substances, that within the limits of their purity can be measured by a simple weighing operation. It is counterproductive for chemistry to mold, stylize and sculpture the balance to be the ultimate analytical instrument, as chemical measurements are mainly invented and practiced to help describe and quantify actions and interactions between substances and are not primarily concerned with "resistance to changes in velocity [3], i.e. mass. The more one relies on methods of measurement that realize chemical measurements by referring to physical units, the more one relies on the quality of preparative chemistry and moves away from analytical chemistry: only very pure substances can in general be characterized reliably in the absence of standards of the entity (analyte) itself.

While a link to another base quantity is surely metrologically useful for the redundancy it provides in the realization of units, at the same time it points to a lack of independence of units that makes the scientific need for seven base units and their independent realization by metrology a dubious matter.

What else can be meant by the "highest metrological qualities" of a result other than that the result has a small uncertainty that will become smaller and smaller with time and experience? As long as nobody is able provide the answer, the concept of primary methods has the questionable advantage of being untestable. However, as we (have to) resort to untestable statements we also depart from experimental sciences. To move in this direction we do not necessarily need National Measurement Institutes. Bu we do need them to stay in close touch with the forefront of good analytical practice, we need them as pools of knowledge and to transfer technology to the bottom of the pyramid. Despite the relevance of ultrapure substances in technology, National Measurement Institutes will have to do analytical research that is concerned with issues other than those related to high-purity substances.

What then is a property that enables a method to be called a primary method, a property not found in a non-primary method? Looking through the list of methods offered by colleagues working on the definition, we find that a large number of them are used in basic laboratory courses in analytical chemistry: gravimetry, coulometry, titrimetry. Undoubtedly, when my students practice these methods, the results do not show the "highest metrological qualities" by any definition. This then brings us back to a point made earlier: the highest metrological quality comes with the operation of procedures, with the care and extent of validation, with the number, the intensity and the quality of cross-checks for interferences or with the operation of independent procedures underpinning the results, with the competence, experience and pride of the analyst in charge. For a good example, see Ellerbe et al. [4].

And the qualities we are looking for are always realized as a matter of degree, these being best reflected by a proper statement of measurement uncertainty. This can also be illustrated by a top-down example from gravimetry: suppose that NIST fails to incorporate in their measurement procedure one correction after another; the quality of results for the gravimetric determination of sulphate as barium sulphate (for a recent account of the procedure applied by NIST, see Vetter et al. [5]) will degrade in a stepwise manner, little by little – first no correction for inclusion of chloride into barium sulphate, then no correction for volatilization, finally no correction for incomplete precipitation.

The results of such a measurement can, of course, also be obtained in the reverse order, i.e. by adding one correction after another as more and more factors that influence the result are recognized. And this actually describes the order in which scientific discovery is made: imagine my students getting more and more practice, having ever better training, learning to do correction on gravimetry by coulometry, ICP-MS and what else might be needed. Undoubtedly, the results they'd produce would also carry less and less uncertainty. Although for obvious reasons I will never attempt to make them as proficient as the colleagues at NIST, the gradual achievement of higher metrological quality can easily be made visible.

It looks then as if we are down to the practice of a method – granted, not all methods are equally well understood as gravimetry after centuries of study. And not all of them will require centuries of study to gain similar prominence. An example of a method that has gained wide acceptance in a much shorter time is isotope dilution. But with all this progress we are still in search of clues to help us identify a primary method of measurement when we happen to come across one, we are searching for at least one outstanding and unique property.

Here is my plea: let's not give up chasing the rainbow, but just so far and not farther.

References

1. ISO (1984) International vocabulary of basic and general terms in metrology. International Organization for Standardization, Genève
2. ISO DIS 17025 (1999) International organization for standardization, Genève
3. Collins Dictionary of the English Language (1981) Collins, London
4. Ellerbe P, Meiselman S, Sniegoski LT, Welch MJ, White EV (1989) Anal Chem 61:1718
5. Vetter TW, Pratt KW, Turk GC, Beck CM II, Butler TA (1995) Analyst 120:2025

W. Wegscheider
Institute for General and Analytical Chemistry, University of Leoben
A-8700-Leoben, Austria
e-mail: wegschei@unileoben.ac.at

Accred Qual Assur (2001) 6:73–75

Alex Williams

Traceability and uncertainty – A comparison of their application in chemical and physical measurement

Abstract Establishment of the traceability and the evaluation of the uncertainty of the result of a measurement are essential in order to establish its comparability and fitness for purpose. There are both similarities and differences in the way that the concepts of traceability and uncertainty have been utilised in physical and chemical measurement. The International Committee of Weights and Measures (CIPM) have only in the last decade set up programmes in chemical metrology similar to those that have been in existence for physical metrology for over a century. However, analytical chemists over that same period have also developed techniques, based on the concepts of traceability and uncertainty, to ensure that their results are comparable and fit for purpose. This paper contrasts these developments in physical and chemical metrology and identifies areas where these two disciplines can learn from each other.

Keywords Traceability · Measurement uncertainty · Chemical metrology · Physical metrology

Introduction

Traceability and uncertainty are fundamental properties of all measurements. Although it is only recently that they have been given a high profile as part of quality assurance and accreditation, their importance have always been recognised by the measurement community. Indeed they were recognised by Galileo, the founder of experimental science. In his famous experiments to study the motion of a ball down an inclined plane, he developed a scale for his measurements of time utilising a water clock. The clock was based on weighing the amount of water flowing out of a large vessel. This enabled him to establish a stable reference for his time measurements and to show that the distance travelled by the ball was proportional to the square of the time. Without this stable reference he would not have been able to establish this relationship.

All measurements are made relative to some scale or standard and are therefore traceable to this scale or standard. The uncertainty on the result will be the uncertainty on the realisation of the scale or standard and the uncertainty on making measurements relative to that scale. The concepts of uncertainty and traceability have developed in different ways in physics and chemistry, leading until recently to quite different approaches to measurements in these two disciplines but now a convergence of approach is emerging.

First I would like to discuss the developments in the concepts of traceability and then in uncertainty in these two areas and show how lessons can be learnt from both that are leading to a common approach.

Traceability

The object of traceability is to enable comparability of measurement results.

This is not just in order to be able to compare results of the measurements on the same sample but also to compare results on different samples to see whether the value of the quantity being measured is larger in one sample than the other. To achieve this, common reference scales must be used in both measurements.

The importance of using common reference scales has been recognised for centuries. For example, in England, King John introduced "consistent measures throughout the land" in 1215. Other countries also had their own measurements scales standards. Many city museums show the standard measures used for trade within the city or local state. As trade widened so did the need for comparability of measurement results and the use of common units widened. The many different measurements scales were harmonised with the introduction of the metric system and the SI units under the Convention of the Metre signed in 1875. An excellent summary of the historical development of units of measurement is given in the NBS Special Publication 420 [1]. Under the Convention of the Metre a hierarchical chain of national and international measurement standards has been developed for the measurement of most of physical quantities.

Chemical analytical measurements were to a large extent left out of these developments. The vast number of chemical compounds and sample matrices made it very difficult to establish a hierarchical system of standards similar to those used in physical metrology. As in physical metrology comparability of results was only achieved locally. For example in the United Kingdom the local control laboratories achieved agreement with each other and with the referee laboratory, the Laboratory of the Government Chemist (LGC), by the development and use of common methods of analysis. International comparability was also achieved in certain trade sectors again by use of common methods, for example use of the Association of Official Analytical Chemists (AOAC) official methods in the food sector. The suitability of the method was checked by means of a collaborative study, which provided parameters that described the spread in results that might be expected from laboratories using this method and gave direct evidence of the comparability of the results. This meant that in general comparability was only possible between results obtained using the same method. This comparability could be extended if suitable reference materials were available and in the (rare) case where the value of reference material was itself traceable to the SI system then a system of traceability similar to that for physical measurements could be established.

Uncertainty

It is both fortunate and unfortunate that both the repeatability and reproducibility of measurement results follow fairly closely the Normal or Gaussian distribution. Fortunate in that a vast number of statistical methods for analysing data have been developed based on the properties of the Normal distribution. But unfortunate, in that has led to the concentration both on the use of the term error and on the errors arising from random effects. This has led to confusion between error and uncertainty and some neglect of the uncertainty arising from systematic effects. When carrying out the statistical analysis the measurement results are often expressed in the following form:

$$x_i = \mu + \varepsilon_i$$

Where ε_i is the error on the ith reading and the expectation value of ε_i is 0 and expectation value of ε_i^2 is σ^2. The values of x_i are taken to be Normally distributed with the mean μ and the standard deviation σ. The values of μ and σ are estimated from the actual readings. Thus although the analysis is carried out in terms of the random errors the data provides an estimate of σ which is the uncertainty arising from random effects. This confusion between error and uncertainty is often added to by referring to σ as the standard error. In addition the statistical analysis is very rarely extended to include systematic errors.

This has meant that for many years there has been confusion in the use of the terminology and an over emphasis on the uncertainty arising from random effects and a neglect of those due to systematic effects. For example at the National Physical Laboratory (NPL) in the United Kingdom in the early 1970s there was an inconsistency on how the accuracy of results was reported on the measurement certificates. In the Radioactivity Section where I worked the measurement certificates stated: "the accuracy of the result is not claimed to be better than ±3 percent". A quite misleading statement since in fact nothing is claimed about the actual accuracy. Faced with the situation I was tasked with some colleagues to pro-lead capacitance can affect the results of AC impedance measurements. There are several effects in other areas of metrology for example the effect of the force on the measuring probe when making the mechanical measurements of length or of phase shifts when making optical measurements. Although these effects may be small they nevertheless can be significant compared with the claimed uncertainty.

In chemical analysis the difference in response the measurement system to a reference material standard of the sample can be very large, indeed in many cases it can be impossible to evaluate. To overcome this standard methods are developed and validated, often by means of collaborative studies and in many cases the performance of the laboratory is checked by participation in proficiency testing schemes. In addition, although the traceability of results is only to the defined method used, the uncertainty on the result can be based on experimental data obtained from validation and collaboration studies. However comparability is often limited to results obtained using the same method.

In contrast methods used in physical metrology are very rarely collaboratively studied neither are the results of the more routine measurements intercompared. The exception being intercomparisons carried out at the highest level by BIPM. The present programme of key intercomparisons will provide data on the agreement between national standards. However this will not be sufficient to ensure comparability of results of routine measurements without some further work to demonstrate that traceability to these standards has been established taking into account any differences between the properties of the standards and the sample being measured. In addition in physical measurements, the uncertainty is not usually evaluated from an experimental study of the method but relies on identifying and evaluating the individual uncertainty components. This means that there is a danger of overlooking a component particularly a component arising from the properties of the sample being measured.

Thus physical measurements would benefit from obtaining direct evidence of comparability by carrying out at least some collaborative studies to check that any difference between the response of their measurement system to the sample of the standard have been correctly accounted for. Proficiency testing schemes could also be more widely utilised. At the present moment undue reliance is placed on the effectiveness of the use of calibrated instruments.

For chemical analysis the need is to develop a hierarchical system, a start on this has been made by the International Committee of Weights and Measures (CIPM), utilising primary methods of analysis. It is envisaged that these primary methods will be used to provide suitable reference materials for using more routine analysis. This will provide a means of achieving traceability to the SI system. However only a limited range of reference materials will be available and it will still be necessary to determine the difference in response of the measure-duce a guide on the statement of accuracy. This guide [2] was published in 1973 and recommended the use of the terms random uncertainty and systematic uncertainty to describe the uncertainty arising from random and systematic effects, respectively. The guide recommended stating the random uncertainty and the systematic uncertainty separately, since at that time there was no agreed way of combining them. The discussion during the preparation of this guide highlighted the problems that had arisen due to the confusion between the terms error and uncertainty. It was mistakenly claimed by some that random uncertainty on a result of one measurement became a systematic uncertainty when that result was used in subsequent measurements. This was clear confusion between the terms random error and random uncertainty and a miss-interpretation of the term systematic uncertainty.

In 1980 the Bureau International des Poids et Mesures (BIPM) published its own recommendations, based on some proposals made by Dr. Müller. These recommendations formed the basis for the ISO Guide "*Guide to the expression of uncertainty in measurement*" [3] with which we are now all familiar. These discussions on uncertainty took place mainly between the national measurement institutes and associated calibration laboratories. To a large extent most testing and analytical laboratories were unaware of these developments. In addition although the general principles for evaluating uncertainty set out in the ISO Guide can be applied to chemical measurements, the examples and to a certain extent the detailed methodology are based on the hierarchical system of standards. This has meant that extra guidance on the evaluation of uncertainty in analytical measurements has had to be developed, which takes into account the way in which comparability is achieved in analytical chemistry and utilises the data obtained from method validation and collaborative studies. This guidance is given in the revised EURACHEM Guide "*Quantifying uncertainty in analytical measurement*" [4].

What lessons can be learnt

At first sight physical systems would appear to have significant advantages. The traceability is achieved by comparison with a hierarchical chain of standards each having its own stated uncertainty. This complies with the requirements for traceability set out in of ISO Guide 17025 where the emphasis is on providing traceable calibrations for the measuring equipment. However this approach overlooks one important point, it does not take into account any difference in response of the measurement system to the standard and the sample being measured. These differences can often be a major if not the largest source of uncertainty and weakest link in the traceability chain.

For example in the case of DC resistance measurements, leakage currents can introduce errors that are large compared to the uncertainty on the calibration of the measuring equipment. Similarly interment system to these reference materials and the samples being analysed. Work on this is already being carried out in a number of institutes and we are seeing the convergence of approach between physical and chemical measurements.

References

1. National Bureau of Standards (NBS) Special Publication 420 (1975) The International Bureau of Weights and Measures 1875 to 1975, US Department of Commerce
2. Campion PJ, Burns JE, Williams A (1973) A code of practice for the detailed statement of accuracy. National Physical Laboratory, HMSO, UK
3. ISO (1993) Guide to the expression of uncertainty in measurement. ISO, Geneva, Switzerland
4. EURACHEM/CITAC (2000) Quantifying uncertainty in analytical measurement, 2nd edn.: http://www.vtt.fi/ket/eurachem/

Presented at the International Conference on Metrology – Trends and Applications in Calibration and Testing Laboratories, 16–18 May 2000, Jerusalem, Israel

A. Williams
Chairman EURACHEM Working Group on Measurement Uncertainty,
19 Hamesmoor Way, Camberley, Surrey, UK
e-mail: alex_williams@camberley.demon.co.uk
Fax: +44-8700-548926

Accred Qual Assur (2000) 5:449–450

Lars Jorhem

Traceability and its role in interlaboratory comparisons (proficiency testing programmes), modeled on trace element determination in biological materials

Abstract Traceability is not always evident in proficiency testing programs, although this is a requirement in ISO/IEC Guide 43-1. The assigned, or "true", value in most programs is not traceable to an independent entity. The test materials should generally be similar in nature to those routinely tested by participating laboratories. This is far from always the case and it is important to realize that if the difference is large, there may be no traceability to the testing program. It is also important that results from participation in proficiency tests are cited when papers are published, in order to enhance reliability/credibility of the published data.

Key words Traceability · Proficiency testing · Elements

Introduction

The ISO/IEC GUIDE 25 [1] stresses, in paragraphs 5.6 and 9.3, the importance of participation in proficiency testing programs or other interlaboratory comparisons, as appropriate and when suitable programs are available. This would imply that such activities are an important part of a laboratory's quality control procedures.

According to ISO/IEC GUIDE 43-1 [2] traceability is: "The property of a result of a measurement or the value of a standard whereby it can be related to appropriate standards, generally international or national standards, through an unbroken chain of comparisons, all having stated uncertainties". It is thus quite clear that traceability is deemed important also in the field of proficiency testing.

The focus of this paper will be on the phrase "an unbroken chain of comparisons". According to the way proficiency tests are normally carried out it is quite possible that this chain might be broken in different ways during the preparation and execution of a proficiency testing program.

Discussion

Traceability of the "true" value in a proficiency testing program

There is no mention of what the interlaboratory comparison or proficiency testing program per se should be traceable to, either in ISO/IEC GUIDE 25 or in ISO/IEC GUIDE 43-1. Independent of whether the "true" or "assigned" value of a proficiency testing program is based on consensus, or based on results from selected (datum) laboratories, it is not traceable to an independent, or known, entity. If the true value is based on consensus, it may be skewed by results from inexperienced laboratories. If this is the case, the results may spread over a very wide range, which may prevent the statistical evaluation procedure from eliminating all outliers. This may result in an over-estimation (or, more rarely, under-estimation) of the true level and consequently give the participants erroneous information about the outcome of the proficiency test. It therefore seems important that the true level is traceable to an independent/separate entity.

That it is of importance that the assigned, or "true", value is as true as possible goes without saying. Most laboratories realize that a bad Z-score (the most common evaluation procedure) will occur from time to time, even within a "well-behaved" laboratory and can accept this even if it is probable that the cause is the true value being a bit off center. For a commercial laboratory, however, working for customers having perhaps little insight in measurement uncertainty and statistics, but with a keen eye on the books, it can be devastating to find that one's result is outside of two or, even worse, three Z-scores. If this result then is based on an erroneous "true" value it may mean a business contract lost for the wrong reason.

The analysis of the testing material by means of an "absolute" technique, e.g., isotope dilution mass spectrometry (ID-MS), would probably be the nearest thing to achieving an absolute or "true" value. This approach is adopted by the International Measurement Evaluation Programme (IMEP) in which the elements to be determined in the sample solution are "certified" by the use of ID-MS or, if this is not possible, it is "assigned" after neutron activation analysis (NAA). Although ID-MS probably is ideal for this purpose, it is perhaps too expensive to be used in programs with more frequent testing rounds and/or several elements.

Another way to establish traceability of the element levels in the test material would be to relate the level found to the level in a certified reference material (CRM) of similar composition and element concentration. The CRM could, for example, be analyzed in parallel to the test material during the homogeneity testing procedure. The result of the homogeneity test would thus be directly related to a certified value. The found and the certified levels of the CRM would then have to be compared using a suitable evaluation procedure, to establish the trueness of the found level. If the outcome is no significant difference, the level in the proficiency-testing sample has traceability to a known entity.

Traceability of analytical results to a proficiency testing program

A laboratory that has achieved accreditation is required by ISO/IEC Guide 25 to take part in proficiency testing activities, provided that a suitable program is available. ISO/IEC GUIDE 43-1 (5.5.3) states: "The test items or materials to be distributed should generally be similar in nature to those routinely tested by participating laboratories".

There will often be a problem with the similarity/suitability of a proficiency test program in relation to the level at which work is usually carried out in a laboratory. If the difference is too great, traceability can no longer be claimed. It may either be the matrix that is not quite agreeable or the concentration of the analyte is at the wrong level.

Biological matrices can vary substantially from each other. Most, however, contain various amounts of interfering elements, salts and oxides, making them potentially problematic to analyze. It may therefore be argued that almost any biological sample matrix, to some degree, is a test of the laboratory's analytical competence.

Samples with too different concentration levels of the analyte pose another type of problem. If the element concentration is too high to be analyzed by the normal procedure, the sample solution is diluted until it can be determined. If this dilution is much larger than what is usually carried out in the laboratory, the matrix might have been diluted to a level where it no longer will have a notable influence on the analyte. It will then have a composition that is quite different from that with which the analyst usually works, and thus become an entirely different analysis. Hence, the result of the analyses normally carried out in the participating laboratory is not traceable to the proficiency-testing program in which it participates.

Is it possible to define how "similar in nature" a specific testing program must be to the analysis carried out in a specific laboratory, in order to provide traceability? Possibly not, but it is important that analysts and quality managers are aware that the problem exists. The value of the proficiency test must not only be seen in light of the Z-score but also its relevancy and to what degree traceability is provided. In this process it can probably be argued, however, that the element concentration is of greater importance than the sample matrix, for the traceability of one's results to the assigned or "true" value of a proficiency test.

Traceability of results in published papers to proficiency tests

Although this perhaps is not traceability in the more "traditional" sense, it is extremely important and should be an integrated part of the efforts to enhance the credibility/reliability of published papers.

Even today, in an age where analytical quality assurance procedures and accreditation are part of most laboratories' daily routines, an often-neglected section in a published paper is where you describe the quality control procedures you have used during the analytical work. It is only natural that the author points out what he/she thinks has been done well. It is equally natural that the weak spots have been suppressed or otherwise disregarded. What procedures the author will describe are thus rather subjective.

Many analysts rely to a very large extent on the analysis of CRMs as the means to guarantee the quality and reliability of the study. This is, however, a procedure that may lead both the analyst/author as well as readers into a false sense of security by over-rating the results. Since the analyst knows the certified level from the outset of the study, it will bias his judgment, probably more or less subconsciously. As a result you rarely see a paper with anything but satisfactory results of the analysis of CRMs. This must be viewed with a certain measure of surprise, since the confidence interval around the certified mean from a statistical point of view is rather difficult to hit with the result of, e.g., a single or duplicate analysis.

Results from a proficiency test, on the other hand, are usually not affected by prior knowledge or personal bias. Satisfactory results from a relevant proficiency test therefore give independent evidence of the quality of analytical results. One problem, however, may lie in the fact that proficiency tests are usually carried out at regular intervals and not in connection with a specific analytical survey. But then again, repeated participation in relevant programs gives a good picture of a laboratory's or analyst's general competence over time.

When results are published in international journals they should therefore be accompanied both by results from relevant proficiency tests as well as relevant CRMs. In this way, results in a publication can be seen as having traceability to an independent entity and possessing reliable analytical quality.

References

1. ISO/IEC Guide 25. General requirements for the competence of calibration and testing laboratories. Third edition 1990. International Organization for Standardization. Genève, Switzerland [Withdrawn in December, 1999 and replaced by ISO/IEC 17025:1999]
2. ISO/IEC Guide 43-1. Proficiency testing by interlaboratory comparisons – Part 1: Development and operation of proficiency testing schemes. Second edition 1997. International Organization for Standardization. Genève, Switzerland

Presented at 3rd EURACHEM Workshop "Status of Traceability in Chemical Measurement", 6–8 September 1999, Bratislava, Slovak Republic

L. Jorhem
National Food Administration, Box 622, 751 26 Uppsala, Sweden
e-mail: lajo@slv.se

RM certification is the whole process of obtaining the property values and their uncertainties, which includes homogeneity testing, stability testing, and RM characterization [5]. ISO Guide 35 [1] requires one to show that the value of such a certified property does not exhibit a systematic error specific to a method or to a laboratory. By widespread opinion, correctness of analytical results is an obvious prerequisite for the RM characterization in contrast to stability and homogeneity studies in which analytical bias is acceptable [5].

In the present paper we would like to discuss primarily the situation when a systematic error in measurement (analytical) results, specific to a measurement method or to a laboratory developing an IHRM, can be overcome. This is possible when a certification of an IHRM is based on a comparative approach providing a characterization of the value carried by the IHRM in comparison to the value carried by the reference material with higher metrological status and sufficiently similar matrix (a CRM).

Accred Qual Assur (2002) 7:122–124
DOI 10.1007/s00769-001-0436-9

Ilya Kuselman
Alexander Weisman
Wolfhard Wegscheider

Traceable property values of in-house reference materials

Abstract The traceability of in-house reference materials (IHRM) is discussed. It is shown that a systematic error in results of a measured value, specific to a measurement method or to a laboratory developing an IHRM, can be overcome if a comparative approach to IHRM characterization is used. A traceability chain of the value carried by the IHRM to the value carried by the reference material with higher metrological status and sufficiently similar matrix (for example, a certified reference material – CRM according to ISO Guide 30) is helpful in such a case. The chain is realized when the IHRM samples are analysed simultaneously with the CRM samples under the same conditions. This and other traceability chains necessary for the IHRM development are examined as the measurement information sources.

Keywords Traceability · Reference materials · Uncertainty estimation · Certification · Comparative approach

Introduction

The traceability of the value carried by a reference material (RM) should be demonstrated by the RM producer [1–3]. The producer shall provide the traceability of results of its measurements to the national or international measurement standards. Where this is not possible, the correlation of results with the values of national or international certified reference materials (CRMs) is required. Ideally, the values of the CRMs should themselves be traceable [2].

More than 220 producers of CRMs throughout the world produce today 12,000–20,000 materials with different matrixes, analytes and properties [4]. However, many testing (analytical) laboratories cannot find suitable CRMs in the market and develop in-house reference materials (IHRMs) themselves. Often IHRMs are developed in a laboratory to conserve the corresponding expensive CRMs. For example, a pharmaceutical company "Chemagis Ltd." produces 30 active pharmaceutical ingredients: steroids, benzodiazepines, antihistamines, hipolipidaemics, blood flow reactants, etc. Only for a few of them – Mometasone Furoate, Fluticasone Propionate and Dobutamine Hydrochloride – are official reference standards for assay supplied by US, British and European Pharmacopoeias with prices of about $ 180 per unit (50–200 mg). Thus, to support its customers Chemagis is forced to develop IHRMs for assay as well as for impurities and related substances of each produced compound. Therefore, certification of such IHRMs that leads to traceable values is very important.

Comparative approach

This approach is based on the transmission of the measurement information from a corresponding CRM to the IHRM [6–9]. The following conditions are required for the IHRM characterization:

1) IHRM and CRM are similar materials,
2) the difference in concentrations of the IHRM and CRM matrix components, which are not under characterization, and corresponding properties of the materials (for example, solubility) does not hinder the use of the same measurement (analytical) method for both the IHRM and CRM,
3) the concentrations of a component under characterization in IHRM and CRM do not differ by more than a factor of two [8].

Test portions in pairs – one of the IHRM and one of the CRM – are analysed by the same analyst and method in the same laboratory and conditions, each pair practically simultaneously. The concentration of the IHRM component under characterization, C_{IHRM}, is compared with its certified concentration, C_{CRM}, in the CRM using the differences in the analysis results $E_i=(C_{IHRM-i}-C_{CRM-i})$ for all pairs: $i=1, 2, ..., n$ ($n \geq 20$). From this comparison the characterized value is calculated as

$$C_{IHRM} = C_{CRM} + E_{avg}, \quad \text{where} \quad E_{avg} = \sum E_i/n. \qquad (1)$$

Obviously, even if C_{IHRM-i} and C_{CRM-i} have an additive systematic error, E_i is free from this error by definition. Additivity of bias is a reasonable approximation for nearly iden-

Table 1 Results of the alcohol determination in the IHRM and CRM LGC5404

i	C_{IHRM-i}	C_{CRM-i}	E_i
1	4.75	5.03	–0.28
2	4.80	5.04	–0.24
3	4.78	5.03	–0.25
4	4.77	5.00	–0.23
5	4.76	5.00	–0.24
6	4.77	5.03	–0.26
7	4.76	5.02	–0.26
8	4.76	5.03	–0.27
9	4.75	5.00	-0.25
10	4.76	4.99	–0.23
11	4.75	5.02	–0.27
12	4.80	5.01	–0.21
13	4.79	5.03	–0.24
14	4.80	5.02	–0.22
15	4.78	5.04	–0.26
16	4.76	5.02	–0.26
17	4.75	4.99	–0.24
18	4.78	5.02	–0.24
19	4.75	5.03	–0.28
20	4.76	5.01	–0.25
Average	4.769	5.018	–0.249
Standard deviation	0.018	0.015	0.019

tical matrices: a multiplicative bias component is assumed negligible at similar concentrations of the analyte in the CRM and IHRM as limited above. Therefore, E_{avg} and C_{IHRM} by Eq. (1) are also unbiased. So, the characterization standard uncertainty [10] is

$$u^2(C_{IHRM}) = [u^2(C_{CRM}) + u^2(E_{avg})]^{1/2} \quad (2)$$

$$u^2(E_{avg}) = \Sigma(E_i - E_{avg})^2/(n-1)n, \quad (3)$$

and $u(C_{CRM})$ is the standard uncertainty of the value carried by the CRM. If the CRM can be selected according to the criteria of ILAC-G9 Guidelines [11] as satisfactory or acceptable, i.e., if $u(C_{IHRM})/u(C_{CRM})>4$, the uncertainty of the value carried by the CRM is negligible. Otherwise it should be taken into account by Eq. (2).

A numerical example of the calculations is shown below.

Characterization of an IHRM of ethanol in water. A reference spirit CRM LGC5404 from LGC with the certified value of the ethanol concentration in water of 5.00±0.03% by volume at the 95% level of confidence is used for the alcohol determination in beer [12]. So, C_{CRM}=5.00% and $u(C_{CRM})$=0.03/2=0.015% by volume (2 is the coverage factor). To prepare an IHRM a suitable ethanol was diluted gravimetrically with water to produce a solution close to 5% ethanol concentration. The solution when thoroughly mixed was suggested to be homogeneous. Twenty test portions in pairs – one of the IHRM and one of the CRM LGC5404 – were analysed in the same conditions using a standard gas-chromatographic method. Results of the analysis are presented in Table 1. The average difference between the results for IHRM and CRM pairs shown in Table 1 is E_{avg}=–0.25% by volume. The value carried by the IHRM according to Eq. (1) is C_{IHRM}=5.00+(–0.25)=4.75% by volume. Note, that the average result of the CRM analysis $C_{CRM\text{-}avg}=\Sigma C_{CRM\text{-}i}/20$=5.02% is biased from the certified value C_{CRM}= 5.00% for 0.02% by volume. It is not influencing the value of C_{IHRM} as far as $C_{CRM\text{-}avg}$ is not used directly in the C_{IHRM} calculation. The standard uncertainty in E_{avg} calculated by Eq. (3) is $u(E_{avg})$=0.004% by volume. So, the characterization standard uncertainty by Eq. (2) is $u(C_{IHRM})$= $[0.015^2+0.004^2]^{1/2}$=0.016% by volume.

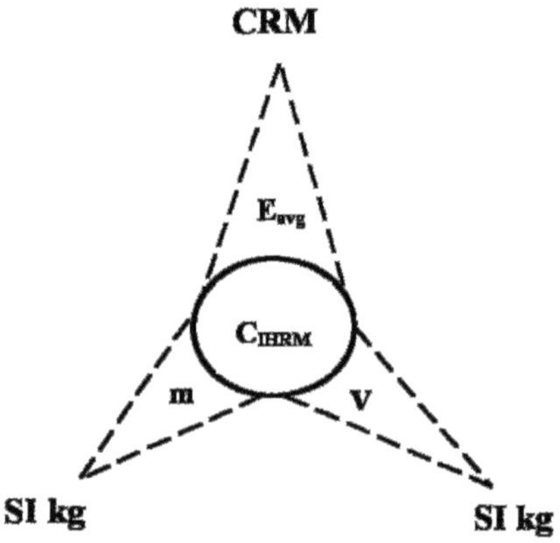

Fig. 1 Traceability chains of a value C_{IHRM} carried by an IHRM. Here m is the mass of the IHRM and V is its volume

Traceability chains

If the first steps of a procedure as described above are weighing and dissolving, a concentration of the IHRM component under characterization C_{IHRM} (the value carried by the IHRM) expressed in % by volume has three traceability chains:

1) of the IHRM mass to the SI kg,
2) of the IHRM volume after dissolving, also to the SI kg, since calibration of volumetric flasks is performed gravimetrically, and
3) of the IHRM carried value comparison to the CRM carried value.

In such a case C_{IHRM} can be shown as a "sun" with three "beams" [13] that are metrological pyramids (traceability chains with minimal uncertainty in the top of the pyramid and maximal in the bottom): see Fig. 1. As a rule, uncertainties in traceability chains to the SI kg are negligible in comparison with the ones from the own analytical process. So, the problem is just how narrow the bottom of the third pyramid is, i.e., how much information is lost in the chain IHRM – CRM.

It is important that the information on a value carried by a CRM is obtained from measurement (analytical) results in other laboratories (not or not only in the laboratory-producer of the IHRM), probably using different methods. Therefore, if a suitable (adequate) CRM is not available, participation of a second laboratory in an IHRM characterization is desirable, even when an unbiased validated method is used, to evaluate a possible bias specific to the laboratory developing the IHRM. Traceability chains of the method form in this case a traceability scheme of the value carried by the IHRM.

Conclusions

It is shown that a systematic error in results of a measured value, specific to a measurement method or to a laboratory developing an IHRM, can be overcome if a comparative approach to IHRM characterization is used. A traceability chain from the value carried by the IHRM to the value carried by the reference material with higher metrological status and sufficiently similar matrix is helpful in such cases. The chain is realized when the IHRM samples are analysed simultaneously with the CRM samples under the same conditions.

If a suitable (adequate) CRM is not available, participation of a second laboratory in an IHRM characterization is desirable, even when an unbiased validated method is used, to assess the laboratory bias. Traceability chains of the method form in this case a traceability scheme of the value carried by the IHRM.

References

1. ISO Guide 35 (1989) Certification of Reference Materials – General and Statistical Principles. 2nd ed, Geneva
2. ILAC-G12 (2000) Guidelines for the Requirements for the Competence of Reference Materials Producers
3. De Bievre P (2000) Accred Qual Assur 5:224–230
4. Zschunke A (ed) (2000) Reference Materials in Analytical Chemistry. A Guide for Selection and Use. Springer, Berlin, p 211
5. Van der Veen AMH, Linsinger TPJ, Schimmel H, Lamberty A, Pauwels J (2001) Accred Qual Assur 6:290–294
6. Shaevich AB (1971) Measurement and Standardization of Chemical Composition of Substances. Publishing House of Standards, Moscow, 280 p (in Russ.)
7. Shaevich AB (1987) Reference Materials for Analytical Purposes. Publishing House "Chemistry", Moscow, pp 147–149 (in Russ.)

8. Pliner YuL (ed) (1988) Metrological Assurance of Materials Composition in Ferrous Metallurgy. Handbook. 2nd ed, Publishing House "Metallurgy", Moscow, pp 157–163 (in Russ.)
9. Pliner YuL, Kuzmin IM (1989) Metrological Problems in Analytical Control of Metallic Products Quality. Publishing House "Metallurgy", Moscow, pp 157–165 (in Russ.)
10. EURACHEM/CITAC Guide (2000) Quantifying Uncertainty in Analytical Measurement. 2nd ed, Teddington
11. ILAC-G9 (1996) Guidelines for the Selection and Use of Certified Reference Materials
12. Kuselman I, Anisimov B, Goldfeld I (2001) Accred Qual Assur 6:107–110
13. Kuselman I (2000) Proceedings of The Millennium International Conference of The Israel Society for Quality. Vol I, Jerusalem, pp 124–129

I. Kuselman (✉)
The National Physical Laboratory of Israel (INPL), Givat Ram, Jerusalem 91904, Israel
e-mail: ilya_kus@netvision.net.il
Tel.: +972-2-6536534
Fax: +972-2-6520797

A. Weisman
Chemagis Ltd., P.O.Box 9091, Tel-Aviv 61090, Israel

W. Wegscheider
Department of General and Analytical Chemistry, University of Leoben, Franz-Josef Strasse 18, 8700 Leoben, Austria

Accred Qual Assur (2002) 7:217–219
DOI s00769-001-0428-9

Linzhen Wang
Qiao Han

Primary reference materials and traceability chain for gas composition

Abstract This article briefly describes research on the development of primary reference gases and the traceability system of gas measurement at the National Research Center for Certified Reference Materials, China.

Keywords Primary reference materials · Traceability system · Intercomparison

Introduction

The gas analysis laboratory of the National Research Center for Certified Reference Materials (NRCCRM) is engaged in gas measurement at the state level. It is responsible for the uniformity of gas composition in China and for maintaining consistency and equivalence with international gas measurements. Our main tasks include: research, development and maintenance of national primary reference materials; research and establishment of precise reference methods for gas analysis; to provide certified reference gas and working reference gas as well as permeation tubes for users; to undertake the verification of gas analyzers and technical advisory services. The focus of our work is the development of research on gas analysis and measurement.

The analysis and measurement of gas components are widely employed in the petrochemicals, electronics, coal, space and instrument manufacture industries, and especially in the areas of environment monitoring, clinical diagnosis and safety assurance which are important to public health and economic development. Hence, government bodies and the departments concerned have been paying great attention to the maintenance, accuracy and uniformity of gas measurement, making gas measurement an important area of chemical metrology.

China began research on gas measurement in the 1960s and made rapid progress by the end of 1970s. We have produced and certified many kinds of certified reference materials and working standard materials including binary or multicomponent organic and inorganic gas mixture, ranging from 1×10^{-6} to 50×10^{-2} (mol/mol). We are continuing our research work on the preparation and measurement of primary reference materials to create a national gas measurement hierarchy, as well as to study the feasibility of the establishment of a traceability system of gas composition measurement on a worldwide scale. Therefore, we have participated in the "key comparison on primary standard gas mixtures" under the auspices of the Comite Consultatif pour La Quantite de Matiere (CCQM) and the success of this program will certainly provide a sound base from which international comparability can be extended.

With reference to the experience gained with other chemical measurements, we designed our research plan on primary reference materials (PRMs) and the traceability system applied in gas measurement. This plan covers the following items:

1. Realize the primary unit of gas composition
2. Develop PRMs and a primary standard device
3. Research and establish reference measurement method with high accuracy
4. Create the traceability system of gas measurement

The realization and maintenance of the value of quantity of gas composition

The way to express the value of quantity of gas composition

For gas mixtures, the preferred way of expressing the composition is in mole fraction (mol/mol), and it is calculated by the following equation:

$$X_i = \frac{n_i}{n_i + \sum n_i} = \frac{n_i}{n} \qquad (1)$$

Where:

XI = mole fraction of component i
nI = amount of substance i
nj = amount of substance j
i, j = signal letter of component, i, j∈[1.p] and i≠j;
p = the total numbers of components

One advantage of selecting the mole is that is provides the amount of substance of the gas composition. In other words, the applicable conditions (pressure and temperature) must not be given and it is not necessary to take additional uncertainty contributions into account.

The method to realize mole fraction – gravimetric preparation

In order to realize the mole fraction, Eq. (1) can be converted into Eq. (2):

$$X_i = \frac{\frac{m_i}{M_i}}{\frac{m_i}{M_i} + \sum \frac{m_i}{M_i}} \qquad (2)$$

Where:

mI = mass of component i (g)
mj = mass of component j (g)
MI = molar mass of component i (g/mol)
Mj = molar mass of component j (g/mol).

By weighing the receptor cylinder, which has been selected and treated, before and after each introduction of component gas and by means of Eq. 2, we are able to get the mole fraction of each component. This method is known as gravimetric preparation. Since mass is one of the seven base quantities and the atomic or molecular weight M can be determined very precisely, the application of gravimetric preparation makes the value of quantity traceable to the mole, the SI base unit in chemistry.

It is necessary to point out that gravimetric preparation is applicable only to mixtures of gaseous, or totally vaporized components which do not react with each other or with the cylinder walls (Reference 1).

Sometimes a multiple dilution method may be used to prepare a final mixture with acceptable uncertainty in typical low concentration of minor components.

National standards of the value of quantity of gas composition

Primary reference gas mixtures

Being the basis of the traceability system, there are certain stringent technical restrictions on the definition of PRMs. They have to meet the following requirements:

a. They must be determined by primary method and traceable to the SI unit.

Gravimetric preparation is an authorized method with high accuracy.

b. The uncertainty and confidence level of PRM should have definite meaning. And the technical specification of the PRM should be advanced in the country or in the world, so that they can be recognized as the base of the traceability system at the highest level.
c. The value of quantity of the PRM must have been proved to be accurate and reliable, and of international comparability by participating inter-laboratory comparison program.

NRCCRM, China has establish many kinds of authorized measurement methods, solved the problem of purity analysis, and completed research on several sets of primary reference gases.

Primary standard facilities

(1) Weighing apparatus – used to weigh the mass of each component introduced into the cylinder. This includes:

a. Balance: capacity – 30 kg, resolution – 1 mg
b. Electric balance: capacity – 16 kg, resolution – 1 mg
c. Weights: first class weights traceable to the national standard.

(2) Gas filling apparatus – used to fill gas component into the cylinder, which consists of the following parts:

a. Vacuum sets – mechanic pump and diffusion pump
b. Vacuum gauge – 10^{-2}~10^{-3} Pa
c. Pressure gauge
d. Valves and pipes.

(3) High pure (99.999%) and pure (99.9) gases.

(4) Other accessories.

Study on the high precise measurement method

In the process of developing PRMs, it is necessary to study and establish measurement methods which are used to analyzed the purity of raw gases and verify the stability of the gas mixture kept in the cylinder. Up to now, NRCCRM has been equipped with several series of analytical techniques including atmospheric pressure ionization mass spectrometer, gas chromatograph, infra-red spectrophotometer with long-path gas cell, chemiluminescent, non-dispersive infra-red, minor O_2 and H_2O analyzer and so on.

Purity analysis of parent gas

Parent gas can be divided into two parts: the minor component and major component. The major component is also known as background gas or diluent gas.

a. Requirements on purity:

The purity of minor gas should be better than 99.9%. Background gases are usually quite stable, for example N_2, Ar and He, and the purity of them is usually better than 99.999% . In addition, in the preparation of PRMs, two kinds of impurities have to be controlled very carefully. The first type are O_2 and H_2O; their existence may lead to a reaction between the components and have an impact on the stability of PRMs. The second type are impurities existing in the background gas which are the same gas as the minor component, which we intend to add in the cylinder. When preparing PRMs with low mole fraction, these kind of impurities should be analyzed accurately. For example, if we want to prepare CO-N_2 of 10×10^{-6} (mol/mol), the CO existing in the diluent gas of N_2 must be determined carefully.

b. Determination of the purity

The determination for the purity of the gas is generally carried out by analyzing the major impurities in a normally 'pure' gas. So the purity of the 'pure' gas is calculated by the next equation:

$$X_p = 1 - \sum_{i=1}^{N} X_i$$

Where:

XI = mole fraction of impurity I, determined by analysis
N = number of impurities likely in the final mixture
Xp= mole fraction 'purity' of the 'pure' gas

After having studied the method above systematically, NRCCRM established a practical analysis procedure which could control the error factors effectively at the same time.

Stability of PRMs

The prepared PRMs must be analyzed regularly to ensure that no significant chemical reactions have taken place and that no absorption or desorption has occurred. This procedure is carried out over a significant period of time to verify the stability in concentrations.

The method employed to evaluate the stability is to use a freshly prepared gas mixture as the calibration gas, and, by means of single-point calibration or linear regression method, determine the Xi (mole fraction of component i) at certain intervals. Then we can figure out the variation factor(expressed in RSD). If the variation factor is within a specific limit, the PRMs are proved stable.

Uncertainty of PRMs

There are a number of sources of uncertainty that influence the final uncertainty for the gas mixtures. For PRMs prepared by NRCCRM, we usually take three sources of uncertainty into account. These sources are listed below:

1. Uncertainty related to the balance and the weight (uc1)

   Which includes:

   a. Resolution of balance
   b. Accuracy of balance
   c. Incorrect zero point
   d. Drift (thermal and time effects)
   e. Instability to draught
   f. Uncertainty in the weight used
   g. Buoyancy effects on the weight used

Table 1 Primary reference gases maintained at NRCCRM

Group	Component	Mole fraction (mol/mol)[a]		
A	CO-N_2	6%;	100×10^{-6};	1000×10^{-6};
B	CO_2-N_2	15%;	100×10^{-6};	1000×10^{-6};
C	NO-N_2		100×10^{-6};	1000×10^{-6};
D	CO_2-N_2		100×10^{-6};	1000×10^{-6};
E Natural gas		E1	E1	E1
	N_2	4.0%	7.0%	14.4%
	CO_2	1.0%	3.0%	0.5%
	C_2H_6	3.0%	9.4%	3.0%
	C_3H_8	1.0%	3.4%	0.5%
	n-C_4H_{10}	0.2%	1.0%	0.1%
	CH_4	(balance)	(balance)	(balance)
F Automobile Emission gas	CO_2	13.5%		
	CO	3.2%		
	C_3H_8	0.2%		
	N_2	(balance)		

[a] nominal value

2. Uncertainty related to the gas cylinder (uc2)
   a. Mechanical handling of cylinder
   b. Buoyancy effects of the cylinder itself
   c. Cylinder temperature differs from surrounding air
   e. Difference of cylinder volume before and after filling
   f. Change of density of air
3. Uncertainty related to the component gases (uc3)
   a. Leakage
   b. Absorption/reaction of component on internal cylinder surface
   c. Reaction between components
   d. Impurities in the parent gases used
   e. Insufficient homogenization
   f. Uncertainty of molecular mass
4. The final uncertainty of the PRMs (U)

   U=2uc with confidence level of 95%; where uc is the combination of uc1, uc2 and uc3which are listed before. For the primary reference gases of NRCCRM, U is in the range of 0.1% to 0.3%.

The intercomparison of PRMs

In order to achieve international multi-recognition and uniformity of the value of PRMs, it is very important to carry out intercomparisons between leading national chemical metrology laboratories, especially in the process of research and development of PRMs. Therefore, in 1992, NRCCRM started its intercomparison work with the Netherlands Measurements Institute (NMI) on CO-N_2, From 1993 to 1999, NRCCRM participated in the intercomparison program between national metrology institutes (NMIs) under the auspices of the CIPM-CCQM – 'Key Comparison on Primary Reference Materials'. The types and contents of the gas mixtures, selected for comparison covered the fields of environment monitoring, detection of automobile emission gas and natural gas. etc., and were representative of the PRMs maintained in each country. In a comparison of 7 types of gas mixture PRMs by 27 groups, the results of 24 groups agreed to +/–1% with other major NMIs in the CCQM comparison (Reference 2). Some unsatisfactory results reflected defects in our work and have motivated us to improve our work in the future.

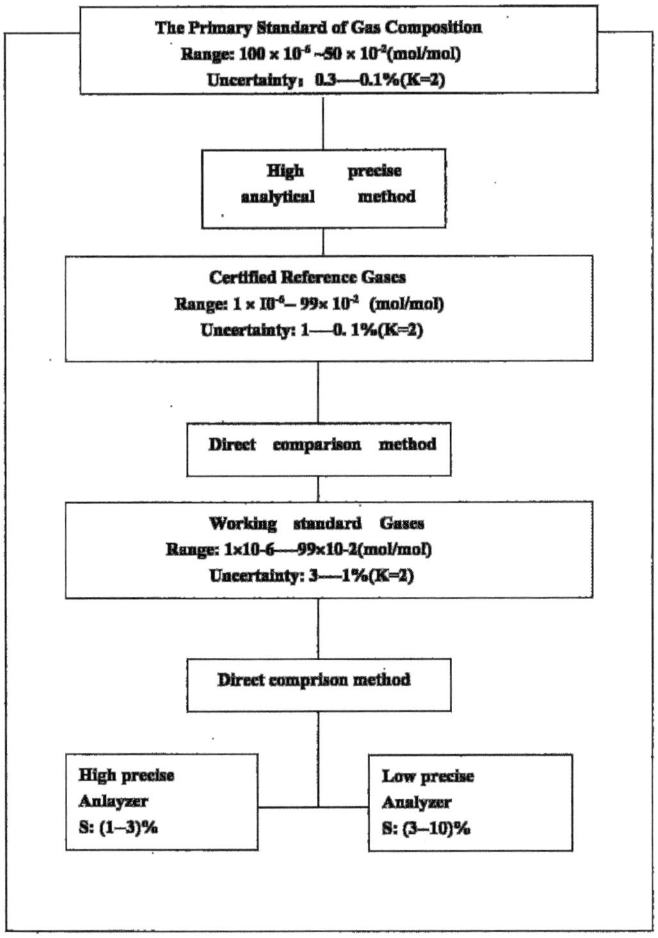

Fig. 1 Traceability system of gas chemical composition

As a result of participating in the intercomparisons, several groups of our PRMs have been set up and validated. Each group is comprised of three cylinders of gas of nearly the same composition. These PRMs are kept in the gas analysis laboratory of NRCCRM.(shown in Table 1).

With reference to the traceability system of other quantities and to meet the demands on gas measurement, we have develop a traceability system (shown in Fig. 1) and hope it will become the basis of the uniformity of gas measurement at the highest level.

References

1. ISO 1998 ISO/CD 6142(1998) Gas analysis – Preparation of calibration gas mixtures – Gravimetric method. ISO, Geneva
2. Anton Alink Metrologia (2000)

Wang Linzhen · Han Qiao
National Research Center
for Certified Reference Materials,
100013 Beijing,
People's republic of China
Tel.: +86-10-64221810
Fax: +86-10-64228404

Accred Qual Assur (2003) 8:475–476
DOI 10.1007/s00769-003-0599-7

Gary Price

Traceability to units

Published online: 9 April 2003

Abstract On the grounds that clear and direct communication is required of us today, it is proposed that traceability be regarded as the ability to demonstrate that measurements are what they are purported to be and that traceability is thus to measurement units rather than reference values per se. It is suggested that such an approach may give greater flexibility in the establishment, maintenance and propagation of traceability, and that accreditation practices are becoming central to the practical establishment of traceability for chemical and biological measurement.

Keywords Traceability · Units · Reference value · Metrology · Accreditation · Scale

Communicative accuracy

The wider society of measurement users is anybody and everybody who makes, or is affected by, a decision dependent upon a measurement. It is literally everybody. The decisions large and small, in their teeming multitude, are made by men and women with a vast variety of knowledge and experience, but seldom directly relevant to the technical issues of the measurement. Through a fog of self- and competing interest, and particular circumstance and need, they must summon sufficient understanding to make informed decisions. They are the audience whenever we write down a measurement result. The dilemmas of communicating to the public are not of course unique to chemical and biological measurement. Warren Weaver, one of the founders of the modern theory of communication, proposed to the scientist struggling with communication problems the concept of communicative accuracy [1]. It rests on the fact that the effective accuracy of a communication depends primarily on the interpretation given to it by the audience. Weaver suggested two conditions for communicative accuracy:

First, taking into account what the audience does and does not know, it must take the audience closer to a correct understanding..... Second, its inaccuracies (as judged at a more sophisticated level) must not mislead....

Both of these criteria must be applied from the point of view of the audience, not from the more informed and properly more critical view of an expert.

Viewed through these criteria, there is considerable room for improvement in our communication practices in chemical and biological measurement. Anyone who wishes to dispute this should contemplate how they would explain to a jury in the face of a persistent and skillful opposing advocate the concepts of the mole and amount of substance as defined in the SI.

Traceable to what?

Traceability is a concept intimately involved in the communication of measurement results and their meaning, but what is the wider society of users of measurement results to make of it? On the one hand they are told of its central importance and the need to have an infrastructure to enable and support it, and on the other hand, confusion, bewilderment and ambiguity as to what it is and most especially, what it is *to*. They are told that traceability is a "property of the result of a measurement or the value of a standard whereby it can be related to *stated references, usually national or international standards*, through an unbroken chain of comparisons all having stated uncertainties" [2] (my emphasis). Yet they find talk of traceability to: pieces of paper or certificates, laboratories or institutions, methods or instruments, pure substances, certified reference materials, reference values, "agreed" reference standards, and the delightfully bureaucratic "appropriate" reference standards. We all know about the multitude of ambiguities attached to the English word "standard", but this is surely ridiculous.

To start with, most of the end points listed are not even the sorts of things that measurements can be traceable to, at least on the face of it. To say that laboratories or substances are properties of measurements per se and to which they may be compared is a simple confusion of logical categories, like saying that a velocity can be compassionate. Indeed, strictly and literally speaking, the only member of the list that lies unambiguously in the universe of possibilities for traceability end points is "reference values" [3].

Not unreasonably it will be objected that this is an overly literal interpretation: that when we say a measurement is traceable to, for example: a laboratory; or a pure substance; or a certified reference material, we are speaking shorthand code for something respectively like: "reference value determined or maintained" by a laboratory; "reference value created by weighing out a sample" of a pure substance; or "reference value stated in an attached certificate" of a certified reference material. It is not an unreasonable approach in specific cases of specific measurement situations when technical peers talk among themselves about specific problems. But as a general explanation to the wider society of measurement users it is both beside the point and misleading, for at least two reasons.

The first reason is that the wider society quite rightly demands transparency and trust in the communication of information essential to such fundamental purposes as production, trade and commerce, health care, environmental policies, and legal judgement. Speaking in codes to your audience is not one of the more notable ways to achieve transparency and trust. The wider society is entitled to ask what is really going on with this linguistic *jiggery pokery*, and whose interests does it serve? More plain speaking and less – not more – obfuscation are what is required of us.

The second reason is that it is incomplete and leaves the traceability chain dangling like a metrological rope trick. On being informed that traceability is to reference values which may be in a variety of material forms, an intelligent but technically unsophisticated measurement user is entitled to say "Yes, I think I understand that, but what's the point? What is all this complex chain of comparison actually for? What is its purpose? And why is traceability such a good way of doing it?" You will generally have no more than 30 seconds to answer before your audience turns their mind to more pressing matters, convinced only of the utter inconsequentiality of the subject.

A small proposal

Allow a modest proposal: "Traceability is the ability to demonstrate that measurements are what they are purported to be". Because measurements are always expressed and communicated in the form of numerical values (with associated uncertainties or equivalent intervals between numerical values, at stated levels of confidence) multiplied by measurement units, it then follows by ineluctable logic that the end point of any traceability chain is simply the units in which the measurement is expressed. What then, is the role of reference values?

There is an older form of words that one still sometimes sees that infers that traceability is to "reliable realisations of the units" in which a measurement is expressed. The proposal here given is a close cousin to that view and is related also to Belanger's approach that traceability is the means to ensure measurements of accuracy sufficient for the purpose at hand [4], an approach that Nicholas and White amended (with close consideration of the evolving role of accreditation) to the view that traceability is the ability to demonstrate the accuracy of a measurement in terms of its expressed units [5].

However those previous views adequately considered neither the complexities of practical chemical and biological measurement nor their highly instrumental nature, nor the variety of measurement units and scales that may be encountered. A more general formulation is that measurement scales are the means by which numbers are assigned to quantities that we may desire to measure and that scales are on the one (theoretical) hand, *defined* by measurement units; and on the other (practical) hand, *realised* in material ways by applying reference values to measurement procedures or instruments (sometimes termed "calibration"). Thus the suggestion being made here is that the construction of a measurement scale is the mediating step between the reference values that may be available to an analyst and the units with which the relevant measurement results are expressed.

One reason to commend the approach suggested is that it gives a proper emphasis to the integrity of the whole measurement, not just the series of comparisons from measurand or analyte to reference values. In actual practice, the uncertainties with which reference values effectively realise units are on occasion omitted from consideration. This is sometimes justified with the claim that, if all is in order, the uncertainty of the reference value is negligible in comparison to that of all operations and influence factors from reference value to measurand. Far from being in order, I wish to suggest that the requirement that reference values have negligible relative uncertainties is a potential barrier to the advancement of practical metrology in chemistry.

Where do reference values come from?

There is one thing that we can definitely say is not a source of reference values and that is interlaboratory comparison. Interlaboratory comparison, proficiency testing and the like are very useful tools, depending on the protocols and intended purposes. They can be used to detect and diagnose problems, evaluate competencies and proficiency, and even possibly propagate traceability, but they can never conjure traceability into existence. If traceability is not independently established, interlaboratory comparison is just as efficient and effective in propagating systematic error.

One answer is that reference values ultimately derive from national measurement institutes, high level laboratories and facilities with the technical capability of making highly accurate primary measurements on carefully prepared materials and these are disseminated down the measurement chain as certified reference materials of progressively larger uncertainties. The Consultative Committee for Amount of Substance (CCQM) defined a primary method of measurement as "having the highest metrological qualities, whose operation can completely be described and understood, for which a complete uncertainty can be written down in terms of SI units" [6]. The central idea was that such a method was in a sense an absolute or defining measurement that could stand alone without reference or comparison to other standards of the same quantity and produce independent reference values. Its essence was the conceptual transparency implied by the terms "completely described and understood" so that a relatively simple equation could describe the measurement and all influence factors could be accounted for and their uncertainty evaluated. It is a very useful idea with large unrealised potentials, but nobody has yet given any cogent explanation of the phrase "highest metrological qualities". Could this phrase possibly be a bit of special pleading in favour of oligarchic (if not monopolistic) supply of reference values? There is certainly no compelling reason in principle to say that a competently applied primary method but of larger than state-of-the-art uncertainty does not still produce a perfectly good reference value for many purposes.

Historically, analytical chemistry more often than not relied on "do-it-yourself reference values", by for example weighing out samples of pure materials, preparing "standard solutions" or creating experimental set-ups where known amounts of a species of interest are generated and the like. These may not have been of the "highest metrological quality" but they were certainly fit for their purpose then. Of course, that was decades ago, when analytical life was very, very much simpler and many of our now commonplace instrumental methods did not exist. But many modern instrumental methods are also potentially highly precise-far more precise than many of their practical field uses may require. Might higher uncertainties in the reference values applied to them still result in overall uncertainties at least adequate to the purpose?

What is contemplated here is a flattening of the practical metrological pyramid, a shortening of the distance between units and their expression, a reduction of the number of comparisons in the traceability chain and a degree of metrological self-sufficiency where laboratories or networks of laboratories could, if it makes sense in their circumstances, construct as directly as possible their own intrinsic realisations of units or reference values, appropriate to their purposes but traceable to units in common with (and hopefully understood by) the rest of the world.

Conclusion

Reference values may be propagated in many ways. Traditional certified reference materials are just one of them. One could for instance imagine electrochemical amount generators, or reference laboratories able to perform matrix-independent reference measurements on samples submitted by field laboratories, the samples then being returned to the field laboratory with a reference value attached. There are many possible paths to metrological virtue. The requirements of communicative accuracy suggest they converge on measurement units. What is essential to all of them nowadays is summarised in the first part of the proposed definition of traceability: "..is the ability to demonstrate.." Accreditation practices, the trust and transparency mechanisms of modern practical measurement systems are I would suggest now central to the establishment of practical traceability for chemical and biological measurement.

References

1. Weaver W (1967) Science and imagination: selected papers of Warren Weaver. Basic Books, New York London, pp 183–184
2. ISO (1993) International vocabulary of basic and general terms in metrology. ISO, Geneva
3. De Bievre P (2002) Accred Qual Assur 7:515
4. Belanger BC (1980) Bulletin OIML 78:21–25
5. Nicholas JV, White DR (1994) Traceable Temperatures: an introduction to temperature measurement and calibration, 1st edn. Wiley, New York, pp 22–25
6. Consultative Committee on Amount of Substance (1998) Report of the fourth meeting. BIPM, Sevres, Paris

G. Price (✉)
P.O. Box 57,
Menai , NSW 2234, Australia
e-mail: gzprice@ozemail.com.au
Tel.: + 61-2-9543 2224
Fax: +61-2-9543 2224

Accred Qual Assur (2003) 8:530–531
DOI 10.1007/s00769-003-0686-9

Ilya Kuselman
Alexander Weisman
Wolfhard Wegscheider

Traceability without uncertainty: current situation in the pharmaceutical industry

Published online: 30 August 2003

Abstract The current situation in the pharmaceutical industry is discussed, when the traceability of measurement (analytical) results to certified values of pharmacopoeial reference standards is required, without evaluating their uncertainties. It is shown that the evaluation of measurement uncertainty is necessary for understanding the level of confidence of the analytical results and their comparability, particularly during preparation and characterisation of the reference standards.

Keywords Traceability · Measurement uncertainty · Reference materials · Quality of analytical results

Comparable analytical results are required in the pharmaceutical industry, not only to avoid duplication measurements, which cost time and money, but also to be sure that the product corresponds to its specifications and is not harmful or even dangerous for a patient. Comparable results obtained in different laboratories at different time can be achieved only if these results are anchored to a common base, i.e. are traceable to this common base, preferably to one widely recognized worldwide [1]. Establishment of traceabilty for an analytical method is based on understanding that:

1. While developing the method, a set of measurement conditions, procedure and a formula for the analyte calculation are optimised,
2. Validation demonstrates that this set of conditions and the calculation are sufficiently complete for the defined purpose,
3. Once these conditions are met, the laboratory needs only to establish traceability, or control of each value in the equation for the calculation, as well as of each of the critical specified conditions [2].

According to the international vocabulary [3], traceability is the property of the result of a measurement or the value of a standard whereby it can be related to stated references, usually national or international standards, through an unbroken chain of comparisons all having stated uncertainties. Uncertainty is a parameter associated with the result of a measurement, which characterises the dispersion of the values that could reasonably be attributed to the measurand/analyte [3, 4].

The concept of measurement uncertainty has been recognized by chemists for many years, and general rules for evaluating and expressing uncertainty in analytical measurement have been formulated in the international guide [5] based on practical experience. The uncertainty estimation, according to the guide, consists of four steps:

1. Specifying measurand/analyte using information given in the relevant standard operation procedure (SOP),
2. Identifying uncertainty sources specified in the SOP, including sources arising from chemical assumptions,
3. Quantifying uncertainty components (from validation data, certificates of calibrations, manufacturer's information etc.) as standard deviations,
4. Calculating combined (overall) uncertainty as a root of a sum of variances associated with the uncertainty components, and then calculating expanded uncertainty multiplying the combined uncertainty by an appropriate coverage factor corresponding to the necessary level of confidence.

The current situation in the pharmaceutical industry is that the traceability of measurement (analytical) results to certified values of pharmacopoeial reference standards is required, without assessing their uncertainties. The problem is, however, that the necessity for uncertainty information follows from the need to ensure that the references used are sufficiently accurate for the purpose and to provide similar information on the analytical result. In dealing with traceability, this is important because (a) the uncertainty of the analytical result cannot be lower than the uncertainty arising from the reference standards in use, which influences the choice of the reference standards and (b) for a given analytical method, achieving low result uncertainty requires control of a larger number of variables [2]. Unfortunately, neither US Pharmacopeia [6] (USP) nor European Pharmacopoeia [7] (EP) define even the uncertainty of certified values of recommended reference standards, in spite of requirements of ISO Guide 35 [8, 9]. In the "Official USP Reference Standards Catalog" [10] one can find only a declaration that all 1350 USP reference standards for pharmaceuticals, excipients, and dietary supplements are established through a process of rigorous testing, evaluation, and quality control. There is also no information on uncertainties of certified values of 1518 EP standards listed in the "European Official Catalog of Chemical Reference Substances and Preparations" [11]. The authors of this catalogue refer to the specificity of pharmacopoeial reference substances that has been officially recognized in ISO Guide 34 [12], where the following is really noted: "... the uncertainty of their assigned values is not stated since it is negligible in relation to the defined limits of the method-specific assays of the pharmacopoeias for which they are used". However, this statement is not efficient, most obviously when measurement results are close to the limits. Moreover, it is in contradiction with the definition of a certified reference material whose certified property values are to be traceable to an accurate realization of the unit in which they are expressed, and also to be accompanied by an uncertainty at a stated level of confidence [3, 13]. As a consequence of this situation, the uncertainty of property values carried by working standards (in-house reference materials), developed in analytical laboratories of the pharmaceutical industry and based on comparison with corresponding reference standards, also cannot be evaluated completely [14, 15]. In this case, the uncertainty of an analytical result at the end of the traceability chain "reference standard → working standard → measurement result" cannot be assessed correctly. This is not good laboratory practice, since knowledge of the uncertainty is crucially important for assessing the quality or compliance of the analytical result [5, 16, 17].

Uncertainties associated with qualitative analysis and with purity assessment, especially at the reference standards characterization, are subjects of increasing attention of the metrological and the analytical communities [18, 19, 20, 21, 22, 23].

Traceability and uncertainty of measurement results are basic technical elements of quality systems in analytical laboratories whose competence is recognized by accreditation according to ISO/IEC 17025 [24]. However, GLP and GMP standards widely used since the 1960s for the quality system assessment in pharmaceutical industry [25, 26] do not include requirements for the measurement uncertainty evaluation.

It is clear today that the issue discussed above should be taken into account by pharmacopoeial committees and regulatory bodies involved in drug quality assurance for harmonization of requirements to analytical results and improvement of their quality, first of all while developing the reference standards.

References

1. Papadakis I, Wegscheider W (2000) Accred Qual Assur 5:388–289
2. EURACHEM/CITAC (2003) Traceability in chemical measurement. A guide to achieving comparable results in chemical measurement. EURACHEM, Teddington
3. BIPM, IEC, IFCC, ISO, IUPAC, IUPAP, OIML (1993) International vocabulary of basic and general terms in metrology. ISO, Geneva
4. BIPM, IEC, IFCC, ISO, IUPAC, IUPAP, OIML (1993) Guide to the expression of uncertainty in measurement. ISO, Geneva
5. EURACHEM/CITAC (2000) Quantifying uncertainty in analytical measurement, 2nd edn. EURACHEM, Teddington
6. US Pharmacopeia XXVI (2003) US Pharmacopeial Convention, Rockville, MD, pp 1966–1967
7. European Pharmacopoeia (2002) 4th edn. Council of Europe, Strasbourg, p 8
8. ISO (1989) Guide 35. Certification of reference materials—general and statistical principles, 2nd edn. ISO, Geneva
9. Van der Veen AMH, Linsinger TPJ, Schimmel H, Lamberty A, Pauwels J (2001) Accred Qual Assur 6:290–294
10. USP (2003) Official USP Reference Standards Catalog. http://www.usp.org
11. European Directorate for the Quality of Medicines (2003) European Official Catalog of Chemical Reference Substances and Preparations. http://www.pheur.org
12. ISO (2000) Guide 34. General requirements for the competence of reference materials producers, 2nd edn. ISO, Geneva
13. ILAC (2000) G12. Guidelines for the requirements for the competence of reference material producers.
14. Kuselman I, Weisman A, Wegscheider W (2002) Accred Qual Assur 7:122–124
15. Weisman A, Gafni Y, Vernik M, Kuselman I (2003) Accred Qual Assur 8:263–266
16. Andersson H (2002) Accred Qual Assur 7:228–233
17. Kuselman I (2000) Rev Anal Chem 19:217–233
18. Hardcastle WA, Ellison S (2002) Expression of uncertainty in qualitative testing. approaches to the problem. Discussion paper. LGC/VAM, Teddington
19. Rios A, Barcelo D, Buydens L, Cardenas S, Heydorn K, Karlberg B, Klemm K, Lendl B, Milman B, Neidhart B, Stephany RW, Townshend A, Zschunke A, Varcarcel M (2003) Accred Qual Assur 8:68–77
20. Roesslein M, Ellison S (2003) CITAC News, p 4
21. Vershinin VI (2000) J Anal Chem (in Russ.) 55:468–476
22. Vershinin VI, Topchij VA, Medvedovskaya II (2001) J Anal Chem (in Russ.) 56:367–373
23. Zolotov YuA (2002) J Anal Chem (in Russ.) 57:677
24. ISO/IEC 17025 (1999) General requirements for the competence of testing and calibration laboratories. ISO, Geneva
25. Garner WY, Barge MS, Ussary J.P (eds) (1992) Good laboratory practice standards. Applications for field and laboratory studies. ACS, Washington DC
26. Miller JM, Crowther JB (eds) (2000) Analytical chemistry in a GMP environment (a practical guide). Wiley, New York

I. Kuselman (✉)
The National Physical Laboratory
of Israel (INPL),
Givat Ram, 91904 Jerusalem, Israel
e-mail: ilya_kus@netvision.net.il
Tel.: +972-2-6536534
Fax: +972-2-6520797

A. Weisman
ChemAgis Ltd.,
P.O.Box 9091,
61090 Tel-Aviv, Israel

W. Wegscheider
Department of General
and Analytical Chemistry,
University of Leoben,
Franz-Josef Strasse 18,
8700 Leoben,
Austria

Accred Qual Assur (2003) 8:247–251
DOI 10.1007/s00769-003-0617-9

A. Zschunke

The role of reference materials in analytical chemistry

Received: 7 September 2002
Accepted: 22 February 2003
Published online: 16 April 2003

Abstract The paper demonstrates a concept and possible models for an international infrastructure of chemical measurements by using reference materials.

The function of reference materials to establish traceability and means of quality assurance is emphasized.

Keywords Amount of substance standards · Chemical composition standards · Traceability · Third party assessment · Primary reference materials

Standards and reference system of analytical chemistry

The Standards of western culture were set by the ancient Greeks in the fifth century before the birth of Christ.

One standard and one measure of men was and still is the Hercules of Ephesus (see Fig. 1).

Human beings need standards and scales to improve their position in life.

Standards make terms conceivable and they make it possible for one to define scales.

The reference system of analytical chemistry comprises terms, standards, scales and the measuring system. The system allows one to formulate analytical problems and to develop analytical strategies.

Results of analytical chemistry are only valid within a reference system. Reference materials are the *standards* of analytical chemistry. They may be characterised for "Identity" (chemical structure) and for "Property values" (specific chemical quantities).

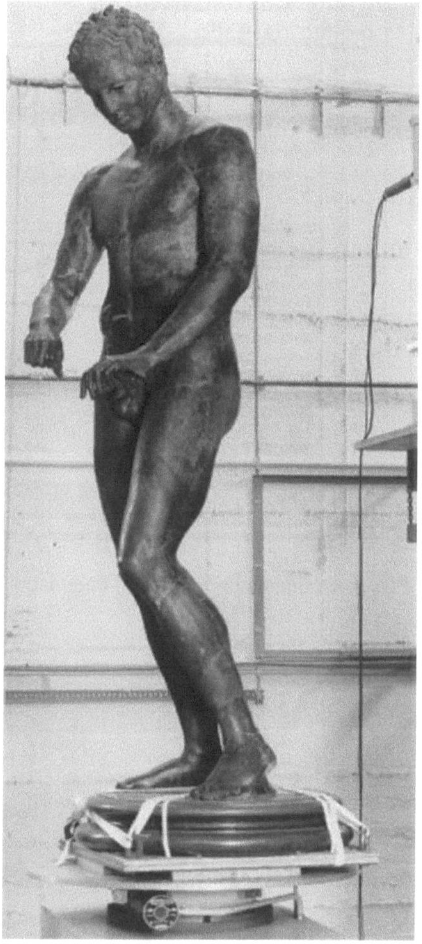

Fig. 1 Hercules of Ephesos – investigated by computer tomography at the Federal Institute for Materials Research and Testing (BAM)

Demands on analytical chemistry

The definition of analytical chemistry was given by the FECS [1] in 1993 and adopted by IUPAC:

– "Analytical Chemistry is a scientific discipline which develops and applies methods, instruments and strategies to obtain *information* on the composition and nature of matter in space and time, as well as on the value of these measurements, i.e. their uncertainty, validation and/or traceability to fundamental standards"

The *information* gain is a step by step procedure in the sequence: information – quantification- localisation – characterisation (Fig. 2). Each step creates macroscopic and microscopic scales based on reference materials or reference methods. In the first step of information gain, the identification, the analyte has to be defined on a macroscopic scale or microscopic scale.

Analytes on the macroscopic scale are compounds, elements, species. On the microscopic scale the identity is given by the electronic, nuclear or molecular structure of the particles [2]. On the macroscopic scale the analyte is in the most cases a sum of different identities.

The detectable analyte (measurand) depends on the selectivity of the analytical procedure (sometimes including the sample preparation).

The full analytical procedure starts with formulation of the analytical problem (Fig.3). However a chemical analysis is never an end in itself. An external need defines the analytical problem and external expertise should participate in assessment and utilization.

Analytical chemistry is a scientific discipline as well as a testing field. The demands on analytical chemistry differ to some extend depending on whether they

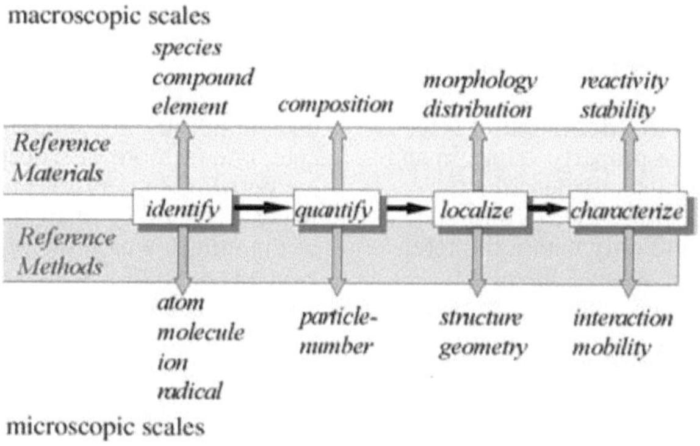

Fig. 2 Information gain in analytical chemistry

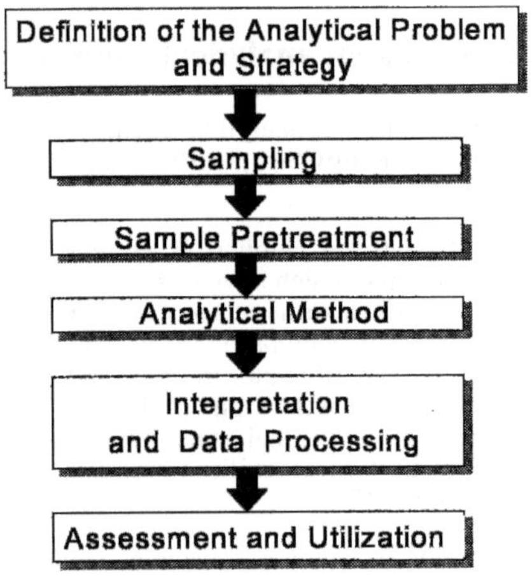

Fig. 3 Step by step procedure of analytical chemistry

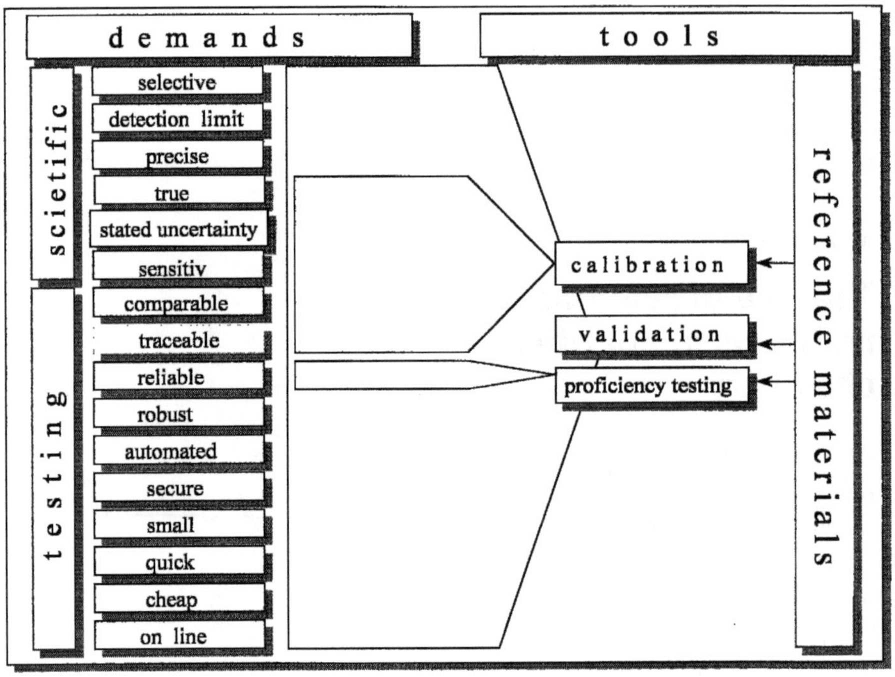

Fig. 4 Demands on analytical chemistry

Name	Symbol	Definition	SI unit	Common unit
1. Fractions				
Mass fraction	w	$w_i = m_i / \Sigma m_i$		
Volume fraction	φ	$\varphi_i = V_i / \Sigma V_i$		
(Chemical) amount fraction	χ	$\chi_B = n_B / \Sigma n_i$		
mole fraction, number fraction		$\chi_B = N_B / \Sigma N_i$		
2. Concentrations				
Mass concentration	γ, ρ	$\gamma_i = m_i / V$	kg/m³	g/L
Volume concentration	σ	$\sigma_i = V_i / V$		
Amount concentration	c	$c_B = n_B / V$	mol/m³	mol/L
Number concentration	C	$C_B = N_B / V$	m^{-3}	m^{-3}
3. Molality		$b_B = n_B / m_{solv}$	mol kg⁻¹	
4. Contents				
Volume content	κ	$\kappa_i = V_i / m$	m^3kg^{-1}	
Amount content	k	$k_B = n_B / m$	mol kg⁻¹	
Number content	K	$K_B = N_B / m$	kg^{-1}	

The use of mole demands the specification of particals (atoms, molecules, ions, radicals, electrons or groups of defined structure) consisting the substance (Sample)

Fig. 5 Quantities describing compositions of mixtures [4]

come from purely scientific sources or arise in the field of testing (see Fig. 4).

The gathered demands have partly overlapping meaning and are not independent of each other.

Securing at the same time selectivity, precision and sensitivity is unachievable for the same reasons as simultaneous securing of reliability, rapidity and cheapness [3].

There is always an optimum in the fulfilment of a certain member of demands that makes chemical analysis fit for purpose.

The measurable quantity describing a composition of mixtures (Fig. 5) is an attribute of a substance that may be distinguished qualitatively (e.g. $CuSO_4$ in water) and determined quantitatively (e.g. mol of analyte in kilogram of solvent).

Function of reference materials

"Reference Materials (RM) are materials or substances on or more of whose *property values* are sufficiently homogeneous and well established to be used for calibration of an apparatus, the assessment of a measurement method, or for assigning values to material."

According to this definition of reference materials the property values in analytical chemistry usually describe the chemical composition [5, 6]. Reference materials are valid only within the reference system of analytical chemistry.

They can be pure chemical substances, blends or synthetic mixtures, simulates or artefacts, spiked or unspiked real-life samples.

The current definition of Certified Reference Materials CRM ("CRM are reference materials, accompanied by a certificate, one or more of whose property values are certified by a procedure which establishes traceability to an accurate realisation of the unit in which the property values are expressed, and for which each certified value is accompanied by an uncertainty at a stated level of confidence" [6]) has the disadvantage that it contains requirements which are valid only for reference materials used as calibrants, i.e. establishing traceability.

However, there are reference materials for other intended uses (validation) which may also be accompanied by a certificate.

In addition, the term "certification" needs clarification. "Certification" in context with CRM means a procedure that establishes the value(s) of one or more properties. It should not be confused with certification as a procedure by which a third party gives written assurance that a product, process or service conforms to specific requirements. Certification in this sense is

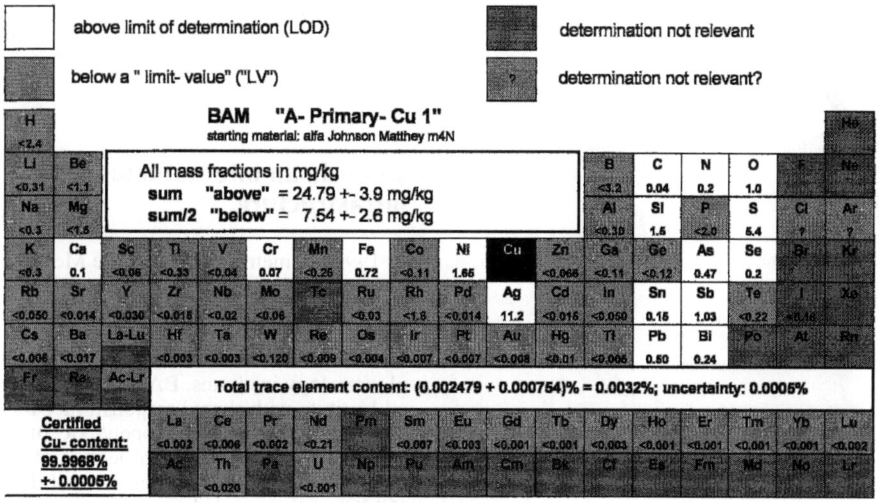

Fig. 6 Trace contents in a primary Cu standard

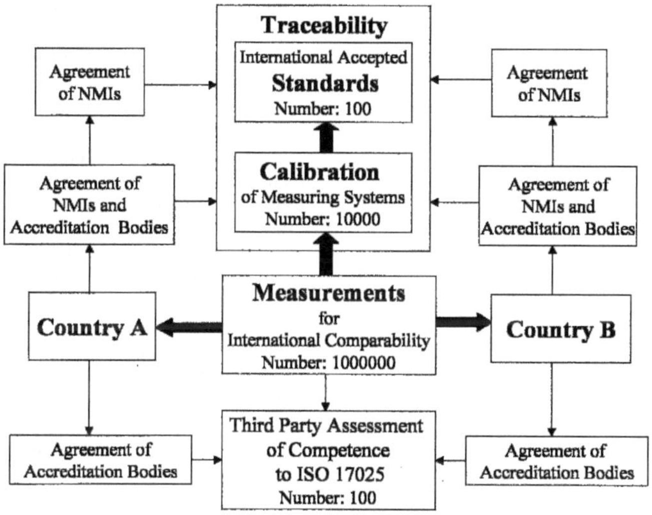

Fig. 7 International comparability by means of traceability on the base of existing infrastructure

a "third party activity" defined in ISO Guide 2 [7].

The assignment of property values to RM is a "first party activity". The quality requirement for reference materials are formulated in ISO Guides 30–35 [6, 8, 9, 10, 11, 12].

Pure elements can be used as *amount of substance standards*, if all impurities are known and determined in stated uncertainty limits. This is done by BAM for copper and some other elements [13].

Examples are BAM A Primary Cu 1 (Cu content 99.9968%±0.0005% mass fraction) and BAM A Primary Fe2 (Fe content 99.987%±0.002% mass fraction).

The limits of determination in Fig. 6 are related to ICP-MS.

The reference materials have the form of pellets, globules, shot, wires or bars. They are intended for use as amount of substance standard (traceability to SI) and are applied for preparation of calibration solutions. They are available only for producer of calibration standards or for national metrology institutes.

The pure element standards can be also used as *chemical composition standards*. The certified properties are in this case the contents of all "metallic" traces at ultra trace level. BAM offers, e.g. BAM B Primary Cu1 with statements (certified values) of the content of 65 trace metal elements. This reference material is suitable for matrix matching in metal analysis, e.g. where using methods of atomic spectroscopy.

The importance of reference materials demands a high degree of confidence in the quality of RM. Customers' confidence can best be achieved by ensuring transparency, reliability and acceptance [6, 14, 15].

In many cases the reliability of measurement is based on reliability of reference materials. Therefore the important role of reference materials is well recognized.

An EA-EUROLAB-EURACHEM working group "Selection and Use of Reference Materials" (EEE-RM) was established in 1996 and since then has met regularly twice a year. EEE-RM was formed to improve transparency in the field of reference materials and consequently to strengthen the confidence in reference materials. In 2000 EUROMET joined the group which now operates under "4E-RM". In 2002 ILAC also joined the group and now a new international structure is approaching. The current status of discussion in the 4E-RM-Group can be illustrated by the following items:

- A thorough third party technical assessment is considered to be an essential component of any RM quality assessment.
- The recommended approach to the quality assessment of RMs is accreditation of producers to ISO/IEC 17025 in combination with ISO Guide 34 [14]. (A reference material producer is a technically competent body [organisation or firm, public or private] that is fully responsible for assigning the certified or property values of reference materials it produces and supplies which have been produced in accordance with ISO Guides 31 and 35.)
- In addition the RM itself can be assessed as a product by an accredited (to ISO/IEC Guide 65 [15]) product certification body. Product certification can be recommended when the RM producer is not a laboratory and therefore accreditation is not possible.
- All kinds of quality assurance of RMs and their production should take into account ISO Guide 34 as the core quality requirements document of the assessment.

Traceability of results

"Traceability is the property of result of measurement or the value of a standard whereby it can be related to stated references, usually national or international standards, through an unbroken chain of comparisons all having stated uncertainty" [16]. Traceability is the proof of trueness of a result. Traceability is also a measure to build up trust in measurement results because it includes a complete documentation of calibration certificates (Figs. 7 and 8).

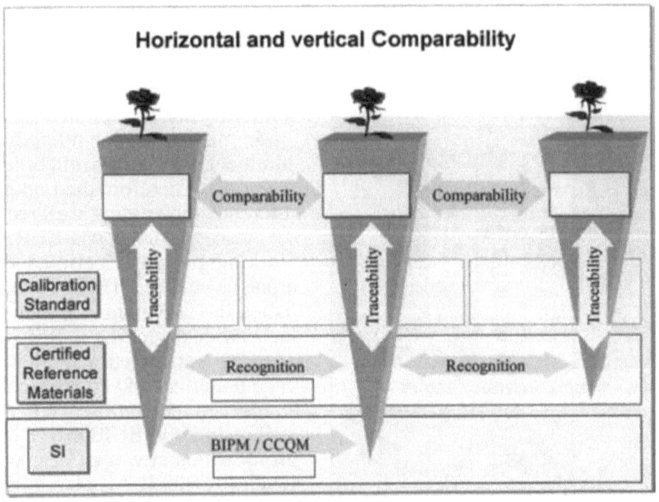

Fig. 8 Horizontal and vertical comparability

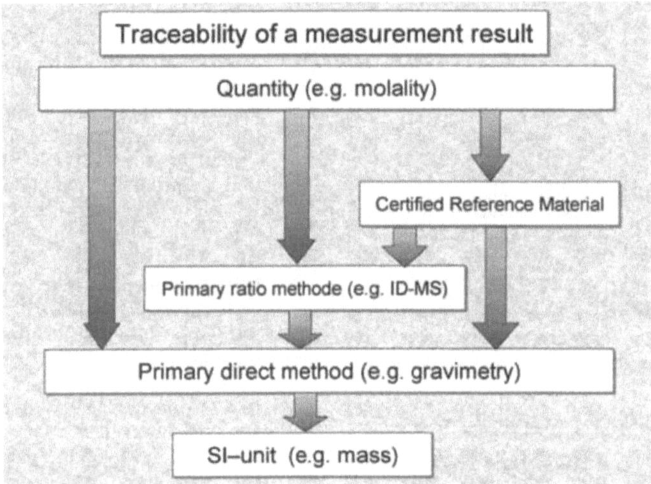

Fig. 9 Primary method concept

The international comparability of analytical results is dependent of the traceability to international standards, preferable to SI-units.

The mutual recognition of international standards can be achieved by agreements between National Metrology Institutes (NMI). Calibration as an important activity in establishing traceability should be done by competent personnel. Therefore the competence should be assessed by a third party. Consequently agreements are also needed for accreditation bodies.

The demand for Certified Reference Materials is much higher than the international infrastructure of metrology institutes can provide. National metrological institutes are often not well enough equipped for efficient production of these materials. An additional infrastructure of reliable and recognised producers of CRM becomes more and more important.

The traceability concept is closely connected with SI, the consistent system of basic units. The final link in the traceability chain to SI is a primary direct method (Fig. 9), i.e. a method having the highest metrological qualities, whose operation can be completely described and understood, for which a complete uncertainty statement can be written down in terms of SI units [17].

A primary direct method measures the value of an unknown without reference to a standard of the same quantity.

In the field of chemical measurement gravimetry and coulometry (as primary direct methods) are usually used to trace to SI units. Certification of reference materials should therefore include primary methods to make property values traceable to the SI-unit.

Towards an international infrastructure

In 1999 the member states of the Meter Convention have signed the Mutual Recognition Arrangement (MRA) on measurement standards and on calibration and measurement certificates issued by national metrology institutes. BAM is a designated Institute with shared responsibility for metrology in chemistry. Appendix C of the CIPM-MRA is a growing collection of the Calibration and Measurements Capabilities (CMC) of the national metrology institutes. The CMC-database is available for everyone on the website of the Bureau International des Poids et Mesures (BIPM) and includes reference materials as well as references methods.

The used methods are proved by key comparisons between the national metrology institutes. For chemical measurements the Comité Consultative pour la Quantité de Matière (CCQM) has been established. The CMC database provides a reliable service for customers all over the world to establish traceability.

In Germany since 2001 BAM and PTB have provided a service regarding national standards for dissemination of traceability. The service elucidates the endpoints of traceability of results of chemical measurements in Germany and is a German contribution to an international system aimed at mutual recognition of national standards.

We distinguish between two kinds of national standards [18]:

- National Primary Standards (NPS): NPS are maintained at BAM and PTB and serve exclusively to link the values of reference materials to SI (e.g. Primary pure elements)
- Primary Reference Materials (PRM): PRM are delivered to clients

The quality criteria for both kinds of national standards are:

- Laboratories certifying both kinds of standards work in accordance with technical standard ISO 17025 and ISO Guides 31, 34 and 35
- Methods of certification are proved by CCQM key comparisons, interlaboratory comparisons linked to CCQM and EUROMET/EURACHEM projects
- Methods of certification are intended to be included in CIPM-MRA Appendix C, Amount of Substance, Calibration and Measurement Capability Declarations

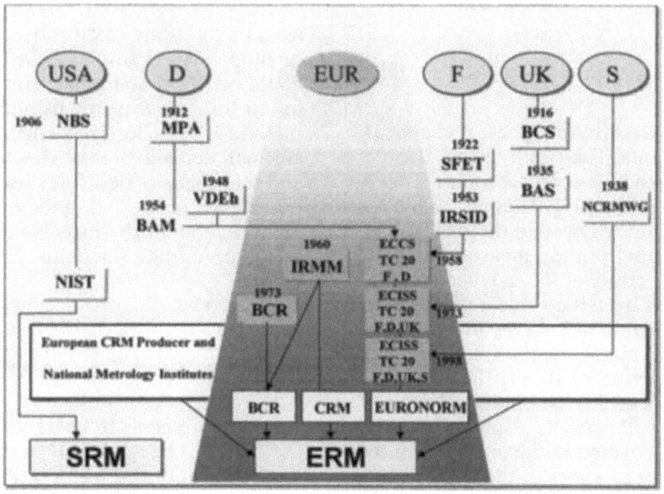

Fig. 10 Towards a global system of recognised reference material

- Certifying laboratories have to prove their competence by successful participation in interlaboratory comparisons [CCQM, IMEP (*I*nternational *M*easurement *E*valuation *P*rogramme)] or PT-schemes [19]

The primary element standards especially the primary pure elements are used for the production of other reference materials (Element-solutions for PTB/Merck and EMPA/Fluka as well as isotopic standards for IRMM [Joint European Project for Primary Isotopic Measurements=JEPPIM]).

Even if the concept of primary element standards has already proved a success it covers presently only a few elements. Only an international division of labor can build up this part of measurement infrastructure. In addition to the initiative of CIPM and the national metrology institutes the leading producers of reference materials should cooperate to establish a worldwide system of recognised reference materials to cover all application fields where international comparability of results and therefore traceability is required. Guidance is given by the ISO Guides 34 and 35 to assure the quality and reliability. The Standard Reference Materials (SRM) Program of NIST is an important part of a global system of recognised reference materials. Additional contributions from Europe are under development (see Fig. 10).

References

1. Federation of European Chemical Societies (FECS), Edingburgh (1993)
2. Zschunke A (2000) Accred Qual Assur 5:441–445
3. Valcarcel M, Rios A (1997) Fresenius J Anal Chem 357:202–205
4. Davis RS (1998) CCQM/98–9
5. Zschunke A (ed) (2000) Reference materials in analytical chemistry. Springer, Berlin Heidelberg New York
6. ISO (1992) Terms and definitions used in connection with reference materials. ISO Guide 30
7. ISO/IEC Standardization and related activities – general vocabulary. ISO/IEC Guide 2
8. ISO (1981) Contents of certificates of reference materials. ISO Guide 2
9. ISO (1996) Calibration in analytical chemistry using reference materials. ISO Guide 32
10. ISO (1989) Use of certified reference materials. ISO Guide 33
11. ISO (2001) General requirements for the competence of reference materials producers. ISO Guide 34
12. ISO (1989) Certification of reference materials – general and statistical principles. Guide 35 (under revision)
13. Matschat R, Czerwensky M, Pattberg S, Heinrich H-J (2002) Phys Stat Sol A 189(1):107–122
14. ISO/IEC(1999) General requirements for the competence of testing and calibration laboratories. ISO/IEC 17025
15. ISO/IEC(1996) General requirements for bodies operating product certification systems. ISO/IEC Guide 65
16. ISO (1993) International vocabulary of basic and general terms in metrology (VIM). ISO, Geneva
17. Milton MJT, Quinn TJ (2001) Metrologia 30:289–296
18. Zschunke A (2001) Reference materials service for chemical measurement in Germany (editorial). Fresenius J Anal Chem 369:6–7
19. ILAC (2000) Guidelines for the requirements for the competence of reference materials producers. ILAC G12

Presented at the Interact 2002 Conference, 21–25 July 2002, Sydney, Australia

A. Zschunke (✉)
Bundesanstalt für Materialforschung und -prüfung,
12205 Berlin, Germany
e-mail: Zschunke35@aol.com

Present address:
A. Zschunke, Rappsweg 115,
04207 Leipzig, Germany

Accred Qual Assur (2003) 8:380–382
DOI 10.1007/s00769-003-0655-4

Meeting ISO/IEC 17025 Traceability Requirements

A New Guide with Worked Examples

Abstract For over ten years Eurachem and CITAC have been working to develop a general strategy for establishing the traceability of chemical measurements. The outcome of this work is a new guide that describes a simple, yet metrologically robust strategy and contains worked examples to illustrate the approach. The guide is at an 'advanced draft' stage and is available on the Eurachem web site. Inter-alia, it will help implement the traceability requirements of accreditation to ISO/IEC 17025. This paper discusses the key issues, coloured by the personal views of the author. The paper was presented at the ILAC Berlin Conference in September 2002.

Keywords Traceability · Chemical reference materials · Accreditation · CITAC · Eurachem

Introduction

Establishing the traceability of measurements to the SI, or, where this is not feasible, to other appropriate measurement standards is a requirement of ISO/IEC 17025. This is because the process of linking measurements to common standards helps ensure the comparability of measurements made in different laboratories, or at different times.

Many of the issues associated with traceability are well-established components of good measurement practice in chemical laboratories. However, some of the formalities are new and, until recently, the views of experts concerning the details of how to achieve the traceability of chemical measurements differed considerably. This has made it difficult to implement the traceability requirements of ISO/IEC 17025 in chemical testing laboratories. In recognition of this problem, CITAC and Eurachem have been working for over ten years to develop a widely agreed strategy applicable to different types of chemical measurements. The result of this effort is a new guide that describes a robust strategy and provides worked examples to illustrate what laboratories need to do to establish the traceability of their measurements. The guide is currently at the 'advanced draft' stage. It has been the subject of an international workshop in Lucerne in June 2002 (1) and is available on the Eurachem web site (www.eurachem.ul.pt). The final document is expected to be published during 2003. The following is a summary of the key issues covered in the guide, coloured by the personal views of the author, who is a member of the working group producing the guide.

Whilst traceability is necessary to achieve comparability of measurement results, it is, of course, not the only requirement of good measurement practice and thus needs to be considered as just one part of the measurement process (2).

The Measurand

An important precursor of valid measurement, and the establishment of measurement traceability, is an adequate description of what is to be measured (the measurand), which includes the measurement units and consideration of the acceptable level of measurement uncertainty (MU). Clearly, if different characteristics are measured, or different measurement units are employed, then different measurement results can be expected. Clarity on this issue can be vital to subsequent decision making. For example, in environmental studies, it may be more important to know the amount of extractable pollutant in a geological material, rather than the total amount of the pollutant. Thus, although self evident when we think about it, it is important to remember that, in addition to making traceable measurements, it is also important to make the right type of measurement.

When we have decided what to measure, often the specification of the measurand can be made using SI units, for example 'the total amount of cadmium in soil, on a dry sample basis'. In this case, the measurand can be described in terms of the amount of substance unit, the mole. In other cases, such as the amount of fat in meat, where the chemical entities vary from sample to sample, it is difficult to describe the measurand in terms of the mole. None the less, the measurement can still be made in SI units, namely, weight/weight units. Sometimes the measurand is highly dependent on the method of measurement and such methods are often called 'empirical methods'. An example of such a method could be 'the amount of extractable cadmium in soil, as measured using standard method XYZ. Clearly the measurement result will depend on the extraction method and this needs to be made clear when reporting the result. All of these measurements can be made traceable to SI, if related to appropriate standards, as discussed below.

Measurement Traceability

For measurements to be traceable to the SI, they need to be made using equipment that has been calibrated using measurement standards, that have themselves been calibrated using higher level standards that are traceable to the SI. Often such measurement standards are obtained from reference or calibration laboratories, simplifying the task of the testing laboratory.

Chemical measurements are invariably made indirectly, by measuring other quantities (such as sample weight, volume of sample solution, signal response from the instrument relative to the response from a series of chemical standards) and calculating the chemical measurement result using an appropriate measurement equation. If the measurement of these 'influence quantities' is carried out using equipment calibrated using SI traceable standards, then the chemical measurement calculated from these results can also be expected to be traceable to the SI. If there are additional quantities, such as time, temperature, pH etc influencing the measurement process, then their effect can often be eliminated by keeping them constant. Where such 'control quantities' have a significant effect, then the measurements used to control them also need to be made using equipment calibrated using SI traceable standards.

This strategy is summarised in Box 1 illustrated by the example in Box 2. It is of course possible to make different types of cadmium in soil measurements. For example, instead of the total cadmium, the amount of cadmium extracted by a specific method, or the amount of cadmium in the sample as received, or dried in a different way could be measured. These are different measurements (measurands) and likely to give different measurement values, but all can be made traceable to SI, including the amount of substance. Hence the importance of adequately defining the measurand, in addition to establishing measurement traceability.

An important related issue is the uncertainty associated with both the measurement standards used to make a measurement, and the uncertainty of the final test result. The value assignment of measure-

Box 1

Strategy for Establishing the Traceability of Chemical Measurements

- ☆ Define the measurand including the measurement units and the acceptable level of MU.
- ☆ Select a method. During method validation establish a valid equation for calculating the result, establish the value of any constants and establish the measurement conditions.
- ☆ During measurement of samples carry out the following:
  - ➢ Measure the variable quantities in the measurement equation using appropriate measurement standards, eg – weight of test portion, volume of test solution, concentration read from a calibration graph, bias correction factor.
  - ➢ Measure any other quantities that need to be closely controlled in order to obtain consistent results, again using appropriate measurement standards, eg – laboratory temperature.
- ☆ Calculate the measurement result, using the measured values and any constants.
- ☆ The calculated measurement value is traceable to the measurement standards employed in making the measurement. Where the measurement standards are traceable to higher level refreences, such as SI, the calculated measurement value is also traceable to those references.
- ☆ The measurement value is traceable to the stated references at a level of MU that depends on the uncertainties associated with the measurement process and the measurement standards.
- ☆ Document and report the traceability by briefly describing the traceability of the measurement standards used to calibrate the key measuring instruments.

Notes

1. The traceability of any constants need to be obtained from the literature (eg – from IUPAC tables) or they will need to be based on traceable measurements made by the testing laboratory, eg – bias correction factors.
2. The traceability of standards or calibrated measuring instruments used for physical measurements, such as weighings, volume measurements etc. can be readily established using measuring devices that have been calibrated using SI traceable standards produced by accredited calibration laboratories.
3. The traceability associated with chemical standards will depend on the purity of the substance used as the RM and on the weights and volumes measured during the preparation and dilution of the standards. Sometimes this traceability is provided by the supplier. However, it is also common practice for laboratories to prepare their own standards, form general purpose chemicals. Where this is done, it is the laboratory's responsibility to establish their traceability as described above.

ment standards, including reference materials (RMs), needs to be at levels of uncertainty such that they are not more than about one third of the uncertainty of the final test result. If this condition is met, the standards will not contribute significantly to the combined uncertainty of the final test result. In practice, the uncertainties of measurement standards are often small compared with other sources of uncertainty in chemical measurement, and hence, whilst important, the standards are not usually a major source of difficulty.

Reference Materials and their QA

None the less, questions may arise concerning the traceability of the values carried by chemical reference materials. The essential requirement is for the traceability of the assigned value to be established at a level of uncertainty appropriate to the final test result. Where the RMs have been obtained from National Measurement Institutes, or from accredited calibration / reference material producers, then the traceability of the standards is assured, typically using the types of measurements described in Box 3. For many pure substance based RMs, the only difficulty is likely to be, how to establish the purity of the starting material. This can be achieved by a combination of direct assay of the substance, and by measuring all the impurities and subtracting these from 100%. When the impurity subtraction approach is used it is important to ensure that <u>all</u> impurities are accounted for. However, since the purity determination only needs to be carried out at a level of uncertainty that is commensurate with the uncertainty required of the subsequent test result, it is not usually a major problem. For example, for a test result uncertainty of U(k=2) = 5%, an uncertainty of 1% in the purity of the pure substance RM used to prepare the calibration standards would not be very significant. In the case of matrix RMs, it has been common to assign property values on a consensus basis, using data derived from one or more 'validated' methods. Although bias may not have been fully evaluated for each method, where results from the different methods agree, it can be concluded that bias is absent. Thus, although traceabilty may not be described in the certificates of such RMs, it has actually been addressed and it is possible for users to make their own judgements on the subject.

As is often the case, where the RMs are obtained from a non-assured source, it is the responsibility of the user to establish their traceability to appropriate references, at an appropriate level of measurement uncertainty. Usually it will be possible to establish traceability to the SI, but if this is not possible, traceability to other references, such as a higher level RM can be achieved. The traceability issues related to 'in-house RMs' are summarised in Box 4.

One of the remaining questions concerns the QA of chemical reference materials and the competence of RM producers. Whilst laboratory accreditation normally requires the use of physical measurement standards that have been produced in accredited calibration laboratories, the situation regarding chemical measurement standards is, so far, much less formal. "General Requirements for the Competence of Reference Material Producers" (3,4) have been available since 1996, but the implementation of accreditation based on these requirements is still in its infancy. There is a need for a more balanced approach to accreditation practice related to measurement standards. For example, should only reference materials produced by accredited producers be used as measurement standards in accredited test laboratories? Or, does the whole system of the QA of measurement standards need to be re-examined?

Conclusions

As indicated at the beginning of this paper, traceability is a requirement of ISO/IEC 17025. However, as also indicated, it has not been feasible to implement this requirement for chemical testing, due to a lack of clarity about how to do it. The advent of this guide changes all that. There now exists a simple, yet metrologically robust guide to help laboratories and accreditation bodies. The guide will help ensure that chemical measurements are made using equipment calibrated using traceable standards, thus ensuring the traceability of

Box 2

Traceability Example – Measurement of Cadmium in Soil

☆ The client requires to know the total amount of cadmium in soil, measured in mg/kg, on a dry weight basis.
☆ The method is based on ICP-MS and the measurement equation is as follows:

$$C = I \cdot \frac{\partial C \cdot V}{\partial I \, m \cdot R} = C_{cal} \cdot V / m \cdot R \text{ mg/kg}$$

Where

C = concentration of Cd in the soil
I = The test sample signal intensity
$C / \partial I$ = The calibration function - ie. the slope of the calibration graph
V = The final volume of the test solution
m = The dry sample weight
R = The method recovery
C_{cal} = Sample concentration read from calibration graph

☆ The traceability of the quantities in the equation can be established as follows:
- ➢ Providing the measurement and calibration conditions are identical, then the traceability issues associated with I and ∂ I will cancel
- ➢ C_{cal} is traceable to SI through the cadmium RM and associated weighing and volume measurements.
- ➢ m and V are traceable to SI through calibrated standards
- ➢ R is a constant and is traceable to SI through a matrix CRM (or by spiking)

☆ The drying temperature and drying time are traceable to the SI through calibrated thermometer and calibrated clock.
☆ Since the contributory measurements are traceable to SI and the measurement equation was shown to be valid, during method validation, the measurement result for Cd is also traceable to SI, at some specified level of measurement uncertainty.

Box 3

Establishing the Traceability of Chemical Reference Materials

☆ The traceability of pure substance RMs and their solutions can be established through the following measurements:
- ➢ Purity determination using a primary meethod(s)
- ➢ Mass measurements using weights traceable to SI
- ➢ Volume measurements using SI traceable devices
- ➢ Atomic weight values. These are fundamental constants available from IUPAC tables

☆ The traceability of matrix RMs can be established through the following measurements:
- ➢ Primary or reference measurements traceable to SI
- ➢ Consensus values based on methods of known bias

Box 4

In-House RMs

☆ Should be related to higher level RMs of the same type, where:
- ➢ The in-house RM MU contributes significantly to the MU of the final test result
- ➢ It is feasible

☆ Where commercial chemicals, or other non-assured materials are the only RMs available, the laboratory needs to:
- ➢ Assess available data
- ➢ Where necessary, characterize the materials

☆ In addition to property value assignment, based on traceable methods, in -house RMs should be stable and homogeneous, at an appropriate level of uncertainty, and stored appropriately

the test results. It is a matter of choice whether to establish traceability to the SI, or to relate measurements to lower level standards, such as reference materials. However, most chemical measurements can, if required, be made traceable to the SI and usually this is worth the small extra effort involved. It will no doubt take time to implement the traceability requirements in chemical laboratories, but as with measurement uncertainty, the scale of the task is less daunting than might appear at first sight. Success will depend on the willingness of laboratories to embrace this element of 'good measurement practice' and on enforcement of the ISO/IEC 17025 requirements by accreditation bodies.

Acknowledgements The author would like to acknowledge the work of the joint Eurachem CITAC working group who are producing the guide, particularly the contributions from the chair, Alex Williams and the secretary, Steve Ellison.

References

1. Eurachem / CITAC Guide, Traceability in Chemical measurement, Workshop Draft, June 2002, www.eurachem.ul.pt
2. Eurachem / CITAC Guide to Quality in Analytical Chemistry: An aid to Accreditation, Edition 2002, www.eurachem.org
3. ISO Guide 34: 2000, General Requirements for the competence of reference material producers.
4. ILAC G12:2000, Guidelines for the Requirements for the Competence of Reference Material Producers.

Presented at the International ILAC/IAF Conference on Accreditation on Global Trade, September 23–25, 2003, Berlin, Germany

Bernard King
Analytical Measurement Consultant,
25 Bolton Gardens, Teddington,
Middlesex, TW11 9AX, UK
Tel./Fax: +44 208 977 2360
e-mail: BKingTeddington@aol.com

Accred Qual Assur (2003) 8:480–482
DOI 10.1007/s00769-003-0683-z

Mike Sargent

UK delivery of traceable chemical measurements in the 21st century: building on the foundation of the VAM programme

Published online: 16 September 2003

Abstract The paper discusses the requirements for achieving traceable chemical measurements in the UK. It is emphasised that success will depend on establishing an appropriate UK chemical measurement infrastructure and encouraging reference and field laboratories to make use of it. The demanding requirements of the BIPM Mutual Recognition Arrangement (MRA) also require a point of focus to link UK reference laboratories into international metrology. Two key factors are described which have provided the UK with the means to meet these requirements and which have established a sound basis on which to build a system of traceable chemical measurements in the 21st century. These two factors are LGC's long-standing role as the UK's national centre for analytical chemistry and the development and delivery over many years of the UK's Valid Analytical Measurement (VAM) Programme.

Keywords Traceability · Laboratory of the Government Chemist · Valid Analytical Measurement · Calibration

Introduction

The UK, like many other industrialised countries, has a long history of governmental action to ensure and underpin the reliability of chemical measurements. It is only in recent years that the emphasis has moved towards traceable chemical measurements but most of the requirements remain essentially the same. Moreover, one UK organisation—LGC, The Laboratory of The Government Chemist—has been at the forefront of this type of activity for over 150 years [1]. In the 1980s LGC's role was extended by the UK government's VAM (Valid Analytical Measurement) Programme [2] and this trend has continued with the recent establishment at LGC of the UK Chemical Calibration Facility which provides a focus for achieving traceability in ISO 17025 accredited laboratories.

LGC was founded in 1842 as The Excise Laboratory with the task of performing the chemical analyses needed to levy excise duty on tobacco, beer and spirits. Towards the end of the 19th century, the UK introduced a range of legislation aimed at safe-guarding the quality of consumer and industrial products, particularly food and agricultural materials. This legislation was enforced by local government laboratories through chemical analysis. To deal with the inevitable disputes regarding the analytical results, the legislation introduced the concept of a *referee analyst* appointed by the government. This role fell to the head of the government's chemical laboratory who was, of course, widely known as the Government Chemist. In 1911, the excise and referee tasks were formally combined with the establishment of the Department of the Government Chemist (Fig. 1).

The government laboratory continued in this form, with a steadily expanding range of chemistry-related activities, for many years and in 1959 acquired the new name of Laboratory of the Government Chemist. Shortly afterwards the LGC expanded its research role and moved into a single new laboratory, but still in its traditional central London location. The first major change, after almost 150 years, occurred in 1988 when LGC was designated a Government agency and moved to a new, purpose-built laboratory at Teddington to the south west of London. At this time LGC took on a number of chemistry-related activities previously carried out at NPL, particularly the Office of Reference Materials, and the DTI (Department of Trade and Industry) VAM (Valid Analytical Measurement) programme was established.

The second, and most fundamental change, in LGC's long history took place in 1996 when it became a private company, partly owned by The Royal Society of Chemistry (RSC). Today the LGC Group has a turnover of well over £50m and more than 500 staff located both at Teddington and other sites in the UK and Europe. LGC remains, however, a key part of the UK National Measurement System and is a signatory to the BIPM (Bureau Internationale des Poids et Mesures [http://www.bipm.org]) MRA (Mutual Recognition Arrangement) through an association with the NPL (National Physical Laboratory). NPL represents the major part of the UK's national measurement institute and undertakes a limited role in chemical measurement, primarily gas and surface analysis. As a private company, LGC has three principal areas of activity:

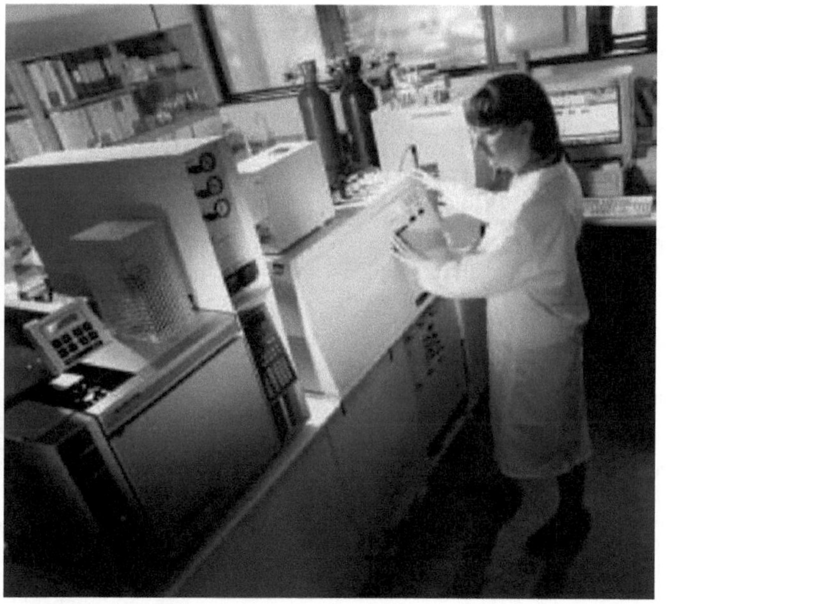

Fig. 1 Gas chromatography combustion isotope ratio mass spectrometry for assay of aqueous ethanol calibration standards. This method offers lower uncertainty than the previous titrimetric procedure and may also be used directly to obtain ethanol reference values in alcoholic beverages

- The national centre for analytical science, a role which includes valid analytical measurement, measurement in support of regulation and innovation in analytical science
- Contract research, analysis and consultancy in the chemicals, environment, forensics, food, health, life sciences and pharmaceuticals sectors
- World-wide distribution of reference materials and standards following acquisition of the Promochem company.

VAM, The DTI programme on valid analytical measurement

VAM was launched in 1988 by LGC and today is a multi-million pound programme which forms part of the UK National Measurement System. The original rationale for such a programme remains true today, not just in the UK but also in most developed countries:

- The analytical measurement sector is large and vital to the UK economy (estimated at approximately £7.5bn of analysis p.a. with around 100,000 analysts in 15,000 laboratories)
- Agreement between laboratories and countries is essential for commerce and international trade to prosper
- The chemical measurement infrastructure is poorly developed in comparison with the infrastructure for physical measurements
- Not all chemical measurements are valid (fit for purpose).

The VAM programme has three main activity areas, namely to develop the chemical measurement infrastructure, to develop the tools needed for better measurements, and to promote the "concept of VAM".

It was recognised from the outset that chemical measurements are global in nature and hence VAM has supported development of both the UK Chemical Measurement Infrastructure and an International Chemical Measurement Infrastructure. Work at the national level initially focussed on establishing VAM working groups (e.g. RM (Reference Materials), PT (Proficiency Testing), training, mass spectrometry), collaborative development of RMs, and developing UK laboratory and knowledge networks in association with the RSC. The recent emphasis on achieving traceability in ISO 17025 accredited laboratories has resulted in the creation of the UK Chemical Calibration Facility. At the international level, VAM enabled LGC to establish the EURACHEM (http://www.eurachem.ul.pt) and CITAC (Committee for International Traceability in Analytical Chemistry [http://www.citac.cc]) organisations and to become a founder member of CCQM (Comité Consultatif pour la Quantité de Matiere). Today, VAM also supports essential participation in the BIPM MRA and EUROMET (http://www.euromet.ie).

The second area of VAM activity, developing the tools needed for reliable analytical measurements, has always represented the most extensive part of the programme. The emphasis lies with the development of reference methods and the production of reference materials and calibration standards. In the early days, distribution of reference materials was seen as a key part of VAM, aimed at encouraging and facilitating their use. Today, however, the importance of reference materials is widely accepted and the distribution function is undertaken as a completely separate, commercial function by LGC-Promochem. A very important aspect of the second area is the collaborative development of validated methods, protocols and guides working in close collaboration with both UK laboratories and overseas organisations.

The "Concept of VAM" has been promoted since the start of the programme as a way of encouraging laboratories, and their managers, to adopt best practice. The message is clear: by adopting six straightforward principles organisations can ensure their results are fit for purpose, demonstrate the validity of data to their customers, and achieve consistency with results obtained elsewhere. These six principles, widely known as "The VAM principles", are:

1. Measurements should satisfy an agreed requirement
2. Use tested methods and equipment
3. Use qualified, competent staff
4. Seek independent assessment of performance
5. Ensure consistency with results obtained elsewhere
6. Adopt QC and QA procedures.

The UK Chemical Calibration Facility

The Facility (Fig. 2) has been established to support the development of traceable chemical measurements by UK analytical laboratories, in particular to meet the requirements of ISO 17025. It will also link UK laboratories into the national and international chemical measurement systems to ensure the future world-wide acceptance of their data. An important role is to provide a single point of focus for a wide range of chemical measurement and calibration services provided by LGC as part of its delivery of VAM.

The principal services include provision of high level calibration services (Fig. 3) and/or standards to suppliers offering secondary services or standards, preparation of standards and reference materials, and limited services direct to field

Fig. 2 Staff of the UK Chemical Calibration Facility at LGC

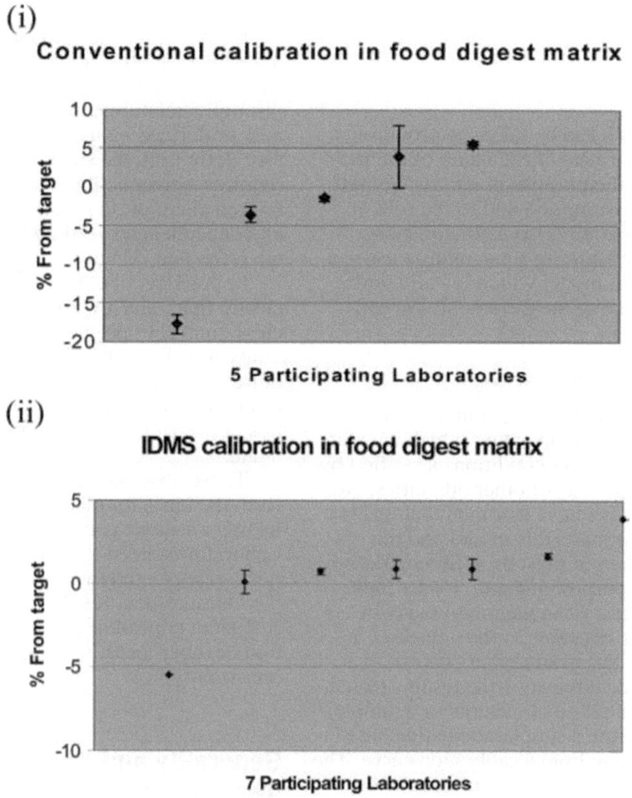

Fig. 3 The diagrams show some of the results of an inter-laboratory trial undertaken with UK expert laboratories based on determination of Cd in an artificial food digest matrix containing a known amount of Cd. The values obtained by the laboratories using (i) their conventional calibration are compared with those from (ii) an "approximate matching" IDMS procedure developed and validated at LGC

laboratories. The latter are necessary primarily in new areas where commercial services have not yet been established. Service provision is in accordance with the BIPM MRA. The initial activities being undertaken include:

- Pure organic substance, including certification or provision of natural or isotopically enriched single substance calibration materials
- Analytical standards, including certification or provision of solution calibration standards
- Preparation and certification of complex matrix CRMs (Certified Reference Material).
- Provision of reference values for matrix samples such as CRMs, PT scheme artefacts, validation samples, etc

Achieving traceability in ISO 17025 accredited laboratories clearly demands more than just provision of suitable calibration materials or values. Hence, traceability demonstrator projects are being undertaken to identify issues, best practice, and key areas to address. It is also necessary to transfer methodology and expertise to UK reference laboratories and to provide guidance on implementation of traceability to UK field laboratories. Methods developed for the UK Chemical Calibration Facility have been transferred to other expert laboratories with the co-operation of the RSC's Analytical Methods Committee (AMC), Sub-Committee on High Accuracy Analysis by Mass Spectrometry (HAAMS). Finally, the Facility assists networking arrangements with key players including the accreditation and regulatory bodies, reference laboratories and field laboratories.

Conclusions

Achieving traceable chemical measurements in the UK depends on both establishing an appropriate measurement infrastructure and ensuring that reference and field laboratories are willing and able to make use of it. The demanding requirements of the BIPM MRA also require a point of focus to link UK reference laboratories into international metrology. Overall, LGC's role for more than 150 years as the UK's national centre for analytical chemistry, together with 15 years of the VAM Programme, have provided the UK with a sound basis on which to build a system of traceable chemical measurements in the 21st century.

References

1. Hammond PW, Egan H (1992) Weighed in the Balance—A history of the Laboratory of the Government Chemist HMSO London ISBN 0 11 515302 0.
2. King B, Phillips G (1991) Anal-Proc (London) Apr 1991 28(4):125-126

M. Sargent (✉)
LGC Limited,
Queens Road, Teddington,
TW11 0LY, UK
e-mail: ms@lgc.co.uk
Tel.: +44-(0)2089437360

Accred Qual Assur (2003) 8:483–485
DOI 10.1007/s00769-003-0661-5

S. L. R. Ellison
A. Williams

Disseminating traceability in chemical measurement: Principles of a new EURACHEM/CITAC guide

Published online: 7 August 2003

Introduction

Traceability to appropriate reference standards provides the units in which results are expressed, and the stability and comparability required for international trade and scientific research. Existing guidance from EURACHEM and CITAC covers measurement uncertainty evaluation [1], method validation [2], and general quality assurance for chemical measurement [3]. However, with the continued rise in importance of traceability to national and international standards, it became apparent that practical guidance on establishing traceability in chemistry was inadequate. In May 2000, therefore, EURACHEM and CITAC established a joint activity to provide guidance on traceability for chemical laboratories. The task was delegated to the then measurement uncertainty working group, whose remit was appropriately increased. At the time of writing, the working group has produced draft guidance which was discussed at an international workshop in June 2002, has been amended following discussion and is circulating for final approval [4].

In this paper, we present the basic principles of the guidance document, and discuss some of the implications.

Basic principle

The basic philosophy on which the guidance is based is most concisely summarised by reference to a mathematical statement about results of measurement, and that is the development which, for brevity, is presented here.

It is assumed that method development and validation have led to a measurement process which can be taken as providing a sufficient estimate of the value of the measurand for the purpose in hand. (We will return this assumption and to the role of validation below). This estimate is obtained by performing a calculation using a mathematical model which is valid under the conditions of measurement. We can then write

$$y = f(x_1, x_2, \ldots x_m)\big|_{x_{m+1}, x_{m+2} \ldots x_n} \qquad (1)$$

that is, y is calculated from input values $x_1 \ldots x_m$ using a relationship f which is valid under measurement conditions specified by the values $x_{m+1} \ldots x_n$ of other quantities. Note that because we have assumed that this is a sufficient estimate, this model and the quantities therein must be sufficient definition for the purpose in hand. Under these circumstances, y can unambiguously be identified as traceable to the values $x_1 .. x_n$.

To place this in a familiar context, in a routine GC determination the result y (usually the concentration of material in a sample) is obtained from a simple calculation based on interpolation from a calibration curve. The curve in turn depends on the values of the concentration standard(s) used to set up the calibration, combined with the values of the mass of sample, dilution volumes etc. x_1 to x_m represent the values of concentration standards used to set up the calibration curve, sample mass, dilution volumes etc. which are used in this calculation of the result.

However, this model is only valid where the signal is free from interference, where prior derivatisation is complete and so on. Freedom from interference is assured by controlling parameters such as GC oven temperature, flow rate, temperature ramp rates etc. to obtain an isolated signal; completion of derivatisation is normally assured by guaranteeing appropriate derivatisation conditions including reagent quantities, concentrations and temperatures. These GC and derivatisation conditions do not generally appear in the calculation itself, yet can clearly have significant effects on the result. These are the types of quantity that are represented by the values x_{m+1} to x_n.

The central tenet of the EURACHEM guidance is simple; if the above equation and associated conditions are sufficient for estimation of the value, all that is necessary for complete traceability to appropriate references is that all the values x_1 to x_n are themselves traceable to appropriate references or are defined values.[1]

[1] "Defined values": for example, unit conversion factors, mathematical constants, or the values of constants used to relate some SI units to fundamental constants.

Note that for the x_i to be considered traceable, we must apply the principle recursively; the x_i must in turn depend, through a relationship like Eq. 1, on traceable or defined values. Viewed as a recursive definition, the statement automatically requires a clear calibration chain—an unbroken chain of *quantitative* comparisons and dependencies leading back to appropriate reference values.

In practice, of course, it is sufficient to ensure that values x_1 to x_n are under sufficient control to provide the required uncertainty in y. For critical quantities, this requires traceable calibration against other reference values. For less critical quantities, less stringent control may be adequate.

These observations, taken together, form the basis for clear guidance on establishing adequate calibration for traceability. Laboratories need to identify those quantities—represented by x_1 to x_n - that affect the measurement results, and then establish sufficient control of their values by calibration or other measures to obtain adequate uncertainty in y.

Generality and "empirical" measurands

If we acknowledge that there may be any number from zero upwards of quantities in $x_1 .. x_{m+1}$ and $x_{m+1} .. x_n$, it is immediately clear that most univariate measurements can be described by this model, which makes it a very practical presentation for a general guide.

This is most striking when we consider the implications for the consideration of "empirical" or "operationally defined" methods (sometimes also called "tests" as distinct from "measurements", due to their dependence on the test method for definition). In other guidance [1] this distinction is discussed in detail in terms of its effect on the definition of the measurand. A typical example is "lead in soil". This statement implicitly means "all the lead present". But many useful and practical measurements involve instead the determination of quantities such as "extractable lead" or "bioavailable lead". Clearly, these are different things, and also differ in general from "total lead". In practice, they are defined most readily by specifying a measurement method suitable for their estimation (leading to the term 'empirical method' in such circumstances, though we prefer the term 'empirical measurand'). Yet in defining our result y as "a value obtained from a calculation *valid under stated conditions*" we immediately accommodate the distinction; the requisite conditions—including any required to specify the method of measurement - *are all specified in* $x_{m+1} \ldots x_n$. Viewed from this perspective, we see that essentially all practical measurements are 'empirical' to some degree. There is no fundamental distinction between 'empirical' and 'rational'; the differences lie only in the number of conditions included in $x_{m+1}..x_m$. We can accordingly write very generally applicable guidance without special rules for 'empirical' versus 'rational' measurements. (It has not escaped us that this view might greatly simplify future guidance on the estimation of uncertainty for empirical measurands).

Implicit models

The presentation of Eq. 1 also applies well where y is implicit; that is, stated as a solution to a set of conditions and not attainable by direct calculation (for example, in least squares fitting to a circular shape, the result for the radius is that value which minimises a sum of squared residuals, and cannot be calculated directly). In these cases, while the function f may not be known explicitly, we can write, for example,

$$g(y,x_1,x_2,...x_m) = 0\,|_{x_{m+1},x_{m+2}...x_n} \quad (2)$$

and then write

$$f = g^{-1}\,|_{x_{m+1},x_{m+2}...x_n} \quad (3)$$

where g^{-1}, represents the solution of g for y. We may not know the exact form of g^{-1}, but since we have a sufficient definition of g and the conditions under which it applies, there is no additional difficulty in identifying the relevant quantities $x_1..x_n$. Since this is the key to achieving 'sufficient' traceability in practice, the presentation can also be extended to these cases.

Sufficiency and method validation

In obtaining this relatively simple development of traceability in practice, we have relied heavily on one key assumption; that is, that the model represented by Eq. 1 is sufficient. This begs a most important question; how is sufficiency demonstrated?.

In the guidance we are discussing here, method validation, among other important functions concerned with adequacy of performance, is recognised as the mechanism used to test this crucial assumption. It answers the question "is our calculation and set of conditions sufficient?" by making experimental tests of the assumptions on which it relies. An overall bias check seeks evidence of significant bias; recovery studies seek evidence of loss of material; linearity checks seek evidence of significant departures from linearity; ruggedness studies seek evidence for the presence of further, specific, effects; precision studies (especially on a broad scale) form a test for the presence of unsuspected additional affects. When all these tests are complete and successful, we can reasonably accept our calculation and set of conditions as sufficient.

Clearly, should validation studies reveal an effect which is not duly accounted for, it is the responsibility of the analyst to correct the deficiency. This may be done by altering the measurement procedure slightly—extending extraction time, for example. Or it may be that the calculation requires an additional correction term, or that the number of quantities $x_{m+1}..x_n$ subject to control may be increased (this is 'method development'). The resulting modified method is then subjected to further validation studies and the cycle repeated. The outcome will be, as required, a calculation valid under specified conditions.

This makes it possible to present a clear and consistent picture of the different roles of method development, validation, and traceability:

- Method development establishes a procedure for obtaining an acceptable estimate of the measurand. This procedure includes an equation that describes how to calculate a measurement result from other measured quantities, and specifies the conditions under which this equation is expected to hold.
- Validation demonstrates that this equation and set of conditions is sufficiently complete for the purpose in hand.
- Establishing traceability, by calibration using appropriate measurement standards, provides appropriate units of measurement and ensures that the values of the measured quantities and the specified conditions are duly related to appropriate standards.

Steps to traceability

On the basis of the principles above, the guidance document identifies the key elements in establishing traceability as

i. Specifying the measurand and the acceptable uncertainty
ii. Choosing a suitable method of estimating the value—that is, a measurement procedure with associated calculation—an equation—and measurement conditions
iii. Demonstrating, through validation, that the calculation and measurement conditions include all the "influence quantities" that significantly affect the result, or the value assigned to a standard.
iv. Identifying the relative importance of each influence quantity
v. Choosing and applying appropriate reference standards
vi. Estimating the uncertainty

Because guidance is already available on many of these steps, the guidance document focuses principally on steps ii to v.

It is not the purpose of this paper to repeat the substance of the guidance in the document itself. However, it is pertinent to consider some items of particular relevance to practical traceability in chemical measurement.

Choice of references for calibration

In practice, many of the conditions of measurement in practical chemical measurement are specified in terms of physical measurements. So, too, are many of the inputs to a given calculation. Though the establishment of traceability in these fields has been far from trivial, it is now essentially a routine matter for laboratories to obtain suitable calibrated equipment for measuring quantities such as length, volume, mass, temperature and time. The problem for most laboratories is related to their chemical reference values for amount of substance measurements.

Where certified materials are available for calibration, these can be used directly or used to prepare working standards, and while the provenance of the materials deserves attention, the problem is straightforward in principle. However, where no certified material exists, it is unclear what references may be used in practice, and laboratories typically use well-characterised but uncertified pure materials. This immediately raises the question of whether this practice can be considered as a suitable means of establishing traceability.

The guidance is unequivocal on this point of principle. Quoting from the draft:

"Chemists have a long history of isolating and purifying such substances, and it is common to find relevant materials of purity sufficient to serve as reference standards. This follows from an almost unique feature of chemical measurement; 100% purity forms a natural reference value, which cannot be exceeded. Coupled with widely available and excellent reference data for atomic and molecular weight, and often with additional data on physical parameters such as density, a high purity material represents a local, practical realisation of concentration units, through conversion of mass to molar quantity. **Calibration with materials of well-established purity is accordingly a valid means of establishing traceability**."

This statement is, of course, not unqualified; it remains important to ensure that materials do indeed have the level of purity sought, that subsequent diluted materials retain the concentrations expected and so on. "Without clear evidence of traceable values of known uncertainty, the adequacy of such a material can only be a matter of care and judgement... . Laboratories (should use) all reasonable checks to confirm reliability of uncertified pure materials."

Matrix reference materials for validation

The principles above lead to an interesting simple 'test' for 'traceability'. Put briefly, if the value is part of the calculation of a result or its operating conditions, the result is traceable to that value[2]. If not, the value does not form part of the traceability chain.

This statement has important implications for the role of matrix reference materials used for validation. A previous position paper [5] on traceability indicates that if a certified material is used to check an analytical method, that material contributes to the traceability of the result. The principles presented above do not admit of that conclusion; in this context, only if the matrix reference material generates a significant correction *which is applied in the course of calculating the result* can we be certain that the result is traceable to the value of the check material.

It is important to be clear on this point; these positions may appear inconsistent, but in practice generate exactly the same expectations of laboratories. Whether or not we speak of traceability of the result to the value associated with a matrix CRM used in validation studies, we regard matrix CRMs as the most appropriate test of reliability available, and wholeheartedly recommend their use wherever practicable. Since validation is seen as essential in the context of the present guidance, matrix CRM use is as important as ever in this paradigm. It is simply seen as important to validation, rather than important to the calibration chain.

Uncertainty

In considering traceability within the VIM framework, measurement uncertainty is an essential topic to address. The importance of uncertainty in achieving quality has already been touched upon; only by considering the uncertainty introduced by each quantity in Eq. 1 above can we decide whether the uncertainty in y is sufficient for the purpose in hand. Uncertainty estimation therefore forms an essential activity in the context of establishing adequate traceability. In addition, of course, a useful measurement result must necessarily be of known uncertainty to allow appropriate interpretation. There is accordingly no doubt within the current guidance that the establishment of uncertainty forms one of the essential steps in providing useful, traceable measurement result.

[2] Noting, again, that y may be rewritten in terms of values on which the x_i depend.

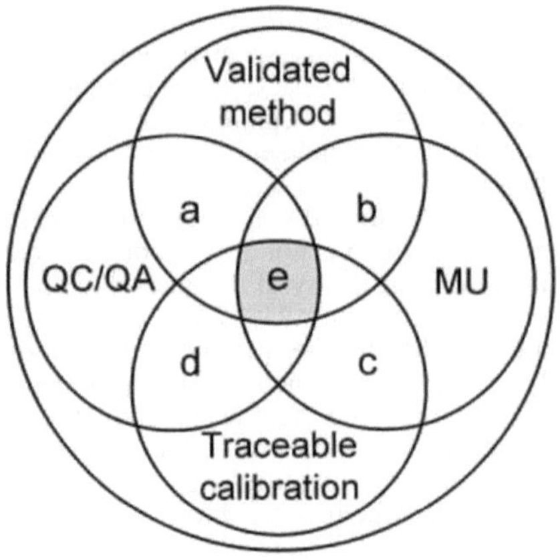

Fig. 1 Elements of analytical quality

A coherent picture

In identifying the concept of traceability most closely with the process of calibration and control for known influence quantities, we risk presenting an incomplete picture of the problem of obtaining reliable results. So it is worth considering whether the picture we present is indeed complete and consistent.

Figure 1 illustrates our presentation schematically. We have identified validation and traceability (represented by the calibration circle) as separate activities in a larger picture, which also incorporates appropriate QA and QC (the third inner circle) and uncertainty estimation. This larger picture is the combination of activities necessary for reliable results; only when all are in place (shaded area e in the centre) can we substantiate a claim that we are presenting a reliable and traceable result. The activities themselves are conveniently considered separately; but it is quite clear that all are essential.

Conclusions

New guidance on traceability presents a simplifying concept of traceability which is appropriate for practical situations. This new paradigm clarifies the relative roles of method development, validation, traceability through calibration, and uncertainty estimation and provides a coherent picture of these activities in the context of wider QA management. The principles are applicable to any situation in which a result is derived from measurements or reference values using a calculation which is valid under stated conditions of measurement, and provide a very general approach to the consideration of whether traceability in a given situation is adequate for its purpose.

Note added in proof The guidance refered to herein was approxed by EURACHEM in May 2003 and CITAC in June 2003.

Acknowledgements Preparation of this paper was supported in part under contract with the UK Department of Trade and Industry as part of the National Measurement System Valid Analytical Measurement (VAM) Programme.

References

1. Ellison SLR, Roesslein M, Williams A (eds) (2000) EURACHEM/CITAC Guide: Quantifying uncertainty in analytical measurement, 2nd edn in English. EURACHEM, ISBN 948926–15–5, LGC Teddington (also available from http://www.eurachem.org/)
2. EURACHEM (1998) The fitness for purpose of analytical methods: a laboratory guide to method validation and related topics. EURACHEM , http://www.eurachem.org/
3. EURACHEM/CITAC (2002) Guide to quality in analytical chemistry: an aid to accreditation. EURACHEM/CITAC, http://www.eurachem.org/, http://www.citac.cc/
4. Ellison SLR, King B, Roesslein M, Salit M, Williams A (eds) (2003) EURACHEM/CITAC Guide: Traceability in chemical measurement. Voting draft March 2003, http://www.eurachem.org/, http://www.citac.cc/
5. CITAC (2000) Traceability of chemical measurement. CITAC, http://www.citac.cc/

Presented in part at the CCQM Workshop on Traceability, BIPM, April 2002

S. L. R. Ellison (✉)
LGC Limited, TW11 0LY Teddington, Middlesex, England
e-mail: steve.ellison@lgc.co.uk

A. Williams
Chairman, Eurachem/CITAC MU and Traceability Working Group,
19 Hamesmoor Way, Mytchett,
GU16 6JG Camberley, Surrey, England

Accred Qual Assur (2000) 5:388–389

CITAC
Co-Operation on International Traceability in Analytical Chemistry

CITAC Position Paper:

Traceability in chemical measurement

Executive Summary

Traceability[1] [1] is a key element in the mutual recognition of testing results. This explains the renewed emphasis on this topic particularly in ISO 17025.

For chemical measurements this involves the need for stated references and a clear uncertainty statement, which should be derived from an uncertainty budget with due regard to the fact that several references, such as amount of substance, mass, volume, time, temperature are generally involved in a single analytical procedure contributing distinct, but different portions to the overall uncertainty.

This uncertainty budget must not only take into account the uncertainties of all the references used in connection with the analytical procedure, but also the uncertainties from the operation of the laboratory procedure as documented in the validation report. The uncertainty from the measurement procedure is frequently much larger than the uncertainties carried by the references.

Background: why traceability

In today's global society comparable results are needed in order to avoid duplicating measurements which cost time and money. The need for mutual recognition – the ability to directly and transparently compare results – explains the emphasis on traceability in ISO 17025 [2]. Comparable results can only be achieved by anchoring them to a common base. In other words we need results traceable to a common base preferably to one with worldwide recognition.

The overall merits of producing and dealing with traceable results of measurement have clearly been acknowledged by the signatories of the Meter Convention whose primary raison-d'être is in fact traceability [3]. It is therefore a central question how this traceability of results is best achieved in chemical measurement. This has to be seen in the light of the two key elements that must be in place for producing traceable values:

- stated and/or internationally agreed suitable references and
- an uncertainty statement for the measurement according to the principles of GUM [4]

whose key role is to enable us to judge the "fitness for purpose" of a result.

Of course, another central question regarding fitness is whether a particular type of measurement is a suitable one for the purpose at all. This is, however, a matter of professional judgement addressed by the choice of an appropriate method.

Merits and added value of traceability for laboratories and customers

The value of traceability for laboratories and customers are in many instances closely related to each other. It has to do with the immediate recognition that an accurate value can only be claimed within the limits of the boundaries indicated in the statement on uncertainty. This helps to avoid over-interpretation of the data and gives a clear view on the limits of validity. Failures in traceability potentially undermine the trust in the professional integrity of analytical chemists. Embarrassing results from these failures could be avoided by paying more attention to the nature and limitations of the traceability of references and of the measurement process itself.

When uncertainty is estimated according to GUM it is given as an interval around the result of the measurement and it is fairly straightforward to decide one or more of the following:

a) Is the upper or lower limit of this uncertainty statement close to a statutory or legal limit, or does it reach beyond such a limit?
b) How much overlap is between the uncertainty statement of similar measurements on the same or another sample?
c) Do the intervals expressing uncertainty of measurements from different laboratories on the same sample overlap?

For most purposes it is less important to have a particularly minute uncertainty, but more pertinent to have a good estimate of the uncertainty for answering questions just as the ones mentioned above.

These and similar questions are important in the self-assessment of a laboratory, benchmarking and establishment of confidence in the working relationship with a customer.

Technical elements of traceability

A laboratory finds itself typically at the end of the traceability chain. Therefore, in order to produce traceable results it must be able to rely on all the references necessary in the measurement process, as well as on method validation [5]. A prerequisite for supplying traceable results to the customers is therefore that the values of all references are themselves traceable to stated references and are accompanied by a reliable uncertainty statement. The technical expertise of the laboratory as established by accreditation then must ensure the proper use and handling of these references and of the samples. This is generally a matter of training and expertise. It is particularly useful if there is expertise in the development and adaptation of analytical procedures, as this is much needed in the obligatory validation procedure.

Determination of amount of substance often requires measurements of different properties, for example: sample mass, on a balance compared to a mass reference; analyte identity by comparison to a reference, perhaps using a spectrometer and a database of known compounds; and analyte quantitation by comparison to a different reference, perhaps a reference material. Each property of the result should be traceable, and each may contribute uncertainty to the reported result. Thus, claims of traceability of a result must include not only a description of the references and uncertainty budgets for comparison to them, but also a description of the scope of traceability.

In most cases in analytical chemistry one faces the situation that the contribution of uncertainties of the references to measurement uncertainty is small relative to those contributions that come from the measurement process itself. Under such

---

[1] Traceability is the property of the result of a measurement or the value of a standard whereby it can be related to stated references, usually national or international standards, through an unbroken chain of comparisons all having stated uncertainties.

circumstances the results can only be improved by improving the analytical procedure.

Laboratories are urged to concentrate on the measurement process they are operating. This involves a thorough validation process leading to valid results including a realistic statement of measurement uncertainty that also duly accounts for the uncertainty of the relevant references.

If validation is exercised with due regard to traceability it must provide sufficient information for the subsequent estimation of measurement uncertainty. In this manner, a traceability chain is established as part of validation.

Traceability of values carried by reference materials

The one key to traceability that must be supplied from outside the laboratory is the traceability of values carried by references, especially by certified reference materials. As these values are also established by measurements, the same features required from analytical laboratories also apply for producers of reference materials. Additionally information on the stability and homogeneity of the reference material in form of an expiration date or by equivalent means is required.

The producers of reference materials must be aware that the values they supply are invariably an indispensable link in the traceability chain. They must implement all procedures necessary to provide evidence internally and externally (e.g. by peer review, laboratory intercomparison studies, etc.) that they have met the conditions required for obtaining traceable results at all times.

Conclusions

- Traceability of results and reference values is a central issue in modern laboratory operation. It is not an end in itself, but serves the purpose of achieving a reliable result.
- Traceability of results can only be claimed if results are accompanied by an uncertainty statement based on traceability of all references, chemical and physical, as well as on procedural contributions to uncertainty.
- A result must be "fit for purpose", thus estimation of measurement uncertainty from uncertainties of references and procedures is added value for laboratories and simple when guidelines are followed.

References

1. International Vocabulary of Basic and General Terms in Metrology, ISBN 92-67-01075-1, International Organization for Standardization, Genève 1993, orders at: http://www.iso.ch/infoe/otherpub.html#Metrology and measurement
2. ISO 17025, General requirements for the competence of testing and calibration laboratories, International Organization for Standardization, Genève 1999, orders at: http://www.iso.ch/cate/d30239.html
3. Mutual Recognition of national measurement standards and of calibration- and measurement certificates issued by National Metrology Institutes, 1999,http://www.bipm.fr/enus/8_Key_Comparisons/mra.html
4. Guide to the Expression of Uncertainty in Measurement, ISBN 92-67-10188-9, International Organization for Standardization, Genève 1993, orders at: http://www.iso.ch/infoe/other-pub.html#Metrology and measurement
5. International Guide to Quality in Analytical Chemistry, An Aid to Accreditation, EURACHEM-CITAC Guide 1, ISBN 0 948926 09 0, London 1995, orders at: http://secure.bookshop.co.uk/ser/serdsp.asp?shop=1&isbn=0948926090&DB=220

Dr. Ioannis Papadakis ✉
(CITAC Secretary)
Institute for Reference Materials and Measurements, Joint Research Centre – European Commission, Retieseweg, B-2440 Geel, Belgium
Tel: +32 14 571 682
Fax: +32 17 571 863
e-mail: ioannis.papadakis@irmm.jrc.be or citac@irmm.jrc.be

Prof. Dr. Wolfhard Wegscheider
(CITAC Chair)
Dept. of General & Analytical Chemistry, University of Leoben, Franz-Josef Strasse 18, A-8700 Leoben, Austria
e-mail: wegschei@unileoben.ac.at

Accred Qual Assur (1996) 1:41–43

John Fleming
Bernd Neidhart
Christoph Tausch
Wolfhard Wegscheider

Glossary of analytical terms*

Introduction

Analytical data play a vital role in our daily lives, with increasing influence on both economy and ecology. The harmonisation of the European market – including the Eastern European countries – and the opening of the international borders for trade and communication have led to serious problems with terminology in analytical chemistry. We can identify the three main reasons that have caused this situation. These can be classified as "linguistics", "semantics", and "acceptance".

Frequent translations of a term through a chain of languages, and the use of terms by non-native speakers, may lead to a misuse of terms followed by grave misunderstandings. In addition, the co-existence of different meanings of terms due to their independent definition by national and international bodies or authorities, together with recommendations given by international organisations like IUPAC, leads to problems of semantics and confusion resulting in reduced acceptance.

* EURACHEM Education and Training Working Group

A strategy on terminology

During the last 5 years, the EURACHEM Education and Training Working Group (E&TWG) has analysed this situation and has developed a strategy which is expected to resolve the dilemma. The first, and most important, step in this concept is to provide a forum which initiates and enables international discussions among experts in the field. The catalyst for these discussions will be a dictionary-like "glossary of terms" which will be published as a series in this journal. Each term in the glossary is provided with a definition (taken from the highest international level, if possible ISO) followed by a scientific description of the meaning of the definition and one or more examples explaining its practical use. In addition, translations of the term into other European languages are given. This structure will facilitate translation of the glossary into other languages, and errors will be minimised if not excluded. The translation will be performed by the E&TWG members, who are experts in the field and native speakers of the respective language, and will finally be published in a suitable national journal.

Feedback will be sought at both national and international levels to enable a dynamic development of the glossary at the highest scientific and linguistic levels possible. This might also include the deletion of existing and the creation of new words, if, in the latter case, the scientific definition and meaning has no linguistic equivalent in a given language. Let us take as an example the term *traceability,* which by definition describes a way to achieve quality (accuracy, comparability) in chemical measurements. The equivalent in German would be *Rückführbarkeit* but the term *Rückverfolgbarkeit* is used as the respective DIN Standard, the linguistic meaning of which is "follow the way (track) back". Consequently, the term *Rückverfolgbarkeit* is part of providing *assurance* of quality and not of *creating* quality. Unfortunately, there is no English word for *Rückverfolgbarkeit.* There are two ways of solving this problem: one is to create a new English word and the other to introduce the German word into the English language.

We are willing to "grasp the nettle" and open the debate on this issue by proposing the term *trackability* to cover this concept.

Discussion forum

It is proposed that the EURACHEM E&TWG should be the catalyst which will promote a wider debate of the issues raised by this glossary of terms. All analytical scientists are urged to contribute to the debate and work towards a consensus on the usage of the key terms covered by the glossary. This debate can be pursued either by corresponding with the editor of this journal or by sending an e-mail message to jwf@lgc.co.uk for consideration by the working group.

Repeatability

Wiederholpräzision (**D, A, CH**); Répétabilité (**F, B**); Repetibilidad (**E**); Ε.παναληψιμ϶τητα (**GR**); Ripetibilitá (**I**); Herhaalbarheid (**NL**); Powtarzalnosc (**PL**); Toistettavuus (**SF**); Ismételhetõség (**H**); **Сходимость** (**RUS**); Repetibilidade (**P**)

Definition

Precision under repeatability conditions.[1]

Description

Repeatability is the closeness of the agreement between the results of independent measurements of the same analyte carried out subject to all of the following conditions:
the *same* method of measurement, the *same* observer, the *same* measuring instrument, the *same* location, the *same* conditions of use, repetition over a *short period of time.*[2]
Independent measurements are made on distinct subsamples of a test material. If possible, at least 8 measurements should be performed.
Repeatability is a characteristic of a method not of a result.

Example

Successive measurements under the above conditions gave eight single results from which a standard deviation is calculated. The standard deviation multiplied by 2.8 gives the repeatability at 95% confidence level.
Suppose that an analyst uses a method for which the repeatability has been established as 2 μg/mL.
If, in a real case, the same analyst reported results of a measurement repeated over a short time interval as 50 and 56 μg/mL, there would be a question over the validity of these results as they are very unlikely to have differed by 6 μg/mL as a result of random variability.

[1] ISO 3534-1 (1993)
[2] International vocabulary of basic and general terms in metrology, 1993, (BIPM, IEC, IFCC, ISO, IUPAC, IUPAP, OIML); ISO central secretariat, 1 rue de Varambé, CH- 1211 Geneva 20

Reproducibility

Vergleichpräzision (**D, A, CH**); Réproductibilité (**F, B**); Reproducibilidad (**E**); Αναπαραγωγιμότητα (**GR**); Riproducibilitá (**I**); Reproduceerbarheid (**NL**); Odtwarzalnosc (**PL**); Uusittavuus (**SF**); Reprodukálhatóśag (**H**); **Воспроизводимость** (**RUS**); Reprodutibilidade (**P**)

Definition

Precision under reproducibility conditions.[1]

Description

Reproducibility is the closeness of the agreement between the results of measurements of the same analyte in distinct subsamples of a test material, where the individual measurements are carried out *changing* conditions such as: observer, measuring instrument, location, conditions of use, time, but applying the same method.[2]

Example

In a laboratory intercomparison samples (e.g. a surface water) were sent to a number of laboratories for determination of e.g. nitrite. Each laboratory reports its results as single values.
The standard deviation from all accepted individual results multiplied by 2.8 gives

the reproducibility at 95% confidence level.
Suppose that the reproducibility of a method has been determined to be x. If two of the laboratories in a real case reported results for subsamples of the same sample which differed by >x there would be a question concerning the quality of performance.
Methods which have a large reproducibility may not be suitable for making valid comparisons in a given real situation. In this case either the method must be improved or another method with a smaller reproducibility must be applied.

[1] ISO 3534-1 (1993)
[2] International vocabulary of basic and general terms in metrology, 1993, (BIPM, IEC, IFCC, ISO, IUPAC, IUPAP, OIML); ISO central secretariat, 1 rue de Varambé, CH-1211 Geneva 20

Traceability

Rückführbarkeit (**D, A, CH**); Tracabilité (**F, B**); Trazabilidad (**E**); Ιχνηλάτηση (**GR**); Riferibilitá (**I**); Herleidbarheid (**NL**); Rastreabilidade (**P**); Jaeljitettaevyys (**SF**); Visszavezethetõség (**H**); Zgodnosc (**PL**); Сходимость (**RUS**)

Definition

The property of a result of measurement whereby it can be related to appropriate standards, generally international or national standards, through an unbroken chain of comparisons.[1]

Description

For each analytical measurement, it should be possible to relate the result of the measurement back to an appropriate national or international measurement standard through an unbroken chain of comparisons. For measurement of weight, this would be the kilogram standard in Paris, or for amount of substance it should be the SI unit, the mole. If calibrated by an accredited body, the balance is an instrument which can provide measures of weight which are traceable to national measurement standards. Instruments for chemical analysis must be calibrated by the use of certified reference materials, or other suitable reference materials.

Example

Determination of lead in water by atomic absorption spectrometry (AAS): The AAS instrument has be to calibrated using reference solutions made up by dissolving known amounts (balance) of a certified reference material (CRM) or a pure substance such as $Pb(NO_3)_2$ in a defined volume of pure water; in the latter case the pure substance has to be compared with a CRM. A calibration graph which covers the concentration range of the analyte in the sample should be prepared.
For more complicated analyses, which might involve extraction and other analytical procedures, the traceability of the result of a measurement can be established by subjecting a certified reference material – with similar composition to the unknown – to the same analytical procedures.
If for example the measurement-standard used has not been compaired with a CRM of the same type the chain of comparison is broken.

[1] ISO 3534-1 (1993)

Trackability

Rückverfolgbarkeit (**D, A, CH**), Relacionabilidad (**E**), Sporbarhet (**NOR**)

Definition

The property of a result of a measurement whereby the result can be uniquely related to the sample.

Description

Each step of an analytical method has to be documented in a way that the result of a measurement can be linked unambiguously to the sample to which it refers.

Example

All samples must be uniquely labelled. All operations performed on a sample must be recorded in a notebook or computer system. Chromatograms, spectra and other instrumental outputs must be labelled with the sample identification.

Track:

Result
⇩
Calculation
⇩
Determination
⇩
Separation
⇩
Preparation
⇩
Storage
⇩
Sampling

Uncertainty of measurement

Meßunsicherheit (**D, A, CH**), Incertitude de mesure (**F, B**), Intercidumbre de la medida (**E**), Αβ βαιότητα τη μ΄τρηση (**GR**); Meetonzekerheid (**NL**); Incerteza da medida (**P**); Mérési byzonytalansá g (**H**); incertezza di misura (**I**); Mittauksen epaevarmuus (**SF**)

Definition

Parameter, associated with the result of a measurement, that characterizes the dispersion of the values that could reasonably be attributed to the measurand.[1]

Description

Uncertainty sets the limits within which a result is regarded accurate, i.e. precise and true.
Uncertainty of measurement comprises, in general, many components. Some of these components may be evaluated from the statistical distribution of the results of series of measurements and can be characterized by experimental standard deviations. The other components, which can also be characerized by standard deviations, are evaluated from assumed probability distributions based on experience or other information.[2]

Example

Overall uncertainty can be estimated by identifying all factors which contribute to the uncertainty. Their contributions are estimated as standard deviations, either from repeated observations (for random components), or from other sources of information (for systematic components). The combined standard uncertainty is calculated by combining the variances of the uncertainty components, and is expressed as a standard deviation. The combined standard uncertainty is multiplied by a coverage factor of 2 to give a 95% level of confidence (approximately).
The uncertainty for the determination of e.g. atrazine in water consists of the calibration of several components of uncertainty, such as the uncertainty of the true content of the atrazine standard, uncertainty from dilution of this standard, uncertainty regarding the loss of atrazine in sampling and storage prior to analysis, as well as that associated with the preconcentration step after correction for recovery.
The result would be expressed as:
1.02 ± 0.13 μg/L

[1] International vocabulary of basic and general terms in metrology, 1993, (BIPM, IEC, IFCC, ISO, IUPAC, IUPAP, OIML); ISO central secretariat, 1 rue de Varambé, CH-1211 Geneva 20

Printing: Mercedes-Druck, Berlin
Binding: Stein+Lehmann, Berlin

MIX
Papier aus verantwortungsvollen Quellen
Paper from responsible sources
FSC® C105338

If you have any concerns about our products,
you can contact us on
ProductSafety@springernature.com

In case Publisher is established outside the EU,
the EU authorized representative is:
Springer Nature Customer Service Center GmbH
Europaplatz 3, 69115 Heidelberg, Germany

Printed by Libri Plureos GmbH
in Hamburg, Germany